The Handbook of Global Agricultural Markets

Also by Luc Nijs

MEZZANINE FINANCING: Tools, Applications and Total Return

SHAPING TOMORROW'S MARKETPLACE: Investment Philosophies for Emerging Markets and a Semi-Globalized World

The Handbook of Global Agricultural Markets

The Business and Finance of Land, Water, and Soft Commodities

Luc Nijs

palgrave
macmillan

First published 2014 by
PALGRAVE MACMILLAN

Palgrave Macmillan in the UK is an imprint of Macmillan Publishers Limited, registered in England, company number 785998, of Houndmills, Basingstoke, Hampshire RG21 6XS.

Palgrave Macmillan in the US is a division of St Martin's Press LLC, 175 Fifth Avenue, New York, NY 10010.

Palgrave Macmillan is the global academic imprint of the above companies and has companies and representatives throughout the world.

Palgrave® and Macmillan® are registered trademarks in the United States, the United Kingdom, Europe and other countries.

ISBN: 978–1–137–30233–5

This book is printed on paper suitable for recycling and made from fully managed and sustained forest sources. Logging, pulping and manufacturing processes are expected to conform to the environmental regulations of the country of origin.

A catalogue record for this book is available from the British Library.

A catalog record for this book is available from the Library of Congress.

To Simon – my best friend for all times
That you may walk through life blessed with joy, equanimity, wisdom, dignity, (com)passion and will do so in good health

Contents

List of Figures

List of Tables

Preface

Agriculture, agribusiness, agrotechnology, agronomy, agroforestry, farmland, nanotechnology, biofuels, and genetically modified crops: the list is endless. In recent years, the agriculture industry has undergone a renaissance. There are very good reasons for that, which inter alia include (1) a constantly growing number of humans living on this planet – at the time of writing in excess of 7 billion; (2) an accelerating shift toward urbanization and aligned living styles; (3) the emergence of a new middle class in many emerging economies around the world, fueling changes in eating and dietary habits; (4) changing weather patterns, regardless of whether they are induced by man or not, that force us to think about how we want to organize our food supply; (5) globalization of agricultural and food-supply chains; (6) the use of agricultural commodities to produce biofuels; and (7) scarcity of water and raw materials involved in the manufacture of fertilizers.[1]

There is also another trend visible on the other side of the fence, a phenomenon that more and more financial investors are seeing: the agricultural space as an investible asset class,[2] or classes I should say, with growing interest in agricultural commodities across the different traditional financial asset classes (stocks, bonds, equities, real estate, land, etc.). This happened from the early 2000s but has been very visible since 2008. This has been fueled by the factors listed above, which have acted as a catalyst accelerating the upward pricing trend, as the fundamentals behind the demand/supply equation for commodities stay positive. Furthermore, monetary policy has had an impact on the pricing of agricultural commodities,[3] something that has become very visible since Ben Bernanke took office at the Federal Reserve. Note that most if not all such commodities are priced in US dollars.

The attraction by investors globally has been hitting the news for years now, with the most exotic stories appearing on the front page. One cannot blame investors for being in search of yield on a continuous basis. But once it (allegedly) started impacting basic necessities, controversy was around the corner. Agriculture has now become a theme that is fully financialized and has absorbed the inner logic of the financial world despite the fact that its own proprietary industry dynamics are still managed in the broader but traditional paradigm following the logic of Mother Nature and the industrial and economic logic of the modern farming organization.

That in itself might sound very logical or justifiable, but it ignores that fact that the financial and industry dynamics often contradict each other. The contemporary financial world is, more often than not, focused on short-term risk-on/off performance and achievements, whereas by contrast the traditional investment dynamics of the agricultural sector are by nature focused on the long term. That conflicting dynamic, including my experiences as an entrepreneur, investor, consultant, and academic, yielded a massive number of thoughts and a variety of experiences that deserved to be logically and thematically dissected. In this way, I ensured that that knowledge and experience will be much more valuable and accessible than the accumulated chaotic potpourri of information as it came to me over almost two decades of professional life.

This book was written with a number of objectives in mind. First, it will provide the reader with a solid understanding of how the agricultural world works globally, highlighting major differences per continent. Second, it will demonstrate how the financial world interacts with the agricultural world – how agricultural products are converted into financial products and assets and how this created a phantom parallel world where each domain lives by its own set of rules but which impact each other, mainly through the market pricing mechanism of agricultural commodities. It allows you to determine how and when agri-investing can become relevant for your portfolio of investments.

But third and maybe most important, it will provide a review of the evolution that is taking place in the wider (agricultural) industry. Relevant questions to be reviewed are (1) How is the industry dealing with (upcoming) scarcity of water, chemicals, etc.? (2) What new technologies are being created to enhance production per unit? (3) What is going to be the impact of the industry's engagement in biofuel development and production? and (4) How is the industry dealing with changing weather patterns, etc.? This leads to follow-up questions about where exactly in the value chain incremental value can be captured by investors. It might even nullify some of the mainstream truths about investing in agricultural land as being the most attractive opportunity in the industry, despite the rising prices of land in recent years.

Ultimately, the agricultural industry is a long-term play, and agricultural commodities are renewable commodities, so their intrinsic logic is somewhat different from other (exhaustible) commodities. It also lacks predictability, particularly in the short term, as Mother Nature can be as unpredictable and psychotic as Mr. Market in the financial world. Maybe even more so.

As I performed research and ordered my thoughts for this book, I became more and more engaged with the topic,[4] not for the sake of enhancing my expertise or deepening my knowledge, but for the sake of us all. This happened because the topic is so intrinsically integrated with who we are as humans and serves some of our most basic needs that keep us going on this planet. That must make you think, for yourself and for all future generations who will have to live with the consequences of our policies and the quality of our decision-making capabilities. Also in this context, the most vulnerable people in the world will be negatively impacted first. Politicians, keeping the next elections in mind and too hungry for their next media appearance, tend to ignore the challenges that this and other industries pose for our long-term prosperity and even survival. This behavior is fueled by austerity-driven dynamics controlling most governments in the Western world.[5]

The upshot is that after having reviewed this book, it should be perfectly possible, depending on the ambitions of the reader, (1) to have developed a thorough understanding of the agricultural industry globally; (2) to develop a personalized in-depth investment strategy covering the wider agricultural space; (3) to determine how one can invest in the agricultural sector and which products are available, including all the do's and don'ts one needs to be aware of in order to hit the ground running in the investment world; (4) to develop a personal educated decision tree regarding the broader challenges, misconceptions and discussions going on about the (possibly) unholy relationship between the financial sector and the agricultural industry; and (5) to be engaged and well-versed in the most important dynamics of and evolutions in the agricultural industry. The book is therefore suitable for both industry and financial professionals, as well as the engaged enthusiast or layperson and the more academically inclined reader.

I have tried to strike a middle ground between practical and theoretical discussions, making the book useful in both an academic and professional context. Every time one considers writing a book, one does so driven by a certain passion for the topic. Nevertheless, every book has its limits of focus in order to be and stay useful, and this book is no exception. Throughout the drafting process, countless decisions had to be made about what topics and data sets to include or leave out. I realized that being fully comprehensive on the data-set front would be fully unrealistic and also undesirable, as it would make the book outdated soon after publication. A balancing act was needed that required some level of discretionary decision making. To compensate for that, I have tried to produce a substantial reading list so that readers who feel particularly attracted by a subtopic or feature discussed will not be left hanging out to dry. I also, where needed, temporarily broadened the space of discussion from agricultural commodities to commodities in general. I have stopped short of making any direct references to firms, funds, research centers or investible equities, bonds, ETFs/ETPs/ETNs, dedicated mutual or hedge funds, or any other possible investment destination. I felt it would compromise the objectivity and "analysis momentum" behind the book and would, for the diligent reader, sugarcoat it with an unacceptable marketing flavor. He or she would be right in having that opinion. Furthermore, if I had done so, I would have had to endlessly reiterate that nothing in this book constitutes (individual) investment advice and would never be fully comprehensive in displaying the possible investment alternatives there are out there globally. There is a right time and place for everything, and there are (more than) enough marketing brochures and investment conferences on the topic where plenty of investment advisors are lined up who can entertain you with what they believe is the right thing to do for you or your organization. Reading this book, however, will make you a worthy, dignified and educated counterparty and conversational partner. That way you can make up your mind yourself about all the well-intended investment advice that undoubtedly will be coming your way! At least it will allow you to consider that the global agricultural space contains many more very promising investment destinations, each with their own very different risk-return profile, than the farmland-grab deals that make headlines in the global media centers.

Underlying this book and its topic is a further ambition. Given my multi-disciplinary and multifocal orientation toward investing and finance over the course of my career, I hope this book contributes to enhancing the quality of interaction between the financial sector and other sectors (in this case the agricultural sector) of the economy. The last few years have proven that these two sectors do not necessarily aim for the same objectives, and this might be at the expense of some of the parties involved. The best financial professionals are those who have a very thorough understanding of the industry they are financing. Likewise, the best industry professional is the one who best knows how to convert the economic activities in the industry or company into sustainable long-term value for which others are willing to pay a price, as it, in their understanding, reflects a certain portable and (hopefully) sustainable value. While trying to shift away from the trade-off paradigm in which financial-sector growth comes at the cost of the real economy and its sectors, or alternatively the real economy can thrive only at the cost of the financial sector, a middle ground is explored that allows the financial sector to lubricate the industry sector according to its needs and proprietary dynamics. This implies no overlubrication with products and services contradicting the dynamics of or unneeded by a certain industry. Additionally, we can only hope that scientists, the corporate world, and regulatory bodies can find a common path in this multiparadigm environment toward achieving the most optimal output given the limits put up by nature.

That will be absolutely key in a globalizing world characterized by interdependence but also trade barriers, import duties, restrictions on exports, and the like. I hope to leave you sufficiently informed about the major topics in the industry to invest in an educated way and see the longevity of the agricultural world and its implications.

Before embarking on the journey of working through the book, consider the massive amount of research that is being done on a daily basis by committed researchers around the world in this field at universities, international and supranational organizations, and industry associations, as well as by commercial organizations. A special word of thanks and acknowledgment is absolutely needed. It would be impossible to name them all here. I therefore ensured that, when dealing with their research, I referred to their names and works as much as possible in the extensive set of notes included in this book. Please consult them whenever possible, as they contain a wealth of information and data that may be valuable to you. Without their research, this book would not have been written. These notes also facilitate and encourage further reading in any of the areas the reader considers worth pursuing. The manuscript was completed on September 30, 2013, and all Internet-related links were last accessed and verified on that date.

Happy reading!

Notes

1. For a first good general reading on the challenges and dynamics facing the agricultural industry in the 21st century, see The Crawford Fund (2012), "The scramble for natural resources: More food, less land?" Editor: Ann Milligan (retrieved: http://ageconsearch.umn.edu/bitstream/152131/2/2012%20annual%20conference%20proceedings%20WR.pdf)
2. Expected is that globally the average exposure to agriculture will increase from 1% in 2012 to 5% over the next five to seven years.
3. The pricing of agricultural commodities can be followed via the monthly FAO food-price index, which can be accessed through http://www.fao.org/worldfoodsituation/foodpricesindex/en. Alternatively, one can turn to Food Price Watch of the World Bank, which can be accessed through http://web.worldbank.org/WBSITE/EXTERNAL/TOPICS/EXTPOVERTY/0,,contentMDK:22838758~pagePK:210058~piPK:210062~theSitePK:336992,00.html.
4. I mean more than I already was.
5. This is the case with the agriculture challenges, but the wider climate-change dynamics are also suffering. For example, in most European countries (including Germany and Spain), subsidies and incentives with respect to solar and renewable energy have largely been abolished (or materially reduced). That this would be caused or justified by pricing dynamics in the industry in recent years (e.g., pricing for solar panels coming down) is intellectually dishonest and not in line with the common financial dynamics of immature industries. More than a century down the road, the immature aviation industry is still heavily incentivized, so it all seems to be a matter of prioritization. Mother Nature does not have cash-rich lobbyists in Washington or Brussels and is always looking for additional naturally inclined enthusiasts.

List of Abbreviations

AIS	agricultural innovation system
APH	actual production history
ARI	agricultural research intensity
BIT	bilateral investment treaty
BROA	bootstrapping real-options analysis
CA	conservation agriculture
CAP	Common Agricultural Policy
CAPM	capital asset pricing model
CBOT	Chicago Board of Trade
CDM	Clean Development Mechanism
CFA	contract farming agreement
CFTC	Commodity Futures Trading Commission
CIT	commodity-index trader
CME	Chicago Mercantile Exchange
CSA	climate-smart agriculture
DCF	discounted cash flow
ECa	electrical conductivity
ECPT	ecosystem carbon payback time
EDR	electrodialysis reversal
EGI	effective gross income
EMH	efficient-market hypothesis
ESMA	European Securities and Market Authorities
ESRB	European Systemic Risk Board
ETF	exchange-traded fund
ETN	exchange-traded note
ETP	exchange-traded product
FDI	foreign direct investment
FI	financial institution
FPIC	free, prior, and informed consent
GHG	greenhouse gas
GIS	geographical information system
GMM	genetically modified microorganism
GMO	genetically modified organism
GWAS	genome-wide association studies
IAEA	International Atomic Energy Agency
ICE	Intercontinental Exchange
IFES	integrated food-energy system
IFRS	International Financial Reporting Standards

IIA	international investment agreements
IIASA	International Institute for Applied Systems Analysis
ILUC	indirect land-use change
IPCC	Intergovernmental Panel on Climate Change
IVSC	International Valuation Standards Council
KIA	Kuwait Investment Authority
KCIC	Kuwait China Investment Company
LEV	land expectation value
LIHD	low-input, high-diversity
LPM	loan policy manual
LRS	loan-rating system
MACD	moving average convergence/divergence
MDG	millennium development goals
MFN	most-favored nation
MIS	management information systems
MPCI	multiperil crop insurance
MSF	multistage flash
NBFI	nonbank financial institution
NCBI	National Center for Biotechnology Information
NCREIF	National Council of Real Estate Investment Fiduciaries
NDVI	Normalized Difference Vegetation Index
NOI	net operating income
NPL	nonperforming loan
NPV	net present value
OECD	Organisation for Economic Co-operation and Development
OLS	ordinary least squares
OTC	over the counter
PG	potential gross income
PPP	public-private partnerships
RAROC	risk-adjusted return on capital
rBST	recombinant bovine somatotropin
REDD	Reducing Emissions from Deforestation and Forest Degradation
REIT	real-estate investment trust
RMA	Risk Management Agency
RO	reverse osmosis
ROA	real-options analysis
rpST	recombinant porcine somatotropin
PGR	plant growth regulator
PPO	pure plant oil
PV	present value
SHIP	Shale Gas Information Platform
SIT	sterile insect technique
SME	small and medium enterprises

SVO	straight vegetable oil
USDA	United States Department of Agriculture
VAR	vector autoregressive
VRA	variable-rate application
VF	vertical farming
WACC	weighted average cost of capital
WNV	West Nile virus

1
Introduction

The agricultural sector is characterized by several dynamics that distinguish it from other economical activities. They could be summarized as follows:

1. High sensitivity to natural conditions

This includes discrepancies among climatic conditions (e.g., rain, temperature) from one season to another that impact the level of agricultural production, especially in perpetual farming areas, as well as in all areas in general. This results in an inconsistent relationship between production inputs (e.g., seeds, fertilizers, plaguing, weeding, hoeing) and the final product, and consequently in production fluctuations that require government interference to restore balance for the benefit of producers, consumers, or both. Man's ability to control the agricultural environment is still limited and varies from one country to another according to the development achieved in biotechnology, which is determined in turn by the location of invention. This explains the slow transfer of these innovations to countries with differences in environmental conditions.

2. The seasonal and biological nature of agriculture

Each group of agricultural crops has its suitable season and can't be planted in other seasons. With advancements in some agricultural activities, such as in the case of greenhouses in plant production from one side, and poultry production in the animal production sector on the other, it has become possible to achieve year round production. The production cycle (the period between starting the production process and getting the final product) is long compared to the production cycle in other activities. In spite of humans' achievements in biotechnology and especially in genetic engineering and hybridization which lead to effective results in innovating new breeds and types with high productivity compared to traditional breeds, shortening the production cycle in agriculture is still difficult to achieve.

Realistic limitations on shortening the production cycle make the task of increased production across time difficult. Growth sources predominantly rely on two sources: (1) horizontal expansion: increased production elements, land, water, work equipment, and production requirements, and (2) vertical expansion: increased productivity of production elements and appropriate application of biotechnology innovations (vertical farming, verticrops; see infra).

3. The local nature of agricultural production

Since agricultural activity is usually attached to climate, soil, and people's traditions and norms (people who work in agriculture), agricultural activity and its advancement requirements are less dependent on external factors, for example the inability to grow palm trees and citrus in northern Iraq, or to have certain areas specialize in growing potatoes.

That leaves an uncertain output relative to other industries. Recent food crisis and growing concerns about global climate change have placed agriculture on top of the international agenda. In 2013 a convergence has been recognized between the dual goals of eradicating hunger and making agriculture sustainable. Achieving these goals will require a significant increase in agricultural investment, but more importantly it will require improving the quality of this investment.

The international financial crisis, which is affecting governments and donors around the world, means that now, more than ever, public resources alone cannot meet the investment needs for agriculture. Governments and donors play a crucial role in catalyzing, channeling, and governing agricultural investment, but private investors – primarily farmers themselves – must be central to any investment strategy for agriculture.

Investors in agriculture can be categorized as public or private and foreign or domestic. The majority of private domestic investors are farmers, and they are by far the largest source of investment in agriculture in low- and middle-income countries and in relative terms also in the developed world. Domestic public investors, primarily national governments, are the next largest source of investment in agriculture, followed distantly by foreign public investors such as development partners and by foreign private investors, such as corporations. These investors – public and private, domestic and foreign – invest in different things and for different reasons. Their investments are often complementary, sometimes overlapping, and are generally not substitutable for each other. Particularly in low- and middle-income countries, the biggest investor is the farmer himself. This has important implications for policy: while public investment remains essential, the focus of investment policy has to shift to facilitating more and better private investment.

Farmers invest to feed their families, to increase and diversify their incomes, and to build their wealth. For farmers, investing in agriculture means giving up something now (such as money, effort or time) in order to accumulate assets or capital that will allow them to increase their productivity and incomes in the future. Farmers and other private investors will invest in agriculture only if the expected returns compensate for the perceived risk and exceed returns from alternative types of investment. The rationale for public investment in agriculture by governments and development partners rests on three interrelated benefits for society that can come from enhancing agricultural productivity: (1) economic growth and poverty reduction, (2) food and nutrition security, and (3) environmental sustainability. For governments and donors, investing in agriculture means allocating scarce public resources to activities that raise productivity in that sector. Agricultural research and market infrastructure are among the most important types of public investment in agriculture. History shows that even though farmers are the largest investors in agriculture, in the absence of good governance, appropriate incentives, and essential public goods, they do not invest enough. Agricultural production is usually seasonal or cyclical in nature and is vulnerable to natural phenomena such as drought, pests, and disease. Producers are often geographically dispersed, and most agricultural products are bulky and perishable. All these factors make agricultural investment risky and highly dependent on the existence of good rural infrastructure, robust input supply and output processing industries, and transparent market institutions and price signals. Appropriate public investment can reduce the risk and increase the profitability of private investment and thus enhance incentives for farmers to invest.

In general, investment involves giving up something today in order to accumulate assets that generate increased income or other benefits in the future. Farmers invest in their farms by acquiring farm equipment and machinery, purchasing animals or raising them to productive age, planting permanent crops, improving their land, constructing farm buildings, and the like. Governments may invest in, inter alia, building and maintaining rural roads and large-scale irrigation infrastructure, assets that generate returns in terms of increased productivity over a long period of time. Governments also invest in other less tangible assets such as the legal and market institutions that form part of the enabling environment for private investment. Determining whether an expenditure, public or private, constitutes an investment can thus be difficult both conceptually and empirically, and in some cases it is not clear cut. Investment is generally defined as activities that result in the accumulation of capital that yields a stream of returns over time. In agriculture, a distinction is usually made between investment and spending on inputs, based rather arbitrarily on the length of time required to generate a return. Thus, planting trees is typically considered an investment because it takes more than a year to generate a return, but applying fertilizer to a maize crop is not considered an investment because it generates a return during the current crop cycle. More important from a conceptual point of view, trees are a capital asset that yields a stream of returns over many years. Even in this seemingly simple case, the distinction may not be clear. If fertilizer use helps maintain and build soil fertility in the long-run, it may also be considered an investment. Similarly, in public expenditures, a distinction is generally made between investment and current expenditures, but again this is not always clear cut, not least because current expenditures are required to maintain the value of capital assets such as roads and other physical infrastructure.

Many of the investments made by governments are called "public goods" because they generate benefits for society that cannot be captured by a private investor. Once a public good has been created, people cannot be excluded from taking advantage of it, and use-by one person does not diminish the ability of others to use it. In technical terms, they are "nonexclusive" and "nonrival." Private investors have little or no incentive to provide public goods because they cannot charge enough to recover the cost of the investment. Examples of important public goods for agriculture include many types of R&D and rural roads and other infrastructure. Other types of public investment, such as building institutions and human capacity, provide less tangible but perhaps even more important public goods for agriculture. What constitutes a public good will depend to some extent on country characteristics and local context, and mixed public/private goods are common in agriculture.

Public investment helps to create an appropriate enabling environment that influences farmers' incentives to invest. It also directly creates other forms of capital that support the development of a thriving agriculture sector. Some types of government investment are specific to agriculture and aimed specifically at enhancing primary production in the crop, livestock, aquaculture, and forest sectors as well as in upstream and downstream activities. These can be referred to as investments in agriculture. Government investment in other sectors can also have a positive impact on agricultural productivity and farm incomes. For example, investments in transport and communications infrastructure, energy, general education, health and nutrition, ecosystem services, market institutions, and broader legal and social institutions all support agriculture and can be considered as investments for agriculture.

From a helicopter view, there are four distinct investment streams to be recognized: (1) domestic private, (2) domestic public, (3) foreign private, and (4) foreign public.[1] Although there are data available for the public capital streams, there is no comprehensive set available for the private investments in agriculture on a global level.

The shape, size, and focus of the investment flows are going to be very different depending on the location of the area in which one intends to invest. In most OECD (Organisation for Economic Co-operation and Development) countries, industrial farming is the dominant form of farming. This refers to the industrialized production of livestock, poultry, fish, and crops. This form of farming is supported by innovation in machinery and methodologies, genetic technology, and techniques for the achievement of economies of scale and assumes global and free trade.

Although the trend is that the agricultural sector has generated more investor attention in recent years, it has been very asymmetric, in that there is a focus on certain assets rather than on the industry in general, but also on the location of investible interest. The trends in agricultural capital stock, agricultural capital stock per worker, and the composition of agricultural capital stock all suggest that investment is seriously lagging in low- and lower-middle-income countries, particularly in Sub-Saharan Africa and South Asia.

Arable land refers to land where crops can be cultivated. This usually refers to the totality of land that is not only already cultivated, but also land that has the potential to be cultivated, such as land where the soil and climate are suitable for agriculture, where there is no existing large-scale human settlement, or where the land is not protected by any land-rights regimen. While land can be made arable by various artifices, arable land in our context refers primarily to land that can be used for production with little or no modification. This is because modifications designed to recapture arable land are often expensive, energy-intensive or politically untenable, and this discussion of arable land is in the context of a future facing greater limitations on non-renewable resources.

1.1 Farmland essentials

Agricultural land (also *agricultural area*) denotes the land suitable for agricultural production, both crops and livestock. It is one of the main resources in agriculture. The standard classification (used, for example, by the FAO – the Food and Agriculture Organization of the United Nations) divides agricultural land into the following components:

- Arable land – land under annual crops, such as cereals, cotton, other technical crops, potatoes, vegetables, and melons; also land left temporarily fallow.
- Permanent crops – Orchards and vineyards (e.g., fruit plantations).
- Permanent pastures – areas for natural grasses and grazing of livestock, such as meadows and pastures.

The first two components – arable land and land in permanent crops – constitute so-called cultivable land. The part of arable land actually under crops is called sown land or cropped land. The term *farmland* is ambiguous in the sense that it may refer, on the one hand, to agricultural land and, on the other hand, to cultivable or even only arable land.

Depending on the use of artificial irrigation, agricultural land is divided into irrigated and non-irrigated land. In arid and semi-arid countries, agriculture is often confined to irrigated land, with very little farming possible in non-irrigated or rain-fed areas.

Agricultural land constitutes only a part of any country's territory, which in addition also includes areas not suitable for agriculture, such as forests, mountains, and inland water bodies. Agricultural land covers 33% of the world's land area, with arable land representing less than one-third of agricultural land (9.3% of the world's land area).

Reasons for arable land scarcity[2] are the following:

- Nearly 2 billion hectares worldwide have been degraded since the 1950s. These 2 billion hectares represent 22% of the world's cropland, pastures, forests, and woodlands (at present some 12% [over 1.5 billion hectares] of the world's land surface is used in crop production. This area represents over a third [36%] of the land estimated to be suitable for crop production to some degree). In particular, Africa and Latin America have the highest proportion of degraded agricultural land. Asia has the largest proportion of degraded forestland, as revenue-poor national governments pursue lucrative policies of deforestation. Degradation is not the only reason for declining levels of arable land. There are a variety of climatic, environmental and human factors that have an effect on available arable land resources.
- Erosion hazard, aluminum toxicity, soil shallowness, and hydromorphy as constraining between 13% and 16% of global arable land area. These soil constraints make a significant portion of land unsuited for the production of crops without serious modification or enhancement.
- Irrigation, deforestation, desertification, terracing, landfills, urban encroachment, and issues surrounding topology and landmass further constrain the availability of arable land.
- Increased investment and diversion of land for biofuel production has further contributed to declining availability of arable land.
- The increasing scarcity of arable land is also complicated by unequal land distribution. Land distribution refers to the concentration of arable land ownership held by a small, wealthy portion of the population.

The remaining arable land is unevenly distributed between regions and countries. An estimated 1.8 billion hectares of potential cropland are located in developing countries. Yet 90% of that 1.8 billion is in Latin America and Sub-Saharan Africa, and half of the total is concentrated in just seven countries (Brazil, the Democratic Republic of Congo, Angola, Sudan, Argentina, Colombia, and Bolivia). There is virtually no spare land available for expansion in South Asia, the Near East, or North Africa. The solutions to addressing the availability of arable land are threefold: (1) the production of more arable land, (2) an increase in the productive capacity of existing cultivated land, and (3) the conservation of arable land in order to prevent degradation. Yield improvements can be added to this list, and all will be extensively analyzed in the chapter on farmland.

1.2 Food wastage[3]

Globally, between a third and a half of all food produced for human consumption is lost or wasted, amounting to around 1.3 billion tons per year.[4] According to a recent study released by the UK-based Institution of Medical Engineers, this figure increases to up to half of all food purchased in Europe and the United States being thrown away without being consumed.

The defining element of the problem is that food wastage occurs at all steps along the food supply chain, with the majority of food wastage in developing nations occurring during the production phase; conversely, in developed nations, an average of around 100 kg of food per person is wasted in the consumption phase. In Figure 1.1, an overview is provided of the sources of waste in the agricultural value chain.

Food waste and food loss are not a new phenomenon. However, they have been lacking a proper definition. A study conducted by the Swedish Institute for Food and Biotechnology (SIK) on behalf of the Food and Agriculture

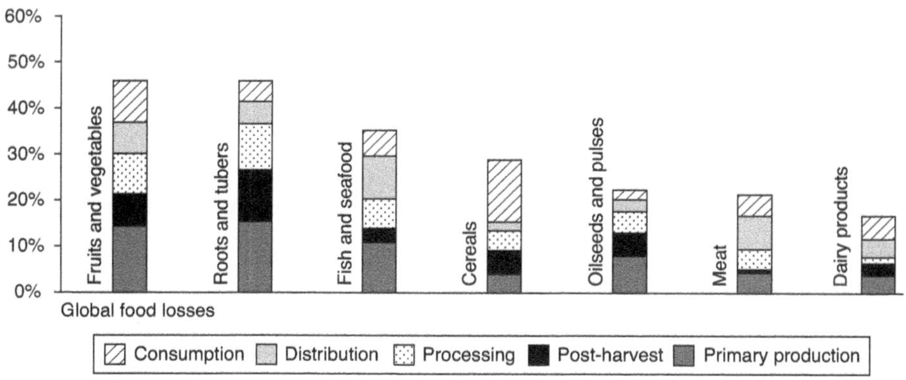

Figure 1.1 Distribution of waste across the agricultural value chain
Source: FAO.

Organization of the United Nations (FAO), "Global Food Losses and Food Waste," draws the distinction between food loss and food waste and provides a definition that will be used throughout this chapter. According to the SIK, the definitions are as follows:

- Food loss measures the decrease in edible food mass (excluding inedible parts and seed) "throughout the part of the supply chain that specifically leads to edible food for human consumption," that is, loss at the production, consumption, postharvest, and processing stages. This definition of loss includes biomass originally meant for human consumption but eventually used for some other purpose, such as fuel or animal feed.
- Food waste is food loss occurring during the retail and final consumption stages due to the behavior of retailers and consumers – that is, the throwing away of food.

The root causes of food wastage differ significantly between developing and developed countries. Broadly speaking, the causes can be split into three distinct categories:

- Production – "from the farm to the factory," it is at this step in the supply chain that the majority of food loss occurs due to either inadequate or inappropriate agricultural processes, usually in developing countries.
- Processing – from the factory to the end retailer, a large element of this stage in the developed world is the overuse of and reliance on excessive packaging.
- Retail – the guilt of the developed world, where we are responsible for as much food wastage as the developing world is for food loss. Covering such issues as sell-by or use-by dates, rejection of produce on aesthetic basis, or simply buying more food than we're prepared to eat, this is the easiest end of the problem to resolve.

Table 1.1 Food loss per continent

Food waste and loss per person per year (kg)	Total	At the production and retail stages	By consumers
Europe	280	190	90
North America and Oceania	295	185	110
Industrialized Asia	240	160	80
Sub-Saharan Africa	160	155	5
North Africa, West and Central Asia	215	180	35
South and Southeast Asia	125	110	15
Latin America	225	220	25

To put these numbers into perspective, Table 1.1 reflects the food loss per continent.[5]

As the above table shows, developed and developing countries differ substantially. In developing countries, more than 40% of losses occur at the post-harvest and processing stages, whereas in the developed world an equivalent amount of food loss occurs at the retail and consumer stages. The total food waste by consumers in developed countries (222 million tons) is almost equal to the entire food production of Sub-Saharan Africa (230 million tons).

1.3 Agricultural policy and international trade

It would be beyond the scope of this book to introduce the details of the different agricultural policies pursued by different countries and regions and how they differ in name and details. The agricultural and food sector is an ideal case for investigating the political economy of public policies. Many of the policy developments in this sector since the 1950s have been sudden and transformational, while others have been gradual but persistent.[6] One of the most notable regional agricultural policies is the European Common Agricultural Policy (CAP),[7] which was developed decades ago and has since then undergone some transformations, although too slowly for the liking of the international community. Each national or regional policy is known for its fair share of beneficial treatment of local producers and products and export incentives.

As shown in Table 1.2, a common set of tools is available to design an agri-environmental policy.[8]

During the last ten years, a major overhaul[9] has been prepared which is now in its implementation phase.[10] The ten key CAP reform points are as follows:[11]

1. Better-targeted income support
2. Improved crisis management principles
3. Green payment for preserving long-term productivity
4. Add R&D investments
5. More competitive and balanced food chains
6. Encouragement of agri-environment initiatives
7. Facilitation of the establishment of young entrepreneurs
8. Stimulating rural employment and entrepreneurship
9. Better addressing of sensitive areas[12]
10. A simpler and more efficient CAP

While agriculture accounts for less than 10% of merchandize trade, high and variable agricultural distortions appear to cause the majority of the cost of distortions to global merchandize trade. Within agriculture, most of the costs

Table 1.2 Common tools in agri-environmental policy

Policy type	Implication
Standard regulation	Standard regulation bans the use of certain (detrimental) inputs and prescribes the use of precautionary measures.
Environmental tax	Input-oriented taxes allow farmers to use the taxed input only when farming can still be profitable with the tax. There may also be output-oriented taxes (e.g., undesired output).
Tradable quotas	Contrary to the environmental tax which deals with price regulation, quotas regulate the quantity of environmental certificates tradable on a special market.
Environmental auctions	An effective solution on a smaller-scale.
Communicative policies	Communicative policies lead to higher uptake levels of the agri-environmental schemes on the production side and improved market transparency on the side of the consumer.
Agri-environmental schemes and measures	AE schemes represent a voluntary instrument and are a mixture of regulatory instruments with economic incentives. They compensate farmers for yield and income loss and higher production costs due to the implementation of environmentally friendly practices.
Cross-compliance	Cross-compliance rules represent an obligatory approach. Noncompliance with certain environmental standards makes farmers ineligible to receive other types of payments, for instance direct payments.
Community-based schemes	The idea behind this instrument is to fund local initiatives aimed at pursuing policy goals at the regional or local level.

appear to arise from trade barriers levied on imports, since these barriers tend to be high, variable over time and products, and levied by a wide range of countries. Negotiations faced a need for balance between, on the one hand, discipline in reducing tariffs and hence creating the market-access gains that were central to the negotiations and, on the other, flexibility in managing political pressures. While the approach of providing flexibility on a certain percentage of tariff lines is seriously flawed, the proposed modalities still appear to provide worthwhile market-access. Better ways appear to be needed to deal with developing countries' concerns about food price volatility while reducing the collective-action problems resulting from price insulation.[13]

Although it has not been the case for a very long time, ultimately agriculture was included in the WTO[14]/Doha Round of negotiations dealing with international trade reform, and this despite the strong political resistance to reforming the industry. Limited results have been booked so far, and within the Doha Round of negotiations it is unclear what can be expected further, if anything. Potentially the comprehensive free-trade negotiations between the United States and the European Union that started in 2013 and which were part of a comprehensive plan to fuel the sluggish global economy will lead to a major leap forward that can function as a stepping stone for a global deal. However, it remains to be seen which exceptions will stand at the end of the day within the context of a full free-trade treaty between two continents.

Not surprisingly, investors' behavior is a function of these (changing) policies. Recently, a study[15] was conducted analyzing investor behavior in the European Union and delivering a comparative analysis among selected EU member states of the investment demand of a sample of specialized field crop farms for farm buildings, machinery, and equipment as determined by different types and levels of Common Agricultural Policy support. The study allowed for the existence of uncertainty in the price of output that farmers receive and for both long- and short-run determinants of investment levels, as well as for the presence of irregularities in the cost adjustment function due to the existence of threshold-type behaviors. The empirical estimation reveals that three investment regimes are consistently identified in Germany and Hungary, across asset and support types, and in France for machinery and equipment. More traditional disinvestment-investment type behaviors characterize investment in farm building in France and the UK, across support types, and in Italy for both asset classes under coupled payments. The long-run dynamic adjustment of capital stocks is consistently and significantly estimated to be toward a – mostly non-stationary – lower level of capitalization of the farm analyzed. By contrast, the expected, largely positive short-run effects of an increase in output prices are often not significant. The effect of CAP support on both types of investment is positive, although seldom significant, while the proxy for uncertainty employed fails to be significant yet in most cases has the expected effect of reducing investment levels.

In general, for decades the world's agricultural markets have been highly distorted by national government policies, but very differently for different commodities. Hence a weighted average across countries of nominal rates of assistance or consumer tax equivalents for a product can be misleading as an indicator of the trade or welfare effects of policies affecting that product's global market. This is especially the case when some countries tax and others subsidize a commodity's production or consumption.[16]

The above results reveal the considerable extent of reform in the agricultural policies of developing as well as high-income countries over the past two decades. Similar studies have been done looking at the expected change in land use post-CAP 2013.[17]

1.4 The agricultural value-added chain

The agricultural value-added chain can be depicted as shown in Table 1.3.

Table 1.3 Agricultural value chain

Resources	Upstream activities	Production	Transport and storage	Processing	Distribution
Land	Machinery	Plantations	Infrastructure	Flour mill	Retail
Water	Fertilizer	Annual crops	Vessels	Sugar refinery	Wholesale
Labor (1) work, (2) intellectual	Plant protection	Animal production	Silos	Slaughterhouse	Export
Climate	Breeding	Aquaculture	(Climate-controlled) Warehouses	Biogas	
Capital	Training	Forestry			
Soil fertility					

1.5 Organic farming[18]

Organic agriculture is a production management system that aims to promote and enhance ecosystem health, including biological cycles and the biological activity of soil. It is based on minimizing the use of external inputs and represents a deliberate attempt to make the best use of local natural resources. Methods are selected to minimize pollution of air, soil, and water. Organic agriculture comprises a range of land, plant, and animal management procedures, circumscribed by a set of rules and limits that are usually enforced by inspection and certification schemes. Synthetic pesticides, mineral fertilizers, synthetic preservatives, pharmaceuticals, genetically modified organisms (GMOs), sewage sludge, and irradiation are prohibited in all organic standards.

1.6 Bio-based economy[19]

Agriculture is playing an increasingly important role in the bio-based economy, providing feedstocks for the production of liquid fuels, chemicals, and advanced materials, such as natural fiber composites for industry. The emergence of green industries provides new opportunities for the rural sector, beyond traditional forestry and the supply of wood. Biological science has the potential both to make incremental efficiency improvements and to bring radical change in a wide range of sectors, including through the use of enzymes, fermentation, and organisms for processes and products in the energy, chemical, pharmaceutical, food, textile, and pulp and paper industries.

The application of biological and material science to agriculture has the greatest potential in the energy, natural fiber composite, and starch sectors.

Much of this potential is already being realized, especially through the rapid growth of the biofuel sector. Currently, ethanol is produced from easily fermentable agricultural feedstocks such as sugarcane, sugar beet, cereal grains, and cassava. Biodiesel is produced from vegetable oil (typically rapeseed, soybean, and palm oils) using a process of chemical modification. The emerging bio-based economy centers on energy efficiency, renewable feedstocks for polymer products, industrial processes that reduce carbon emissions, and recyclable materials. Natural fibers exemplify these attributes. For example, growing 1 ton of jute fiber requires less than 10% of the energy used in the production of competing polypropylene. The sustainability of a rapidly growing bio-based agricultural economy, especially one that relies on liquid fuels, has generated the "food-versus-fuel" debate. The links between bio-industries and food security are complex and multifaceted. Ensuring the sustainable development of bio-sectors becomes challenging when efforts are made to capture the potential benefits for development. For instance, the rapid growth and sheer scale of the biofuel sector has potentially negative implications for all four dimensions of food security – availability, access, stability, and utilization – as it may increase competition for land and water resources. For a more detailed discussion, see the chapter on biofuels.

1.7 Vertical farming

From an investor point of view, the agricultural sphere is not necessarily the most attractive investment area. The dynamics and high volume of input variables make it a very volatile environment characterized by high levels of uncertainty on top of the long payback periods (in many instances). Mother Nature is not the best and most trustworthy business partner one can imagine. Combined with demographic and other societal issues, the list of complications is very long. To name a few, (1) over 7 billion people need to be fed, (2) climate change will alter the agricultural landscape, (3) many agricultural commodities are not exhaustible relative to their hard-commodity counterparts, (4) food-borne illnesses are on the rise, (5) drinking water is becoming scarce in many parts of the world, (6) more crops are failing due to plant pathogens and insect pests, (7) half the world goes to bed hungry every night, and (8) half the world harbors 80% of the intestinal worm infections. Many of these issues will have a negative impact on crop production. Wheat rust and rice blast account for 125 million tons of impacted production. The 2012 drought in the United States caused an estimated 20 billion US dollar loss in crops, the monsoon in Asia arrives much earlier than normal and is much more aggressive, deforestation caused to a large extent by cattle ranging and only in minor ways by small- and large-scale agriculture and logging all lead to reduced agricultural production. The share of deforested land that ultimately is used for extensive agriculture has diminished from 80% in the 1980s to just above 40% in 2010. And last on the list, the loss due to flooding, storage and shipping, extended winter periods, and the expansion of desert areas all magnify the uncertainty related to agricultural investing.

Consequently there is a need for more controllable alternatives. Controlled urban indoor farming is one of those initiatives. In an urban environment, the scarcity of landmass leads to the idea to engage in farming using the height of a skyscraper to cultivate crops in a controlled environment. The concept is not only about more stable food production but also the potential for horticulture, vaccination and the development of better-quality seeds and crops (and reduced carbon emissions). Although vertical farming (VF) projects have been developed in many parts of the world, the quest is now for a way to turn it into an affordable alternative through increasing volume and size and combining best practices in making the projects more efficient.

1.7.1 The concept and characteristics

Currently, traditional agriculture makes it difficult to achieve profitability, distances customers from their food, and hurts the environment. The supply chain for food and the typical disconnect between the customer and where their food is produced leads to low profit margins and an environmentally unfriendly model going from industrial farm to transportation to wholesaler/distributor to retailer to customer. The sky or vertical farm combines horizontal and vertical dimensions to magnify the space available for an urban farm. Different levels of integration are then possible: (1) the you-pick system, where in the same building where crops are cultivated, people live, shop for these products, and wine and dine using the crops cultivated there; (2) The farmers-market model where humans would not be living in the same building space and would therefore have to travel to the urban farm to shop, wine and dine, etc.; (3) the education lab, where universities can conduct classes and training regarding agricultural matters; and (4) the export model where shopping and wining and dining are disconnected from the urban farm as well as living environments of citizens, demonstrating a deteriorating level of integration when moving through the different models indicated.

Vertical farming can therefore be described as the cultivation of plants and animal lives within a skyscraper greenhouse. In theory, the idea of a vertical farm has been around as long as the Hanging Gardens of Babylon, one of the Seven Wonders of the Ancient World. Pilot projects have been around since as early as 2009, and the first commercial vertical farm became operational in 2012 and is located in Singapore, counting three stories.

The advantages can be identified as follows (most being derived from scaling up hydroponic [growing plants using mineral–nutrient solutions in water without using soil] or aeroponic [the process of growing plants in an air or mist environment without the use of soil or other aggregate medium] growing methods):

- It brings farming closer to the urban consumer, which is expected to account for 80% of the global population by 2050 which is expected to be over 9 billion people. This will reduce the need for the expansion of arable land globally and the increase in crop yields, which has become somewhat technologically constrained.
- It not only can produce more crops (higher yield) but also do so year round. All-season farming has the potential to increase the productivity of crops by a factor somewhere between 3 and 40 times depending on the type of crop, given a similar amount of farmed surface.
- In the case where crops are sold in the same infrastructure where they are cultivated, they don't need transportation, thus avoiding spoilage, infestation and energy consumption. Spoilage and infestation often create a loss of up o 30% of total production in many emerging countries.
- The use of dwarf versions of certain crops that are smaller but richer in nutrients would allow for richer crop yields than when using their default versions.
- The protection of crops against the background of changing climate patterns and all sorts of weather anomalies is increasingly important. Since vertical farming happens in a controlled environment, crop production would be insulated from weather events.
- Efficiency and conservation of resources: the efficiency ratio is estimated to be up to 20 times that of outdoor farming. It further reduces the need for additional arable land and reduces issues such as deforestation and overall pollution. As mentioned, the reduction of the food supply chain will save on transportation and refrigeration costs as well as the fossil fuels used in conventional plowing, planting, and harvesting farm machinery.
- The controlled crop environment reduces the need for pesticides, herbicides, and fungicides.
- It halts the extinction of certain animal and vegetative species due to mankind's quest for more farmland in precious unspoiled parts of the world.
- It is less damaging to the health of agricultural workers due to nonexposure to chemicals, infectious diseases, and injuries caused by the use of powerful (motorized) agricultural equipment.
- Vertical farms have the potential to become partly energy independent through the use of methane digesters, which could deliver energy to the project by transforming organic waste into biogas, which to a large degree is composed of methane.

In short, the concept further embodies:

- a groundbreaking opportunity for the high-efficiency growing of quality fresh food in city, urban, and rural environments in any climate;
- minimum skill levels for consistent product delivery due to the integration of proven industrial-grade computer monitoring and control systems;
- optimum growing conditions 365 days a year, resulting in many more crop rotations per year than previously possible with traditional farming;
- a sustainable and effective solution to increasing concerns in many regions regarding security of the food supply and its quality, adverse impacts on agriculture from climate change, and the diminishing availability of arable land and traditional water resources; and
- many worthwhile commercial advantages to both the growers, wholesalers, and consumers, including
  - super reliable production – no crop losses from weather, pests or wildlife;
  - a sterile product – no prepack washing required;
  - reduced carbon emissions – typically no refrigerated transport from farm to retail location;
  - consistently high-quality product – no pesticide, fungicide or herbicide used;
  - the ability to "tune" crop characteristics to specific market demands and expectations;
  - production that can be established in close proximity to any client or market location; and
  - rapid flexibility of crop variety being grown to meet seasonal market opportunities.

Additionally, high-efficiency LED lighting used by VF systems matches specific photosynthetically active frequencies to greatly enhance plant growth for minimum energy inputs. VF systems also optimize daylight and night-time durations, overcoming the seasonal limitations that occur in traditional outdoor farming.

Plants must have quality air for optimum growth when they breathe in CO_2 and breathe out oxygen. VF systems control air temperature and humidity as well as UV sterilize the air for maximum plant protection against yeasts, moulds, and other harmful bacteria.

Optional CO_2 enhancement further increases natural plant growth in several of the VF systems developed around the world.

Water used by edible plants must be free of heavy metals, harmful bacteria, and algae, and yet rich in oxygen and essential minerals and nutrients. As well as UV sterilization, VF systems use magnetic fields to neutralize the bonding ability of minerals and reduce their size to less than four microns, which stops algae growth, improves plant health through faster mineral uptake, and allows plants to thrive even when using higher-salinity bore water. Water used in VF systems is automatically monitored and controlled to be rich in oxygen, with an exact balance of biologically active microbe-based nutrients as well as the correct temperature and pH level.

The vertical farming model has led to the development of even smaller-scale micro-vertical farms as close to the urban consumer as their kitchen.

SmartGreens, as they are often referred to, often have more than 1.5 m^2 of growing area built into a modern cabinet no bigger than a normal refrigerator. These SmartGreens give everyone the freedom to have an organic garden in their own kitchen or pantry, with all the benefits of saving money and serving family and friends the healthiest fresh greens, sprouts, and herbs available. Chefs in restaurants and resorts are able to use SmartGreens to grow microgreens, garnish herbs, and salad greens specifically to suit planned menus and to serve only premium fresh-picked produce as ingredients in their dishes.

The patented SmartGreens technology uses energy-efficient LED lighting and biologically active plant minerals to provide a simple, reliable, and cost-effective way to grow a wide range of salad greens, herbs, sprouts, and microgreens all year round. SmartGreens is connected to a standard power point, filled with water, and the selected seeds dropped into individual pots in the growing trays. A simple touch of the LCD touch screen, and SmartGreens does the rest. Many different plant types can be grown in the SmartGreens at the same time. Individual leaves or entire plants can be picked as required. Many plants can even be cut two or three times before needing to be replanted.

1.7.2 The remaining challenges

a. Financial sustainability

It has been argued that the intended cost benefits of reduced transportation outweigh the operational costs of running a vertical farm, including the additional costs of lighting, heating, and power. The economic and ecological benefits of running a vertical farm are largely built around minimizing "food miles," although recent research suggests that transportation is only a minor contributor to the economic and environmental costs of supplying food to urban populations.[20]

In case the power needs of the vertical farm are met by fossil fuels, the environmental benefit might be lost. That would even be true under low-carbon-producing conditions relative to normal farmland exploitation.

The reality, however, is that convincing financial models have never really been presented showing evidence one way or the other and taking into account all variables, including the often very expensive office rent in major cities and hubs around the world.

b. Energy use and efficiency

During the summer, or sunny season, sunshine will cover a vertical platform from a steep angle, thereby producing much less daylight for indoor crops than under normal natural circumstances on a farmland. Additional artificial light is needed to produce economically viable crop yields. Stress testing has been performed on financial models with respect to the cost of artificial light required (they used the cost of normal light bulbs instead of LED lights, which can reduce the cost by 50% to 70%). Nevertheless, the results are somewhat disturbing as some estimates point to an additional cost of almost 10 US dollars[21] per loaf of bread produced.

Besides the lightning, heating, and cooling costs, there is the complicating matter – both technically as well as financially – of plumbing and elevator systems to distribute food and water throughout the structure. One possible solution for this is to build an anaerobic digester that will allow the farm to operate off the energy grid (the biogas opportunity mentioned earlier in this chapter) and additionally will recycle waste from nearby businesses.

c. Pollution and other environmental aspects

Typical greenhouse models are known to create more greenhouse gases than field production, largely due to a higher energy cost per produced unit. Given the additional cost components mentioned before, the risk of additional carbon emissions is significant.

As plants acquire, under normal circumstance, nearly all their carbon from the atmosphere, vertical farm growers commonly supplement CO_2 levels at three to four times the rate normally found in the atmosphere. This increase in CO_2, which has been shown to increase photosynthesis rates by 50%, contributes to the higher yields expected in vertical farming. It is not uncommon to find greenhouses burning fossil fuels purely for this purpose, as other CO_2 sources, like from furnaces, contain pollutants such as sulphur dioxide and ethylene, which significantly damage plants. This means that a vertical farm will require a CO_2 source, most likely from combustion, even if the rest of the farm is powered by "green" energy. Also, through necessary ventilation, much CO_2 will be leaked into the city's atmosphere.

Greenhouse growers commonly exploit photoperiodism in plants to control whether the plants are in a vegetative or reproductive stage. As part of this control, growers will have the lights on past sunset and before sunrise or periodically throughout the night. Single-story greenhouses are already a nuisance to neighbors because of light pollution; a 30-story vertical farm in a densely populated area will surely face problems because of its light pollution.

Hydroponic greenhouses regularly change the water, meaning there is a large quantity of water containing fertilizers and pesticides that must be disposed of. While solutions are currently being worked on, the most common method of simply spreading the mixture over a sufficient area of neighboring farmland or wetlands would be more difficult for an urban vertical farm.

1.7.3 Complementary technology used in vertical farms

Previously, the concept and advantages of vertical farms were described. In executing their business model, vertical farm operators use a variety of techniques to grow crops, enhance yields, and attempt to become or stay financially sound.

The most commonly used techniques in vertical farming are as follows:

- Greenhouse dynamics: cultivating crops and plants indoors mainly under a glass or plastic roof. It heats up because incoming visible solar radiation (for which the glass is transparent) from the sun is absorbed by plants and crops inside the building. Air warmed by the heat from hot interior surfaces is retained in the building by the roof and wall. In addition, the warmed structures and plants inside the greenhouse re-radiate some of their thermal energy in the infrared spectrum, to which glass is partly opaque, so some of this energy is also trapped inside the greenhouse. The primary heating mechanism of a greenhouse is convection.
- Composting: refers to the process of organic matter decomposed and recycled as fertilizer or soil amendment. Modern-day methodical composting is a multistep, closely monitored process with measured inputs of water, air, and carbon- and nitrogen-rich materials. The decomposition process is aided by shredding the plant matter, adding water, and ensuring proper aeration by regularly turning the mixture. Compost is quite often rich in nutrients.
- Grow light: artificial (often electric) light source designed to stimulate plant growth by emitting an electromagnetic spectrum appropriate for photosynthesis. Grow lights either attempt to provide a light spectrum similar to that of the sun, or a spectrum that is more tailored to the needs of the plants being cultivated. Outdoor conditions are mimicked with varying color temperatures and spectral outputs from the grow light, as well as varying the lumen output (intensity) of the lamps. Depending on the type of plant being cultivated, the stage of cultivation (e.g., the germination/vegetative phase or the flowering/fruiting phase), and the photoperiod required by the plants, specific ranges of spectrum, luminous efficacy, and color temperature are desirable for use with specific plants and time periods.

- Folkewall: a construction with the dual functions of growing plants and purifying waste water. This technique makes an efficient use of space by fulfilling two essential functions: vertical plant growing and purification of gray water. The advantages are a better use of gray water, more efficient use of the area, purification of the percolating water, and a lower cost of housing.
- Aero-/hydro-/aquaponics: hydroculture subsets focusing on growing plants and crops by using mineral–nutrient solutions in water without soil (hydro), in an air or mist environment without the use of soil (aero) or by combining conventional aquaculture with hydroponics in a symbiotic environment.[22]
- Phytoremediation: involves the treatment of environmental problems (bioremediation) through the use of plants that mitigate the environmental problem without the need to excavate the contaminant material and dispose of it elsewhere. Phytoremediation consists of mitigating pollutant concentrations in contaminated soils, water, or air with plants able to contain, degrade, or eliminate metals, pesticides, solvents, explosives, crude oil and its derivatives, and various other contaminants from the media that contain them.

Despite the wide variety of techniques and piloting cases worldwide using this concept, the fundamental investment question relates back to the (financial) sustainability of the concept and the different models applied. No convincing (financial) models or properly executed projects are readily available that would allow the concept to go mainstream in the near- to midterm future.

1.8 Precision agriculture

Precision agriculture (PA) can be defined as a suite of technologies (yield monitor, GPS, VRT, and GIS) that promote improved management of agricultural production by accounting for variations in crop performance in space. It is also sometimes called "precision farming," "site-specific management" or "information-intensive farming." A yield monitor, the cornerstone of PA, is a sensor installed within a header that enables continuous measurement of yield during harvest. It must be connected to a GPS to generate maps. A geographical information system (GIS) is a software package that allows organization, analysis, and illustration of maps, images, and points. A global positioning system (GPS) is a positioning system based on an array of 24 satellites that orbit the earth. This is a key technology on which precision agriculture depends since it enables precise spatial location within paddocks. Variable-rate technology (VRT) is a new technology that enables agricultural inputs such as fertilizers to be varied continuously within or between paddocks. It comprises one or more of the following components in a spreader or sprayer: variable-rate capacity, GPS, and/or a digital map control.[23]

1.8.1 Applications

Precision agriculture (PA) technologies can enhance the productivity of irrigated agriculture. It is expected that growers will adopt new technology only when it solves a problem in an economical way.

a. Use of mechanized technology and farm management

Technological innovation in agriculture has been the basis for the outstanding gains in productivity that have been achieved.[24] In the area of energy use, significant progress was made after the transition from animal-drawn implements to mechanized operations. Seeking higher levels of field capacity, modern agriculture has experienced a process of using ever-increasing farm equipment size. This trend has promoted management systems that rely on using production inputs in a uniform way across the field. Uniform management has the advantage of being simple; however, it disregards the spatial and temporal variability within the field that impacts crop yield. As the average farm size has increased due to land consolidation, there has been an unintended loss of efficiency in the use of energy-intensive production inputs. Uniform management, in order to be sustainable, needs to deal with the ever-increasing cost of energy inputs as well as its environmental footprint.

b. Application

Advances in agriculture technologies over the last 20 years point to a new paradigm in which two views are combined:

i. The use of energy-intensive inputs (i.e., chemical fertilizers and pesticides) based on the right amounts, at the right time, and in the right place. This type of management is commonly known as "site-specific management."
ii. Technology used to perform the conventional pattern of mechanical operations but at higher levels of efficiency. This is the case of tractor auto-guidance.

This new paradigm defines the context of PA, a new production system with efficiency in mind. PA technologies are, to some extent, alternative ways to do the same operations while saving on resources such as time, labor, fuel, fertilizers, water, and so forth. Because resources can be utilized more efficiently, the use of PA technologies can also help in reducing the environmental footprint of agriculture.

Table 1.4[25] shows how information-intensive technologies in precision agriculture are an integral component of modern farming (traditional operations versus PA upgrade).

Table 1.4 Information-intensive technologies in precision agriculture

Traditional operations	PA technology	Benefits	Observations
Chisel plow	Tractor steering control	• Keep consistent spacing between runs. • Makes possible strip tillage in conservation systems and drip-irrigated fields.	• Light-bar, steering assist, or hands-free options. • High levels of accuracy are generally not required.
Leveling	RTK elevation monitoring	Yearly maintenance of field slopes without expensive laser leveling.	Requires high level of accuracy.
Disking	Tractor steering control	• Avoid excessive overlap between runs (more land covered). • Optimizes turning pattern.	• Light-bar, steering assist, or hands-free options. • High levels of accuracy are generally not required.
Row listing	Tractor steering control	• Repeatable row position year after year. • Avoid guess row. • Increase speed of operation. • Improve timeliness of operations prior to planting. • Optimize turning pattern. • Reduce operator's fatigue.	• Makes possible day and night shifts. • Steering assist, or hands-free options. • Requires high level of accuracy. • Recommend using stabilizers to reduce implement drift.
Bed shaping/ mulching	Tractor steering control	• Improves bed uniformity, straight pattern, and spacing between runs.	• Steering assist, or hands-free options. • Requires medium-high levels of accuracy. • Recommend using implement stabilizers to reduce drift.

Planting	• Tractor steering control • Variable-rate seeding	• Deliver more seed in germination-trouble areas. • Drills: avoid overlap between runs (saving seed). • Row planters: straight seed line (for close cultivation) and spacing between runs.	• Steering assist, or hands-free options. • Requires high level of accuracy. • Variable application rates can be achieved through prescription map or manual adjustment.
Border disc	Tractor steering control	• Straight berms to create uniform-width borders. • Repeatable position within season and year after year.	• Steering assist, or hands-free options. • Requires medium-high levels of accuracy.
Spraying (of chemical pesticides and liquid fertilizer)	Tractor/sprayer steering control Rate control with speed compensation and variable-rate spraying Boom section control Boom height control	• Directional control is particularly useful in flat-ground applications. • Avoid excessive overlap between runs. • Reduce chemical drift. • Avoid spraying outside field boundaries.	• Steering assist, or hands-free options. • Requires medium to high levels of accuracy. • Variable application rate through prescription map or manual adjustment. • Single/multiple chemicals.
Side-dressing	Tractor steering control Rate control with speed compensation and variable-rate application	• Automatic dispensing of fertilizer according to prescription. • Tracking actual amounts of fertilizer used. • Increased speed of operation.	• Controller changes rates automatically according to zones and changes in speed. • Can be combined with other operations in the same pass (i.e., cultivation).
Cultivation	• Tractor steering control • Implement steering control	• Allows cultivation close to the seed line. • Increase speed of operation. • Optimizes turning pattern. • Reduce operator's fatigue.	• Can be combined with other operations in the same pass (i.e., side-dressing).
Harvest	• Harvester (hay swather, grain combine, cotton picker) steering control • Yield monitors	• Improves turning pattern to avoid incomplete swaths. • Records differences in yield with the field.	• Provides yield information that can be used to detect trouble areas, delineate management zones, evaluate variety performance, etc.

As can be derived from the table, virtually all operations in modern farming can be improved with the use of PA technologies. Also included are some of the benefits of adopting these technologies and important observations. It is worth noting that the most frequent PA technology adoption is tractor auto-guidance, which has been commercially available for the last ten years. The high rates of adoption of tractor auto-guidance[26] demonstrate that this technology is perceived by growers as a significant improvement on mechanized operations. The table also denotes one trend, which is the integrated use of technologies. It can be observed that steering control can be used in combination with rate controllers and other hardware (such as global positioning systems [GPS], hydraulic valves, and pumps) to increase the functionality of the same power unit (tractor or sprayer) and allow multiple operations to take place on the same pass through the field (i.e., cultivate and side-dress at the same time). As we will see (infra), this is possible through the use of vehicle-mounted computer displays that have exceptional computing and electronic communication capabilities for monitoring and control functions.

c. Precision agriculture equipment and software[27]
   1. GPS receivers

Several kinds of GPS receivers are available for growers to use on the farm. The most basic type corresponds to handheld receivers, sometime used for recreational purposes, but still these are quite useful for navigation and georeferencing field samples. Generally, these receivers have low levels of accuracy (about three to five feet), but are inexpensive solution alternatives. Next are machine-mounted GPS receivers with external antennae whose purpose is to provide positioning information that enables a variety of functions such as guidance, variable-rate application, and yield monitoring. These receivers range in accuracy from 12 inches to less than an inch depending on the correction source. Generally speaking, medium levels of accuracy can be obtained with free-of-charge correction sources available through satellite or ground stations. For high accuracy, GPS signal correction can only be attained from fee-based sources or by installing an on-site base station. Growers should keep in mind that cost and accuracy go hand in hand, and the need for a specific accuracy level depends on the application.

   2. Machine steering control

Early developments in machine guidance are the "light-bar" systems. These are pieces of hardware mounted in the windshield of the tractor/sprayer/

harvester cab in direct operator line of sight. With proper configuration, these units guide the operator who visually observes a series of green/red LED lights to indicate the machine's position relative to the desired direction. Light bars provide a good entry-level option for steering control, but limitations are evident when compared to the more sophisticated "hands-free" systems currently available. "Steering assist" systems are devices that attach to the steering wheel – or steering column – to take over the directional machine control. One advantage of these systems is portability because the unit engaging the steering wheel can be easily moved from one machine to another. Steering assist systems provide an excellent option to auto-guidance, but their accuracy is lower than the top-of-the-line autopilot systems that interface with the machine steering through the hydraulic system. These "hands-free" autopilot systems generally use the most precise GPS receivers that provide consistent signals of less than one inch accuracy. Directional control at this level requires extra navigation hardware and motion sensors such as gyroscopes and accelerometers.

In power-demanding operations such as tillage, autopilot systems allow growers to save fuel and time by optimizing the number of passes into the field. Moreover, autopilot makes tractor operation easier, reducing operator fatigue and allowing the driver to pay more attention to the quality of the work done by the tractor/implement system. Proper use of autopilot results in more productivity that is seen in more acres covered during the operator's shift. During the growing season, mechanized operations for planting, weed and pest control, fertilization, and harvest can be performed with high-efficiency by integrating a variety of PA technologies such as steering control, variable-rate controllers, yield monitors, etc.

3. Multifunction computer displays

The integration of measurement and control systems on the machine is done through cab computer display. Computer displays installed in the cab can combine a variety of functions such as guidance, variable-rate application, and land leveling. These systems are very good options for farmers using the same tractor for a number of different applications. The display is a human–machine interface communicating with peripheral devices (e.g., servo-hydraulic valves, GPS receivers, rate controllers, speed sensors) to enable monitoring and provide control for a variety of functions. Multifunction computer displays (such as the Ag Leader Integra, John Deere GreenStar, and Trimble FMX) are available for tractor systems that are capable of a variety of functions, such as automatic guidance, yield monitoring, variable-rate seeding, spraying, side-dressing, and ground elevation monitoring.

4. Rate controllers

Rate controllers are devices designed to control the delivery rate of chemical inputs such as fertilizers and pesticides, either liquid or granular. These rate controllers monitor the speed of the tractor/sprayer traveling across the field, as well as the flow rate and pressure (if liquid) of the material, making delivery adjustments in real-time to apply a target rate. Rate controllers have been available for some time and are frequently used as stand-alone systems, but the newer computer displays installed in the machine cab usually are capable of communicating with these rate-control devices to enable higher functionality such as memory storage for as-applied and prescription map based rate changes.

5. Soil and plant sensors

Sensor technology is an important component of PA technology, and their use has been widely reported to provide information on soil properties and plant fertility/water status. Adamchuk[28] provided a comprehensive list of current sensors as well as desirable features for new sensors to be developed in the future. One of the most popular ways to characterize soil variability is surveying the field with soil-apparent electrical conductivity (ECa) sensors that collect information continuously when pulled over the field surface. Because ECa is sensitive to changes in soil texture and salinity, these sensors provide an excellent baseline to implement site-specific management.

An area of intense study is the use of active-light canopy reflectance sensors to infer a crop's nutritional status by estimating plant size, biomass, and photosynthetic activity and use this information to activate rate-control systems to dispense fertilizers and/or plant growth regulators (PGR). There are published extension bulletins on the use of these spectral sensors for real-time nitrogen application in wheat,[29] corn,[30] and cotton.[31] The operational principle of active-light spectral sensors is based on measuring the % of light at certain wavelengths that bounces back after modulated light is directed to the crop canopy. Vegetation indices are computed based on the amounts of reflected light from the green, red, and near-infrared portions of the electromagnetic spectrum. Sensor-based management is an emerging area with very high potential to improve fertility and PGR management through real-time variable-rate applications.

6. Yield monitors

Current technology for yield monitoring includes the mechanical harvest of grains and cotton. In the case of grains, yield is continuously recorded by measuring the force of the grain flow as it impacts a sensible plate in the clean grain elevator of the combine. Some cottonseed yield-monitoring systems use optical sensors installed in the mechanical picker chutes to measure cotton flow rate as harvested cotton travels from the header to the basket. A recent development of a cotton mass flow sensor works on the principle of transmitting beams of microwave energy and measuring the portion of that energy that bounces back after hitting the stream of cotton flowing through the chutes. In all yield monitors, GPS receivers are used to record the location of yield data and create yield maps.

Other yield-monitoring systems include devices used in forage crops to keep track of weight, moisture, and other information on a per-bale basis. There are few examples of yield monitors for specialty crops, which are less developed at this time.

7. Precision irrigation in pressurized systems

Recent developments are being released for commercial use in sprinkler irrigation by controlling the irrigation machines' (e.g., pivots and linear-move) motion with GPS-based controllers. In addition to motion control, wireless communication and sensor technologies are being developed to monitor soil and ambient conditions, along with operation parameters of the irrigation machines (e.g., flow and pressure) to achieve higher water application efficiency and utilization by the crop. O'Shaughnessy and Evett[32] have done extensive research looking closely into ways to increase the efficiency of these irrigation machines. These technologies show remarkable potential, but further development is needed before they become commercially available.

8. Software

Applying precision agriculture technologies will frequently require the use of software to carry out diverse tasks such as display–controller interfacing, information layers mapping, on-the-go and postprocessing data analysis and interpretation, farm accounting of inputs per field, and many others. Generally speaking, these software packages come at affordable prices, are user-friendly, and are easy to learn with some training. There are different options for acquiring software licenses depending on the intended use and the users' skill. These packages fall within three categories: generic GIS software, specialized farm-management software, and proprietary software from machinery dealers. All are excellent options for PA farm management and record keeping to keep up with the needs of modern, information-intensive farming systems.

This is a rapidly evolving field, with new technologies coming to the market at short intervals.[33] One of the areas in which the evolution took place rapidly is the evolution of spatial analysis where in a few years' time the following evolution took place:

1. "Eyeballing" of printed maps (nonquantitative) to indentify patterns. However, the human brain is good at finding patterns even when there are no patterns.
   - Subjective and misleading
   - GIS as the ending point
2. Numerical analysis (quantitative)
   - Averaging by district, zone, county, other
3. Statistical analysis (quantitative)
   - Regression, ANOVA
4. Spatial statistical analysis (quantitative)
   - Adapted from epidemiology, criminology, geography, regional economics, agriculture

1.8.2 What does reality tell us?[34]

In commercial practice the implementation of precision agriculture (PA) has in common the use of spatially-aware technologies made possible through the use of global positioning systems (GPS). Most commonly this includes the use of vehicle guidance to reduce overlap in application of agricultural chemicals; reduced traffic associated with tramlining to reduce compaction and operator fatigue; shielded spraying of pesticides in row crops; yield monitoring; variable-rate technology (VRT) for application of agricultural chemicals, especially fertilizer; and within-paddock zone management for agricultural operations. Although PA technology has been available for more than a decade, it has been estimated that only a limited number (around 3–5%) of grain growers are using some form of the technology. One of the chief reasons for low adoption of PA is the reluctance of farmers to invest many thousands of dollars in PA without knowing if the technology will return a profit. A number of studies have reported the economic benefits of tramline farming and guidance for chemical application. Few studies have examined the value of variable-rate technology and zone management.

The adoption of precision agriculture technologies has been uneven, both geographically and temporally. The economic theory of induced innovation predicts that new technologies will be developed and adopted where they make more efficient use of the scarcest productive resources. Indeed, adoption of precision agriculture technologies has been fastest where labor is costly but land and capital are relatively less costly. Where precision agriculture is being adopted, the uneven adoption rate is tied to normal cycles for replacing the expensive machinery in which many precision agriculture technologies are embodied. Equipment replacement decisions are affected by many factors exogenous to the farm, such as bank interest rates and commodity prices. Adoption is likely to continue in labor-scarce, land-abundant countries, with rates of adoption accelerating when commodity prices are high and interest rates low.[35] Although spatial precision agriculture encompasses four key information technologies, farmers tend to use it in one of two major ways. The four PA technologies include location determination (via the global positioning system, GPS), computerized geographic information systems (GIS), computer-guided controllers for variable-rate application (VRA) of crop inputs, and sensing technologies for automated data collection and mapping. The GPS and GIS technologies underpin both of the major PA practices that farmers have begun to adopt. One of these is nutrient management; it involves spatially referenced soil sampling, often linked to VRA fertilizer spreading. The other is yield monitoring, usually tied to yield mapping. Agricultural technologies can be viewed as means by which farmers seek to achieve their production objectives. Farmers have many objectives, including risk management, quality of life, and environmental stewardship. But for the majority of farmers who rely on agricultural income, expected profitability is the sine qua non – they must earn enough to stay in business. In attempting to produce profitably, farmers are constrained by limited access to essential productive resources such as land, labor, equipment, buildings, and management knowledge. Profitable farming calls for using these resources up to the point where the cost of additional resource use is no longer compensated by the value of the resultant gain in output. The profitability appeal of VRA input control has been the potential to tailor input use site specifically, increasing it where justified by expected yield gains or reducing inputs where the costs exceed the expected benefits.[36]

1.8.3 The economics of precision agriculture

There have been studies that attempt to quantify the economic benefits of PA. They examined six case study farms from the Australian wheat belt.[37] They did not confine their analysis to VRT alone but also considered benefits to guidance and reduced traffic. The farm case studies covered a range of agro-climatic regions (Mediterranean, uniform, and summer-dominant rainfall patterns); cropping systems (wheat-lupin, wheat-canola, and winter and summer crops); farm sizes (1,250 to 5,800 ha cropping program); soil types (shallow gravels to deep-cracking clays); and production levels (average wheat yields from 1.8 to 3.5 t/ha). Among the examined farmers, all had invested in guidance and were practicing some form of variable-rate management of fertilizer. However, only some were using auto-steer and tramlining. One was using Normalized Difference Vegetation Index (NDVI), and another the GreenSeeker technology for in-season nitrogen management.

As such, the data set covered the range of likely situations confronting practitioners of PA in the Australian wheat belt.

Each grower was interviewed, and information was collected on area of cropping program, crops grown, area of the cropping program to which PA technologies were applicable, average cropping gross margin, PA equipment purchased, included date and cost, management actions associated with PA technology implementation, the estimated reduction in overlap for tramlining/guidance, the rates of fertilizer applied in each zone for zone management, areas of management zones in each paddock, rates of fertilizer applied for uniform zone management, yield in each management zone, and growers' opinion of nonmonetary benefits of PA. Standard economic analysis were applied including gross margin calculations and discounted cash-flow analysis. The level of capital investment in PA varied from $55,000 to $189,000, which is typically at the medium to high end of investment for Australian grain growers. When expressed as capital investment per hectare cropped, it varied by a factor of three from $14 to $44/ha. The estimated annual benefits from PA ranged from $14 to $30/ha, and consequently the investment analysis showed that the initial capital outlay was recovered within two to five years of the outlay, and in four out of the six cases within two to three years.

For all farmers, they were able to quantify benefits of variable-rate fertilizer management, ranging from $1 to $22/ha across the six farms. On a per-paddock basis, benefits ranged from –$28 to +$57/ha/year. Variation in monetary benefits from farm to farm could be explained by (1) whether or not starter fertilizer was being varied and not just nitrogen topdressing, and (2) the degree of within-paddock yield variation. The methodology for estimating the benefits of VRT requires further testing on paddock-scale data where yields and fertilizer rates are recorded for uniform and VRT-managed strips.

Benefits due to reduced overlap of spraying were typically in the order of 10% savings on spraying costs. Other benefits nominated by farmers and estimated by us were less fuel use, less soil compaction, less hired labor required, and more timely sowing. Intangible benefits listed by farmers were the ability to conduct on-farm trials, increased knowledge of paddock variability, increased confidence in varying fertilizer rates, and better in-crop weed control due to shielded spraying.

All farmers were highly literate in the use of computers, GPS technology, and variable-rate controllers; they routinely tested soil and kept good farm records. All invested considerable time in setting up their system in the beginning (with considerable teething problems in some cases), but ongoing labor demands were minimal. Some did not use a consultant, while others placed heavy reliance on consultants for zone definition, yield-map processing, and variable-rate map production. The researchers also found that, while a number of farmers are trialing VRT in test strips within paddocks, it seems that very few have taken the jump into full commercial implementation of VRT on their farms. This study was the first of its kind to estimate the economic benefits of precision agriculture in a commercial context. It demonstrated that Australian grain growers had adopted systems that are profitable, were able to recover the initial capital outlay within a few years, and also saw intangible benefits from the use of the technology. While the results here did go some way toward informing the debate about the profitability of PA in general, it also illustrates that the use of, and benefits from, PA technology varies from farm to farm, in line with farmer preferences and circumstances.

1.8.4 The future of precision agriculture

Precision agriculture is here to stay. Not that it is a one-stop-shop solution for all the challenges lying ahead. Nevertheless, precision agriculture, just like biotechnology, will help to advance the optimization of time and resources in a way never done before. It is therefore fair to say that there will be two trends driving the dynamic forward from here: (1) data and data analysis will be more and more automated, and software and analysis will be performed in the "cloud," and (2) although the value of single-farm data is finite to farmers, value can and is exponentially derived from pooling experiences and data sets in communities.

1.9 The rise of industrial agriculture[38]

Industrial farming is a form of modern farming that refers to the industrialized production of livestock, poultry, fish, and crops. The methods of industrial agriculture are techno-scientific, economic, and political. They include innovation in agricultural machinery and farming methods, genetic technology, techniques for achieving economies of scale in production, the creation of new markets for consumption, the application of patent protection to genetic information, and global trade. Industrial agriculture can predominantly be found in developed economies and increasingly worldwide. Industrial agriculture views the farm as a factory with "inputs" (such as pesticides, feed, fertilizer, and fuel) and "outputs" (corn, chickens, and so forth). The goal is to increase yield (such as bushels per acre) and decrease costs of production, usually by exploiting economies of scale. It implies modern farming methods that depend on synthetic fertilizers and pesticides, large amounts of irrigation water, major transportation systems, factory-style practices for raising livestock, and machine technology. In short it refers to a technique that involves or implies the following:

- Heavy dependence on synthetic inputs
- Concentration of production
- Vertical integration of producing and marketing food
- Spectacular increases in productivity

It depends on expensive inputs from off the farm (e.g., pesticides and fertilizer), many of which generate wastes that harm the environment, use large quantities of nonrenewable fossil fuels, tend toward concentration of production, involve vertical integration, and are associated with many environmental and public health concerns. It traces back to the first industrial revolution and the way it changed how the world produces goods. It created new innovations including the development of new energy sources, new crop varieties, and the use of fossil energies and energy to extract synthetic fertilizers and pesticides. It further improved yields (efficient production and cheap food), reduced labor needs, opened export markets, and replaced draft animals. The increased yield was impressive: in 1920 the average US farmer produced 30 bushels of corn per acre – that increased to 134 bushels by the end of the century (a 350% increase). It has led to a situation where agricultural output outpaced population growth and consequently helped to mitigate hunger and starvation and improve balanced diets. Agricultural production across the world doubled four times between 1820 and 1970 to feed a global population of 1 billion human beings in 1800 and 6.5 billion in 2002. During the same period, the number of people involved in farming dropped as the process became more automated.

It has created a situation where in the developed world households spend about 10% of their disposable income on food. In the developed world, that is still 50% or more of global GDP. In terms of agricultural operations and organizational capabilities, large corporate-controlled industrialized organizations dominate the industry (both production and sales). Those corporations control genetics, the manufacture and distribution of seed, fertilizers, pesticides, and machinery; manage through contracts; and control the processing, distribution, and increasingly marketing and retail.

The industry became more globalized, allowing export markets to emerge or at least for countries to become self-sufficient. The industry accounts for significant percentages of country GDP, often as large as 30% (e.g., Brazil).

Although there are major achievements in place, there are still niche areas of possible improvement. Further, there are some questionable aspects to industrial farming. Besides the fact that only 10% of total revenues go to farmers (the rest goes to marketing and input firms), there are implications, most of them emerging on the health and environmental side as well as in our economy and in societies in general.[39]

These externalities are not included in the retail price or the analysis of real productivity. Some of these externalities are as follows:

- Depletion of resources – for example, fossil fuel, water, soil, and biodiversity
- Pollution of resources by the products of fuel combustion, pesticides, and fertilizers
- Economic and social costs to communities – for example, lost property values

The environmental impact is significant and includes the following:

- Approximately two-thirds of water use worldwide is devoted to irrigation and therefore water is consumed at unsustainable rates.[40]
- Worldwide, aquifers are being depleted for irrigation faster than they can be replenished.
- Soil is eroding much faster than it can be replenished.
- Biodiversity is inevitably affected by both the scope and methods of agriculture.
- Monoculture, the growing of a single plant species in one area, usually the same type of plant year after year, affects biodiversity among both plants and animals.
- Industrial agriculture is based on the maximum yield of a few types of crops grown in monocultures.
- Nitrogen and phosphorous from fertilizers, pesticides, and agricultural waste.
- Synthetic chemical pesticides and fertilizers pollute soil, water, and air.
- Crops absorb only one-third to one-half of the nitrogen applied to farmland.
- Over 1,600 chemicals are used in the manufacture of pesticide – most have not been tested.
- Worldwide, 6 million tons of pesticides are used per year.
- Human health and poisonings; long-term effects on the immune, reproductive, and nervous systems; increased cancer risk.
- Agricultural runoff pollutes ground and surface waters.
- Agriculture accounts for 28% to 40% of all land use worldwide. Of this, 31% is in crop production and 69% in managed pasture.
- Deterioration of rural communities.
- Larger agriculture has driven out small producers and impacts communities.
- There are fewer farm families to support local schools, churches, public institutions, and retail businesses.

- Local communities are bypassed in purchasing production inputs/marketing products.
- Quality of life issues, for example, air quality and odors.

The total cost of industrial agriculture is often difficult to estimate as there are several contributing factors that interact, and the amount of potential harm is unlimited and often unknown at this stage. Studies have attempted to put a number on it and arrived at a bandwidth ranging from USD 7 to 18 billion annually.[41] With the rise of industrial farming came the extensive use of agricultural econometrics,[42] scalability became a keyword,[43] and foreign direct investments in agriculture are now a driver of economic growth.[44]

1.10 Irrigation and irrigation techniques[45]

1.10.1 Introduction

As water scarcity is a central theme and 70% of global water consumption is used for agriculture, the topic needs no real introduction. Lots of thinking and experimenting have been going into the area to improve the way irrigation can be developed while avoiding waste and improving efficiency.

An irrigation system consists of canals and structures to convey, regulate, and deliver water to users. Two basic types of irrigation systems exist: open-canal systems and pressure-piped systems. A pressure piped irrigation system is a network installation consisting of pipes, fittings, and other devices properly designed and installed to supply water under pressure from the source of the water to the irrigable area. The basic differences between traditional surface irrigation and piped irrigation techniques are as follows:[46]

- The water flow regime: With traditional surface methods, the size of the stream should be large, while in pressure piped irrigation systems, very small flows, even 1 m^3/ha, can be utilized.
- The route direction of the flow: With traditional surface methods, the irrigation water is conveyed from the source and distributed to the field through open canals and ditches by gravity following the field contours. The piped system conveys and distributes the irrigation water in closed pipes by pressure following the most convenient (shortest) route, regardless of the slope and topography of the area.
- The area irrigated simultaneously: With traditional surface methods, the water is applied in large volumes per unit of area, while piped irrigation systems distribute the water at small rates over a very large area.
- The external energy (pressure) required: Traditional surface gravity methods do not need external energy for operation, while piped irrigation systems require a certain pressure, two to three bars, which is provided from a pumping unit or from a supply tank situated at a high point.

Piped irrigation systems are classified according to the pressure required for operation, the method of delivering water to plants, and the type of installation.

1. Pressure

The pressure of the system is the maximum water pressure required for normal system operation and encompasses (1) the friction losses in the piping network from the control station to the distal end of the system, (2) the pressure required at the emitter, and (3) the difference in elevation (plus or minus). Systems can be classed as follows:

- Low-pressure systems, where the pressure required is 2.0 to 3.5 bars
- Medium-pressure systems, where the pressure required is 3.5 to 5.0 bars
- High-pressure systems, where the pressure required exceeds 5.0 bars

2. Water delivery method

The water delivery method is the way the water is distributed to the plants. Systems can be classified into the following categories:

- Sprinkler (overhead) irrigation. The water is delivered in the form of raindrops precipitated over the entire area. There are many variations of this method in terms of the discharge and diameter coverage, the height of the water jet above ground (overhead, under the foliage), the type of sprinkler mechanism (fixed or portable "hose-move"), microsprinklers (low-capacity water emitters), minisprinklers per tree, and so forth.
- Surface irrigation (furrow, basin, border, etc.). The water is delivered to the field plots direct from the main or submain pipelines through hydrants and is spread all over the area, or it is side applied. Either the entire field is flooded (basin irrigation), or the water is fed into small channels (furrows) or strips of land (borders).
- Microirrigation (localized irrigation) by drippers, sprayers, bubblers (a localized, low-pressure, solid permanent installation system often used in tree groves), microjets, and so forth. The water is delivered to the plants without being spread over the entire area but by being applied in low rates to a limited soil surface area around the plants. Drip irrigation is the most advanced irrigation method with the highest application efficiency. The water is delivered continuously in drops at the same point and moves into the soil and wets the root zone vertically by gravity and laterally by capillary action. The planted area is only partially wetted. In microirrigation, fertilizers can be applied through the system with the irrigation water directly to the region where most of the plant's roots develop. This process is called fertigation and is done with the aid of a special fertilizer apparatus (injectors) installed at the head control

unit of the system, before the filter. The element most commonly applied is nitrogen. However, applications of phosphorous and potassium are common for vegetables. Several techniques have been developed for applying fertilizers through the irrigation systems, and many types of injectors are available on the market. There are two main techniques: the ordinary closed tank and the injector pump. Both systems are operated by the system's water pressure.
- Low-cost hose irrigation: mainly used in developing economies. Water is applied to basins and furrows in plastic hoses which are portable, "hand-move," and can be extended in various directions. When one furrow or basin has been filled up with water, the hose is moved manually to the next one and so on.

The water delivery method and the kind of the water emitter are the main characteristics of a piped irrigation system. In many cases they influence and specify the other characteristics (pressure and type of installation) and performances, such as the flow capacity of the system and the duration of application. The flow capacity of a system is the water flow (in cubic meters per hour or liters per second) given or designed to meet the irrigation requirements of the irrigable area at peak demand. It is inversely proportional to the duration of application. Where designed, it is usually the minimum permissible in order to economize on pipe size and other equipment. The duration of application is the time required for the completion of one irrigation cycle.

3. Types of installation

Systems can be classified as

- Solid installations (fixed systems), where all the components are laid or installed at fixed permanent or seasonal positions.
- Semi-permanent installations, where the mains and submains are permanent while the laterals are portable, hand-move, or mechanically moved.
- Portable installations, where all the component parts are portable.

The suitability of the various irrigation methods, that is, surface, sprinkler, or drip irrigation, depends mainly on the following factors:[47]

- Natural conditions
- Type of crop
- Type of technology
- Previous experience with irrigation
- Required labor inputs
- Costs and benefits

1.10.2 Improved irrigation systems and sustainability of irrigation

Population and economic growth, changing social values about the importance of water quality and the environment, and water-right claims will continue to drive growing demand for water resources. Energy-sector growth is expected to significantly increase water demands for an expanding biofuels sector, utility-scale development of solar power, innovation in thermoelectric generating capacity, and commercial oil-shale and deep shale natural gas development. At the same time, projected climate change – through warming temperatures, shifting precipitation patterns, and reduced snowpack – is expected to reduce water supplies and increase water demand across much of the world. These trends are placing greater pressure on existing water allocations, heightening the importance of water management and conservation for the sustainability of irrigated agriculture. How well irrigated agriculture adapts to growing constraints on water, particularly by increasing its water-use efficiency, will directly affect the economic health and sustainability of the sector. A recent US study[48] found the following:

- Irrigated agriculture, which accounts for 80% to 90% (globally that is about 70%) of consumptive water use in the United States, represents a significant share of the value of US agricultural production.
- Irrigated farms accounted for 54.5% ($78.3 billion) of the value of all crop products sold and contributed to the farm value of livestock and poultry production through animal forage and feed production.
- Nearly 57 million acres were irrigated across the United States in 2007, or 7.5% of all cropland and pastureland.

While substantial technological innovation has increased the efficiency of irrigated agriculture over the past several decades, significant potential exists for continued improvement. At least half of irrigated cropland acreage across the United States is still irrigated with less efficient, traditional irrigation application systems. In addition, most irrigators do not make use of the more efficient on-farm water-management practices that conserve the most water. Less than 10% of irrigated farms use advanced on-farm water-management decision tools, such as soil- or plant-moisture sensing devices, commercial irrigation-scheduling services, or computer-based crop-growth simulation models. The sustainability of irrigated agriculture may depend partly on the willingness and ability of producers to adopt irrigation "production systems" that more effectively integrate improved water-management practices with efficient irrigation application systems.

1.11 Conservation agriculture[49]

Conservation agriculture (CA) is an approach to managing agro-ecosystems for improved and sustained productivity, increased profits, and food security while preserving and enhancing the resource base and the environment. CA is characterized by three linked principles, namely,

- Continuous minimum mechanical soil disturbance
- Permanent organic soil cover
- Diversification of crop species grown in sequences and/or associations

CA principles are universally applicable to all agricultural landscapes and land uses with locally adapted practices. CA enhances biodiversity and natural biological processes above and below the ground surface. Soil interventions such as mechanical soil disturbance are reduced to an absolute minimum or avoided, and external inputs such as agrochemicals and plant nutrients of mineral or organic origin are applied optimally and in ways and quantities that do not interfere with, or disrupt, biological processes.

CA facilitates good agronomy, such as timely operations, and improves overall land husbandry for rain-fed and irrigated production. Complemented by other known good practices, including the use of quality seeds and integrated pest, nutrient, weed, and water management, CA is a base for sustainable agricultural production intensification. It opens increased options for integration of production sectors, such as crop-livestock integration and the integration of trees and pastures into agricultural landscapes.

Conservation agriculture provides knowledge and tools to enable farmers to achieve acceptable profits from high and sustained crop production levels while at the same time conserving resources and protecting the environment. CA methods enhance natural biological processes above and below the ground by reducing interventions such as mechanical soil tillage to an absolute minimum. They also ensure that application of external inputs, such as agrochemicals and mineral or organic nutrients, does not interfere with or disrupt biological processes.

Conservation agriculture offers a powerful option for meeting future food demands while also contributing to sustainable agriculture and rural development. CA methods can improve the efficiency of input, increase farm income, improve or sustain crop yields, and protect and revitalize soil, biodiversity, and the natural resource base.

Conventional "arable" agriculture is normally based on soil tillage as the main operation. The most widely known tool for this operation is the plow, which has become a symbol of agriculture. Soil tillage has in the past been associated with increased fertility, which originated from the mineralization of soil nutrients as a consequence of soil tillage. This process leads in the long-term to a reduction of soil organic matter. Soil organic matter not only provides nutrients for the crop but is also, above all else, a crucial element for the stabilization of soil structure. Therefore, most soils degrade under prolonged intensive arable agriculture. This structural degradation of the soil results in the formation of crusts and compaction and leads in the end to soil erosion. The process is dramatic under tropical climatic situations but can be noticed all over the world. Mechanization of soil tillage, allowing higher working depths and speeds, and the use of certain implements like plows, disk harrows, and rotary cultivators have particularly detrimental effects on soil structure. Excessive tillage of agricultural soils may result in short-term increases in fertility but will degrade soils in the medium term. Structural degradation, loss of organic matter, erosion, and falling biodiversity are all to be expected.

Soil erosion resulting from soil tillage has forced us to look for alternatives and to reverse the process of soil degradation. The logical approach to this has been to reduce tillage. This led finally to movements promoting conservation tillage, and especially zero tillage, particularly in southern Brazil, North America, New Zealand, and Australia. Over the last two decades the technologies have been improved and adapted for nearly all farm sizes, soils, crop types, and climatic zones. There is still experience to be gained with this new approach to agriculture.

Application experience has shown that these techniques, summarized as conservation agriculture methods, are much more than just reductions in mechanical tillage. In a soil that is not tilled for many years, the crop residues remain on the soil surface and produce a layer of mulch. This layer protects the soil from the physical impact of rain and wind, but it also stabilizes soil moisture and temperature in the surface layers. Thus this zone becomes a habitat for a number of organisms, from larger insects down to soil-borne fungi and bacteria. These organisms macerate the mulch, incorporate and mix it with the soil, and decompose it so that it becomes humus and contributes to the physical stabilization of the soil structure. At the same time, this soil organic matter provides a buffer function for water and nutrients. Larger components of the soil fauna, such as earthworms, provide a soil structuring effect producing very stable soil aggregates as well as uninterrupted macropores leading from the soil surface straight to the subsoil and allowing fast water infiltration in case of heavy rainfall events.

Keeping the soil covered and planting through the mulch will protect the soil and improve the growing environment for the crop.

The processes carried out by the edaphon, the living component of a soil, can be called "biological tillage." However, biological tillage is not compatible with mechanical tillage, and with increased mechanical tillage the biological soil structuring processes will disappear. Certain operations such as moldboard or disc plowing have a stronger impact on soil life than others, as for example chisel plows. Most tillage operations are, however, targeted at loosening the soil, which inevitably increases its oxygen content, leading in turn to mineralization of the soil's organic matter. This inevitably leads to a reduction of soil organic matter, which is the substrate for soil life. Thus agriculture with reduced or zero mechanical tillage is only possible when soil organisms are taking over the task of tilling the soil. This, however, leads to other implications regarding the use of chemical farm inputs. Synthetic pesticides and mineral fertilizers have to be used in a way that does not harm soil life.

As the main objective of agriculture is the production of crops, changes in pest and weed management become necessary with CA. Burning plant residues and plowing the soil is mainly considered necessary for phytosanitary reasons: to control pests, diseases, and weeds. In a system with reduced mechanical tillage based on mulch cover and biological tillage, alternatives have to be developed to control pests and weeds. Integrated pest management becomes mandatory. One important element to achieve this is crop rotation, interrupting the infection chain between subsequent crops and making full use of the physical and chemical interactions between different plant species. Synthetic chemical pesticides, particularly herbicides, are, in the first years, inevitable, but they have to be used with great care to reduce the negative impacts on soil life. To the extent that a new balance between the organisms of the farm ecosystem, pests and beneficial organisms, crops and weeds, becomes established and the farmer learns to manage the cropping system, the use of synthetic pesticides and mineral fertilizer tends to decline to a level below that of the original "conventional" farming system. Burning crop and weed residues destroys an important source of plant nutrients and soil improvement potential. The phytosanitary motives for burning and plowing can better be achieved by integrated pest-management practices and crop rotations.

Conservation agriculture, understood in this way, provides a number of advantages on the global, regional, local, and farm levels. It provides a truly sustainable production system, not only conserving but also enhancing natural resources and increasing the variety of soil biota, fauna and flora (including wildlife), in agricultural production systems without sacrificing yields at high-production levels. As CA depends on biological processes to work, it enhances the biodiversity in an agricultural production system on a micro as well as macro level. No-till fields act as a sink for CO_2, and conservation farming applied on a global scale could provide a major contribution to control air pollution in general and global warming in particular. Farmers applying this practice could eventually be rewarded with carbon credits. Soil tillage is, among all farming operations, the single most energy consuming and thus, in mechanized agriculture, the most air-polluting operation. By not tilling the soil, farmers can save between 30% and 40% of time, labor, and, in mechanized agriculture, fossil fuels as compared to conventional cropping.

Soils under CA have very high water-infiltration capacities, reducing surface runoff and thus soil erosion significantly. This improves the quality of surface water, reducing pollution from soil erosion, and enhances groundwater resources. In many areas, it has been observed after some years of conservation farming that natural springs that had dried up many years ago started to flow again. The potential effect of a massive adoption of conservation farming on global water balances is not yet fully recognized. Conservation agriculture is by no means a low-output agriculture and allows yields comparable with modern intensive agriculture but in a sustainable way. Yields tend to increase over the years, with yield variations decreasing.

For the farmer, conservation farming is mostly attractive because it allows a reduction of production costs; a reduction of time and labor, particularly at times of peak demand such as land preparation and planting; and in mechanized systems a long-term reduction in the costs of investment and maintenance of machinery.

Disadvantages in the short-term might be the high initial costs of specialized planting equipment and the completely new dynamics of a conservation farming system, requiring high management skills and a learning process by the farmer. Long-term experience with conservation farming all over the world has shown that conservation farming does not present more or less but different problems to a farmer, all of them capable of being resolved. Particularly in Brazil, the area under conservation farming is now growing exponentially, having already reached the 10 million hectare mark. Also in North America the concept is widely adopted.

In summary:

1. The principles of conservation agriculture

CA offers farmers an array of practices, but at its core are three interlinked principles that can be applied in a variety of combinations to meet the needs of resource-poor farmers:

- Continuous minimal mechanical soil disturbance
- Permanent organic soil cover
- Diversified crop rotations of annual crops and plant associations of perennial crops

Conservation agriculture is more than a zero-tillage-based cropping system. Farmers following CA principles use low-cost tools and equipment and traditional crop varieties without herbicides or herbicide-tolerant varieties.

2. CA addresses development challenges

CA addresses some of the development challenges in the following areas:

- Agricultural production: CA has tremendous potential for achieving sustainable yield increases by improving the growth conditions for crops and the efficiency of input.
- Natural resource base: CA reverses soil degradation processes and builds up soil fertility by facilitating better infiltration of rainwater and enabling the recharge of groundwater, which reduces erosion and leaching and, in turn, water pollution.
- Biodiversity: CA conserves and enhances biodiversity in the field.
- Labor shortage: CA eliminates power-intensive soil tillage, thus reducing the drudgery and labor required for crop production by more than 50% for small-scale farmers. For mechanized farms, it reduces fuel requirements by 70% and the need for machinery by 50%. This is especially important for households affected by HIV/AIDS, where children or the elderly have responsibility for farm labor.
- Climate change: CA reduces crop vulnerability to extreme climatic events. In drought conditions, it reduces crop water requirements by 30%, makes better use of soil water, and facilitates deeper rooting of crops. In extremely wet conditions, CA facilitates rainwater infiltration, reducing the danger of soil erosion and downstream flooding.
- Livelihoods: CA gives farm families opportunities to improve their livelihoods. Farmers who adopt CA no longer need to spend time tilling and can use that time in other ways, such as on-farm processing, which adds value to their production. In southern Brazil, the income-generating opportunities of CA have reversed rural–urban migration.

CA offers an alternative to plow-based agriculture and the drudgery of relying on the hand hoe, but to be successful, farmers need a regular supply of reduced-tillage equipment and seed stock for cover crops. Thus, policy should encourage local private-sector entrepreneurs to manufacture and maintain CA equipment and to identify and market multifunctional seed stock.

2
The Investible Agriculture Space

Bringing the different pieces of the agricultural puzzle together would yield a list that would invariably include the following:

- Growing world population
- Lower expected production growth for most crops
- Limited availability of (additional) arable land
- Improved infrastructure to unlock new arable land where possible
- Increased affluence and changing dietary habits
- Need for additional (bio)technologies and fertilizers to improve crop yields
- Water scarcity
- Unknown impact of climate change
- Need for newer business models and supply chains
- Impact of agriculture on climate change
- Enhanced sustainability of the agricultural sector
- Impact of biofuel production

Closing the emerging production gap between demand and supply for agricultural output, two potential solutions can be framed: (1) raise productivity (closing the yield gap) and (2) bring additional land into production. The major industry trends expected to drive agricultural productivity growth over the coming decades include greater irrigation, increased nitrogen fertilizer utilization, and higher yields for crops. While there has been investment in the agricultural sector over the past few decades, mostly in Europe and the United States, it has only had marginal impact on agricultural growth on a global scale.[1]

Commercializing agricultural production across the globe will require tremendous investment in the following sectors:[2]

1. **Irrigation** remains a key input for raising productivity across a broad range of lands. A GIS system helps us discover areas where irrigation has improved

yields and identifies areas for further improvement (e.g., deployment of more efficient irrigation systems).

2. **Fertilizer** – optimizing application rates provides an opportunity to raise productivity. Analytical maps show us where fertilization is improving yields and the geographic patterns of application. By improving the efficiency of applied fertilizer and water, the potential associated water-quality problems and emissions of nitrous oxide could be decreased.
3. **Agricultural equipment** now includes high-performance spraying, harvesting, and construction equipment, as well as smaller tractors and upgraded utility vehicles. Precision farming products that optimize efficiency in farming operations are becoming more prevalent in developed economies. They include integrated machines with advanced guidance systems, display options, and comprehensive information management.
4. **Farm commercialization** using sophisticated management, technologies, and inputs results in greater and more consistent output, and it has the potential to raise agricultural productivity across the globe. For this to occur, significant investment must be made into the infrastructure of an agriculturally productive region. Spatial information can help policy makers identify where infrastructure development investments should be placed in order to integrate farms at all production levels. The large-scale financing required can be insured using agricultural risk-management solutions in meteorologically challenging areas to permit the investment required to increase crop production significantly. Additionally, attention must be paid to raising the competence levels of farmers in adopting and using these technologies and management practices. This area of investment is typically under the responsibility of the state or national extension services. We believe that this is a strong area of investment opportunity for private capital as well.
5. **Leading companies in supply-chain management** between developing countries and emerging economies capitalize on the dismantling of inefficient and bureaucratic agricultural systems and putting into place transparent financial and efficient logistics networks.
6. **Land Supply Response**:

   Intensification: Raising the productivity of existing croplands through technology developments as listed above will help close the production gap. Using spatial analysis tools will help identify where the production gaps can exist and are most likely to be closed.

   Extensification: In order to meet the growing production needs beyond the improvements of productivity, additional land must be brought into production. Many land-use types that currently exist have the potential to provide incremental production growth without expanding into forested land. They include multiple cropping on existing lands, improving degraded crop and pasture lands, bringing nonforested abandoned croplands into production, and converting productive pastures into biofuel production.

Biofuels, a Special Case: Current feedstocks of first-generation biofuels (e.g., corn, soy, sugarcane) compete with land needed for current and future food production. Generally, it is the view that while sugarcane is likely to remain viable, other agrofuels will not be seen as acceptable for biofuels as new feedstocks used in second-generation technologies, such as lignocellulose and algae, are developed and commercialized. If we are both to feed a growing, more affluent population and to use alternatives to fossil fuels, innovation and collaboration are needed across many industry sectors. Even though biofuels remain a small percentage of overall fuel consumption, as consumption increases, techniques that can be used to provide substitute feedstocks for second-generation biofuels that do not compete with current or future land will be required.

Additionally, degraded croplands and pastures, rehabilitated nonforested abandoned croplands, and multicropping on productive lands will also be essential.

At the same time, water is becoming increasingly stressed, and fertilizers currently emit greenhouse gases. In this context, many farmers, markets, and governments have looked at the potential for biotech, or in some case genetically modified organisms (GMO) or biotech crops, that can increase yield with lower water and fertilizer inputs. Also policies and dynamics of increasing international trade and trade patterns for agricultural commodities will impact the investible dynamics. Policies that promote global free trade will do more to raise productivity and thereby promote agricultural and economic development in emerging economies in the long run. Policies such as renewable-fuel standards, which encourage "second-generation" biofuel mandates; development of infrastructure such as transport, ports, telecommunications, energy, and irrigation facilities; and fostering management skills, labor supply, and capacity to use modern technology (e.g., global positioning systems) will all help raise productivity.

In conclusion, the investible space can then be depicted as follows in Table 2.1.

Agriculture has significantly impacted our planet, and in addition to supplying food, fuel, and fiber, it has had profound effects on essential conditions for human health and well-being, such as water quality and supply, atmospheric composition, biodiversity, and regional temperature and precipitation. With respect to greenhouse gases, agricultural practices such as deforestation and the use of certain fertilizers currently account for about 30% of

Table 2.1 The agricultural investible space

Area	Improvement	Type of investment
Machinery	More efficient farming and harvesting	Manufacturing, technology, R&D
Primary production	Enhancement of volume in the right location embedded in the right infrastructure, animal and fish meal replacement	(In)direct farm investments but also aquaculture and fisheries due to severely constrained D/S from reduced supply (climate change), increased demand (preteine, omega), and commercial overfishing
Irrigation	Increased productivity, lower consumption (often combined with fertilizer distribution)	Technology, R&D, manufacturing/infrastructure
Cultivation land/land expansion	Increased land volume for production	Private and public land, different business models for development, cooperative, supply-chain models
Fertilizers	Increased productivity	Technology, R&D, mining, natural resources
Commercialization of farms	Increased productivity, reduction of GHG emissions,[a] precision technology, newer business models to include smallholders	Logistics, equipment and machinery, manufacturing, infrastructure, business models, supply-chain management, education, consolidation, cooperative models
Biofuels	Variety of crops depending on region and type of regulation in place re bio/renewable energy (corn, soy, sugarcane, ligno-cellulose, algae)	Manufacturing, forests, R&D, supply-chain management
Biocrops	Better seed quality (seed trait genetics), crop protection, drought and disease resistance, lower water and fertilizer requirements	Biotech, agronomics
Water techniques and infrastructure	Better and lower usage, availability, desalination, waste recycling, lower usage per acre, infrastructure for facilitation and unlocking unused or unfit land	R&D, manufacturing, technology, natural-resources management

Climate change	Improving resistance and creating adaptation	Initiatives across the board
Commercialization of farms and new business models	Improved efficiency of farms and farming groups, consolidation	Education, supply-chain models, new business models,
Infrastructure and transport/fuel efficiency	Unlock unused farmland, reduce and facilitate cost of production and transport, reduction of GHG emissions	Infrastructure, technology, R&D, land-management techniques, storage capacity/transportation (logistics in general)
Downstream activities along the supply chain	Improve efficiency, environmental impact, unlock hidden value	Processing, storage, transportation/logistics, protein production, animal protein protection
Education, policy, and best practices (facilitator or enabler)	Adoption of technology, management, and best practices (mostly developing economies)	Smaller farmers or family businesses

Notes: [a] According to McKinsey & Company's greenhouse gas abatement cost curve analysis, agriculture represents about 14% of global greenhouse gas emissions (6.2 Gt CO_2-eq./year), with a growth trajectory of about 1%. This includes emissions from soils (in the form of nitrous oxide), methane from livestock, rice cultivation, and manure management, as well as other agricultural practices such as open burning of crop residues. Estimates of the abatement potential from agriculture are actually quite significant (4.6 Gt CO_2-eq./year, or 12% of global abatement opportunities) and represent a decline by 60% of sector emissions. These abatement levers include improved pastureland management, restoration of degraded lands, enhanced cropland management, and reduced methane emissions from livestock.

global greenhouse gas (GHG) emissions. In some cases, agricultural practices are unsustainable,[3] using soil and water resources much faster than they are replenished, thereby reducing the capacity to grow crops in the future.

2.1 World population and the demand/supply story

As a result of the agricultural and industrial revolutions, compulsory vaccinations, and continuing improvements in sanitation and medicine, world population has expanded for the past seven centuries, even in the face of large population losses due to world wars and disease. The world's population has steadily increased since the peak of the Bubonic Plague (Black Death) during 1348–1350, with its highest annual rate of increase occurring in 1963 – 2.2%. The United Nations Population Division predicts that by 2050 the world's population will be between 8.1 billion and 10.6 billion people, with a median estimate of 9.3 billion.[4] The world's human population does not grow linearly, but rather geometrically (i.e., 1, 2, 4, 8, 16, etc.), which explains the fivefold increase in population from 1.2 billion to 6.1 billion during the 20th century. Rapid population growth is predicted to continue for the first half of the 21st century, with rates of growth declining during the latter half of the century.[5] Despite the projected stabilizing of population by the end of this century, rapid growth is expected in the next 40 years and will likely place a huge burden on global resources, and the agriculture sector in particular. Population is growing most rapidly in the least developed and developing nations of the world, while population levels in the most developed nations are expected to remain flat or even decrease. Over the next 40 years, 97.5% of the projected increase will occur most markedly in Africa and Asia. By 2050, Asia is projected to be home to 56.4% of the population of the world and Africa to 24.7% of the world. If predictions materialize, just fewer than 80% (79.8%) of the entire world population will be living in Africa and Asia alone. Africa's population will increase by 159.7% by 2050 from 2010 levels. But that is not where the party ends. By 2050, also Central America's population is projected to increase by 68.5%, the Middle East's by 58.3%, and Oceania's by 50.1%, while the population of the European Union will decrease by 3.5%.[6]

Those data then need to be combined with the significant urbanization dynamic the world is experiencing. In 2009, 3.4 billion people lived in urban areas, and this is predicted to reach 6.3 billion by 2050. In fact, between now and 2050, urban areas of the world will be absorbing *all* of the world's population growth while also absorbing immigration from rural areas as well. Also longevity and gender are an argument in the mix. Due to better health care, access to education, and economic growth, life expectancy has risen across all socioeconomic groups across all areas of the world. The most rapidly growing age group worldwide is those over 80 years old. In contrast to the developed world, the younger generation under 25 make up often between 30% and 50% of the population of developing societies.

In terms of demand/supply, Table 2.2 provides the following balance:

Demand/supply models have been developed (MIT,[7] Potsdam, etc.) to assess the impact, demonstrating that with a climate change policy constraint and minimal deforestation, biofuels would have to specifically displace pasture lands in order to increase food and feed production to meet demand. However, the challenge to the global equilibrium models is that they underestimate the complexities of local land-use change and patterns of productivity.

But what is the impact on returns of all this? Agricultural returns are influenced by two main drivers: prices and productivity. Given what was said about population growth, changing dietary habits, and growing affluence around the world,[8] it is expected that prices will rise, even as the supply-side response creates higher efficiency. Historically, returns in agriculture have shown low correlation to equity markets. This is true of investments in production itself but less so for publicly listed agroconglomerates. The combination of land-price appreciation plus increases in productivity has led to long-term investment returns. Agricultural commodity prices are often volatile in the short term, displaying large spikes in prices. Input prices, such as fuel and fertilizers in the developed world, are often drivers of these spikes. Yet, in the long-term, in real terms, these commodity prices have remained relatively flat. But that has been changing in recent years and will continue to do so.

Table 2.2 Demand/supply balance in agriculture

Demand	**Supply**	**What will/can hold us back? (i.e., constraints)**
Growing world population	Expansion of land, unlocking unfit land	Land ownership, land grabbing, and consolidation within industry (in case not properly managed)
GDP per capita is growing	Fertilizers	Environmental issues: nitrogen, salinization, pests, etc.
Increased consumption (in kcal) per capita	Better irrigation and production techniques	Carbon emissions and pollution
Increased demand or regulatory dynamics for biofuels	Biotechnology (plant breeding and genetics)	Deforestation
	Animal and production standards	Draining of natural resources overall
	Insect and pest control	Effects of climate change
	Machinery and equipment	Geopolitical influences
	More efficient food systems, supply chain, infrastructure, and business models	Trade barriers/export constraints
	Water conservation and techniques (desalination, availability)	Water scarcity and availability

Traditional economic models of price movements of commodities show responses to these pressures by following the classic cobweb interaction of supply and demand out of equilibrium. In many areas of agriculture, supply can adjust to demand shocks within a growing season or two by reallocating land to different crops, optimizing inputs, and following other agricultural management practices.[9] Despite the overall statements about rising food prices scoring headlines, it should be noted that reality is more nuanced across regions and crop types. This does not take away that the overall picture will continue to look constrained. Higher prices will lead to more (much-needed) investment, which will take some of the burden away on the supply side. In both price cases, flat or rising, the strong investment returns from agriculture will have to come primarily from increases in productivity. Technological and management improvements can lead to higher productivity rates. And as production per unit of input land increases, revenue will increase, thus producing higher investment returns.

2.2 Agriculture and climate change

The changing patterns of climate change can potentially have severe impacts on agricultural productivity both ecologically as well as economically. Due to shifting precipitation patterns, movement of insect populations, more dramatic shifts in daily temperature regimes, and other large-scale ecological changes that come with climate change, agriculture is the economic sector that will be most severely impacted by climate change.[10] I can refer to the many reports of the Intergovernmental Panel on Climate Change (IPCC) evidencing the impact of, for example, rising temperatures on crop prices. Climate change impacts agriculture through potential variation in temperature regimes. These regimes can alter yield patterns and cause increased crop damage, soil erosion, and overall land degradation. These changes will require improved land management, along with adjustments to planting dates and crop variety using dynamic planning tools for crop relocation. Studies have been conducting demonstrating how rising temperatures will impact crops and eventually crop prices.[11] As indicated, agriculture contributed to climate change through the production of GHGs (see above). This amount is divided roughly evenly between CH_4 and N_2O (about 45% each), with CO_2 from fossil-fuel combustion and electricity use accounting for the remaining share. At the activity level, the largest agricultural source is soils management (40% of the sector total), where emissions result from particular tillage and cropping practices, such as fertilizer application. The second-largest source is methane emissions from livestock (27% of the agriculture total), which is a by-product of the normal digestive process of cattle and other livestock. Other important agriculture sources are wetlands rice cultivation (CH_4) and manure management (CH_4).

Agriculture also contributes to CO_2 through land clearing and the burning of biomass.[12]

The agricultural sector has to respond through a variety of actions (adaptation and mitigation):

- Development of new agrotechnologies, water-management systems, and farm practices, but also investment in infrastructure and logistics to facilitate the development of these adaptive agricultural systems (i.e., adaptation) and replacing agrofuels with cellulosic-derived biofuels that do not compete with food for land.
- Ability to store and cycle CO_2 and to provide potential offset markets for the trading/management of carbon through technologies and management practices such as no-till cropland management, planting of perennials, and the development of biochar resources.[13]
- Offset market models[14] include practices such as no-till agriculture, winter cover crops, and some biofuels mandates. The duration of offset contracts has also caused some criticism as farmers are used to spot markets, such as 6 to 12 months, whereas in the offset market, farmers would need to sell carbon credits on a 5- to 10-year carbon-contract basis. In that time, costs of production will vary, such as planting costs, fertilizers, and use of land.
- Short- and long-term adaptations to reduce GHGs.[15]

The right policy mix and business models to invest climate smart are heavily dependent on local circumstances, policies, and regulations.[16]

2.3 Agriculture and biofuels

The production of biofuels has created many heated debates in recent years. It has been blamed for partly causing the spike in food prices in the 2008–2009 period.[17] Others took a more nuanced view and balanced the arguments as they were reviewed in the earlier part of this chapter. Since the price spikes of 2008, commodity and biofuel prices have both fallen (with the necessary variations over time), illustrating their tight correlation with the price of oil and other primary energy products. The use of agricultural crops as a biofuel feedstock is a fast-growing segment, but one that is still less than 5% of total agricultural production. Of these feedstocks, crops used to produce ethanol account for about 90% of the diverted agricultural products, while 10% are used in biodiesel. It is important to note that production of first-generation biofuels generates many co-products that are used in other industry sectors. For example, bagasse from Brazilian sugarcane plants is used to power the ethanol conversion process, and dried distillers grains from corn ethanol are used as livestock feed.[18] As per capita caloric consumption increases, the efficiency of each sector in providing consumable calories will be of the utmost importance. The overall increase in demand puts additional pressure on our agricultural system in the short term, increasing the likelihood that marginal lands will be put into production and cause further GHG emissions and other environmental damage. Producing low-carbon fuels is a global priority, but diverting food crops for conversion into biofuels is not seen as a long-term strategy. Unless second- and third-generation biofuels are introduced, increasing biofuel demand will lead to further competition for land used to grow food.[19] For further analysis I refer to the chapter on biofuels.

2.4 Getting the most out of our farmland

As mentioned before, in order to increase agricultural productivity, we need to consider both land use and crop yields. In order to properly assess the potential of our existing or potential farmland, a (geographic information system) system is used. The majority of the analyses consider only the possibility of increasing production on lands currently used to grow crops or to pasture animals. Using complementary data sets to show other land uses (e.g., forests, urban areas) and their ecological integrity (degradation status, carbon sequestration potential), it is possible to identify lands not currently part of the agricultural system that could be converted with minimal environmental cost. In general, these lands are highly productive tropical pastures, abandoned agricultural lands, or degraded lands available for rehabilitation. While extensification is possible, intensification of land currently in production is also possible through increasing yields and multiple cropping. I refer to the chapter on farmland for further analysis. Only when a diverse mix of management, technology, and foresight are applied do we seem to be able to solve one of the greatest challenges we face in the next century. As mentioned, production is a function of yield and harvested area. Maximum theoretical yields are very difficult to obtain, and yields usually stagnate around 80% of their potential maximum. There are myriad factors that can decrease yields, such as genotype, nonoptimal solar radiation, temperature, plant population, water deficiency, nutrient deficiency or imbalance, diseases, insect pests, and weed competition.[20] The most recent performance per crop can be found in the outlook chapter.

Given the environmental risks (e.g., erosion, reduced water quality, CO_2 emission) of expanding agriculture onto marginal lands, it makes sense to first concentrate on improving yields on currently used cropped land. The capacity of currently cropped lands to increase productivity needs to be assessed in order to find the best opportunities for growth. Based on available data sets,[21] it is possible to calculate the current range of yields for each crop. Yields should be similar between locations with similar climate, but in most cases they are not, due to differences in management (irrigation, fertilization, mechanization, seed technology, etc.).

The range of yields for croplands with similar climates can be grouped together for analysis. The average and maximum yields of a crop can be computed and compared across the growing region. The "production gap" can then be calculated for each climatic region to quantify the difference between the higher-producing regions and the lower-producing regions.[22] Regions can then be identified that are most likely to be able to close the production gap with appropriate technology and with minimal environmental constraints. This analysis differs from previous production-gap analyses by using actual yield data as a constraint for future production, not theoretical maximum yields, which are rarely achieved. Calculating the production gap between current production and 90% of maximum production[23] for areas with a similar climate and comparing current production with a 90% maximum production projection for a given crop in a given region, aggregated globally, the global production system is falling behind the growing demand for 2030 and 2050. Closing the production gap for lands currently in production can be accomplished through changes in management and technology from the plant to field scale. But even reaching the assumed 90% maximum output levels will be a challenge in many parts of the world, in addition to constraints that may be faced. For example, environmental problems arising from misuse of chemical fertilizers, irrigation, and machinery can cause land degradation and irreversible ecosystem damage, such as loss of genetic diversity, while other problems arise from ill-defined property rights and urbanization, water stress,[24] and CO_2 emissions. Nevertheless, switching land use and expanding land supply are going to be important elements in the global solution mix. However, the analysis will need to go beyond improving yields to include land and food prices as well as the types of farmland. A holistic conceptual framework has been developed to bring the arguments together.[25]

2.5 Agricultural R&D

In order to support the increasing demand, R&D in the agricultural space will focus on drought-resistant crops, better seed quality, and the like. Much R&D has been going into areas such as crop-yield improvements, sustainable fertilizers, smart irrigation, biological controls, erosion control, and precision agriculture (overall the categories are agricultural technologies and biofuel technologies, including algae biodiesel, biogas, biomass, and cellulosic ethanol). However, they are still very minor relative to what is needed.

2.6 Collateral inputs required

Besides the basic ingredients for the solution to the food challenge, there are the collateral inputs. Two major categories in this respect are as follows:

- IT in order to measure, monitor, and verify components of the agricultural ecosystem. Both on-farm and remote monitoring systems that track crop productivity and constantly changing environmental variables will need to be developed and integrated into the agribusiness economy (GPS).
- Patent landscapes to observe and prioritize investment opportunities.

2.7 Water usage and efficiency

The extent of irrigation has grown markedly, but it can cause problems such as water-quantity depletion (aquifer or surface water), erosion and loss of nutrients from runoff, and soil salinization. Only a fraction of farmland is irrigated (+/– 20%), but a significant portion of crop output comes from irrigated land (40–45%). The agricultural industry accounts for a significant portion of the water withdrawals (70–85% depending on year assessed and regions analyzed).[26] R&D will have to focus on increasing production from the same amount of water or maintaining the same production from less water. While it is clear that irrigation is essential to productive agriculture in areas without sufficient rainfall, there is significant room for improvement in this area. IT applications help to identify where areas of improvement are located. Water conservation and irrigation techniques (such as drip-irrigation and precision-irrigation techniques) are critical, especially for water-stressed regions of the world. Within that spectrum can also be named microirrigation systems and fertigation – the application of fertilizers through an irrigation system.

Drip irrigation has the following benefits compared with normal (flood) irrigation techniques:

- High water savings (up to 60%) compared to normal techniques due to avoidance of evaporation, surface runoff, and percolation.
- Irrigation efficiency double the normal amount (80–90%).
- Consequently less is spent on labor, fertilizers, and pesticides.
- Fertilizer efficiency is high and supply is regulated (no leaching).
- Water control is regulated, and yield increases 50% to 100% compared to flood techniques.
- Capital cost per surface volume is significantly lower but will also depend on crop spacing.

More details are to be found in the chapter on water resources.

Other production-enhancing techniques are as follows:

- (Advanced) fertilizers

In areas of low soil fertility, fertilizers are used to supplement plant nutrition. In the last decade, somewhere between 160 and 250 million tons of the three major fertilizers (nitrogen [N], phosphorus/phosphate [P205], and potassium/potash [K20]) were placed on agricultural fields. Improper timing or overfertilization can lead to serious problems that negatively affect air, water, land, and human health. Nitrogen fertilizer has the most wide-ranging potential consequences and solutions.[27]

Nitrogen pollution from the United Kingdom and Germany is estimated to cost one-third of the value of the crops and livestock produced there.[28] Large consumers are Southeast and South Asia as well as North America. Based on the Haber-Bosch process, about 50% of the world population depends one way or another on nitrogen.[29] Although it can do a lot of good helping the world to eat, once it leaves the fields it is often at the center of many discussions regarding health-related and other problems.

It is implicated in environmental problems, from "blue babies" syndrome due to nitrate in drinking water, to "dead zones" in the ocean, to global warming due to nitrogen-containing GHGs.[30] The grand challenge presented here is to decrease the negative externalities of fertilizer use while increasing the availability of low-impact, highly efficient fertilizers in low-productivity areas. In that context, one of the challenges is that the use of fertilizers is very region specific and follows certain patterns and needs to be judged within the context of different management techniques to optimize and raise yields. This leads to the conclusion that we shouldn't apply nitrogen fertilizer on farmland across the board. This is highly inefficient, since excess nitrogen use produces diminishing yield increases and can cause dramatic fertilizer loss via nitrate leaching. Furthermore, such leaching has led to nitrogen pollution in many inland and marine ecosystems. We need to manage trade-offs between agricultural production on the land, drinking-water safety, and the health and productivity of fisheries.[31] Foley also indicated a diminishing marginal utilization rate. Adding additional fertilizers is less likely to increase yields significantly as this ceiling is reached, but it can result in more pollution-causing fertilizer runoff. In areas approaching their yield ceilings, the trade-offs between potential yield increases and environmental degradation from nutrient pollution must be weighed. For example, yield of corn in the United States has risen from 1.6 t/ha to 8.6 t/ha between 1930 and 2001, largely due to increases in the use of nitrogen fertilizers. While the Corn Belt has enjoyed increases in yield, the rate at which this has been occurring has decreased from 3.4% per year to 0.78% per year, as the yield ceiling has been reached. Also within the context of climate change, fertilizers play a role. When soils are waterlogged for a period of time, nitrate from nitrogen fertilizer is converted to nitrous oxide by a process called denitrification. It has been estimated[32] that nitrous oxide losses from fertilizer are only about 1.25% of applied nitrogen fertilizer, but nitrous oxide has 310 times the global warming potential of CO_2 per molecule.

A further 30 megatons (Mt) per year of CO_2 could also be removed from the atmosphere by plants as a result of sound agricultural and forestry management practices. Solutions to the fertilizer efficiency challenge include increasing nitrogen-use efficiency. This can be accomplished through changes in management (conservation tillage, timing of fertilizer application) and also through fertilizer technology and genetic improvements of crops. Fertilizer companies have conducted field trials showing increased nitrogen-use efficiency in particular varieties of crops, including testing of GMO crops. Promising tests have shown that some crop varieties demonstrated high yields using significantly less nitrogen fertilizer than conventional varieties. Improvements were equivalent to using two-thirds less nitrogen fertilizer than conventional varieties to generate similar yields. This type of genetic improvement is an area of active investment today.[33] Potential mechanisms to achieve a lower overall use of nitrate include using split application, nitrification inhibitors, and controlled-release fertilizers to match nutrient availability with crop needs over time.

Other strategies to reduce soil nitrogen loss include improved timing of nitrogen fertilizer application, better use of soil and plant testing procedures to determine nitrogen availability, switching to use of variable-rate nitrogen fertilizer applications and other more effective nitrogen fertilizer application methods, application of nitrification or urease inhibitors, and use of nitrogen fertilizer sources that are suitable for local environmental conditions. By improving the efficiency of fertilizer and water, emissions of nitrous oxide could be significantly reduced.[34] Research has still been focusing to a large degree on protein development (aced, shell, and fermentation technologies). We could consider this shift therefore in an early-stage process.

- Agricultural machinery and equipment

To meet some of the challenges, agricultural equipment is becoming more advanced, with improvements in high-performance spraying, harvesting, and construction equipment, as well as smaller tractors and upgraded utility vehicles. Precision-farming products that optimize efficiency in farming operations are becoming more prevalent in developed agricultural economies. They include an integrated machine with components for increased accuracy, advanced guidance systems and display options, as well as comprehensive information management. Some advanced machines now allow low soil disturbance. Machines that are able both to plant seed and apply fertilizer can fertilize more accurately and reduce input costs and reduce environmental impact. Other innovations include reduced engine emissions, improved fuel efficiency, controls over nitrogen oxide emissions, and the ability to work around the retained crop residues in reduced tillage systems. Agricultural machinery that can increase output, use less fuel, and cause less soil compaction is highly desirable.[35] A key driver in the process is that investments should drive the cost down so that the technology will be available to all layers of the commercial landscape.

- Infrastructure

To allow for unused land to become cultivated, it is required that the necessary infrastructure be put in place. A GIS system can then help to determine where those investments are needed or provide the most benefits. In cases where agricultural productivity is high and it is only a matter of logistics to store and transport the produce from one region to another, business models and refocusing supply-chain management, as well as converting inefficient and bureaucratic state-run agricultural systems, are part of the equation. Some of these initiatives get involved in the establishment of transparent financial and efficient logistics networks, including prepayments for farmers, guaranteed off-take at market prices before the planting season, and redeployment of underutilized legacy processing assets.[36]

- Farm commercialization

Farm commercialization – or using more sophisticated farming techniques – will drive the input/output relationship and will drive productivity growth. The relationship between fertilizer application, irrigation, and crop output is closely linked, and the rate of commercialization will drive other input growth rates. This process will be more pronounced in developing economies than in the Western world where consolidation and modern farming practices have been used for decades. There is however a trade-off in this effort, as maximizing output needs to be moderated by the need to protect environmental resources. However, many developing economies are pursuing agricultural policies that favor food self-sufficiency, including China, Russia, and other nations. It is the commercialization of the factors of production that will raise productivity. Since the commercial farm is typically much larger, it often integrates food/fuel/waste elements and enjoys significant yield and cost advantages.[37] Commercialization is not only about size and economies of scale. It is about optimizing professionalism that takes into account the downside of each initiative. The pace of farm commercialization, growth in fertilizer demand, and the growth of other agricultural chemicals go hand in hand to a certain degree.[38]

- Agricultural risk management and insurance

Sound risk-management practices can help manage the volatility of prices, yields, government policies, and foreign markets. State-sponsored insurance schemes have been in place for decades but have not always provided the right incentives or coverage. With the exception of hail insurance, most crop insurance has involved heavy subsidies to mitigate the expense of the premiums. For example, both the United States and Canada have three forms of subsidy: (1) a direct premium subsidy, (2) a subsidy in the delivery costs, and (3) some form of government sharing for the most catastrophic risk. The world experience with multiple-peril crop insurance has been particularly troublesome because the amount paid by the farmer is typically a fraction of the total cost of delivery and underwriting for this form of insurance.[39] The farmer often pays only 30% of the total insurance cost. Obviously no country can sustain a crop insurance program characterized by problems that result in extremely poor actuarial performance. This will be discussed in extenso in the chapter on risk management and insurance.

- Biofuels

As discussed above, feedstock production for biofuels competes for land needed for present and future food production. This puts significant stress on the system. There is not a one-stop-shop solution, but a mix of initiatives is required to turn the tide. Khosla[40] presents an example rubric named CLAW, in which future biofuels must (C)ost less than gasoline, have low to no additional (L)and use, and have limited impacts on (A)ir quality (e.g., carbon emissions) and limited (W)ater use. Cropping intensification is always favored over extensification, and there are many areas around the world that can support more than a single crop harvest. The Global Agro-Ecological Zones project discussed earlier led by the International Institute for Applied Systems Analysis (IIASA) mapped the multiple-cropping potential of rain-fed production by computing the length of the growing season as defined by the seasonal patterns of temperature and precipitation. In parts of the humid tropics, it is possible to have three continuous crops; in temperate regions, it is possible to grow winter wheat or other cover crops during the fallow cycle of summer crops. It is always preferable to plant a low- to no-input crop that also helps to rebuild soil carbon and structure, instead of leaving bare soil that is prone to erosion.[41] Where intensification is not an option, lands not currently in production or not suitable for food production can be used to grow second-generation biofuel feedstocks. One category of land that can be used is degraded lands. This is possible largely because the perennial plants that produce second-generation biofuels generally do not require tillage and can be sustained on otherwise erodible land[42] How much degraded land there is around is not only not very often measured, but the outcomes are very diverse and range from 350 to 2,000 million hectares of land. Most of this degradation occurred because of poor agricultural practices.[43] While degraded lands might not be suitable for food production, there are opportunities to rehabilitate the land and bring it back into production or to plant crops that can tolerate subprime growing conditions. Another potential land resource for second-generation biofuels production is highly productive tropical pastures. While pastures currently support livestock populations, the potential energy that is captured by tropical grass is transferred very inefficiently to livestock, and it may be in the societal best interest to harness the production of these lands for second-generation biofuel feedstocks. If biofuel feedstock production were to displace livestock in tropical areas, it would be crucial to prevent forest clearing in other areas, as that would negate any carbon benefit of the biofuels.[44] First-generation biofuels are high-input and low-diversity systems (e.g., maize, sugarcane, soybeans). But research suggests using low-input, high-diversity (LIHD) native perennial grasses (i.e., second generation) as a feedstock. Second-generation biofuels production includes feedstocks such as switchgrass, jatropha, miscanthus, and fast-growing trees like poplar. As opposed to annual crops, perennial crops build up below-ground carbon in roots and the soil, which gives them the potential to store carbon. Calculations[45] show that LIHD biomass yields "51% more usable energy per hectare from degraded infertile land than does corn grain ethanol from fertile soils" if converted using integrated gasification and combined-cycle technology with Fischer-Tropsch hydrocarbon synthesis (IGCC-FT). Globally, if LIHDs were planted on 500 million hectares of abandoned or degraded land, it could displace approximately 13% of liquid fuels for transportation and 19% of global electricity consumption – but it will require massive investments in infrastructure.

Third-generation biofuels based on waste, ligno-cellulose, and algae, with an even smaller land footprint, are now well under way and used for the production of biodiesel and hydrogen. Dr. Michael Boot at the Technical University Eindhoven (and others around the world) are making tremendous efforts at commercializing third-generation biofuels (in his case ligno-cellulose based fuels, which is a waste element of paper pulp). The third-generation biofuels distinguish themselves as being noneatable alternatives compared to the first two generations.

In the story about biofuels, it is important to apply the so-called life-cycle analysis to the production of biofuels: that is, biofuels were once thought to be a climate-friendly energy source, but recent analyses[46] have accounted for the CO_2 emissions from land conversions to grow biofuel crops, including the expansion and shifting of agriculture needed to fulfill biofuel mandates. The ecosystem "carbon payback" time (ECPT) is an index that gauges the length of time it would take for agricultural biofuels to offset the CO_2 emissions from clearing carbon-rich ecosystems. It is required to prevent putting new biofuel crops into areas of forestland that store large amounts of carbon.

- Biotechnology crops and distribution

Although traditional breeding has improved massively over the years and decades, generating new varieties of crops with desirable traits entails crossing and recrossing plants over several generations, and selective breeding whereby thousands of genes are combined has delivered great results. However, interbreeding, that is, the crossing of totally different species in order to achieve a desired trait, cannot happen in nature and is typically not considered an evolutionary pathway. Therefore, scientists have attempted to use genetic modification to create desirable traits that don't currently occur in nature. That is the removal of specific genes from one organism and their insertion into another species. This is referred to as a transgenic or genetically modified organism (GMO), or what we call biotech crops.[47] Stronger plant varieties, a key element of raising productivity, have been developed through traditional crossbreeding and biotechnology. The development of the crop-science market has included herbicides, insecticides, fungicides, biotech, and other agro applications. Biotech crops have the potential to contribute to the closing of the production gap through improving yields with lower inputs of water and fertilizers in some cases. Some developed traits can be used to enhance a plant's nutritional value, while others are used to strengthen crops and boost yields by improving a plant's built-in resistance to disease, pests, or harsh climates. Advances in crop genetics and farming technologies have helped keep pace with growing agricultural demand.[48] In the case of biotech applications to biofuels, efforts have been directed for example to pursuing a direct microorganism approach to the breakdown and conversion of plant materials into ethanol. In nature, a live organism "feeds" on plant materials by producing enzymes to convert the plant material into energy for itself. Through genetic-engineering techniques, such organisms can be modified to degrade the plant material much faster than normal, with the energy released used to drive a direct biochemical pathway to biofuel production.[49] Within biotech crop research, companies are using molecular biology techniques to efficiently and effectively mine the genetic library for useful traits. By combining technologies, such as molecular markers, with other breeding tools, the pace of research can be significantly increased. Major research is focused in the areas of grain yield and quality, pest control, disease resistance, and lipid enhancements to increase the oil and fatty acid contents of crops. And although not cheap at this stage, firms that help improve seed quality are in high demand. After all, the cost of seeds is one of the largest input costs for farmers around the world.

Overall, the demand for GMO seed is growing. In the short run, GMO seed and other superior germplasms show significant potential to increase food production. In the medium term, more regions will adopt GMO seed, driving demand higher. And in the long-term, GMO seed could be a primary driver of higher productivity.[50] But there is a downside to all of this: in a recent study by the Union of Concerned Scientists, they concluded that the promise has not yet been met by increased yield, that research and development is very expensive, and that the benefits of this research have not reached poor populations.[51] Policy and regulation will be needed to guide the biotech field to maturity while protecting public health and safety.[52] Moreover, the advent of the Green Revolution in agriculture, when biotech crops first arrived on the scene, sparked controversy over problems associated with farmer indebtedness, control over seed distribution, and overproducing the capacity of a given landscape which causes water scarcity as well as salination of soils. The promise of increased yields and reduced costs must be met with skepticism until it can be demonstrated in third-party-documented field trials. Although the sensitivity of this issue and the accompanying complex ethical and social equity issues can easily be acknowledged, it seems unthinkable given the challenges that are ahead in the agricultural sphere that farmers and governments will not do everything they can and consider every option they have available, including biocrops, to mitigate the negative impact of the trends indicated at the beginning of this chapter.

In each and every aspect discussed, policy and regulation can and should play a role. However, agriculture is one of the most restricted markets when it comes to international trade, and some misguided policies and lack of investments are leading to underproduction. From an investment point of view, it is crucial to understand that policies and regulation can impact the risk/return dynamics of agro-investments to a large degree. The primary goals of any agricultural policy must be to protect food security and promote economic development. Trade barriers are an often-used weapon especially in periods of drought or just high food prices, although a free-trade scenario will do more to raise productivity and thereby promote agricultural and economic development. The global food system should be a free- and fair-trade system that fosters growing the right crop, on the right soil, in the right climate, with the right technology and management practices.[53] Issues such as GHGs, biofuels, and their adequate product life-cycle costs would benefit from meaningful policies. That includes renewable-fuel standards.[54] Our global agricultural system has integrated the production of food, feed, and fuels, which requires careful balancing between the different interests involved.

All the measures discussed can be achieved in concert with good land-use planning. Good policy and fair but open trade will be elements in the mix to get the most out of our limited agricultural resources.

The remainder of this book will help us to flesh out the dynamics of some of the issues and opportunities discussed. It also serves as a benchmark to understand that in most cases the underlying reality is not as black and white as the mainstream media, policy makers, and lobbyists make it sound. Most of the topics are in flux, although with varying degrees of intensity – they are therefore, just like this book, a work in progress. The options, however, are limited, as malnutrition comes at a high cost to society and requires integrated action across sectors. Or as the FAO state and food agriculture report of 2013 indicates, *Maintaining the momentum of growth in agricultural productivity will remain crucial in the coming decades as production of basic staple foods needs to increase by 60 percent if it is to meet expected demand growth.*[55] It is in this context that meaningful and balanced allocation of capital is required to arrive at a justifiable risk/return rate, while monitoring how much risk should be taken with what levels of capital committed to each of the opportunities that will present itself for many decades to come.

3
Climate Change and Agriculture

The severity and pace of climate change in the 21st century is presenting an unprecedented challenge.[1] Current global surface temperatures are now about 0.6°C higher than the average for the last century. This increase is consistent with model predictions of the effects of rising atmospheric concentrations of carbon dioxide (CO_2) and other GHGs, which are (potentially) a result of human activity. Also in line with the same model simulations, the observed warming is greater at higher latitudes – particularly in the northern hemisphere, where most landmasses are located – than in the tropics. At the same time, extreme-temperature events are becoming more frequent, causing increasing damage to ecosystems, agriculture, and human health. Such worrisome trends will intensify in this century if emissions of anthropogenic GHGs continue to follow a business-as-usual scenario, with global atmospheric surface temperatures predicted to rise by at least 4°C by 2100. Moreover, the hydrological cycle will strengthen because of increased rates of evaporation from land and sea surfaces. As a result, rainfall may increase in the tropics and at higher latitudes, while decreasing over large continental interiors, with critically water-scarce areas of the world expected to become drier and hotter. More frequent climate extremes will increase the incidence and intensity of droughts and flood events worldwide. Finally, a sea-level rise will put millions of people at risk, presenting a significant challenge for rural, low-lying areas in many poor, developing countries. The future availability of water to match crops' water requirements will be particularly critical for agricultural production in arid and semiarid regions,[2] and in the southern, drier parts of Europe and Northern America. Runoff and groundwater recharge are both likely to decline dramatically in these areas. An excess of water may become a problem in specific areas. Where rainfall volume increases and becomes more intense, and where significant glacier melting alters water storage dynamics, a greater proportion of runoff may occur as flood flow. In particular, the expected loss of glaciers will have large impacts on some river systems, changing the availability of surface water for irrigation and the amount of groundwater recharge. The poorest and most food-insecure regions around the globe will be the most vulnerable under climate change: first, already scarce land and water resources will likely become even more scarce; and second, insufficient technical and financial means will make adaptation to the new climate very difficult.[3] But the problem is wider. Most primary agricultural projects are economically and financially (very) sensitive. The consequence is that even the slightest change in temperature or weather conditions overall has the potential for turning a decent and economically viable project into an absolute no-go. The stakes are high and the challenges numerous, as will be demonstrated in the remainder of this chapter. I will not, however, deal with the causes of climate change. That deserves a book on its own and is the subject of continued debate, although the Intergovernmental Panel on Climate Change (IPCC) in its most recent report tries to close the door[4] on the debate surrounding whether climate change is human induced or not.[5] Nevertheless, the perception among farmers of the relevance of climate change seems limited.[6]

3.1 Greenhouse gas emissions[7]

GHG emissions from agriculture, including crop and livestock production, forestry, and associated land-use changes, are responsible for a significant fraction of human-induced emissions – up to 30% globally, according to the Intergovernmental Panel on Climate Change. GHG emissions presented in this section are available from the newly launched FAOSTAT emissions database.[8] GHG emissions from agriculture are dominated by non-CO_2 gases such as methane (CH_4) and nitrous oxide (N_2O) arising from crop and livestock production and management. These include emissions from the following categories: enteric fermentation, manure-management systems, synthetic fertilizers, manure applied to soils, manure left on pastures, crop residues, rice cultivation, cultivated organic soils, and burning of crop residues. These emissions have increased by an average of 1.6% per year since 1990. Adding savannah burning and degraded organic soils under pasture, which are not currently available in FAOSTAT, total agricultural emissions in 2010 were estimated at five to six Gt CO_2/year. This equals 10% of all anthropogenic GHG emissions.[9] Previous data sets suggested it accounted for 14% of global GHG emissions, or 25% if agriculture-driven deforestation is included. This makes agriculture as big a contributor to climate change as the energy sector. In the EU-25, the food system was estimated to contribute 31% of total GHG emissions, or 40% if the hotel and restaurant sector was included.[10]

In particular, emissions from the cultivation of organic soils are receiving increased attention because of their importance in terms of absolute carbon-stock changes when peat lands are drained and degraded, thus becoming a significant source of GHG emissions (CO_2 and N_2O). More generally, changes in carbon stocks and ecosystem functions linked to human activities such as land-use change and land management determine the GHG emissions and removals that are reported by countries for IPCC's Land Use, Land-Use Change, and Forestry (LULUCF) categories. Activities that increase terrestrial carbon stocks over time lead to removal of CO_2 from the atmosphere, while activities that decrease total carbon stocks lead to net CO_2 and non-CO_2 emissions. Global GHG emissions from net forest conversion decreased from 3.6 Gt CO_2/year to 2.6 Gt CO_2/year during the period 1990–2010. It requires us to reflect on the discussion regarding food waste and the impact on GHGs.

Analyses performed show that avoidable food waste in, for example, the United States exceeds 55 million metric tons per year, nearly 29% of annual production. This waste produces life-cycle greenhouse gas emissions of at least 113 million metric tons of CO_2 annually, equivalent to 2% of national emissions, and costs $198 billion.[11]

3.1.1 The specifics of greenhouse gases and agriculture

Agriculture contributes to the release of the three main GHGs: carbon dioxide (CO_2), methane (CH_4), and nitrous oxide (N_2O) – the last two are the two most potent GHGs in terms of global warming potential. Compared to other sectors, agriculture contributes disproportionately heavily to emissions of methane and nitrous oxide, possibly in the 40–60% range.[12] About half of the GHGs come from deforestation followed by fertilization (emissions of nitrous oxide from soil), ruminants (and methane production by ruminant animals – enteric fermentation – see infra), and burning. GHG emissions are usually expressed in mass of CO_2 equivalent.

There are wide regional differences in GHG emissions generated by agriculture. About 75% of total agricultural GHG emissions are the result of agriculture in low-and middle-income countries, and this share is expected to increase in the future. In the period 1990–1995, these countries experienced the most rapid increase in agricultural GHG emissions, 35%, whereas industrialized countries as a group showed a decrease of 12%.[13] The main sources of agricultural emissions also vary geographically.[14] As expected, emissions from biomass burning and rice production come almost exclusively from less industrialized countries. Emissions from manure are more evenly spread geographically. Agriculture-driven GHG emissions are expected to increase in response to population growth and income growth in developing countries as a result of increased consumption of meat and dairy products. Consequently, the world woke up to the idea that targets needed to be set to reduce these emissions. Hence the Kyoto Protocol (created in 1997 and entered into force in 2005) and the CDM (Clean Development Mechanism) were created.[15] Further UN climate conventions in recent years in Durban, Cancun, and Copenhagen have not so far yielded a legally binding (only loose unilateral and voluntary commitments) follow-up for the Kyoto Protocol that will expire in a few years' time.

3.1.2 The way forward

Given that there is clearly a strong link between agriculture and climate change, deep thinking has been going into what should be the way forward for agriculture given the challenges ahead.[16] Solutions should be area and crop specific, but also holistic to include changing landscapes, food security, water and soil management, sustainable crop and livestock production, and forestry and aquaculture to ultimately arrive at an all-inclusive sustainable value chain for the agricultural sector to include climate-smart crop-production systems (including genetic resources) and water, energy, and soil management. It will also impact the way agriculture is financed.[17] This all relates to investments for adaptation,[18] to foster resilience building, climate proofing, and qualitative transformational change, which will cover agronomy, integrated nutrient management, tillage management, agroforestry, and water management. This will have to happen as part of a triangulation as shown in Table 3.1, which starts

Table 3.1 Adaptation triangulation in combating climate change

Type of adaptation action		Examples
Addressing the adaptation deficit	Resilience building	• Livelihood diversification to reduce poverty in the context of climate variability • Crop insurance, seasonal forecasting, and other agricultural innovations, including irrigation • Early-warning systems for disaster risk reduction
Adapting to incremental changes	Climate proofing	• Upgrading of drainage systems to accommodate greater runoff due to more intense precipitation • Adapting cropping systems to shorter growing seasons, greater water stress, and heat extremes (e.g., through crop substitution, irrigation, and new strains) • Improving disaster risk reduction systems to cope with more frequent and severe extremes
Adapting to qualitative changes	Transformational change	• Phased relocation of settlements away from areas at existential risk from sea-level rise • Shifts in emphasis in large-scale economic activity away from areas/resources threatened by climate change (e.g., away from water-intensive agriculture, climate-sensitive tourism, and high-risk marine resources, to less sensitive activities) • Transformation of agricultural systems from unsustainable (under climate change) intensive rain-fed or irrigated agriculture to lower-input (e.g., pastoral, agro-pastoral) systems

Source: See comprehensively, FAO (2013), "Climate-smart agriculture sourcebook" (retrieved: http://www.fao.org/docrep/018/i3325e/i3325e.pdf), note 8.

from changes in the agricultural system having an impact on food production (systems) and leads to impacts on yield variability and exposure to extreme-weather events.[19]

Research[20] indicates that while there is considerable potential to capture synergies between productivity, stability, and mitigation from changes in agricultural systems, there is also considerable potential for trade-offs to arise in specific contexts. For a fully comprehensive overview of the suggested food production and resilience impacts of changes in agricultural production systems as well as the impacts on yield variability and exposure to extreme-weather events, refer to the research mentioned.

3.2 Capturing carbon and reducing emissions in agriculture

The main categories[21] for achieving a climate-friendly agriculture by both sequestering carbon and reducing emissions are (1) enriching soil carbon (for instance through using perennials), (2) promoting climate-friendly livestock production systems, (3) minimizing the use of inorganic fertilizers, and (4) restoring degraded lands and preventing deforestation – which largely occurs for agricultural purposes.

3.2.1 Increasing carbon content in soil[22]

Soil is made of minerals, water, air, and organic materials. Organic matter originates from dead plant, animal, and microbial matter as well as living roots and microbes. These organic materials retain air and water in the soil. They also provide nutrients for plants and soil fauna. Agricultural practices that manage organic matter, and thus build and conserve soil carbon (as plant residues and manure) instead of depleting it, also provide soils in the long-term which are rich in carbon and require fewer chemical fertilizers. The following directions are being considered:

- Managing soil cover and residues. This can be achieved using the following techniques:
  - Avoid bare soils

    Bare soil is susceptible to erosion and nutrient leaching, and its soil carbon content is very low. Crops grown during fallows (sometimes called green manure) can be plowed into the soil while green or shortly after flowering in order to increase its fertility by adding nutrients and organic matter. Grasses and cereals make a good choice because they leave a large amount of residues on the soil surface, they decompose. slowly due to a high carbon/nitrogen ratio, and they improve the soil faster due to their aggressive and abundant rooting system.

- Use nitrogen-fixing cover crops
  Cover crops such as legumes enrich the soil with nitrogen through a symbiotic relationship with bacteria. They convert biologically unavailable atmospheric nitrogen gas N_2 to biologically available mineral nitrogen NH_4^+. When the system has stabilized, cover crops with an economic function such as livestock fodder may be used as cover crops.
- Composting refers to the decomposition of food and plant waste into dark organic matter.

- Increasing yield
  A higher yield will increase the amount of carbon that is sequestered by the plant and released into the soil either during growth or when plant residues are incorporated into the soil. Yield increases may also lead to lower cropland requirements. Apart from using chemical fertilizers, crop yields can be increased by several means.
  - Livestock manure or compost may be used (fertilizers, organic, or not).
  - Improved locally adapted crop varieties. Crop varieties could be improved to increase yields in several ways, for instance greater efficiency in water or nutrient use. Generating varieties and breeds that are tailored to ecosystems and the needs of farmers is crucial.
  - Intercropping. A judicious combination of compatible crops will increase yields on a given piece of land – for instance, a deep-rooted crop with a shallow-rooted one, or a tall crop with a shorter one requiring some shade. The introduction of leguminous species into grassland can increase yields, thus resulting either in higher productivity for the same input or a lower amount of fertilizer required. Intercropping also promotes biodiversity by providing a habitat for a larger variety of insects and soil organisms than in a single-crop environment.
  - Water management. The effectiveness of irrigation depends on water availability, its cost to farmers (dependent on water price and level of subsidies for water and energy), and energy requirements. Water harvesting is a key feature of sustainable water use.
- Reducing tillage. This leads to fewer emissions and increased carbon storage.
  In order to improve crop growing and uproot weeds, the soil is normally tilled, or turned over. This tilling, however, exposes anaerobic microbes to oxygen and suffocates aerobic microbes, which releases carbon dioxide.
  - No-till is often practiced in farming systems, combining it with other soil conservation practices such as crop rotation and green manure crops. For instance, conservation agriculture achieves minimal soil disturbance by combining no-till, permanent organic soil cover, and crop rotation.

These techniques result in healthier soil, enhanced carbon sequestration, and decreased erosion, as well as reduced use of water, energy, and labor. This brings benefits in terms of productivity/profitability and sustainability, including mitigation of climate change. The latter occurs by preventing carbon dioxide from escaping from the soil during tillage, by returning carbon to the soil by decomposition of the crop residues maintained on the surface, and by fewer passes over the field with fuel-driven machinery.

- Farming with perennials

Increasing the use of perennial crops, shrubs, and trees provides an important way of mitigating climate change by storing carbon in soil while crops are growing.

Indeed, unlike annual grains, perennial grasses retain a strong root network between the growing seasons so that a large amount of biomass remains in the soil instead of being released as GHG. Their large roots also help hold soil organic matter and water together, which reduces both soil erosion and GHG emissions. The need for annual tilling, seed-bed preparation, and application of agrochemicals is reduced, which further reduces emissions. Perennials are also beneficial in terms of enriching the soil and having a more conservative use of nutrients. Indeed, their roots allow them to support microorganisms and other biological activity as well as accessing nutrients and water in larger volumes of soil. The latter also makes them suitable for cultivation[23] in areas considered marginal.

Shifting from producing annual to perennial grains is not without challenges: the breeding of perennials takes longer, and their seed yields tend to be lower. However, researchers have already developed perennial relatives of cereals (rice, wheat, and sorghum), forages, and sunflower that are nutritious and good tasting. Further research in this area will support the switch toward high-carbon agricultural systems. Finding perennial substitutes for livestock feed is particularly promising given that one-third of global cereal production is used to feed livestock. Large areas of land are being converted to grow biofuel crops – which often have a net negative impact on GHG emissions once cultivation, fertilization, and fossil fuels use are factored in. Therefore growing perennial biofuel crops could offer a significant opportunity for biofuels of the second or third generation.[24]

- Agroforestry

Planting trees in crop fields and pastures provides another way of increasing carbon storage in agriculture. Agroforestry was practiced traditionally for agriculture in forest and woodland ecosystems. It is nowadays being introduced into subsistence and commercial systems. The trees may provide products (fruits, nuts, medicines, fodder, firewood, timber, etc.), benefits to farm production (increased crop fertility through nitrogen fixation from leguminous tree species, reduced risk through diversification), and ecosystem services (habitat for pollinators, improvement to microclimate, and reducing pressure on deforestation by meeting demand somewhere else).

- Rice management

  Cultivated wetland rice soils emit significant amounts of methane. These emissions can be reduced by various practices such as draining wetland once or several times during the growing season, adjusting the timing of organic residue additions, producing biogas, or keeping the soil as dry as possible between seasons.

- Decomposition, biochar, and silicates

  Provided it takes place within the soil and not on the surface, decomposition of plant matter increases carbon content in the soil.

  Another way is biochar, a charcoal not used as fuel: burning biomass in a low-oxygen environment results in a stable solid rich in carbon content that can be used to lock carbon into the soil. It keeps carbon in soil for a long time and slowly releases nutrients, as already realized by Amerindians some 2,000 years ago.[25] So, planting fast-growing trees in degraded areas (keeping in mind competition for water with other crops), converting them to biochar, and adding to the soil provides a way of taking carbon from the atmosphere and turning it into a slow-release organic fertilizer benefiting plants and the soil. It is a promising option for carbon emission offset payments.

  Some researchers propose to mitigate GHG by spreading calcium or magnesium silicates, such as olivine, as fine powder over land areas in the humid tropics.[26] Through chemical reaction with the silicate and water, CO_2 is transformed into bicarbonate, which washes down to the oceans where it ultimately precipitates as carbonate. On top of mitigating climate change, this process may bring additional benefits, enriching soils with mineral nutrients and reversing the acidification of soils, rivers, and oceans. However, before proceeding to large-scale deployment, more needs to be understood about the environmental impact, particularly in ecosystems adapted to acid conditions, in terms of biodiversity, etc.

3.2.2 Making livestock systems climate friendly[27]

Livestock production generates massive amounts of GHGs in various ways. Methane is produced from the fermentation of feed in the ruminant's stomach;

manure releases methane and nitrous oxide. Land clearing for pasture and feed crops, soil degradation, and the consumption of fossil fuels also result in the release of GHGs.

Curbing global consumption of meat and dairy is likely part of the solution to mitigating climate change, but a number of strategies also exist at the level of livestock production. The influence of grazing intensity on emission of GHG gases is not well established, apart from the direct effects of the number of animals on emission levels. Grass-fed ruminants have a higher methane production per kilogram of product than grain-fed ones, but their raising is associated with reduced fuel consumption and increased potential for carbon sequestration, as well as landscape management and biodiversity conservation.

- Rotational grazing for soil protection

Allowing vegetation to regenerate after grazing protects the soil from erosion and helps maintain its organic matter and carbon. Livestock productivity can even benefit from this practice. Rotational grazing is practiced successfully in the United States, Australia, New Zealand, Europe, and Africa (South and East).[28]

- Innovative feeding for reduced emissions

Innovative feed mixes, with increased starch content for instance, make digestion easier, in turn reducing methane production. Advanced techniques directly tackle the methane-producing microbial organisms present in the stomach by replacing them with other bacteria or by affecting their action through vaccines. These sophisticated approaches, still in development, will probably be expensive and raise ethical issues. They are potentially useful in large-scale intensive livestock operations.

- Manure management: A promising technology for millions of farmers

Manure is both a major source of methane (400 million of CO_2 equivalent according to the FAO) and an opportunity to produce biogas. In a biogas digester, anaerobic microbial action breaks down manure (or food waste) into biogas (methane) – which can be burned for heat or electricity – and sludge, a potential fertilizer.

For example, the US government is already providing subsidies to large dairies and pig operations to invest in anaerobic digesters. Numerous households in the developing world use manure to produce biogas for their cooking needs. Low-income producers need financial help at the outset, but investing in a biogas digester has proved worthwhile, with benefits for both the climate and well-being through access to energy.

a. Fertilizer management

Soils with nitrogen fertilizers release nitrous oxide, the most potent greenhouse gas. Fertilized soils release more than 2 billion tons of CO_2-equivalent GHGs every year.[29] Around 160 million tons of inorganic fertilizers (chemical nitrogen, phosphate, and potash) are used worldwide, mostly in industrial countries and in irrigated regions of developing countries.

- Substituting for inorganic fertilizers

A number of practices, such as composting, use of manures, or rotations with legume crops, allow for substituting or minimizing the need for inorganic fertilizers by increasing soil organic matter while capturing carbon from the atmosphere.

- Improved fertilizer application methods

Nitrogen applied in fertilizers (as well as manures) is not always used efficiently by crops. Timing of application can be optimized to when nitrogen is least susceptible to loss (often just prior to plant uptake). Using remote-sensing techniques, through precision agriculture for instance, allows farmers to understand in-field variability in order to fine-tune chemical usage (as well as water usage and sowing density).

b. Conserving and restoring forests and grassland

Deforestation and forest degradation at large account for about 20% of global GHG emissions and climbing. This share is the second largest after the energy sector (almost on par) and is higher than that of the entire global transportation sector. It is driven by agricultural expansion, conversion to pastureland, infrastructure development, destructive logging, fires, and so forth.[30]

- Reducing or avoiding deforestation

The share of deforestation and forest degradation driven by agriculture is estimated at 75% globally and varies regionally between 65% and 80% when one adds up the areas cleared for the purposes of small-scale agriculture, commercial crops, and cattle ranching.[31] Between 1980 and 2000, in the tropical zone, around 55% of new cropland area came from primary forests, and 25% came from secondary forests.[32]

- Yield increases and forest conservation

Forests and grasslands are an important reservoir of carbon (and biodiversity) on top of being an important part of the water cycle, locally and sometimes globally. Protecting them with the carbon they hold could have a massive impact in reducing emissions from land-use changes. One way to support their conservation may be to promote a judicious use of fertilizers when required and not to minimize the use of inorganic fertilizers at the expense of the forest. For instance, in "The Hidden Climate Costs of Chocolate," the Climate Change Agriculture and Food Security Partnership reports on serious damage to the West African Guinean rainforest when farmers who grow cocoa for a living seek to maintain or increase their incomes by burning adjacent forests, creating ash-fertilized soil conducive to growing cocoa as well as oil palm and cassava. If fertilizers or arable land had been accessible to farmers, significant deforestation could have been avoided.[33] Increasing agricultural yields may be an important lever to avoid emissions through deforestation.

- Incentives required: REDD, product certification, and land tenure rights[34]

As opposed to many of the climate-mitigating practices discussed above, protecting large areas of standing natural vegetation often provides fewer short-term financial or livelihood benefits for landowners or managers. Since it may even reduce their income or livelihood security, it is often required to provide stakeholders with incentives for conserving natural habitats.

One approach is to raise the economic value of standing forests and grasslands. This can be done by improving markets for sustainably harvested products or by paying land managers directly for their conservation value. Initiatives are under way to address the significant methodological, institutional, and governance challenges related to the implementation of a REDD mechanism.[35]

A second incentive for conservation is product certification, for instance the initiative by the International Finance Corporation for palm oil, soy, sugarcane, and cocoa. A third approach is to secure local tenure rights for communal forests and grasslands: local people thus have an incentive to manage them sustainably and protect them from illegal commercial logging, land grabs for agriculture, and the like. The burning of forests, grasslands, and agricultural fields can be an effective way of clearing and rotating plots for crop production, weed control, and soil fertility improvement. However, excesses are harmful to the environment and can be tackled via better regulatory enforcement (in the case of large-scale ranchers or commercial crop producers) or via investments in sustainable production linked to fire control for small-scale, community producers.

- Management of peat land

Apart from driving deforestation, agriculture results in places in the destruction of peat land, a potent carbon sink (containing on average about ten times more carbon than other soils[36]). Peat land is a wetland that over many centuries accumulates acidic peat, a deposit of dead plant material, mostly carbon. To be used for agriculture, the soils are drained, which aerates the soil and promotes decomposition, and emissions of carbon and nitrogen dioxides ensue. Peat areas are also very sensitive habitats, of high importance for biodiversity.

- Restoring degraded areas

A large amount of agricultural lands have been degraded by excessive disturbance, erosion, loss of organic matter, salinization, and acidification. Restoring part of these degraded areas can be a winning proposition from multiple perspectives of climate, economic development,[37] and ecosystems services, particularly with regard to wildlife habitat and watershed functions. (Regarding the latter, poor vegetative cover limits the capacity to retain rainfall in the system or to filter water flowing into streams and lakes).

Among the various land uses, croplands contain the lowest concentration of carbon, apart from deserts and semideserts. The reversion of cropland (see Table 3.2) to another land use, for instance grassland, is one of the most effective ways to reduce emissions and increase carbon sinks (through less soil disturbance and reduced carbon removal – especially since no harvest is taking place). On highly degraded soils, some cultivation or reseeding may be needed.

3.2.3 Low-emission farming systems[38]

Some farming systems and practices have been using or introducing some of the practices described above.

- Organic farming[39]

According to the FAO, lower GHG emissions for crop production, enhanced carbon sequestration, and lower-input of fossil-fuel-dependent resources give organic agriculture considerable potential for mitigating and adapting to climate change. Life-cycle assessments show that emissions in conventional production systems are higher than those of organic systems.[40] Although the

Table 3.2 GHG reduction at producer level

Practice	Comment
Replace inversion plowing with no-tillage (NT) and conservation agriculture (CA).	Possibility to lock up 0.1 to 1 t carbon/ha/year, cut CO_2 emissions by over 50% by reducing fossil fuel in plowing. If another 150 million ha of rain-fed cropland is converted to NT/CA by 2030, 30 to 60 Mt carbon/year could be taken up during the first few years. Benefits may be offset if increased pesticides or machinery are used.
Increase yields.	To increase carbon sequestered during plant growth and incorporated in residues, by improving water management, fertilization, and crop varieties.
Adopt mixed rotations with cover crops and green manures to increase biomass additions to soil. Minimize summer fallows and periods with no ground cover to maintain stocks of soil organic matter.	Avoid burning of residues. Values of secondary crops are critical in building financial returns and optimizing GHG per output value.
Use soil conservation measures to avoid soil erosion and loss of soil organic matter. Apply composts and manures to increase stocks of soil organic matter.	Soil carbon can be built with the use of soil additives, including silicates and biochar.
Improve pasture/rangelands through grazing, vegetation, and fire management to reduce degradation and increase soil organic matter.	Avoid overgrazing; select livestock to optimize yields and GHG performance; link with improved livestock practice.
Cultivate perennial grasses (60% to 80% of biomass below ground) rather than annuals (20% below ground).	Includes restoration of arable land to grassland and possible changes of livestock systems.
Adopt agroforestry in cropping systems to increase aboveground standing biomass. Convert marginal agricultural land to woodlands.	
Restore natural vegetation.	End use of woods also affects GHG by using wood for energy production or to replace energy-intensive materials such as steel, aluminum, and concrete. Can also be used for second-generation biofuels.
Conserve fuel and reduce machinery use to minimize fossil-fuel consumption. Use biogas digesters to produce methane.	It is estimated that restoring Australian rangelands (covering 70% of landmass) could absorb at least half of national annual GHG emissions.

Source: J. Wright (2010), "Feeding nine billion in a low emissions economy", review for the Overseas Development Institute and Oxfam. UK Government Office for Science (2011), "Foresight: The future of food and farming", London and Deutsche Bank Research. Quoted in C. Schaffnit-Chatterjee (2011); ibid., note 15, p. 13.

potential of organic farming to feed the world is still being debated, there is little dispute about the fact that many farmers can maintain yields while using significantly less inorganic fertilizer, with major benefits to the environment, the climate, farmers' health, and the local economy (higher prices/revenues, higher labor requirements).[41]

- Agro-ecology

Agro-ecology is the result of the convergence of agronomy and ecology. It is highly knowledge intensive, based on techniques that are based on farmers' knowledge and not delivered top-down.

Agro-ecology has been proven to raise productivity at the field level, reduce rural poverty, and contribute to improving nutrition, as well as contributing to adapting to and reducing climate change.[42]

- Conservation agriculture

This system[43] involves using reduced tillage, permanent organic cover, and crop rotations. It is increasingly used and promoted by the FAO, especially in developing and emerging economies. One of its drawbacks is that its adoption is associated with significant use of herbicides and may lead to water contamination. Over time, however, soil-cover practices tend to prevent weed emergence and allow reduced use of herbicides. There is also little documentation on the potential of conservation agriculture to feed the world.

- Integrated food-energy systems

Integrated systems combining various types of crops or trees and crops increase and diversify production. Thus they minimize risks and enhance the economic resilience of farmers. Integrated crop and livestock systems also increase efficiency and environmental sustainability of both systems as the waste product of one component serves as a resource for the other: manure increases crop production, and crop residues and by-products are used as animal feed. Integrated food-energy systems simultaneously produce food and energy.[44] One method combines food and energy crops on the same plot of land such as agroforestry systems growing trees for fuelwood and charcoal. Another method is based on the use of by-products/residues of one type of product to produce another.[45]

- Urban and peri-urban agriculture

The share of the global population living in cities is currently around 50% and is expected to rise to 70% by 2050.[46] On top of improving food security and employment opportunities for the urban poor, urban agriculture contributes to reducing emissions by cutting down on transport. Up to 15% of the world's food is produced by urban agriculture, and 70% of urban households in developing countries participate in agricultural activities, according to the FAO.[47] Competition for land and issues related to tenure rights are major constraints on the development of agriculture. Environmental impact and food safety (due to use of wastewater and organic material) are also concerns to be addressed.

3.3 The consumption side of the story: climate-aware consumption behavior

It is a given that the world's population will continue to increase in the next decades and will need to feed itself. It may, however, not be a given that consumption patterns will stay similar to what they are and cannot be changed toward a more sustainable use of resources. It is important to question demand rather than systematically consider it as an exogenous variable.

3.3.1 Reducing food loss

Recent studies[48] identified that around one-third of food produced for human consumption is lost or wasted globally. On top of representing a massive waste of resources exacerbating food insecurity, this means that the GHG emissions occurring along the food chain could have been avoided, as well as the significant amount of methane emissions resulting from food rotting. On a per capita basis, much more food is wasted in the industrialized world than in developing countries. In all regions, food is mostly wasted early in the food supply chain. In medium- and high-income countries, a significant amount of food is wasted at the consumption stage, including food still suitable for human consumption. Food is also wasted due to quality standards that reject food items that are not perfect in shape or appearance. The causes of food loss and waste in low-income countries are mostly associated with financial, managerial, and technical limitations in various areas: harvesting techniques, storage and cooling facilities, infrastructure, and packaging and marketing systems. Both private and public sectors have a role to play in investing in these areas. Food wasted in industrialized countries can be reduced by raising awareness among food industries, retailers, and consumers and by putting safe food that is presently thrown away to good use, for instance by directing toward people who need it.

3.3.2 The link between mitigation, biofuels, and food consumption

Some biofuel systems have net positive effects for GHG emissions, but many first-generation biofuels do not contribute to lowering GHG emissions while reducing the area of land available to grow food. A significant amount of bioethanol is produced from sugarcane (in Brazil), sugar beets, and grain crops (maize in the United States, wheat in Europe), and biodiesel is derived from vegetable oils (rapeseed, palm, soybean, or sunflower).[49]

Second-generation biofuels from cellulosic material (leaves, wood, green waste) are promising and expected after 2020 (sufficiently scalable to be commercially relevant). They also provide an opportunity to be combined with livestock production associated with lower methane emission (through feeding on less gas-producing by-products). The use of aquatic algae for biofuel production would be combined with carbon uptake. These third-generation biofuels are still under development. Well-informed policies are needed in order to minimize detrimental competition for land and water, including speeding up the development of second- and third-generation biofuels.

3.3.3 Modifying diet preferences[50]

Evidence suggests that a climate-friendly diet could be achieved through substituting animal proteins with vegetable proteins (such as pulses) and favoring consumption of food that is locally produced in season (to reduce both transport and cold storage).

The issue of transport[51] in a context of low GHG emissions is not straightforward. The concept of whole life-cycle emissions (life-cycle analysis) is very appealing but challenging to evaluate. Food multinationals are showing increased interest in measuring their carbon and water footprint, but it is important to develop comparative standards which are able to guide consumer choice and to provide policy options. Raising consumers' awareness is key.

Internalizing the environmental costs into the market price of high-carbon foods like meat is likely an effective way to reduce emissions but needs to be associated with measures ensuring adequate nutrition for all, through vegetal proteins for instance. Such a carbon tax attached to food items may raise fairness issues and is likely to meet strong resistance from industry and politicians.

Skeptics do not consider significant changes in food consumption behavior to be a realistic option. However, there is evidence of a slow but growing awareness – both at the individual and collective levels – of the impact of one's actions on the environment and one's health. Many governments are implementing measures to promote sustainable consumption, be it water pricing, labeling,[52] or what have you.[53] The ecological footprint of food is significant. For example, one kilogram of beef produces 16 Kg CO_2 eq., 15.5 liters of water footprint, and uses on average 7.9 m^2 of land. Or 1 liter of milk produces 10.6 Kg CO_2 eq., 1.0 liters of water footprint, and uses on average 9.8 m^2 of land.[54]

The impact of agriculture on climate change cannot, given the above, be underestimated.[55] Although there is, as discussed, a broad pallet of initiatives possible, there are many implementation[56] hurdles. A number of issues make broad implementation challenging, particularly the following three: (1) monitoring, (2) the sheer number of GHG sources and mitigation practices, and (3) agricultural and trade policies biased toward other agricultural systems.

3.4 Measuring and monitoring climate impact

There is no recognized rigorous methodology for assessing agriculture-related emissions, sequestration, and storage, and this is a serious constraint to including agricultural GHG emissions in offset schemes. Scientists are rapidly developing methodologies for assessing carbon balances for specific components of land use such as soil organic-matter enrichment, conservation tillage, agroforestry systems, and so on. This will allow an indirect measure of carbon content in a field by measuring adoption of specific practices whose average impact has been validated for a particular agro-ecosystem. What is still missing is an integrated approach to landscape-level carbon accounting that would reflect diverse land uses and practices, enabling verification of whether a climate benefit in one component of the landscape is undermined by increased emissions in another component.[57]

3.4.1 Fragmentation of GHG sources and mitigation practices[58]

A high level of fragmentation at several levels complicates the scaling up of climate-friendly practices. There is indeed a large variety of land uses, of sources and patterns of emissions across ecosystems, and of practices to reduce emissions or sequester carbon, and a large number of players: in particular, implementing climate-friendly agricultural techniques means working with hundreds of millions of smallholder farmers and requires community-scale planning, both technically and organizationally.

In building the case for investing in farming and land use for climate-change mitigation, three issues are often addressed: permanence (whether emission reductions will be too short-lived), additionality (whether climate benefits will be greater than those expected in the baseline conditions), and leakage (whether emission reductions will only be offset by increases elsewhere).[59] Most of the climate-friendly agricultural practices described in this chapter have proven successful in pilot studies or initiatives within a particular landscape. In order to make a difference to the world's climate, institutional investment, capacity building, and finance are required in order to scale up the institutional models and activities that have proved able at the local level to overcome these challenges.

3.4.2 Policies biased against other types of agriculture

A number of market distortions and market structures act as disincentives to transitioning toward sustainable agricultural practices.

- Agricultural subsidies in developing economies prohibit change in the sector

The significant subsidization, in developed countries, of agricultural production and of agricultural exports to developing countries comes at a high cost to the latter countries. By restricting developing-country producers' access to markets, their capacity to grow and to afford the shift toward sustainable, climate-friendly agricultural production on a large scale is limited.[60] The international WTO Doha Round as well as an appropriate way of pricing carbon could be instrumental in phasing out these subsidies.[61]

- Other adverse effects of national policies

Also, in other parts of the world, farmers are paid for production and agricultural inputs. This is mostly the case in the United States and Europe, but high subsidies are also found in Japan, India, China, and other countries. Often linked to production, these payments and pricing policies of agricultural inputs usually lead to overuse of pesticides, fertilizers, water, and fuel or encourage land degradation.[62]

3.5 Market concentration and sustainability

The high concentration of the corporate food system is also a challenge in scaling-up sustainable agricultural practices. A handful of powerful companies dominate the global agricultural input markets of agrochemicals, seeds, and biotechnology. Their immediate interest is obviously to sell their products, thus to maintain an external-input-dependent, mono-culture-focused, and carbon-intensive industrial approach to agriculture. With regard to seeds, industry consolidation tends to result in a narrower choice for farmers and loss of access to some varieties. Due to their size, the major food processors and retailers also influence global food supply chains into sourcing from scale-focused monocrop production at the expense of diverse multicropping and integrated livestock and crop farming systems.[63] Partnering with smallholders is now increasingly taking place to ensure supply in a resource-constrained world.[64]

3.6 Climate-smart agriculture

Climate-smart agriculture (CSA), as defined and presented by the FAO at the Hague Conference on Agriculture, Food Security, and Climate Change in 2010, contributes to the achievement of sustainable development goals. It integrates the three dimensions of sustainable development (economic, social, and environmental) by jointly addressing food security and climate challenges. It is composed of three main pillars:

1. Sustainably increasing agricultural productivity and incomes
2. Adapting and building resilience to climate change
3. Reducing and/or removing greenhouse gases emissions, where possible

CSA is an approach to developing the technical, policy, and investment conditions to achieve sustainable agricultural development for food security under climate change. The magnitude, immediacy, and broad scope of the effects of climate change on agricultural systems create a compelling need to ensure comprehensive integration of these effects into national agricultural planning, investments, and programs. The CSA approach is designed to identify and operationalize sustainable agricultural development within the explicit parameters of climate change.

CSA[65] is not a single specific agricultural technology or practice that can be universally applied. It is an approach that requires site-specific assessments to identify suitable agricultural production technologies and practices. This approach

1. addresses the complex interrelated challenges of food security, development, and climate change and identifies integrated options that create synergies and benefits and reduce trade-offs;
2. recognizes that these options will be shaped by specific country contexts and capacities and by the particular social, economic, and environmental situations where it will be applied;
3. assesses the interactions between sectors and the needs of different involved stakeholders;
4. identifies barriers to adoption, especially among farmers, and provides appropriate solutions in terms of policies, strategies, actions, and incentives;
5. seeks to create enabling environments through a greater alignment of policies, financial investments, and institutional arrangements;
6. strives to achieve multiple objectives with the understanding that priorities need to be set and collective decisions made on different benefits and trade-offs;
7. should prioritize the strengthening of livelihoods, especially those of smallholders, by improving access to services, knowledge, resources (including genetic resources), financial products, and markets;
8. addresses adaptation and builds resilience to shocks, especially those related to climate change, as the magnitude of the impacts of climate change has major implications for agricultural and rural development;

9. considers climate-change mitigation as a potential secondary co-benefit, especially in low-income, agricultural-based populations; and
10. seeks to identify opportunities to access climate-related financing and integrate it with traditional sources of agricultural investment finance.

As Schaffnit-Chatterjee points out,[66] there are essentially four approaches to lower emissions:

1. Market incentives: in the form of grants, subsidies, levies, carbon taxes, or carbon cap-and-trade schemes,[67] or, more broadly, payments for environmental services or in general pricing the negative externalities.
2. Regulation: mandatory emission limits, potentially associated with higher production costs and market adjustments.
3. Demand pressures: driven by consumer choice (requiring active and informed consumers and accurate and trusted information such as labeling[68] or certification).
4. Voluntary measures: driven by corporate social responsibility.

The balance between incentives and regulation will depend on the availability of financial resources for creating incentives and supporting enforcement.[69] Agriculture as an industry can play a significant role, but aligned efforts are needed coming from different directions.[70] There is no single approach to climate mitigation since the optimal path will take into account the whole system and depend on the context. Beyond GHG levels, it is essential to also consider the amount of food produced, the quantity of inputs required, food prices, ecosystem services, animal welfare, and more. Climate change finance is considered essential in achieving all these objectives. To date, we consider climate change finance to be in its infancy, and no serious measurements are possible or have been attempted to quantify the investments that have been made so far in this space.[71] A further challenge is to do so within the boundaries of responsible agro-investment principles (RAI) and responsible farmland investment principles (PRI).[72]

3.7 Mitigation projects[73]

The entry points to the projects[74] are all linked to unsustainable or low-productivity land-management practices, having caused land degradation. The projects' direction is generally toward improving agricultural or forestry practices by introducing new techniques, as well as rehabilitating degraded soils. Agroforestry is a practice that is part of many of these GHG-mitigating projects, because it contributes both above- and belowground to carbon sequestration, as well as providing other important benefits. Conservation agriculture is another technique that plays an important role in climate-friendly projects, as well as organic production techniques and bioenergy. Some of the projects from this inventory have a carbon-market orientation as they aim to generate carbon credits to be sold on markets. These are all at different stages. In addition, there are several projects aiming to reduce the release of GHGs by introducing new techniques and providing carbon sinks but that do not foresee generating carbon credits from these activities. It makes eminent sense to adopt climate-smart farming practices with important synergies between productivity, adaptation, and mitigation. The environmental services provided by farmers should be remunerated in some way. The specific demands of agricultural production, the need for investments for improved farming practices, the slow process of the accumulation of carbon, and the time lag for increased productivity all create a challenge for financial mechanisms that could facilitate the transformation of current agricultural systems to climate-smart agriculture. Similarly, the sheer number of the farming units and their generally relatively small size and modest amount of carbon accumulated per hectare all call for innovative rethinking of financing systems.

3.7.1 The specific case of livestock and GHGs[75]

Animal production is a significant source of greenhouse-gas emissions worldwide. Depending on the accounting approaches and scope of emissions covered, estimates by various sources[76] place livestock contribution to global anthropogenic GHG emissions at between 7% and 18%. The current analysis was conducted to evaluate the potential of nutritional, manure and animal husbandry practices for mitigating methane (CH_4) and nitrous oxide (N_2O) – that is, non-carbon dioxide (non-CO_2) – GHG emissions from livestock production. These practices were categorized into enteric CH_4, manure-management, and animal husbandry mitigation practices. Emphasis was placed on enteric CH_4 mitigation practices for ruminant animals (only in vivo studies were considered) and manure-mitigation practices for both ruminant and monogastric species.

In evaluating mitigation practices, the use of proper units is critical. Expressing enteric CH_4 energy production on a gross energy intake basis, for example, does not accurately reflect the potential impact of diet quality and composition. Therefore, it is noted that GHG emissions should be expressed on a digestible energy intake basis or per unit of animal product (i.e., GHG emission intensity), because this reflects most accurately the effect of a given mitigation practice on feed intake and the efficiency of animal production.

3.7.2 Enteric CH_4 mitigation practices[77]

Increasing forage digestibility and digestible forage intake will generally reduce GHG emissions from rumen fermentation (and stored manure), when scaled per unit of animal product, and are highly recommended mitigation practices. For example, enteric CH_4 emissions may be reduced when corn silage replaces grass silage in the diet. Legume silages may also have an advantage over grass silage due to their lower fiber content and the additional benefit of replacing inorganic nitrogen fertilizer. Effective silage preservation will improve forage quality on the farm and reduce GHG emission intensity. Introduction of legumes into grass pastures in warm-climate regions may offer a mitigation opportunity, although more research is needed to address the associated agronomic challenges and comparative N_2O emissions with equivalent production levels from nitrogen fertilizer. Dietary lipids are effective in reducing enteric CH_4 emissions, but the applicability of this practice will depend on its cost and its effects on feed intake, production, and milk composition. High-oil by-product feeds, such as distiller's grains, may offer an economically feasible alternative to oil supplementation as a mitigation practice, although their higher fiber content may have an opposite effect on enteric CH_4, depending on basal diet composition.

Inclusion of concentrate feeds in the diet of ruminants will likely decrease enteric CH_4 emissions per unit of animal product, particularly when above 40% of dry-matter intake.

The effect may depend on type of "concentrate," inclusion rate, production response, impact on fiber digestibility, level of nutrition, composition of the basal diet, and feed processing. Supplementation with small amounts of concentrate feed is expected to increase animal productivity and decrease GHG emission intensity when added to all-forage diets. However, concentrate supplementation should not substitute high-quality forage. Processing of grain to increase its digestibility is likely to reduce enteric CH_4 emission intensity. Nevertheless, caution should be exercised so that concentrate supplementation and processing do not compromise digestibility of dietary fiber. In many parts of the world, concentrate inclusion may not be an economically feasible mitigation option. In these situations, improving the nutritive value of low-quality feeds in ruminant diets can have a considerable benefit on herd productivity, while keeping the herd CH_4 output constant or even decreasing it. Chemical treatment of low-quality feeds, strategic supplementation of the diet, ration balancing, and crop selection for straw quality are effective mitigation strategies, but there has been little adoption of these technologies.

Nitrates show promise as enteric CH_4 mitigation agents, particularly in low-protein diets that can benefit from nitrogen supplementation, but more studies are needed to fully understand their impact on whole-farm GHG emissions, animal productivity, and animal health. Adaptation to these compounds is critical, and toxicity may be an issue. Through their effect on feed efficiency, ionophores are likely to have a moderate CH_4-mitigating effect in ruminants fed high-grain or grain-forage diets. However, regulations restrict the availability of this mitigation option in many countries. In ruminants on pasture, the effect of ionophores is not sufficiently consistent for this option to be recommended as a mitigation strategy. Tannins may also reduce enteric CH_4 emissions, although intake and milk production may be compromised. Further, the agronomic characteristics of tanniferous forages must be considered when they are discussed as a GHG-mitigation option. There is not sufficient evidence that other plant-derived bioactive compounds, such as essential oils, have a CH_4-mitigating effect. Some direct-fed microbials, such as yeast-based products, might have a moderate CH_4-mitigating effect through increasing animal productivity and feed efficiency, but the effect is expected to be inconsistent. Vaccines against rumen archaea may offer mitigation opportunities in the future, although the extent of CH_4 reduction appears small, and adaptation and persistence of the effect is unknown.

3.7.3 Manure-management mitigation practices

Diet can have a significant impact on manure (feces and urine) chemistry and therefore on GHG emissions during storage and following land application. Manure storage may be required when animals are housed indoors or on feedlots, but a high proportion of ruminants are grazed on pastures or rangeland, where CH_4 emissions from their excreta is very low and N_2O losses from urine can be substantial. Decreased digestibility of dietary nutrients is expected to increase fermentable organic matter concentration in manure, which may increase manure CH_4 emissions. Feeding protein close to animal requirements, including varying dietary protein concentration with stage of lactation or growth, is recommended as an effective manure ammonia and N_2O emission mitigation practice. Low-protein diets for ruminants should be balanced for rumen-degradable protein so that microbial protein synthesis and fiber degradability are not impaired. Decreasing total dietary protein and supplementing the diet with synthetic amino acids is an effective ammonia- and N_2O-mitigation strategy for nonruminants. Diets for all species should be balanced for amino acids to avoid feed-intake depression and decreased animal productivity. Restricting grazing when conditions are most favorable for N_2O formation, achieving a more uniform distribution of urine on soil, and optimizing fertilizer application are possible N_2O-mitigation options for ruminants on pasture. Forages with higher sugar content (high-sugar grasses or forage harvested in the afternoon when its sugar content is higher) may reduce urinary nitrogen excretion, ammonia volatilization, and perhaps N_2O emission from manure applied to soil, but more research is needed to support this hypothesis. Cover cropping can increase plant nitrogen uptake and decrease accumulation of nitrate, and thus reduce soil N_2O emissions, although the results have not been conclusive. Urease and nitrification inhibitors are promising options to reduce N_2O emissions from intensive livestock production systems, but they can be costly to apply and result in limited benefits to the producer.

Overall, housing, type of manure collection and storage systems, separation of solids and liquid, and their processing can all have a significant impact on ammonia and GHG emissions from animal facilities. Most mitigation options for GHG emissions from stored manure, such as reducing the time of manure storage, aeration, and stacking, are generally aimed at decreasing the time allowed for microbial fermentation processes to occur before land application. These mitigation practices are effective, but their economic feasibility is uncertain. Semipermeable covers are valuable for reducing ammonia, CH_4, and odor emissions at storage but are likely to increase N_2O emissions when effluents are spread on pasture or crops. Impermeable membranes, such as oil layers and sealed plastic covers, are effective in reducing gaseous emissions but are not very practical. Combusting accumulated CH_4 to produce electricity or heat is recommended. Acidification (in areas where soil acidity is not an issue) and cooling are further effective methods for reducing ammonia and CH_4 emissions from stored manure. Composting can effectively reduce CH_4 but can have a variable effect on N_2O emissions and increase ammonia and total nitrogen losses.

Anaerobic digesters are a recommended mitigation strategy for CH_4, they generate renewable energy, and they provide sanitation opportunities for developing countries, but their effect on N_2O emissions is unclear. Management of digestion systems is important to prevent them from becoming net emitters of GHG. Some systems require high initial capital investments, and as a result their adoption may occur only when economic incentives are offered. Anaerobic digestion systems are not recommended for geographic locations with average temperatures below 15°C without supplemental heat and temperature control. Lowering nitrogen concentration in manure, preventing anaerobic conditions, and reducing the input of degradable manure carbon are effective strategies for reducing GHG emissions from manure applied to soil. Separation of manure solids and anaerobic degradation pretreatments can mitigate CH_4 emission from subsurface-applied manure, which may otherwise be greater than that from surface-applied manure. Timing of manure application (e.g., to match crop nutrient demands, avoiding application before rain) and maintaining soil pH above 6.5 may also effectively decrease N_2O emissions.

3.7.4 Animal husbandry mitigation practices

Increasing animal productivity can be a very effective strategy for reducing GHG emissions per unit of livestock product. For example, improving the genetic potential of animals through planned crossbreeding or selection within breeds, and achieving this genetic potential through proper nutrition and improvements in reproductive efficiency, animal health, and reproductive life span, are effective and recommended approaches for improving animal productivity and reducing GHG emission intensity. Reduction of herd size would increase feed availability and productivity of individual animals and the total herd, thus lowering CH_4 emission intensity. Residual feed intake may be an appealing tool for screening animals that are low CH_4 emitters, but currently there is insufficient evidence that low-residual-feed-intake animals have a lower CH_4 yield per unit of feed intake or animal product. However, selection for feed efficiency will yield animals with lower GHG emission intensity. Breed difference in feed efficiency should also be considered as a mitigation option, although insufficient data are currently available on this subject. Reducing age at slaughter of finished cattle and the number of days that animals are on feed in the feedlot by improving nutrition and genetics can also have a significant impact on GHG emissions in beef and other meat-animal production systems.

Improved animal health and reduced mortality and morbidity are expected to increase herd productivity and reduce GHG emission intensity in all livestock production systems. Pursuing a suite of intensive and extensive reproductive-management technologies provides a significant opportunity to reduce GHG emissions. Recommended approaches will differ by region and species but will target increasing conception rates in dairy, beef, and buffalo, increasing fecundity in swine and small ruminants, and reducing embryo wastage in all species. The result will be fewer replacement animals, fewer males required where artificial insemination is adopted, longer productive life, and greater productivity per breeding animal.

Concluding, improving forage quality and the overall efficiency of dietary nutrient use is an effective way of decreasing GHG emissions per unit of animal product. Several feed supplements have the potential to reduce enteric CH_4 emission from ruminants, although their long-term effect has not been well established, and some are toxic or may not be economically viable in developing countries. Several manure-management practices have significant potential for decreasing GHG emissions from manure storage and after application or deposition on soil. Interactions among individual components of livestock production systems are very complex but must be considered when recommending GHG-mitigation practices. One practice may successfully mitigate enteric CH_4 emissions but increase the fermentable substrate for increased GHG emissions from stored or land-applied manure. Some mitigation practices are synergistic and are expected to decrease both enteric and manure GHG emissions (for example, improved animal health and animal productivity). Optimizing animal productivity can be a very successful strategy for mitigating GHG emissions from the livestock sector in both developed and developing countries.

3.8 Forest conservation[78]

Forest conservation can help mitigate the effects of climate change and facilitate adaptation to new climatic conditions. It plays a particularly important role for the water balance. The clearing and overexploitation of "old" woodland leads to accelerated desertification and exacerbated flooding. International climate negotiations aim to foster the protection and sustainable use of forests. The clearing of woodland releases enormous amounts of greenhouse gases, which exacerbates climate change. These emissions account for nearly 20% of the global total. Compounding the problem, an area such as the Amazon rainforest produces its own rain. If too much woodland is cleared, local water cycles are thrown out of balance. As forests have the capacity to absorb water, rapid runoff of rainwater is prevented. Also, roots keep the soil in place. If this is hampered by timber lumbering or the burning of forests, floods are increasingly likely to occur and/or to have catastrophic consequences. On top of that, forests help clean the water, which is particularly important in the catchment area of drinking-water reservoirs. Now, one might think that these negative processes could be stopped by reforestation per se. But that is not automatically the case. Studies have shown that reforested land requires considerable amounts of water, especially in the first few decades, which means that, for instance, rivers in the vicinity may see their water levels drop or even dry up completely. Eucalyptus plantations and other rapidly growing trees for timber production, in particular, can make local water shortages worse by using up large volumes. Hence, plantation farming only makes sense in regions with sufficient precipitation.

In this context, emissions trading may play a decisive part. The REDD+ (Reducing Emissions from Deforestation and Forest Degradation) program planned by the UN is designed to avoid woodland clearance by enabling emissions certificates to be generated through forest conservation. Sustainable use of the forests concerned, and ideally also of nontimber products, can continue to take place nonetheless. For one thing, this creates an incentive to protect forests and/or to use woodlands in a sustainable fashion. For another, the local population does not lose this source of income. If commercial use of the forests in question were to be prohibited altogether, this conservation measure could increase pressure in other areas. In some places, forests are already being used traditionally to produce nontimber products such as fruit, nuts, or latex. In many cases, forest products can be marketed profitably. Support from smallholders and more sustainable land use by large farms can also help to ease the pressure on the last remaining primeval forests. Some technical issues still need to be solved and abuse of the instrument prevented; for example, replacing natural forests with plantations should be avoided at any cost, as forest conservation must be given the highest priority.

In general, climate change will affect forest conditions (area, health and vitality, and biodiversity), allowing increases in growth rates in some areas while endangering the survival of species and forest communities in others. Temperature, availability of water, and changes in seasonality may all become limiting factors, depending on geographic area, original climatic conditions, species diversity, and human activities. Most commonly, these changes will affect the frequency and intensity of fires and insect pests and diseases, as well as damage done by extreme-weather conditions, such as droughts, torrential rains, and hurricane winds. In some cases, this may lead to expansion of forest areas; for example, temperate forests are expected to spread poleward.

In other cases it may lead to reduction of forest areas, such as in the northeast Amazon region, where forest dieback is expected to reach enormous proportions due to reduced availability of water, in combination with unsustainable land-use practices. Provision of forest ecosystem services and goods will be altered by these changes, posing a number of new challenges to forest managers. In some areas, responses to climate change will affect the demand for forest products, for example, increased demand for forest-based fuels as a substitute for fossil fuels. Societies react to their perceptions of the actual and potential impacts of climate change on ecosystems, as well as to changing requirements related to forest production and trade, by developing policies and legislation. A global survey by FAO found that, although most forest managers are aware of and concerned about climate change and its potential impacts, only few have clear ideas on how to prepare for and react to it. Monitoring of changes is possibly the activity that would add the most burden to forest-management activities, since to date few effective and cheap ways to monitor changes have been found and implemented. It is nevertheless important for future forest-management operations, as it is mainly through monitoring that forest managers will be alerted to changes early on. In addition, several of the opportunities that are currently being discussed in relation to climate change, such as payment for ecosystem services, require monitoring to identify and measure services rendered. A range of management activities will contribute to maintaining or increasing the adaptive capacity of forests. They include, among others, actions oriented to maintaining forest health and vitality (e.g., by application of appropriate silvicultural treatments and by fire, pest, and disease management) and to conserving or enhancing biodiversity in forests (e.g., by effective management of forest conservation areas, enhancing connectivity between forest areas). Many of these management actions also contribute to climate-change mitigation through reducing emissions from forests, conserving forest carbon, or enhancing forest carbon sinks. Forest carbon management offers the potential for some immediate financial benefits. However, so far only a few people have benefited from these opportunities. Accessing international financial mechanisms and voluntary carbon markets has proven to be difficult and cumbersome due to the requirements to measure carbon and show both additionality and permanence of the carbon stock. This may improve as existing mechanisms are modified and new ones are developed. In addition, new international opportunities for financial and technical support for climate change adaptation are emerging.[79]

Four broad strategies have been identified for a sustainable future:

- Plant trees and invest in ecosystem services.
- Promote small and medium forest-based enterprises and gender equity.
- Use wood for energy; reuse and recycle wood products.
- Enhance communication and coordinate development.[80]

3.9 Wider investible climate change space

In Table 3.3 a schematic overview is provided regarding the opportunities in the wider climate change sphere (i.e., to include agriculture in the strict sense).[81] The real challenge is to come up with solutions that mitigate the impact of climate change and/or adapt to changing climate conditions while meeting sustainability standards.[82] The key will be to move beyond carbon systems and regulatory mechanisms and toward a market-based system of payments for ecosystems and environmental services.

There are also direct implications coming from climate change regarding the risk profile for investors in agriculture.[83]

Table 3.3 Investment opportunities related to climate change

Asset class	Sector	Investment strategy	Risk/volatility
Equity	• Clean tech	Broad/diversified	Medium
	• Energy efficiency/ energy management	Focused/broad	High/medium
	• Agriculture	Focused	High
	• Water	Focused/diversified	Medium
Fixed income	Climate change/ carbon tilt	Broad and diversified	Low
Clean Technology/ Agricultural Technology	Clean tech/agri-tech	Focused	High
Infrastructure/real estate/land	Renewable energy, natural gas, water, agriculture, timber	Broad and diversified Focused	Medium to low

4
Agricultural Risk Management and (Crop) Insurance

Risk management in agriculture is important on several grounds: even if reducing farming risk does not always improve farmers' welfare, failure to manage risks has direct repercussions on farmers' incomes, market stability, and potentially food security. The two main risks faced by farmers – yield volatility and price volatility – are expected to rise due to changing weather patterns and tightening demand/supply fundamentals. Further, particular features such as the Common Agricultural Policy (CAP) of the European Union has been undergoing reforms that have significantly reduced the extent of market interventions. Tighter budgets, environmental factors, and trade considerations have resulted in an increased market orientation of the CAP. All in all, as market effects are gradually introduced, more of the actual risk will be in the hands of farmers/investors. This chapter therefore needs to be read in conjunction with the chapter on agricultural financial markets.

4.1 Agricultural risk and risk management

Agricultural risks can range from independent (for example, localized hail losses or an individual farmer's illness) to highly correlated (for example, market-price risk or widespread drought). Managing risks in agriculture is particularly challenging, as many risks are highly correlated, resulting in whole communities being affected at the same time. Clearly, given the widespread nature of resultant loss, financial recovery is particularly difficult and challenging.[1] For governments, the fiscal implications of social-safety-net payments or the rebuilding of damaged infrastructure can be serious. For insurers, sudden losses suffered by a large number of policyholders places a strain on their reserves and financial stability. For farming communities, there is often no other option than to sell assets, normally at distressed prices. However, the impacts of realized agricultural risks are not peculiar to farmers alone. There are also the companies and service industries that supply the farmers, and the processing and logistics companies that move the produce from the farm to the markets (that is, the wider supply chain).

If one has to break down the portfolio of agricultural risks, one could use the following categorization:[2] (1) risks prevalent at the farm level and (2) risks prevalent within supply chains. Price risk could well be added to that list. That particular topic, being financial in nature, is covered in the financial chapter (Chapter 11). For a more comprehensive and technical overview of the risks and risk-management techniques applied to agriculture, I refer to the appendix of this chapter.

4.1.1 Risks prevalent at the farm level

Risks faced by farmers vary depending on location, climate, and production systems.[3] A good summary could read as reflected in the list below:

- Weather risks: rainfall, hail, extreme-weather events
- Biological risks: Pests, diseases, contamination
- Price risks: Price and demand/supply volatility
- Labor and health-related risks: illness, death, injury, incapacity to work
- Political and regulatory risk: political turmoil, political and policy changes, overall disruption of markets

They all invariably lead to a reduction in yield, price, or loss of pricing power. Additionally, farmers face "constraints" that cause them not to be able to either improve or increase their production and revenues. Examples of such constraints are limited access to finance, dislocation from markets, poor access to inputs, lack of advisory services and information, and poor infrastructure (for example, irrigation or rural roads).[4] These constraints are generally worse in low-income countries, where public goods and private-sector service delivery are often poorly developed.

The importance of noting the difference between a risk and a constraint is that often the latter are a function of the former. For example, improving access to finance for farmers in developing countries would improve if the potential financiers were able to be assured that the risks inherent with agricultural production had been managed, thereby reducing their repayment risk.

However, many possible constraints are often not driven by one underlying risk alone. Building forward on the "access to finance" example, it can be observed that even if the underlying weather risk is managed through the purchase of an insurance product or the installation of irrigation, this still leaves the financier running a number of risks. For example, the farmer may simply sell the product and not repay the bank, or prices may fall to such an extent at harvest that the revenue is insufficient to repay the loan amount, or perhaps the crop is destroyed by locusts and there is no crop left to sell at the due repayment date.[5]

The understanding of the existence of multiple risks in agriculture is critical as the apparent management of one major risk leaves stakeholders (although rarely the farmers) with the impression that the overall risk profile has been managed. However, this is often not the case; even when farmers and their partners have managed their own direct risks, indirect risks can cause losses. For example, an outbreak of aflatoxin in maize in a given country may lead to the imposition of an import ban by potential buyers. Even though farmers and the supply chain they are involved with may have managed this risk well and their maize is aflatoxin free, they will suffer from the country's market access restrictions.[6] Even if a farmer has managed contamination risks in her own basket of goods, should the processor fail to control its crop collection or processing activities correctly, the farmer may well suffer due to the exclusion of the processor from the market (there being no other buyer for the farmer's output). Conclusively, consideration of risk throughout a supply chain enables a more comprehensive assessment and management of risks.

4.1.2 Risks prevalent with supply chains

The underlying objective of agricultural supply-chain management is to provide the right products (quantity and quality), in the right amounts, to the right place, at the right time, and at a competitive cost. Logistics and communications are embedded in all of these flows, and poor logistics and communications are often major constraints that can exacerbate underlying risks in many agricultural supply chains.

An overview of the agricultural supply-chain risks[7] is provided in Table 4.1:

A different categorization could be (1) production risk,[8] (2) price and markets risks, (3) regulatory risks, (4) technological risk, (5) financial risk, and (6) human resources risk.[9] In fact, many categorizations can (and do) exist and cover using different anchor risk patterns in the total agricultural risk spectrum.[10]

Apart from being categorized according to their sources, risks can be classified according to the frequency of occurrence of certain negative events and the magnitude of their impact. Risk management starts with decisions on the farm and at the household level: which outputs to produce, how to allocate land, which inputs and techniques to use. Diversification of activities on- and off-farm normally contributes to reducing risk. The level of the farmer's integration in the food-supply chain also affects the degree to which the farmer is impacted by price volatility. Vertical integration – when the farm controls a commodity across two or more levels of activity – typically reduces risks associated with variation in quantity or quality of inputs (backward integration) or outputs (forward integration). Vertical integration is more common in the livestock sector (integration backward into feed manufacturing) or in

Table 4.1 Agricultural supply-chain risks

Risk typology	Examples
Extreme/changing weather patterns	Rainfall (excess/deficit), winds, hail, flooding, droughts, hurricanes
Environmental and biological aspects	Pests and diseases; contamination affecting food safety, natural resources/environment, or production and processing
Market related	Changes in supply or demand that impact domestic or international prices of inputs or outputs; changes in demand for quantity or quality attributes, food-safety requirements, or timing of product delivery; changes in enterprise or supply-chain reputation and dependability
Logistics/ infrastructure	Changes in transportation, communication, or energy costs; degraded or undependable transportation, communication, or energy infrastructure; physical destruction, conflicts, or labor disputes affecting transportation, communication, energy infrastructure, and services; remoteness from ports/ transportation hubs
(Operational) management	Poor management decisions; poor quality control; forecast and planning errors; breakdowns in farm or farm equipment; use of outdated or below par quality seeds; lack of preparation to change product, process, or markets; inability to adapt to changes in cash and labor flows
Policy design and institutions/ government	Uncertain monetary, fiscal, and tax policies; uncertain regulatory and legal policies or enforcement; uncertain policies on trade, market, or land and tenure systems; governance-related uncertainty; weak institutional capacity to implement regulatory mandates
Political	Security-related risks; uncertainty associated with sociopolitical instability within a country or in neighboring countries; interruption of trade due to disputes with other countries; nationalization or confiscation of assets
Perishability of goods	

the fresh-vegetables sector (integration forward into sorting, assembling, and packaging). The level of interconnectedness is also important: it is of relevance, for instance, for insurance or policy purposes whether only a few farms are affected (idiosyncratic risk) or whether it is a large number of farms (systemic risk).

4.1.3 Risk assessment

Identifying risk is good, but assessing (and quantifying it) is paramount. Risks (and their impacts) are assessed by quantifying three main variables: hazard, vulnerability, and exposure.

- Hazard is the categorization of the type of risk being considered – for example, weather, price, pest, policy, or market. The quantification of the hazard is then undertaken by assessing three subvariables: (1) frequency: How often or likely is the risk to occur? (2) severity: What are the likely fiscal impacts of such a risk if it occurs? (3) spatial extent: How widespread would the impact of the risk be – one person? one village? one country?
- Vulnerability: is an estimation of what the impact of the realized risk would be given the assets affected by the event and taking into account the current ability to manage the impact.
- Exposure is the identification of the location of crops, livestock, and farm holdings that may be directly impacted by the hazard. Interdependency in the supply chain leads to indirect exposure for other parties.

Risk modeling is used to assess assumptions and variables and their impact on estimated (financial) loss. It should be noted, almost evidently, that agricultural risk assessment is particularly dependent on the relationship between the timing of the loss event and the agricultural calendar. Risk assessment in agriculture is further complicated by the fact that vulnerability is heavily influenced by many local variables, such as soil type, crop varieties, cultural practices, irrigation, and drainage. Local access to information and intelligence is key.

4.2 Risk-management systems and models[11]

Ex ante risk management in the agricultural sector will be based on the following:[12]

- Mitigation/reduction
- Transfer
- Coping
- Avoidance/prevention (although difficult)

As always in risk management, two of the key questions are (1) what is the probability of impact and the economic damage it may cause? and (2) who is the best party involved who can manage that risk at the lowest cost possible (in fact the party that is best positioned to manage that risk is the one who often can do so at the lowest cost possible)? A reflection on these issues with respect to the agricultural industry can be found in Table 4.2.

It is clear that a distinction needs to be made between measures that aim to create and foster the management of risk by markets on an ex ante basis and the management of risks by government (particularly emergency humanitarian relief, compensation for catastrophic events, and reconstruction of

Table 4.2 Risk-allocation grid in agriculture

Probability/ economic damage vs. best party in supply chain	Buyers/ community	Markets/ innovation	Government
Low	Crop diversification, food-stock buffer	New innovation, improvement of seed quality	Irrigation infrastructure, research, (weather) data systems
Moderate	Labor diversification, risk pooling	Savings	
High	Sale of assets, migration	Formal lending, risk-sharing contracts over parties involved in supply chain	State-arranged lending facilities
Miscellaneous	Sharecropping, water-resource management	Insurance	State-sponsored insurance, disaster relief

public goods), normally on an ex-post basis. Facilitating the use of market-based approaches can reduce the needs and scope for government interventions and thereby decrease the costs incurred by government in ex-post coping activities.

4.3 Risk-management tools for farmers

4.3.1 Diversification

The idea behind diversification is that returns from various enterprises or activities are not perfectly positively correlated. A favorable result in one enterprise may help to cope with a loss in another enterprise. Diversification thus reduces overall risk. However, there may be a trade-off with cost-increasing effects linked to this strategy, for example, higher cost for additional equipment and costs in terms of foregone economies of scale which would alternatively be achievable with higher degrees of specialization. Lack of managerial expertise and market outlets, as well as climate, soil quality, or the availability of water, may limit the opportunities for diversification of farm activities.

4.3.2 Contracts and vertical integration

In a *marketing contract*, a farmer agrees to sell a commodity at a certain price to a buyer before the commodity is ready to be marketed. The farmer retains full responsibility for all production-management decisions. The contracts can take many forms. They can be based on a fixed price, or alternatively they may depend on the development of the commodities futures price. The latter type of contract does not eliminate price risk completely. Price risk is zero only in those cases where the exact price to be paid to the producer upon delivery is fixed. The (opportunity) costs borne by the farmer result from forgoing the opportunity of achieving a higher price on the open market. Besides reducing risk, contracting provides the farmer with an opportunity to differentiate his products from mass production and to draw an economic rent from this. Intermediate and final consumers may be willing to pay a higher price for a contractual guarantee of a certain quality or trait. Contracting allows distributing this additional value ("quality premium") along the production chain.[13] However, the premiums agreed on in most of the quality-related contracts are not fixed at a specific value but rather depend on market conditions. A further advantage of marketing contracts in the crops sector is related to time management. During the busy harvesting season, time does not have to be allocated to selling the product.

Production contracts typically give the contractor (the buyer of the commodity) considerable control over the production process. These contracts normally specify the production inputs to be used, the quality and quantity of the final product, and the price to be paid to the producer. The contracts vary in the degree of control exercised by the contractor. Apart from other possible advantages for the farmer,[14] the contract partially shifts price risk to the processor. On the downside, the farmer depends to a large extent on one buyer, thus incurring a risk of losing his only outlet following contract termination. In the crops sector, the development of biotechnologies has recently led to an increase in production contracts. If segregation or identity preservation develops, then crops, be they genetically modified or not, will increasingly be grown and sold under contracts. In the United States, some agricultural commodities are produced almost 100% under contracts. Contracts are now the primary method of handling sales of many livestock commodities, including dairy, hogs, broilers, and turkeys, as well as of major crops such as sugar beets, tobacco, fruit, and processing tomatoes. In recent years, we have seen dramatic shifts toward contracts and away from spot markets.

However, spot markets still govern nearly 60% of the value of agricultural production and remain an efficient way to produce and distribute many products. This is especially true for more generic products for which differentiation is less important to the final consumer. The use of contracts is closely related to farm size. Farms with at least $1 million in sales have nearly half of their production under contract.

A *vertically integrated* firm retains ownership control of a commodity across two or more levels of activity. There are many complex reasons to vertically integrate; risk reduction is only one of them. Vertical integration helps to reduce risks associated with a variation in the quantity and/or quality of inputs (in the case of backward integration) or outputs (in the case of forward integration). Vertical integration is more common in the livestock sector, with integration backward into inputs (feed manufacturing) or in the fresh-vegetables sector, with forward integration into sorting, assembling, and packaging.

The alternative for farmers would ultimately be to accept the spot prices in the open market at any given point in time (I ignore the option for the farmer to manage the risk that comes with accepting spot prices from a financial point of view; see below).

Therefore the options of the farmer, based on our discussion so far, for bringing his products to the market can be summarized as shown in Table 4.3.

Why would farmers want to use contracts instead of spot markets? Data of the USDA (United States Department of Agriculture) focuses on two explanations for the shift that has been unfolding for some time now (starting in the mid-1990s). Contracts may be seen as a device to limit price and income risks (risk-sharing approach), or they may be regarded as a means to reduce the costs of using spot markets to arrange transactions (transactions-cost approach). Contracts can obviously reduce certain risks[15] for the farmer, but then again

Table 4.3 Options for bringing agricultural products to the market

Mechanism	**Control over production decisions**	**Pecuniary compensation for farmers**
Spot prices	Farm operator controls assets and production decisions.	Farmer receives price for farming output, negotiated at time of sale just prior to delivery.
Production contract	The contractor holds control over some of the assets and some production decisions. The contract determines the output, delivery time, quantities, and other essential elements.	Farmer is paid a fee for farming services in the production of the commodity.
Marketing contract	Farm operator controls production decisions and assets. The contract determines the output, delivery time, quantities, and other essential elements.	Farmer receives a price for the farming output, negotiated before or during the production of the agricultural commodity.
Vertical integration	A single holistic firm controls assets and production decisions in farming and subsequent processing stages.	Farm operator-manager is compensated for skill and time.

there are many ways to reduce risks, and many contracts appear not to be targeted at risks. The transactions-cost approach demonstrates that contracts can be designed to improve incentives to lower production costs and deliver products with specific attributes. They can also facilitate coordination among stages of production – speeding adoption of new technology; improving information flows; managing quality, uniformity, and delivery; and enhancing access to credit. If transaction costs are important, then contracting can lead to improved productivity and higher product quality. Increased contract use creates several types of concerns for producers. Contracts may lead to unanticipated new risks for producers. Under some conditions, they can allow buyers to exercise market power, reducing prices received by producers. And as more production shifts to contracts, reductions in spot-market volumes can raise spot-market costs.

Some contracts commit producers to long-term investments that will support production for a particular buyer. If contracts give producers only short-term purchase commitments, they will face new risks from contract cancellation or buyer failure. Moreover, many contracts specify fees for producer services rather than market prices. Without reliable market information on fees and services, producers can find themselves at a bargaining disadvantage with contractors.

Critics presume that contracts create market power for buyers and reduce farm prices. Under the right conditions, contracts can be designed to limit the entry of potential rivals into concentrated markets. They can also be designed to limit the intensity of price competition among existing rivals or to expand buyer profits through price discrimination (by targeting lower prices at some sellers who have few alternatives). However, the success of such actions depends on the precise terms of agricultural contracts, the structure of the agricultural markets involved, and the responses of rival buyers. In particular, contracts that aim to create market power generally require highly concentrated markets with limits on entry by rivals, and they frequently need to have existing rivals adhere to similar contracts. Because contracts are often used in concentrated markets, there may be cases in which contract terms do allow buyers to exercise market power. However, since contracts can also lead to enhanced productivity and improved responsiveness to consumer demand, it is important for policy responses to target only those contracts that extend market power without offsetting gains in efficiency.

In summary,[16] Table 4.4 displays the magnitude and types of risk transfer and risk allocation under the different types of contracts.

The USDA and other organizations worldwide have long provided agricultural market information to the public to facilitate smooth operation of the spot market. However, spot prices are relevant only to the extent that they provide information about the cost and value of products moving through the

Table 4.4 Types of risk transfer and allocation

Contract type/sources of risk	Price risk		Yield risk	
	Output	Input	Common	Idiosyncratic
Marketing contracts				
Flat price – determined fixed price before harvest.	Some or all, depending on share of expected output under contract	No	No	No
Basis – fixed difference from uncertain future price.	Very little – only basis component of price risk	No	No	No
Production contracts				
Absolute performance – contractor provides almost all inputs except labor, facilities, and infrastructure.	Almost all of the risks	Almost all of the risks	Some – depends on contractual arrangements and incentives built in	Some – depends on contractual arrangements and incentives built in
Relative performance – fee is based on relative performers to other farmers; rest is the same as under absolute performance.	Almost all of the risks	Almost all of the risks	Almost all of the risks	Few

whole system. The expansion of contracting may have reduced the value of traditional price reporting to producers and may consequently have raised the costs of using spot markets.

In summary, contracts are likely to govern a growing share of agricultural production going forward because of a number of (obvious) reasons. First, demand for differentiated agricultural products to meet specific consumer preferences should continue to grow, and such products are generally produced under contract. Second, pressures will mount to ensure traceability of products for health and consumer concerns, and contracts provide one way to ensure traceability. Third, pressure to reduce environmental degradation associated with agricultural production will likely result in upgraded production technologies and require tighter management of production systems through contracting. Finally, large farms account for sharply growing shares of agricultural output. Contracting is closely associated with farm size, and contract use can be expected to grow along with the increase of large farms.[17]

4.3.3 Hedging risks via options and futures

The reduction of (pricing) risk via financial instruments is a very important mechanism that links the agricultural industry to the financial industry. It therefore deserves a significant place in the chapter on financial dynamics of the agricultural industry where we will be dealing with the topic extensively. Given its role as a risk-management tool, I feel obliged to at least provide a short overview of the dynamics and relevance of hedging practices within the risk-management arena.

Nevertheless, before talking about how to hedge (price risk), we need to analyze the dynamics of price risk.

A. The dynamics of price risk

Price risk refers essentially to variability in output prices and input prices.

1. Input price risk

In crop agriculture, input price risk has been given considerably less attention in the literature and has been considered less substantial than output price risk and yield risk. It normally does not translate into return variability of the same magnitude.[18] Moreover, the time window of input price risk is shorter: fertilizer and input costs are usually incurred within a few months of the onset of production, whereas the uncertainty around output price and yield usually remains for at least six months. However, the magnitude of input price fluctuations can be significant, and there is no system in place to hedge against input prices (although there is the possibility of storing some items on the farm). Input price risk is still significant and may be overlooked. Variability in fuel and fertilizer prices appears to be the main component of input price variability in crop production, partly because fuel and fertilizer amount to most of the input costs in conventional agriculture, and partly because, as commodities themselves, they are subject to price fluctuations like all other commodities. These variabilities are expected to increase, in line with increased volatility of energy prices. As to the livestock sector, input costs amount predominantly to feed costs. The following discussion on output price risk for crops thus covers most of the issues pertaining to input price risk for the livestock sector.

2. Drivers and types of output price variability

Output price risk arises due to the biological lag inherent in agricultural production. Obviously producers must make production decisions months (even years for tree crops) before they have a product to sell, before the actual crop prices are known. During this period, output prices may change dramatically in response to shocks in supply and demand. This may put farmers in a difficult situation if commodity prices decrease drastically during the production and marketing cycle.

Many factors contribute to price changes, as witnessed during the food-commodity price spike of 2007–2008. Income and population growth, rising energy prices, and subsidized biofuel production have contributed to surging consumption of agricultural products. At the same time, productivity and output growth have been impaired by natural-resource constraints, under-investment in rural infrastructure and agricultural science, farmers' limited access to agricultural inputs, and weather disruptions. While speculation has been mentioned as a driver of price increases, the issue has been heavily debated, but there is no conclusive evidence that speculation drove prices up – at least in the long run (see more on this topic in the chapter on financial speculation). The consumption of cereals had also been consistently higher than production in previous years, which had reduced stocks. Stocks play a critical role indeed through their cushioning effect: low levels of stocks are associated with high price volatility.[19] Macroeconomic factors also impact the volatility of agricultural prices, for example, variability in inflation rates, exchange rates, and interest rates.

The effect of shocks on the agriculture and food system is compounded by low elasticities of both supply and demand. Indeed, since food is an essential product with no substitute (apart from alternative foods), demand responses to price increases are typically low (until, for the poorest of all, it translates into hunger). Supply responses are also very low short term, until production decisions can be made for the next season or more land brought into or taken out of production.

Price variability is complex in nature and is often captured as price volatility, which can be defined in several ways. It refers to movements in prices of a certain periodicity: the period can be a day, a month, a season, or a year. Volatility can also be considered as being made up of high- and low-frequency components, which may be useful to distinguish.[20] A common measure of volatility is the coefficient of variation (ratio of standard deviation to its average value) of a historical price series with a given frequency. Historical volatility may also be defined as the annualized standard deviation of the percentage change in daily settlement prices. Some prefer to focus on the uncertainty component of volatility: given that some variability can be predicted (on the basis of seasonal variation or business cycles, for instance), they are concerned about not overstating the degree of uncertainty. Therefore, they restrict volatility to the unpredictable price movements. There are also measures of implicit volatility, which represent the market's expectation of how much the price of a commodity is likely to move.

Price transmission from global markets to domestic markets is affected by the level of trade and the extent to which the trade environment is managed domestically. For markets more open to trade, volatilities in domestic prices are close to those on international markets. So a price shock to a crop like wheat tends to spread globally, and the magnitude of price risk for a commodity will tend to be similar for producers worldwide.

The relationship between price fluctuations and trade levels is complex. A closed market, although less affected by external shocks – for example, a global recession – is highly vulnerable to an internal shock, for example, a drought. An open market is obviously directly affected by the instability of global markets, but it can share its risk worldwide, which has a stabilizing effect on prices. Future price volatility will depend on a number of factors, including variabilities in supply and demand and responsiveness to these variabilities (elasticities of supply and demand).

The supply/demand balance will be key. The increasing demand (for food, feed, and fuel) combined with long-term resource-scarcity trends (of water, arable land, and energy) points toward a tight balance. This is a particular concern at times when global stocks are low and likely indicate a rise in volatility, although it is difficult to distinguish between price trend and volatility. (See the last chapter on the outlook for the global agricultural markets for further details.)

In the context of climate change, extreme-weather events are expected to be more frequent. This and other factors mentioned affecting yields will contribute to increased price volatility.

Moreover, food prices and oil prices are linked increasingly closely, due to the use of agricultural crops for energy production and agricultural production's reliance on energy inputs. A potential increase in volatility in energy markets (also in a context of tight supply) would likely spill over into food markets. Technological progress and trade liberalization, by allowing faster and greater responses to shocks, are expected to reduce the volatility of food prices downward, but this is unlikely to be enough to balance out the other factors. In short, the short- to medium-term outlook is mixed and volatile, with some downward pressure on a number of commodities (see outlook chapter).[21]

3. Hedging instruments and application

On spot or cash markets, prices are set for goods that are immediately available. Production and marketing contracts as well as futures markets add the time dimension to these markets. They allow users to hedge their price risk by locking in the price of a commodity they wish to purchase or sell at a future date. Futures market contracts differ in three important respects from production and marketing contracts: First, they are standardized in terms of contract terms, and thus they can be more easily traded. Second, they are traded in organized exchanges under rules and regulations. And third, while production and marketing contracts usually involve physical delivery of goods at maturity, in futures markets this is the exception. Trading in futures enables shifting risk from a firm that desires less risk (the hedger) to someone who is willing to accept the risk in exchange for a profit. A hedger can be a farmer, a trader, or a processor who wants to "lock in" a price for a commodity he is going to sell or buy on the spot market at a later stage. Futures contracts help to protect against price risk, as futures and cash prices converge against the end of the delivery period, resulting in losses and gains in the two markets offsetting each other.

A farmer would usually use the instrument as follows: At the beginning of the growing or feeding period, he sells a futures contract, thus entering an obligation to sell after the end of the production cycle a certain quantity of a commodity at a certain price. The futures position ("short futures position") is held until the product is ready for sale. Before the end of the delivery period of the futures contract, the farmer typically buys an offsetting futures position ("long futures position"). The commodity itself is then traded through its usual marketing channels. Losses (gains) in the expected value of the output due to price changes tend to be offset by gains (losses) in the value of the futures position. Thus, the farmer is able to "lock in" a certain price for his commodity.

An option gives the holder the right (without incurring the obligation) to take a futures position at a specified price (called the "strike" price) at a specific time. As opposed to futures, options giving the right to sell a futures contract do not constrain potential profits resulting from increasing prices. Such options give security against the implications of decreasing prices. The costs of achieving this security are materialized in the price to be paid for the option.

Hedging largely eliminates price risk but leads to what is called basis risk. Basis is the technical term for the difference between the price of a commodity on the spot market and the price of the corresponding futures contract. The variability of this difference introduces a new category of risk associated with futures markets. The basis reflects differences in price across space (transportation costs, exchange rates), time (storage costs), and quality (difference between actual quality and standard quality specified in the futures contract). The effectiveness of hedging on futures markets may also be reduced by yield variability. Uncertainty about effective yields makes it not advisable to sell a futures contract for the total expected output because of the potential obligation to buy additional quantities in order to fulfill a mature contract. Generally the effectiveness of hedging to reduce risk diminishes as yield variability increases. The effectiveness of hedging furthermore depends on the correlation between yield and price for a given commodity. If there is a strong negative correlation between yield and price (a "natural hedge," with yield and price fluctuations offsetting each other), hedging becomes – because of the cost associated with it – inefficient. With increasing trade liberalization, this negative correlation will decrease.

Trading future contracts goes as far back as 1730 where rice futures were traded in Japan, followed by grain futures in the 19th century on the CBOT (Chicago Board of Trade). Many other exchanges have emerged since.[22] In the US grain markets, the largest groups of hedgers in this day and age are the grain merchants and processing firms. As indicated, its use is still modest compared to other risk-management tools. In terms of volume of futures traded, that is definitely true since the arrival of financial investors in the agricultural space.

A formal, though summarized, product-by-product review of investment types includes the following:

- *Forward contracts*: A simple instrument available to deal with price risk is a forward contract (also called cash contract). Farmers and buyers of agricultural output agree in advance on the terms of delivery regarding quantity and price (either fixed or in line with futures prices). With this type of contract, the farmer foregoes the opportunity of achieving a higher price on the open market but partially shifts price risk to the processor. There are two broad categories of forwards. A production contract usually gives the buyer of the commodity (a processor) considerable control over the production process. The farmer also depends largely on only one buyer and incurs the risk of losing his outlet after the contract ends. In a marketing contract, the farmer retains full responsibility for production management and has an opportunity to differentiate his products from mass production. Inherent to these cash contracts is counterparty risk: the risk for the producer that the buyer may not show up or not pay on time (and for the buyer that the producer may not deliver the commodity as called for in the contract). These default risks are considered reasonable, and cash-forward contracting is used extensively in the United States, offered by the commodity vendors. Time management is another advantage of these contracts: during the busy harvesting season, farmers often do not have enough time to sell their products. On the downside, it may require high transaction costs to find the potential buyer and negotiate the terms of the forward contract.[23]
- *Futures contracts*: A futures contract is a forward contract traded on an organized exchange (and not over-the-counter), standardized in terms of quantity, quality, and delivery time and location. (Futures markets were originally developed in the Middle Ages to meet the needs of farmers and merchants). A farmer hedging his price sells a futures contract when planting his crop, but he typically does not deliver the commodity at the end of the contract. He buys rather a futures contract for the same delivery date, thus undoing his position, and sells later on the cash market. (For example, if a farmer, through his/her broker and trader, sells a corn contract in May for December

delivery, his or her position may be offset by buying a December corn contract at any time before the end of the delivery period in December.) Indeed, the major motive in trading futures is to hold a temporary position for hedging, not to physically deliver (or acquire) a commodity. It is also cheaper for producers to deliver through normal channels.

By using a futures contract, the farmers reduce their risk but retain basis risk, measured by the difference between the cash price and the futures price. Buyers and sellers of futures are also required to make margin deposits with their brokers to guarantee their respective commitments. On top of this initial margin, there is a variation margin: if the price of the contract moves against the farmer, he gets a margin call from his broker to post additional margin to cover the loss so that a minimum margin is maintained. This maintenance margin is usually somewhat lower than the initial margin, which is typically 5% to 10% of the underlying value of the contract. All in all, hedging involves costs that appear modest compared with the risk reduction for most farmers. The basic link between the prices of the futures contract and the underlying cash market is the *cost of carry* (i.e., storage plus insurance) until the expiration of the contract. According to this link, supply and demand factors affecting prices in cash markets should be transmitted to the futures markets so that futures and spots converge at the end of the settlement period. Historically, the relationship between the cash and futures markets has been fairly constant with predictable seasonal variation. However, futures prices at the expiration period have lately tended to be well above spot prices in the United States, leading to basis risk. No problem regarding convergence has been observed on European futures markets for food commodities, but a debate is ongoing regarding the causes and consequences of the lack of convergence. Futures markets are recognized as an effective way for farmers to reduce price risk on top of sending important signals regarding future prices. Their use provides certainty of income for food producers, and certainty of costs for processors, distributors, and retailers. Some governments encourage farmers to use futures markets, mostly by providing information and technical advice (e.g., Mexico provides subsidies to farmers who buy commodity options on US futures markets).

- *Options*: Options on futures provide a further possibility to cover price risk for some commodities. Options give the right to sell a futures contract or to buy one, at a strike price. Thus options truncate the probability distribution of price at the strike price and provide protection against adverse price movements (low spot price for sellers/put holders, high spot price for buyers/call holders). At the same time, an option allows the option holder not to exercise it and to profit from favorable movements (high prices for put options and low prices for call options). Farmers can use put options to create a floor price for their produce.

According to a recent OECD report,[24] the literature is not conclusive about the effectiveness of options contracts. This probably has to do with the cost to buy and with the fact that options limit the downside without affecting the upside: benefits of risk aversion are hard to quantify.

There are (still) huge differences in trade volumes between the North American and the European exchanges. This can be historically explained. The main obstacle for the establishment of futures markets (including markets for derived products) in Europe has been the CAP-induced relative price stability for major commodities (resulting in little interest from both the farmers' as well as the speculators' side). It was only after the 1992 CAP reform that price volatility increased. As mentioned, with the CAP providing security, farmers have had little incentive to turn toward market-based risk-management tools to deal with price risk; rather, the expectation has been that the consequences of adverse events on farmers' incomes will be borne by taxpayers. Another reason is the higher level of consolidation of farming in the United States, especially for the big commodities: in the EU, grains are still often grown by small undercapitalized farms, which are much less inclined to use financial innovation. Larger farms tend to use more risk-management strategies in general, but this holds especially true for hedging. Additionally, Americans have a longer history of trading soft commodities on exchanges. It takes time to adopt financial innovation. As a result, the main futures markets for agricultural commodities are located in the United States, the benchmark for several commodities (especially wheat and corn) being the Chicago Mercantile Exchange (CME), which absorbed the Chicago Board of Trade in 2007. In the EU, the main agricultural contracts are traded on Euronext in Paris (milling wheat, rapeseed, corn) and Liffe (London International Financial Futures Exchanges) (cocoa, coffee, sugar, feed wheat). There are also a few minor futures markets (e.g., in Germany, Hungary, and the Netherlands). Commodity traders are more prone to use futures markets for hedging and price discovery on European-based exchanges today than they were before successive reforms of the CAP significantly reduced guaranteed prices.

More recently, traded volumes have been coming down across the board. The number of futures and options traded on US exchanges fell 13.2% in 2012 to 7.0 billion contracts from 8.1 billion in 2011, according to statistics compiled by the Futures Industry Association.[25] The number of futures and options traded on exchanges around the globe fell 15.3% in 2012 to 21.2 billion contracts from 24.9 billion in 2011. The total number of contracts traded in 2012 was the lowest since 2009 and reflected double-digit declines across Asia-Pacific, Europe, and North America, according to statistics compiled by the Futures Industry Association. Volume in the Asia-Pacific region fell by 23.4% to 7.5 billion from 9.8 billion contracts. North American volume fell by 11.9% to 7.2 billion from 8.2 billion contracts, and volume in Europe fell by 12.5% to 4.4 billion from 5.0 billion. By sector, contracts based on individual equities fell by 8.4% to 6.5 billion contracts from 7.1 billion; equity-index volume plunged by 28.5% to 6.0 billion from 8.5 billion contracts, and interest-rate-contract volume fell by 16.0% to 2.9 billion from 3.5 billion.

Whereas futures and options have significant advantages and potentially have an important role to play in the future, they also have a number of limitations. They only deal with short- to medium-term risks of market downturns (as other risk-management tools like insurance), and their use requires considerable investment in know-how and infrastructure.[26] In line with that, futures markets have a number of limitations of their own:

- Basis risk (the risk that the change in the price of a hedge may not match the change in the price of the asset it hedges) can be a serious obstacle to some farmers' participation. Trading costs may be an issue as well, although futures contracts incur lower transaction costs than forward contracts. Some governments have provided subsidies toward the cost of futures contracts (e.g., Mexico provides subsidies to farmers who buy commodity options on US futures markets).
- Derivative markets are not available for all commodities. However, this is changing rapidly and therefore hardly an obstacle.
- The quantities specified in the contracts may not be within the scale of many smaller producers. But financial intermediaries such as brokers can provide help here, or an organization of farmers trading as a group.
- Derivatives markets are still relatively new, and there is still a lack of understanding. More information and training are necessary. Governments may encourage farmers to use futures markets by providing information and technical advice.
- Since the food-price spike in 2008, there is the overall concern of financial investors or speculators perverting the market, in particular the price-discovery function of commodity markets. This topic will be extensively covered in the relevant chapter.[27] There is also no clear-cut distinction between the various actors on commodity futures markets: the hedgers who want to protect themselves from price risks (e.g., farmers and agribusiness buyers) and the speculators/investors who want to make a profit from bets on future price movements – who are necessary counterparts for hedgers. (Since hedgers' positions usually do not cancel each other out, the functioning of the futures requires a party willing to accept the risk in exchange for an expected profit.) In fact, there is rather a

continuum between pure risk avoidance and pure speculation. This brings of course the possibility of positions distorting the normal functioning of the market, for instance due to their sheer size. Such speculation can have grave consequences for farmers and consumers and is, in principle, unacceptable.

The use of derivatives is likely to grow in Europe with the further decrease of public-market-management tools. Public intervention may also encourage this trend by promoting training on derivative products, by ensuring availability of information, and by having an appropriate supervision regime in place. Smaller operators should have the opportunity to join forces through risk-sharing cooperatives.

4. Mutual funds or mutual stabilization funds

Mutual (stabilization) funds provide a way for a group of producers to share risk. A loss incurred by a member will be fully or partially compensated through the collected money available in the fund according to predefined rules (often with an additional collection from participants). Established on private initiative, these are mainly set up either for a specific sector or a specific region.

The difference between a mutual fund and a mutual insurance scheme (also a nonprofit cooperative based on self-help) is the legal nature of the institution. For mutual insurance there is a legal title of compensation, and the premiums are calculated on an actuarial basis (as opposed to a fixed amount independent of risk). The advantage of these risk-sharing cooperatives is that farmers often know each other, which reduces moral hazard and adverse selection. Drawbacks are limited resources, especially in the fund's early years, and interconnectedness: a farmer may at the same time incur a loss and have to contribute to the fund to cover other farmers' losses. Reinsurance or cooperation with other regions may help counter this problem

4.4 Agricultural risk layering[28]

Traditional subsidized agricultural insurance programs are not sustainable in the long run, especially in low-income countries that face fiscal constraints. Multilateral institutions promote a proactive, strategic approach for the financial management of agricultural production risks. This financial risk–management model deals with residual risks that cannot be mitigated, that is, risks that remain after cost-effective risk-mitigation techniques (e.g., irrigation, pest treatments) have been successfully implemented. The country agricultural risk financing model is based on four pillars and aims to offer countries an operational template to implement sustainable agricultural insurance. Agricultural risks can be financed with farmers' self-retention, private financial markets, governments, and international donors through an appropriate layering of risks.

The *bottom layer of risk* includes high-frequency but low-consequence risks that affect farmers and/or herders from a variety of mostly independent risks. These losses may be caused by inappropriate management decisions and are thus exposed to moral-hazard and adverse-selection problems. They should be retained by the farmers/herders and financed by individual savings/credit.

The *mezzanine layer of risk* includes less frequent but more severe risks that affect many farmers/herders at the same time (e.g., locust swarms, hail). The private-insurance industry has demonstrated its ability to cover these losses.

The *top layer of risk* includes low-frequency but high-severity risks such as major droughts or floods. These catastrophic risks, by definition, are not well documented, and their probable maximum loss can be very large. Innovative financial products, backed by governments, may offer new risk-transfer opportunities. Table 4.5 provides an overview of the different risk layers.

Table 4.5 Different risk layers

Institutional capacity building • Data management • Regulatory and supervisory framework • Information and education • Technical expertise	**Financial management of agricultural production risk**	**Agribusiness segmentation** • Social vs. commercial insurance • Traditional farming sector • Emerging farming sector • Commercial farming sector
Agricultural risk financing • Risk layering • Product development • Agricultural insurance pool • Blending insurance and rural finance		**Agricultural risk assessment** • Probabilistic catastrophic risk-assessment model • Actuarially sound rate making
	Risk transfer/ mitigation →	
Catastrophe insurance coverage		Reinsurance market
Base insurance coverage	**Loss/frequency**	Insurance markets
On-farm risk mitigation, savings, credit		Farmers

4.4.1 Insurance

The agricultural sector is a major economic sector and a critical source of livelihood in many developing countries. Agriculture is particularly exposed to adverse natural events, such as insect damage and poor weather conditions, that negatively impact production. The economic costs of major climatic disasters may increase further in the future due to climate change. Farmers and herders have developed risk-management strategies to cope with these adverse events, sometimes with government assistance. Agricultural insurance is one financial tool that agricultural producers can utilize to mitigate the impacts of unpreventable risks and other related macroeconomic variables.[29]

Although agricultural insurance has been offered in some industrialized countries for more than a century, the agricultural sector remains underserviced in middle- and low-income countries. Since the late 1990s, however, dwindling public support to agricultural producers in emerging markets has led to a renewed interest in agricultural insurance. The development of agricultural risk-modeling techniques and the emergence of insurance pools and index-based insurance contributed to a revisiting of the potential role of agriculture insurance in emerging economies. Insurance is one of the tools that farmers and other stakeholders can use to manage risks that are too large to manage on their own (risk layering). Part of the risk is transferred to another party, who takes it in return for a fee (or premium).[30] But insurance should always be complemented with other risk-management practices through risk sharing and diversification. Insurance can assist farmers in accessing new opportunities by improving their ability to either borrow money or raise additional credit lines.

4.4.2 Definition and categories

Crop insurance is taken out by farmers and others to protect themselves against either the loss of their crops due to natural disasters, such as hail, drought, and floods, or the loss of revenue due to declines in the prices of agricultural commodities.

Many agricultural risks are considered to be in-between risks that are neither perfectly independent nor perfectly correlated (systemic), for example, yield and price risk. Insurance solutions for agriculture also have to include provisions to deal with problems of asymmetric information (moral hazard and adverse selection). The different insurance categories in the insurance sector are as follows:

- Yield insurance
- Catastrophic losses
- Price insurance

- Revenue insurance (which essentially is a combination of price and yield insurance)
- Income insurance
- Reinsurance

Two basic models for reinsurance dominate:

- Proportional reinsurance (quota share arrangements): The insurer and reinsurer share premiums and risk. The reinsurer assumes an agreed-upon, fixed percentage of all the insurance policies written by a direct insurer. The quota determines how premiums and losses are distributed between direct insurer and reinsurer.
- Nonproportional reinsurance:
  a. Excess of loss: the reinsurer covers up to a certain amount any part of a loss resulting from a single catastrophic event that exceeds an agreed deductible.
  b. Stop-loss: the reinsurer covers up to a certain amount any part of a total annual loss that exceeds an agreed deductible.

The three general categories of crop insurance are called crop-yield insurance, crop-revenue insurance, and index-based crop insurance.

1. **Crop-yield insurance**: There are two main classes of crop-yield insurance.
   - *Crop insurance* (aka damage-based indemnity insurance) is crop insurance in which the insurance claim is calculated by measuring the percentage damage in the field soon after the damage occurs. The damage measured in the field, less a deductible expressed as a percentage, is applied to the preagreed sum insured. The sum insured may be based on production costs or on the expected revenue. In case the damage cannot be determined directly after impact (occurrence of loss), the assessment can be delayed until later in the season. Damage-based indemnity insurance is best known for hail (often called crop-hail insurance) but is also used for other named-peril-insurance products (such as frost and excessive rainfall).
   - *Multiperil crop insurance* (MPCI; aka yield-based crop insurance): Coverage in this type of insurance is not limited to just one risk. Usually multiperil crop insurance offers hail, excessive rain, and drought in a combined package. Sometimes additional risks such as insect- or bacteria-related diseases are also offered. The problem with multiperil crop insurance is the possibility of a large-scale event. Such an event can cause significant losses beyond the insurer's financial capacity. To make this class of insurance, the perils are often bundled together in a single policy

(also because it is often difficult to determine the cause of a loss) called a multiperil crop-insurance policy. Yield-based crop insurance is coverage in which an insured yield (for example, tons/ha) is established as a percentage of the farmer's historical average yield. The insured yield is typically between 50% and 70% of the average yield on the farm. If the realized yield is less than the insured yield, an indemnity is paid equal to the difference between the actual yield and the insured yield, multiplied by a preagreed value. MPCI coverage is usually offered by a government insurer, and premiums are usually partially subsidized by the government. In the United States, for example, the US Department of Agriculture is known to have implemented the earliest MPCI program in 1938, and the Federal Crop Insurance Corporation has managed this multiperil insurance program since then. Additionally, the Risk Management Agency (RMA) has been active in calculating premiums based on individual risk factors since 1996.[31]

2. **Index-based crop insurance,**[32] which can be subcategorized as follows:
   - *Area-yield index insurance*: the indemnity is based on the realized average yield of an area such as a county or district, not the actual yield of the insured party. The insured yield is established as a percentage of the average yield for the area. An indemnity is paid if the realized yield for the area is less than the insured yield regardless of the actual yield on a policyholder's farm. This type of index insurance requires historical area-yield data.
   - *Weather index insurance (WII)*:[33] the indemnity is based on realizations of a specific weather parameter measured over a prespecified period of time at a particular weather station. The insurance can be structured to protect against index realizations that are either so high or so low that they are expected to cause crop losses (e.g., too much or too little rainfall). An indemnity is paid whenever the realized value of the index exceeds a prespecified threshold or when the index is less than the threshold (too much rainfall versus too little rainfall). The indemnity is calculated based on a preagreed sum insured per unit of the index.[34]

A farmer or grower may desire to grow a crop associated with a particular defined attribute that potentially qualifies for a premium over similar commodity crops, agricultural products, or derivatives thereof. The particular attribute may be associated with the genetic composition of the crop, certain management practices of the grower, or both. However, many standard crop-insurance policies do not differentiate between commodity crops and crops associated with particular attributes.

The investment opportunity lies to a large degree in the developing-markets space where access to agricultural insurance is generally very limited. Insurers have tended to concentrate on urban and industrial risks and therefore often do not have networks in rural areas. One of the challenges for insurers is that where agricultural insurance can be provided, risks (weather, pest, disease, and so on) are often highly spatially correlated and therefore financially difficult for insurers to manage. Hence the necessity for a decent risk layering as outlined above, featuring three key but distinct layers:

- *Self-retention*: risk retention (by the farmer) is needed for manageable, smaller, frequent risks that have to be mitigated by the farmer using standard farming practices. Rearrangements for the payment of principal loan amounts or interests due are part of this category as well.
- *Transfer of (market) risk*: Insurance is best suited to infrequent but severe events through insurance and reinsurance based on risk appetite and capital availability.
- *Market-failure layer*: Extreme losses from extremely rare, highly catastrophic events are not suitable for commercial insurance. For these types of losses, governments or the broader international community may be needed to aggregate and transfer this risk layer out of the domestic economy to international markets. This is also known as the "government intervention layer," as the fiscal responsibilities for reconstruction or such interventions as social safety nets lie with the affected government.

3. (Weather) index-based insurance[35]

Index-based agricultural insurance relies on the measurement of an objective and independent parameter that is highly correlated with the actual loss incurred by a farmer/herder. Under parametric index insurance, payouts are based solely on the measurement of a particular parameter (for example, rainfall at a named meteorological station) according to an agreed payout scale (established in the insurance policy) related to the rainfall actually recorded at a specific meteorological station. Under aggregate-index insurance, payouts are based on an index developed from the aggregated statistics of farm production or yield in specified districts (e.g., area crop-yield estimates for crops, livestock mortality estimates for livestock). Cost-effective agricultural insurance programs usually combine traditional indemnity-based insurance (e.g., named-peril insurance) with innovative index-based insurance solutions.[36]

From a weather risk-management standpoint, and as already briefly highlighted, there are two main types of risk to consider. These relate to (1) sudden, unforeseen events (for example, windstorms or heavy rain) and (2) cumulative events that occur over an extended period (for example, drought). This has significant implications with respect to the (assessment of the) relationship between yield and weather events. Short-duration extreme-weather events (such as hail, windstorm, or heavy frost) can cause devastating direct damage to crops in the fields. Assessment of these damages can be immediately observed. On the other hand, while the final outcome of cumulative events can be devastatingly obvious, much of the damage already occurred during an earlier stage of crop development. However, correlations of the weather event and damage are often difficult to model, except for the most extreme events. In the case of cumulative rainfall deficit (drought), the best correlations exist for rain-fed crops grown in areas where there is a clear sensitivity of the crop to deficits in available water, and clearly defined rainy seasons. Less-clear relationships are found in areas of higher and more regular rainfall or less-clear seasonality, or where other influences, such as pests and disease, are important causes of crop losses. If partial or full irrigation is in place, the relationships become much less strong. Rain-fed production in the tropics (where rainfall is higher and less seasonally marked) is an example where correlations may be less easy to establish. However, droughts are also a feature of tropical crop production, where both floods and droughts can occur in the same year. Going short, generalizations are intrinsically risky.

Interest in using index-based agricultural insurance has grown in recent years, particularly with respect to addressing the systemic component of agricultural production losses (e.g., widespread drought). Index-based insurance offers advantages over traditional insurance relying on individual losses, such as lower monitoring and loss-adjustment costs and a more transparent indemnity structure. This type of insurance also faces some challenges, however, such as basis risk caused by the imperfect correlation between the actual loss and the index, which make it cost effective only for specific crops, perils, and geographical areas.

While risk is a factor that affects all areas of human endeavor and activity, risk in agriculture is particularly problematic. Agriculture's reliance on natural conditions (rain, temperature, sunlight, etc.) and lack of ability to either control them or effectively mitigate when they are insufficient or too abundant means that things go wrong in agriculture on a very regular basis (high frequency). The fact that these phenomena tend to affect large areas invariably means that when things do go wrong, a lot of people are affected (covariance). Finally, due to the nature of farming systems in developing countries, when things go wrong it can often result in big losses (high impact). Most insurance products are able to provide you coverage from loss at attractive premiums because risks rarely materialize (low frequency), they normally happen to only a very few of the insured (idiosyncratic), and oftentimes losses are low and below the threshold of the actual insurance (low impact). Before managing agricultural risk, it must first be assessed and approaches identified that can be used to most efficiently and practically manage it. Although insurance is often the first thing that comes to mind, it is often also the least attractive or practical approach. Risk management mainly consists of three types of activity – mitigation, transfer, and coping. Mitigation covers a number of traditional activities. From irrigation to hybrid seeds, from vaccinations to pest control, many agricultural activities are really the application of risk-management approaches. Many vets would not consider themselves to be risk managers, but in effect, that is a large part of what they do. Transfer is the simple act of passing one's financial risk to a third party, who is prepared to accept it because they charge a fee (or premium) for the service. Crop insurance and price hedging are the most common forms of transfer in agriculture, but they are often very expensive.

Index insurance is a simplified form of insurance in which indemnity payments are based upon values obtained from an index that serves as a proxy for losses rather than upon the assessed losses of each individual policyholder. The sum insured is normally based on production cost on an agreed-value basis (fixed in the policy in advance), and payouts are made based on a preestablished scale set out in the insurance policy. In order for the underlying index to be a sound proxy for loss, it has to be based upon an objective measure (for example, rainfall, wind speed, temperature) that exhibits a strong correlation with the variable of interest (in this case, crop yield). Additionally, the weather variable that can form an index must possess the following properties: (1) observable and easily measured, (2) objective, (3) transparent, (4) independently verifiable, (5) reported in a timely manner, (6) consistent over time, and (7) experienced over a wide area. Given these requirements, weather indexation is most applicable to highly correlated risks, such as drought and temperature. Localized (independently occurring) risks, such as hail or fire, do not lend themselves to index insurance. Indemnity payouts are made in accordance with a schedule laid out in the policy itself. For example, insurance for rainfall-deficit coverage can imply a payout structure whereby a full payout can occur as long as actual rainfall is below a certain threshold normal for the environment (let's say 50 mm of rainfall). From there onward, the payout structure is downward sloping to end at 0, the moment the actual rainfall equals a normal rainfall season (and normal distribution of that rainfall over the relevant period). This last comment is key for making the payout structure relevant. If not, the payout structure would be too simplified to capture the true correlation between rainfall and crop-yield loss. The timing, not just the amount of rainfall during the various growth phases of a plant, is very important for satisfying the soil water balance and therefore the ultimate yield. Dry spells, or deficits over the main phases of crop growth, can cause yield loss, even if cumulative seasonal rainfall is adequate. Commonly index-product designs use several phases of measurement during the crop season (typically three phases for grain crops), each with their own thresholds and limits of the weather parameter. Any given development of weather indexes has to be based in reality using data collection of terrestrial-based monitoring systems (weather stations), the latter often not being available in emerging economies. The importance of the different risks within the index can be summarized as follows:[37]

- Rainfall deficit: High
- Rainfall excess: Medium
- High temperature: Medium
- Low temperature: Medium
- High wind speed/direction: High/Low (infra)
- Hours of sunshine: Low

Weather-index insurance has the flexibility to be designed for different types of clients or for clients with different needs. It can be designed on a microlevel for particular farmers (for rainfall for example), on a macrolevel (a risk-pooling basis) for risks covering a whole area (hurricanes in the Caribbean), or somewhere in the middle (called the mesolevel), covering risks that a dedicated area is exposed to with abnormal distribution patterns (El Niño – induced excess rainfall[38]). In these cases, the insurance provides liquidity to continue operations and to ensure repayment of existing loans, although there is no (or reduced) income from the current harvest season.

The question has been raised, given the dynamics of weather-index insurance, whether this still qualifies as an insurance or whether it might more accurately be called a financial derivative product (the value being derived from the index). While the two instruments feature different regulatory, accounting, tax, and legal issues, the risk-transfer characteristics and benefits are similar. From a practical point of view, therefore, the discussion is academic and less relevant.[39] Weather-index insurance products are most meaningful when used to cover extreme risks, as they cannot solve or seriously address other agricultural constraints. The role of coverage is to improve efficiency or unlock further potential. It requires, in order to work efficiently, that other essential aspects are in place (e.g., access to finance). A key element in the mix that should be emphasized is the potential one with agricultural finance. Without linking these insurance programs explicitly to finance (such as bundling the insurance with agricultural production loans or inputs), a weather-index insurance initiative will often find that many farmers (in developing economies) lack both the capital to pay the insurance premium and sufficient incentive to use scarce resources on risk management. The model therefore works best when it integrates weather-index insurance with access to finance and supply-chain development. Therefore the advantages of the product group can be summarized as follows (although it will undergo changes as it matures):

- *Reduced risk of adverse selection*: This is due to asymmetry of risk whereby the farmer will take insurance if it is believed that there is a higher level of risk, putting the insurer at risk. The advantage of index insurance is that farmers subscribe based on the terms, conditions, and payout scale for all farmers in their defined area, virtually eliminating the adverse-selection problem for insurers.
- *Reduced moral hazard*: Under normal insurance positions (individualized), farmers may be able to influence the claim (by exacerbating physical losses) through their behavior. With index insurance, farmers have no ability or incentive to influence the claim, since payout is based on an independent and exogenous weather parameter, independent of farmers' behavior.
- *Field-loss assessment eliminated*
- *Reduced information requirements and bureaucracy*
- *Facilitation of reinsurance (lower premium cost)*
- *Transparency and educed disputes*
- *Facilitating access to financial services*

However, there are also a number of remaining challenges that could impact its ability to become a mainstream product:

- *Basis risk*: this is the difference between the payout as measured by the index and the actual loss incurred by the farmer.[40] Because no field-loss assessment is made under index insurance, the payout is based entirely on the index measurement and may be either higher or lower than the actual loss.
- *Data availability*: for underwriting and pricing purposes and for the recording of the parameter(s) for payout calculations during the period of insurance, as well as historical yield data to assess risk, improve design, and price the product, if the weather index is to serve as an accurate proxy for loss.
- *Integrity of weather station*
- *Need for farmer/insurer/regulator capacity building and education*: for insurers, this is a new type of insurance product, so they require substantial technical assistance in designing contracts and indexes and extensive capacity building to enable them to undertake product development on a sustainable basis.
- *Limited diversification in product group*: currently limited product options for different weather risks. The majority of WII products have been designed for rainfall risk, which is not necessarily the most serious or prominent weather risk in many areas. There is limited experience with insuring other types of weather risk (and the interaction of perils).

- *Research, local adaptation, and scalability*:[41] involves the analysis of weather data and interpretation of it in relation to the specifics of the crop to be insured. Correlations need to be carried out between the weather data and historical yield data in order to find good index parameters.
- *Limited truly universal application*: It can be an effective instrument, but not for all crop types, cropping systems, or hazards. Where crop type or climate show complex and multiple factors affecting crop damage or loss, as may occur in humid climates or where pest and disease are dominant causes of loss, indexation with weather-index insurance may be problematic. In such circumstances, an area-yield index product may be more applicable.

Based on a verifiable and independent measurement of a variable that impacts crop development, it is argued that index insurance can potentially reduce insurance premiums and make insurance accessible to more farmers. The cost of insurance is made up of two components. First, the underlying risk based on the frequency, severity, and extent of impact of loss (known as pure risk). Second, there are also large costs involved in administering and implementing insurance – mainly individual risk assessment and loss adjustment. Together, specifically in agriculture, these costs are very high, and therefore premiums are excessive for most farmers (hence most agricultural insurance schemes are subsidized). Index-based insurance, at least theoretically, can cover many farmers in a given area without the need for loss assessment and adjustment or the need for excessive paperwork and individual risk assessment when writing a policy. This can reduce some of the administrative and implementation costs, and it also has the potential to reduce payouts where loss is due to factors other than natural ones (e.g., farmer malfeasance or poor farming practices).

Pre-feasibility work is important, as initiating index insurance schemes is often very challenging. The very nature of an index-based product creates the chance that an insured party may not be paid when they suffer loss and/or that they may receive a payment when they have suffered no loss. Known as basis risk, this phenomenon is a particular problem for index products. It is frequently caused by the fact that the measured variable (e.g., rain) at the measurement site differs from that on the individual farmer's field or that the complex rainfall data and crop-model calculations fail to accurately capture yield losses. Of course, there is also the problem that index products tend to only capture one variable, and loss can be caused by a number of factors (e.g., a farmer is covered for drought, but he lost his crops to disease). There is a "duality" to basis risk – insurers who have received premiums from farmers who have suffered losses, but are not paying due to a lack of triggering of the index, face serious reputational problems. They run the real risk that such incidents can negatively impact other products that they sell in the market (e.g., auto insurance, life insurance). Arguably, part of the reason for the failure for scale-up on the index product by commercial insurers is directly linked to this duality of basis risk. There is a lack of clarity as to the regulatory and legal status of index-based products in nearly all jurisdictions. Clearly index products do not align with the traditional definition of insurance, as they do not indemnify actual loss, and a policyholder does not actually have to have an insurable interest before they purchase an index-based contract.

We will not embark on the journey of analyzing the technical aspects of weather-index insurance models as that would justify a book on its own.[42]

The value of crop insurance is clear, bridging to the next topic and the investible universe:

- Crop insurance can serve as collateral for operating loans. This improves producers' access to credit. It also helps protect the portfolios of lenders, which protects the infrastructure of rural communities.
- Crop-insurance gives producers the confidence to forward-price a larger percentage of their expected yield, which, over the long term, leads to greater profitability.
- Because producers pay a part of the premiums, crop-insurance protects the interests of taxpayers.
- Because crop insurance companies put their own earnings at risk in order to contribute to indemnity payments when disasters occur, taxpayers are, once again, protected.[43]
- The way crop-insurance APH (actual production history) is measured is subject to debate in an era with biofuels.[44]

Every view should be balanced, and therefore a highlight of the other side of the coin is warranted.[45]

The basic notion is this: crop insurance doesn't work in isolation but goes hand in hand with the direct-subsidies program in most countries. Even before analysis, one identifies the potential issue, that being the fact that if not properly aligned, crop insurance will serve not as a safety net but as an income enhancer for farmers and can have perverse effects that result from the way these (often government run) programs are managed. Not only goes a large part of the subsidies/insurance yield to the investors behind the farmer rather than to the farmer itself. The combination also induces farmers to take more risk, even up to a level of risk similar to before the subsidies were provided. You ultimately can't take risk out of an industry. Risk subsidies are different than price subsidies, however. Under the current subsidy structure, those with the highest risk and those in the highest-risk regions gain the most.[46] Recently, the crop-insurance program in the United States (worth 117 billion USD in product coverage) has come under renewed attack. Under the insurance program, the US taxpayer subsidizes the majority of premiums paid by farmers, covers much of the administrative costs tallied by insurers to run the program, and guarantees that all losses are covered.[47]

4. Crop-insurance savings accounts

Crop insurance is a critical risk-management tool for farmers to protect against yield and revenue losses, to smooth income over time, and to remain a viable operation after catastrophic events. However, designing crop-insurance instruments that achieve broad participation among farmers at a low cost to governments has proven to be a formidable challenge. Agricultural production and prices are highly volatile, and the correlation between historic and future outcomes is quite limited due to weather variability, unforeseen pest problems, frequent changes in technology and market globalization. As a consequence, it is difficult for both the insurer and the producer to accurately assess the level of yield and revenue risk associated with a particular farm operation. As exemplified,[48] the inability of either party to ascertain what the actuarially fair premium is within a reasonable margin of error limits farmer participation unless the overall premium levels are highly subsidized. As a result, achieving broad participation in crop-insurance programs has proven costly to governments. Given the pressing need to bring most budget deficits under control over the next decade, it is important to explore the viability of less costly alternatives to help farmers manage their yield and revenue risks.

A second unresolved issue that has affected most crop-insurance programs for many years has been complaints from farmers, producer organizations, and legislators about the rating structure not being fair across crops, cropping systems, and geographical regions. This could also be a result of the insurer's inability to accurately determine what the actuarially correct rates are in each particular situation and the producers themselves not being able to ascertain what they should be paying.

In short, given (1) anticipated government budget constraints to fund the expansion of crop insurance that would be needed to make it effective as the sole source of government income support for all agricultural producers, (2) concerns about whether an expansion of this program as currently structured would meet the risk-management needs of all producers, and (3) the random variations in the allocation of the available support funds due to premium estimation inaccuracies at the farm, county, state, and regional levels, it is perhaps time to reconsider alternatives that could provide a reliable and more equitable safety net for at least some agricultural producers at a much lower cost to the government.[49]

4.5 Involvement of governments

Reading the previous part undoubtedly raises questions about the role of governments in agricultural risk management, or more specifically, what should not be the role of governments in dealing with agricultural risks. From the looks of it, one can conclude that it is a contentious debate, as food security is a highly visible public matter, involving public and even national security. An efficient risk-management system for agriculture will preserve the standard of living of those who depend on farming, strengthen the viability of farm businesses, and provide an environment that supports investment in the farming sector. The OECD has recently (late 2011) thrown their 50 cents into the debate by indicating that in their understanding,[50]

- **Government policies should take a holistic approach to risk management**, assessing all risks and their relationship to each other, and avoid focusing on a single source of risk such as prices. Governments can help farmers to assess and manage their own risk by providing information and training.
- **Agricultural risk-management policies should focus on catastrophic risks** that are rare but cause significant damage to many farmers at the same time. Contingency plans should define in advance the procedures, responsibilities, and limits of the policy response.
- **Subsidized insurance** is one way of providing disaster assistance, but it tends to crowd out the development of private-insurance markets and has not been successful in preventing additional ad hoc assistance being granted after the event.
- **Facilitating good "start up" conditions** – information, regulation, and training – should be the primary role for the government in the development of market-based risk-management tools such as futures, insurance, and marketing contracts.

Government policies should not provide support to deal with normal risk. Managing normal risk should be the preserve of farmers themselves. Minimum intervention prices or payments that are triggered when prices or returns are low may even be counterproductive as they tend to induce riskier farming practices (driving out farmer-specific strategies). If there are policies in place that are linked to production and price risks, farmers will not see any need to develop their own strategies, nor will private-sector, market instruments develop in the absence of demand. Government payments, as well as production and price-linked policies, affect farmers' risk exposure, but they also change farmers' risk-management behavior. Governments have often assumed that the answer to farming risk lies in stabilizing prices. In fact, by doing this they may actually increase the variability of income and have the opposite effect. For farmers in many OECD countries, years of low prices are also years of better yields. Statistical analysis shows that this correlation matters because it reduces income variability. Price stabilization undermines this natural risk-reduction link. Price policies may actually induce riskier behavior on the part of farmers because the guaranteed price lets them specialize in products or methods that are not well adapted to their particular circumstances. The higher the intervention price, the larger the probability of the net effect being an increase in income variability for some farmers. Price interventions will isolate farmers from underlying market fundamentals such as high prices that signal a negative supply shock or low prices that signal oversupply. Governments end up carrying the entire burden of risk management at high cost to consumers and taxpayers because their actions have crowded out the efforts of farmers themselves and the private sector.

4.6 Agricultural risk management and government policy

Recent years have helped us to relearn what we forgot during the last two decades, and that is that markets, although performing well in the long run, tend to have periodic moments in time where the market behaves irrationally or at least not in a sufficiently balanced way to ensure the continuity of mainstream operations, public goods, and the basic needs of societies. Hence the need for the legislator to provide a robust and stable framework within which the market can allocate capital efficiently, providing transparency and a backstop in case the market gets ahead of itself one way or the other. The question of to what extent and depth that regulatory grip should be on the market is beyond the scope of this book and often yields much controversy and many opinions.

A missing market for risk coverage – in addition to other reasons such as the existence of monopolies, externalities, and the "public good" character of certain products – can explain why markets may fail in their role to provide an efficient allocation of resources. Several risk-management tools are offered on the market, even without any government intervention (the ones discussed). So far, however, these developments remain quite limited. Several reasons have been identified why markets for risk-management products may not be available or may be incomplete. As a result, the market for risk coverage clears at less than socially optimal quantities of risk sharing.[51]

Reasons can be found both on the supply and the demand side.[52] On the demand side they include the following:

- Know-how to make use of certain risk-management tools (e.g., use of futures markets) can be acquired only at high start-up transaction costs and is, therefore, not always available to farmers.

- Farmers perceive the risks they face as being smaller than they actually are. This phenomenon is called "cognitive failure" and can result from insufficient information or misjudgment. Events of low probability that are associated with high potential losses (catastrophes) are very likely to be neglected in individual decision making.
- Even if farmers do not underestimate the risks they face, they might count on other safety nets, including off-farm income or government support programs (disaster aid), in case of significant losses and as a consequence might not use the risk-management tool available.

On the supply side, the reasons vary from product to product:

- Insurance contracts might not be offered on the market because the conditions for insurability, such as independence of risk and symmetry of information, are not sufficiently fulfilled. Each violation of these conditions increases the marginal cost of insurance and reduces market supply.

  Agricultural risks are often natural-hazard risks, which means that they are not independent risks. Only if risks are independent and an insurance company can cover a large number of exposure units can it accurately predict average future losses. Normally natural-hazard risks are correlated up to a certain degree, meaning that many exposure units (farms) incur losses at the same time, making losses for the insurance company highly variable. The higher the variance around expected loss cost, the greater the need for including a reserve load in the premium. A particular challenge is the insurance of catastrophic risks, which are infrequent but of high consequence, making it necessary to add catastrophic loads to the premium. These factors lead to premium rates that are much higher than those for largely independent risks.
- Because of potentially big losses incurred in covering agricultural natural-hazard risks, an insurance company will normally try to obtain reinsurance, which involves additional transaction costs. US evidence shows that reinsurance can be very expensive, especially after catastrophes have happened. Furthermore, agricultural reinsurance markets tend to have limited capacity. Agricultural risks are only reinsured if they fit into the portfolio of reinsurance companies and if the expected profit is high enough. If the coverage of agricultural risks yields lower returns than other sectors, less capacity is attributed to the agricultural sector. The capacity of the industry is also limited by virtue of the fact that for the highly specialized reinsurance market, entry barriers are quite high.
- For a successful establishment of futures and options markets, several conditions have to be fulfilled (as discussed above). Farmers need to adopt the necessary know-how. In addition, it has to be recalled that futures markets

can emerge only under the condition of substantial price variability, a sufficient number of traders and speculators, and products with standardized grades of well-defined quality.

Apart from market failure, there are two other possible reasons for government intervention in markets for risk-management products:[53]

- By setting the legal framework for farm operations, the government influences to a considerable extent farmers' risk exposure, for example by creating the legal framework for animal disease control and veterinary measures. To create an incentive for the government to set up an effective risk-minimizing framework, it can be argued that it should be financially co-responsible for potential losses.
- Concerning reinsurance, government intervention has also been justified by economic reasons. The government might be able to provide reinsurance cheaper than the private sector, as it is able to raise more and cheaper capital than the private sector to cover potentially high losses.

The overall objective of government involvement in agricultural risk management can be to help reduce farmers' risk exposure (through diversification, sanitary and phytosanitary measures, price safety nets) or help them cover losses resulting from risk exposure. Governments can either help to prevent/cover big losses or contribute in general to preventing/covering losses, whatever their size.

From a policy instrument point of view, there are basically four potential fields of action for governments:[54]

- First, the setting up of the necessary legal framework for risk reduction (e.g., sanitary and phytosanitary legislation) or the establishment of markets for risk-management tools. For example, competition policy should take into account that risk pooling between insurance companies (which may result to a certain extent in reduced competition) might be necessary to cover catastrophic risks.
- Second, governments could help to create the necessary conditions for developing markets providing risk-management tools (market facilitation). For example, they could provide information and training on risk and risk-management tools to farmers.
- Third, policy could consider lowering the costs of risk-management tools. As regards insurance, subsidies could be justified if an insurance product – because of the systemic nature of the risk it covers or problems of asymmetric information – would be otherwise unaffordable for farmers. Possibilities include subsidies for insurance and reinsurance premiums. Subsidies could

also be justified on a temporary basis to encourage the development of market solutions.

- Fourth, governments can also provide risk coverage themselves, for example, by providing disaster-aid payments, public insurance and reinsurance, or a specific safety net. Such approaches could be justified if private markets for risk coverage for market-failure reasons (as elaborated on above) cannot sufficiently develop to provide necessary protection for all categories of farmers (missing or incomplete markets).

Beyond these mainstream tools, other more specific tools could be considered: tax policies, disaster aid, and other safety-net policies. In order to stabilize markets and prices in particular, governments can and will engage in any of the following policy actions:

- Trade intervention measures.
- Export subsidies (although a country's import/export restrictions may bring some short-term relief to selected farmers/domestic consumers, economic analysis clearly shows that their overall impact on the domestic economy and on the rest of the world is negative[55]).
- (In)direct public price support intervention: Fixed-price purchases are applied only for certain products (soft wheat, butter, and milk powder) and in quantities determined in advance. Price support has been found to have low effectiveness, in that the price reduction it achieves is mostly captured by other agents of the full supply chain (input suppliers, processors and distributors, landowners), but little by the farmers it is supposed to help.[56] International market stabilization policies tend to not be very effective when judging the past. Following the 2007–2008 food-price crisis, numerous proposals have been put forward regarding the establishment of reserves, including an internationally coordinated strategic-reserve system for food grains. The creation of a virtual reserve is an innovative plan proposed by von Braun and Torero[57] in order to moderate prices: participating countries would commit funds to intervene, if needed, in the grain futures markets (by selling contracts to deliver the commodity) until futures and spot prices have stabilized. Maintaining a strategic reserve involves dealing with the challenges of determining optimum stock levels and coping with the uncertainties the reserve may cause in the market place.
- Direct payments to farmers: These payments are deemed effective at increasing farm welfare but are not as effective at reducing income variability as subsidizing risk-market instruments is. Further, direct payments in their current form are biased toward big producers and tend to favor intensive agriculture.

- Ad hoc payments or calamity funds. The main advantage of these funds over ad hoc aids is that they avoid major distortions to the government budget. However, these public aids have a major caveat: their perverse effect on farmers' risk appetite. Indeed, if farmers are aware of an existing safety net, they have less incentive to reduce either the risk of losses or the farm's exposure beforehand. As a result, their actions and choices (disease control, crop selection) may not be as judicious as if they were bearing the full consequences of any adverse event.

4.7 Crop insurance: a growing global business opportunity?[58]

Farming is an inherently risky business including uncertain weather conditions, market shifts, and other events beyond a farmer's control that affect farm yields and commodity prices, thus creating variability in farm revenue. Crop insurance has developed in most of the major agricultural territories in the world. Most programs are focused on yield insurance; however in recent years some countries have introduced revenue insurance. Solutions need to include the variety of needs of individual stakeholders as part of the process. These include producers, processors, the insurance company, and ultimately the public and the government.

They key drivers behind crop insurance are as follows:

- Commodity prices
    - Increasing oil prices
    - Changing global dynamics
    - Biofuels
- Increasing farmer premium subsidies
- Improving farming technology
- Attitude to risk
- Access to credit
- Significant weather events

In most developed nations the following insurance products are readily available:

- Crop hail
- Specified peril
    - Excess moisture
    - Frost etc.
- Multiperil crop insurance
    - Yield guarantee

- Crop-revenue insurance
    - – Yield and commodity price risk
- Indexed insurance
    - – Area yield
    - – Weather

However, in other parts of the world the situation is often more complicated. The majority of solutions are supported by government focusing on the (1) concern regarding food security, (2) ability to budget for future loss, and (3) high cost to the producer. Consequently the key challenges to the development and sustainability of any program are (1) quality of data, (2) cost to the producer, (3) infrastructure, (4) technical expertise, now and in the future, and ultimately the (5) cost to the taxpayer.

The size of the insurance market in 2013 in terms of collected insurance premiums was (in billions of USD):

United States	12.1	Canada	2.1
Europe	2.4	India	1.1
China	3.1	Latin America	1.0

The agricultural insurance penetration levels vary greatly from over 85% in the United States to often below 30% in some developing economies. Crop and livestock insurance is becoming an important tool to manage agricultural risk globally. Government support is critical if a national insurance program is to be effective, affordable, and sustainable. Most national agricultural insurance programs are still developing, and in light of this, market penetration is relatively low. What is particularly important to understand is that the crop-insurance sector is one of the few sectors within the global insurance space that at the industry level generates consistently over the years an underwriting profit – this in contrast to many other insurance areas like Property/Casualty and life. The key is to develop an understanding of how the mechanisms of crop-insurance policies are implemented on a national level and how opportunities consequently manifest themselves.[59] For example, the level of risk experienced by EU farmers is very heterogeneous, varying from country to country and according to farm type and size. The development of agricultural insurance in each country is linked to the risk level but also to the policy supporting the insurance system.

Single-risk insurance (mostly hail) is well developed in Europe. Usually, private companies insure only hail and fire, and as the government increases its involvement in insurance, more comprehensive coverage is provided. Yield insurance provides coverage against all the main climatic hazards, but plant diseases and plagues are usually not covered. This requires ascertaining the cause of the loss, as opposed multiperil crop insurance in the United States, where damages are simply calculated as the difference between guaranteed and actual yield. The European insurance system is associated with higher loss-adjustment costs but no moral hazard.

The insurance system in the EU-27 varies considerably from one country to the other in terms of market penetration, premium rates, loss ratio (ratio of indemnities to premium), and level of subsidies. However, most countries share the feature that there are few market players, with one or two companies dominating the very specific sector of agricultural insurance. Promoting competitiveness could result in lower insurance costs and better access for farmers.

In comparison, insurance in the United States is provided by 17 private companies, working in agreement with the USDA. About 45% of field crops' production value is insured. The average premium rate is close to 9%, much higher than in Europe, mostly because of a wider coverage through revenue or yield insurances as opposed to single-peril insurances. Government support to insurance (premium subsidies, funds for administrative costs, reinsurance) amounts to 72% of the total premiums (66% in Canada, 32% in the EU). Some pan-European studies have been undertaken[60] analyzing various aspects and implications of potentially implementing an EU-wide system of agricultural insurance, according to the different types of insurance presented above.

The current insurance level in the EU is generally insufficient to smooth major income reductions in bad years, although many member states have worked on further developing insurance in the last few years (by increasing the risks covered and by covering new crops and livestock). An indication of this gap is the amount of ad hoc aids still transferred by governments: over EUR 900 million (probably strongly underestimated by lack of information).

Crop-yield insurance is most likely unviable without government support, and subsidizing crop-yield insurance appears effective in reducing income variability, if limited;[61] otherwise the crowding-out effect of other tools becomes predominant. The use of insurance is also prone to numerous undesirable effects such as excessive risk exposure, misreporting, information asymmetries, and rent seeking by insurance companies. Index-based insurance products have relative advantages and are best suited for homogeneous areas, where farms have correlated yields. Given the heterogeneity of climates and geography in many European countries, they may be more useful for reinsurance at the aggregate level than at the farm level. But banks may get more involved, providing farmers with ways to manage basis risk.

The EU may opt for further supporting insurance as one way to stabilize farmers' income. At the same time, settling for a homogeneous common insurance system does not seem advisable given the high diversity of risks and socioeconomic backgrounds in the EU. The EU can also encourage the development of national systems by facilitating the composition of databases, providing public reinsurance, and establishing a common regulatory framework.[62]

Appendix

Risk analysis and management in agriculture

1. Introduction

Risk is uncertainty that matters and may involve the probability of losing money, possible harm to human health, repercussions that affect resources (irrigation, credit), and other types of events that affect a person's welfare. Uncertainty (a situation in which a person does not know for sure what will happen and the probability can't be properly modeled) is necessary for risk to occur, but uncertainty need not lead to a risky situation. For an individual farmer, risk management involves finding the preferred combination of activities with uncertain outcomes and varying levels of expected return. One might say that risk management involves choosing among alternatives for reducing the effects of risk on a farm, and in so doing, affecting the farm's welfare position. Risk management typically requires the evaluation of trade-offs between changes in risk, expected returns, entrepreneurial freedom, and other variables. Some risk-management strategies (such as diversification) reduce risk within the farm's operation, others (such as production contracting) transfer risk outside the farm, and still others (such as maintaining liquid assets) build the farm's capacity to bear risk. Managing risk in agriculture does not necessarily involve avoiding risk, but instead it involves finding the best available combination of risk and return given a person's capacity to withstand a wide range of outcomes.[63]

The four main aspects of risk management involve (1) identifying potentially risky events, (2) anticipating the likelihood of possible outcomes and their consequences, (3) taking actions to obtain a preferred combination of risk and expected return, and (4) restoring (if necessary) the firm's capacity to implement future risk-planning strategies when distress conditions have passed.[64] Different farmers confront different situations and structural characteristics, and as explained in this report, their preferences toward risk and their risk-return trade-offs have a major effect on decision making in each given situation. A large, industrialized operation, for example, may hire marketing expertise to directly use hedging and options, while a small family farm may prefer to forward contract with other parties better able to hedge directly. Understanding risk in farming is important for two reasons.[65] First, most producers are averse to risk when faced with risky outcomes. Someone who is risk averse is willing to accept a lower average return for lower uncertainty, with the trade-off depending on the person's level of risk aversion. This means that strategies cannot be evaluated solely in terms of average or expected return, but risk must also be considered. Second, identifying sources of uncertainty helps farmers and others address the most important strategies for mitigating risk and aids in circumventing extreme outcomes, such as bankruptcy. In practical terms, farmers seem to be most concerned about the potential decrease in crop yields and livestock production, evolution of commodity prices, the ability to adopt new technology, incoming laws and regulation, and the ability to anticipate new consumer preferences.

In the main chapter we already discussed the risks inherent to farming and the way they can be classified. One particular risk needs further attention: the identification and measurement of price and yield risk. Yield risk varies regionally and depends on soil type, climate, the use of irrigation, and other variables. In contrast, price risk for a given commodity depends on such factors as commodity stock levels and export demand.

2. Measuring and pricing (agricultural) risk

Risk must be quantified in order to evaluate whether various risk-management tools and strategies are effective in achieving producers' risk-reduction goals. The measurement of uncertainty involves estimating the probabilities of future outcomes. Estimates may be made, for example, of the probability of yield less than 130 bushels per acre, the probability of price falling below $8.00 per bushel, or the probability of revenue less than $900 per acre. More generally, one would like to estimate the joint probability distribution of yield, price, and revenue so that one might, for example, specify the probability of revenue falling below any specified level. By observing historical data sets, the attempt is made to separate random variability from systematic variability. Quantifying yield randomness generally involves summarizing what is known about deviations from expected yields, as measured by trends. The degree of randomness is reflected in the width of the distribution and in the number of observations that are distant from the mean. Often the curve is not symmetrical, as the lower tail is longer than the upper tail. This so-called negative skewness is typical of yield distributions. This shape occurs because devastating weather can cause very significant yield declines (as low as zero), while very good weather is likely to only moderately boost yields above trend due to the physiological limitations of the plant. For many purposes, a single number is a more convenient measure of randomness (or dispersion) than an entire distribution. The most widely used measures of randomness are the variance and its square root, the standard deviation. Variance is the average squared deviation from the mean, or trend. By using the variance of deviations from trend, a large part of the systematic variation is removed.

One problem with the variance and standard deviation is that they are difficult to interpret without knowing the level or magnitude of the underlying variable. A variance of 15 bushels, for example, has quite different implications for the tightness of the distribution when the mean yield (adjusted for trend) is 55 bushels per acre than when it is 170 bushels. As a result, proportional variability – or variability relative to the mean – is often measured to facilitate comparisons. The most commonly used measure of relative variability is the coefficient of variation, which equals the standard deviation divided by the mean. The variance (or alternatively, the standard deviation or coefficient of variation) is a good measure of variability for approximately symmetric, bell-shaped distributions.

It fully describes the variability in a normal distribution, which is a particular bell-shaped mathematical distribution that closely approximates many observed distributions. Farmers, like other decision makers, are fundamentally concerned with randomness in future events, not the distribution of past outcomes. They are concerned about the probabilities of outcomes to be observed in the future and the effects of these outcomes on their economic welfare. The probability associated with any given outcome indicates the strength of one's belief that such an outcome will occur, ranging from zero (which represents no possibility) to one (representing absolute certainty). Two sources of information about such probabilities are available: logic and experience.

While yield expectations before planting generally follow trends, price expectations often fluctuate substantially from year to year depending on commodity stock levels, export demand, and other factors. Futures-price quotes serve as useful proxies for price expectations for commodities traded on futures exchanges. Price randomness can be estimated by measuring futures-price quote changes from one trading date to another.[66] Price variability or risk can be measured using ratios of successive prices, P_t / P_{t-1}, instead of differences, $P_t - P_{t-1}$. This includes a number of benefits: (1) there is no need to adjust for inflation unless the inflation over the covered period has changed significantly, (2) it is a unit measure and allows for comparison across commodities, (3) using ratios allows the comparison of volatilities estimated over time intervals of different lengths. Futures quotes provide a vehicle for observing price volatility changes over the growing season.

Prices for agricultural commodities at the national or world level tend to be high when yields are low, and vice versa, because total demand for food changes only moderately from year to year, while supply can fluctuate considerably due to weather in major producing countries.

Consumers bid up the price for crops in short supply, while crops in abundant supply clear the market only at low prices. When two variables, such as price and yield, tend to move in opposite directions, they are said to be negatively correlated.

The ways that farmers reduce risk were already discussed in this chapter.[67] They can be categorized as follows:

A. Economic or operational risk management
- Enterprise diversification
- Off-farm income generation
- Vertical integration
- Production contracts
- Marketing contracts

B. Financial risk management
- Financial products to hedge risks (options, futures, forwards)
- Maintaining financial reserves and using leverage[68]
- Liquidity[69]
- Using leasing constructions[70]
- Insuring crop yields and crop revenues

For the technical discussion, it can be referred back to the relevant chapters on financial markets and this chapter's main text. What follows are a number of numerical examples of how these (financial) techniques are being used. This section provides a detailed analysis of the effectiveness of several risk-management tools on income uncertainty within the year (intrayear income risk), income uncertainty between years (multiyear income risk), and farmers' average returns. The many different options available for managing income risk lead to questions about their effectiveness across different producing regions and about how they can best be combined to reduce producers' risks. A producer's choice among strategies is particularly complicated when both price and yield (output) risk are present – the case for a farmer with a growing crop in the field. In this situation, the degree to which strategies, such as forward contracting or hedging, reduce income risk depends on yield variability, the correlation between price and yield, and whether or not the crop is insured.

A primary use of futures[71] involves shifting risk from a firm that desires less risk (the hedger) to a party who is willing to accept the risk in exchange for an expected profit (the speculator). Also, hedgers with opposite positions in the market trade with each other, and speculators with opposing views of the market may also trade. Because contracts are standardized, the only issue to be negotiated at trading time is price. Enforcing contract terms is a key function of the exchanges where trading occurs, and guaranteeing contracts is a key function of the exchange clearinghouse.

Most futures contracts are offset by opposite trades before delivery time, with each party to the transaction selling (or buying) a futures contract that was initially bought (or sold). For example, if a farmer (through his or her brokerage house and its trader on the Chicago Board of Trade) sells a corn contract in May for December delivery, his or her position may be offset by buying a December corn contract at any time before the end of the delivery period, which is about December 20. Such an offset usually occurs because the major motive in trading futures is to hold a temporary position and then trade for money, not to physically deliver or acquire a commodity. Most hedgers offset because making or taking delivery on futures would be more costly than delivering through normal channels, while speculators generally do not want to own the actual commodity. Because futures contracts are commitments to trade in the future, actual delivery and payment are not required until the contract matures. However, both buyers and sellers are required to make margin deposits with their brokers to guarantee their respective commitments. Because the margin deposit is small (typically 5% to 10% of the underlying value of the contract), speculators (who provide liquidity) are attracted to the market. The exchanges set minimum margins by contract, which can be raised by brokers to provide the protection they deem necessary. Using the December corn contract as an example, and assuming an $8.00 per bushel price quote, a cattle feeder who buys one contract (5,000 bushels) makes a $40,000 commitment. With a 10% margin, the feeder must post $4,000 with his or her broker. A "margin call" occurs when the price of the contract moves against the trader, say to $7.90 in this example. When a margin call occurs, the producer must post additional margin with his or her broker to cover the loss and restore the deposit.

Similarly, when the price moves favorably for the trader by a specified amount, money can be withdrawn from the margin deposit. Because futures prices reflect values of commodities at future delivery points, the local cash prices confronted by farmers usually vary from futures contract quotes at a given point in time. The differences between futures and cash prices are termed "basis" and reflect differences in price across space (due to transportation costs), time (which are associated with storage costs), or quality (such as differences in protein premiums for wheat). The basis is calculated as the difference between the cash price (at a given location and at a given point in time) and the futures price (associated with a specified exchange and contract month).[72] The basis varies over time and reflects only transportation costs and quality differences as the contract reaches maturity. As seen below, hedging largely eliminates price-level uncertainty, but not basis uncertainty, which generally has a smaller variance. Two categories of hedging exist: "long" hedging (where a futures contract is purchased) and "short" hedging (where a futures contract is sold). Either type of hedge involves holding a futures position in anticipation of a later transaction in the cash market, and in both cases, the futures position is opposite to the cash position.

Because futures and cash prices tend to move up and down together, losses and gains in the two markets tend to offset each other, leaving the hedger with a return near what was expected (the initial futures price plus the end-of-period basis). Thus hedging helps protect the business from changes in price levels. Farmers may choose to hedge in many different situations, including the following:

- *Storage hedging*: Farmers or merchants who own a commodity can protect themselves from declines in the commodity's price by short hedging. This involves selling futures contracts as the commodity is harvested or acquired, holding the resulting short futures position during the storage period, and buying it out when the cash commodity is sold. Losses (gains) in the value of the cash commodity due to unexpected price changes will be largely offset by gains (losses) in the value of the futures position, leaving the owner of the commodity with approximately the expected return from storage.
- *Production hedging*: Crop and livestock producers can protect themselves from declines in prices of expected outputs by short hedging. This generally involves selling futures contracts at the beginning of or during the growing or feeding period, holding the resulting short futures position until the product is ready to sell, and buying the futures as the output is sold. Losses (gains) in the value of the output due to unexpected price changes tend to be offset by gains (losses) in the value of the futures position. However, yield variability reduces the risk-reducing effectiveness of hedging for crop growers and generally makes it inadvisable to sell futures equal to more than one-half to two-thirds of the expected crop.
- *Hedging expected purchases*: Livestock feeders anticipating the purchase of corn or feeder cattle can protect themselves from price increases by long hedging. This involves buying corn or feeder cattle futures contracts to match anticipated requirements and selling the resulting long futures positions as these inputs are purchased on the cash market. Increases (declines) in the cost of feeders or feed due to unexpected price changes will be partly offset by gains (losses) in the value of the futures position, leaving the feeder with approximately the expected costs of inputs. Feeders' overall price risks may be further reduced by selling futures on prospective outputs.

To better understand the importance of basis risk in hedging, consider the example[73] of a corn producer with irrigated acreage who is considering the pricing of his growing crop. Because the producer irrigates and faces few other natural perils, he knows the size of his crop with a great deal of certainty and is concerned only with price risk. If the farmer does not hedge, his risk is solely associated with the harvest cash price (P2), which can also be calculated as the harvest futures price (F2) plus the harvest basis (B2). Thus, the farmer's net return in a cash-sale at-harvest situation (Ru) can be calculated as the cash price (F2 + B2) at harvest multiplied by actual production (Y2), minus production costs (C): $Ru = [(F2 + B2) \times Y2] - C$.

Suppose now that the producer places a short hedge (for example, sells a futures contract) to reduce the risk of a price decline and a lower sale price for his growing crop. The expected final net return at harvest (Rh) is based on the cash price at harvest (F2 + B2), and the profit or loss associated with the farmer's futures market position (F1 – F2). The farmer's actual level of production is designated as Y2 in the following equation, and the quantity hedged is $h \times Y1$, where h is the hedge ratio and Y1 is expected production: $Rh = [(F2 + B2) \times Y2] + [(F1 - F2) \times (h \times Y1)] - C$.

Assuming that output is known with certainty at the time the hedge is placed, and that actual production equals the quantity hedged (for example, $Y2 = h \times Y1$), gives the following (example also applies to a storage hedge, where output is known with certainty):

$$Rh = [(F2 + B2) \times Y2] + [(F1 - F2) \times Y2] - C,$$

or

$$Rh = [Y2 \times (F1 + B2)] - C.$$

This last equation indicates that the price component of the farmer's net return depends on the futures price at the time the hedge is placed plus the harvest basis. Because the futures price is known with certainty at planting, and output is known with certainty in this example, the only risk faced by the farmer is the risk associated with the harvest basis. Thus, price-level risk is eliminated by this anticipatory hedge, and the only risk faced by the grower is basis risk (the uncertain nature of B2). The existence of basis risk is a key factor distinguishing the risk associated with futures hedging and the use of many types of cash-forward contracts (see previous section).

When a producer enters into a "flat price" forward contract with his or her local elevator, for example, the basis risk he or she faces is zero. In addition, forward contracts are generally less standardized than futures contracts, and specific terms may vary across elevators. Physical delivery to the local elevator at harvest is generally required, and no margin calls exist when cash-forward contracts are used. An outright comparison[74] between a flat-price cash-forward contract and a futures hedge is provided in Table 4.6.

Using a numerical example[75] to illustrate hedging, suppose the corn producer discussed earlier wishes to reduce his income uncertainty by selling a futures contract at planting time. Because the farmer irrigates his corn crop, he is not concerned about yield risk, and the hedge quantity is assumed to equal actual output.

Table 4.6 Comparison between cash-forward contract and a futures hedge

Characteristic	Flat-price cash-forward contract	Futures hedging
Competitiveness of price	Depends on margin taken by elevator	Yes
Basis risk	No	Yes
Default risk	Some	No
Ease of recontracting or offset	Depends	Yes
Physical delivery	Yes	Seldom
Margin calls	No	Yes

Basic outlook: A wheat producer wishes to reduce his income uncertainty by selling a futures contract at planting time. He is not concerned about yield risk (he irrigates, etc.) and hedges a quantity equal to actual output, which he offsets at harvest time.

Futures price per ton at planting time is EUR 200 per ton. He expects a harvest basis of EUR –10, giving an expected cash price of EUR 200 plus EUR –10, or EUR 190 per ton. Two scenarios are illustrated in the table below: a EUR 15 price decrease between planting and harvest, and a EUR 15 price rise. In both cases, the realized harvest basis is EUR –10, as expected. With hedging, the return per ton is EUR 190 in both cases. It can be calculated as (1) the futures price at planting time plus the harvest basis or (2) the cash price. The effects of hedging on a wheat grower's return per ton can be assessed as follows:

	Price decrease scenario	Price increase scenario
Cash price expected at harvest	190	190
Cash price realized at harvest	175	205
Futures price at planting	200	200
Futures price at harvest	185	215
Futures return to the producer	15	–15
Net price realized with hedging	190	190

With use of futures contracts: A wheat producer has a 200-hectare farm. His return from a cash sale at harvest (with no short hedge) ranges from EUR –42,400 to EUR 70,000 across the six scenarios. Now the farmer decides to hedge in the spring. He anticipates that his output will not likely fall below 800 tons and considers his optimal hedge to be 16 contracts (at 50 tons per contract). At harvest time, he lifts his short hedge by buying back his futures contract and sells his cash crop in the marketplace. He receives the proceeds from the cash sale of the crop (less production costs of EUR 110,000) plus the gains or losses associated with the futures transaction (less commission charges of

EUR 480). With hedging, his returns spread only between EUR –14,480 and EUR 29,520, and their standard deviation is one-third of what it is without hedging. The inputs into the model are as follows:

- 200 ha farm with yields of 6, 5, and 4 tons per ha
- Cost of production: EUR 110,000
- Producer sells 800 tons of futures or 16 contracts
- Producer pays EUR 30 per contract for brokerage: EUR 480 in total

The returns to a cash sale at harvest using a futures hedge can then be summarized as follows:

	Production	Harvest cash price	Spring futures	Harvest futures	Revenue from crop	Net return from cash sale at harvest	Net return with hedging	Difference in return
Scenario	Tons	Eur. Per tons	Eur. Per ton	Eur. per ton	Eur.	Eur.	Eur.	Eur.
1	1,200	65	125	75	78,000	-32,000	7,250	39,250
2	1,200	150	110	160	180,000	70,000	29,520	-40,480
3	1,000	70	130	80	70,000	-40,000	-480	39520
4	1,000	165	125	175	165,000	55,000	-14,520	-40,480
5	800	85	145	95	65,800	-42,000	-2,480	39520
6	800	170	130	180	136,000	26,400	-14,480	2,720
Net average return						6,167	5,687	-40,480

As can be seen by this example, the estimation of hedging amounts and risk reduction is much more complicated in the presence of yield risk. Generally the effectiveness of hedging in reducing risk diminishes as yield variability increases and the correlation between prices and yields becomes more negative. Although hedging can reduce income uncertainty for many farmers, it never completely eliminates such uncertainty. In addition to the considerations discussed previously in this section, hedging involves possible costs for interest forgone on margin deposits and for bias in futures prices. These costs generally are small relative to the value of the positions taken, but they partly offset the risk-reducing benefits from hedging.

A commodity option gives the holder the right, but not the obligation, to take a futures position at a specified price before a specified date. The value of an option reflects the expected return from exercising this right before it expires and disposing of the futures position obtained. If the futures price changes in favor of the option holder, a profit may be realized either by exercising the option or selling the option at a price higher than paid. If prices move so that exercising the option is unfavorable, then the option may be allowed to expire. Options provide protection against adverse price movements, while allowing the option holder to gain from favorable movements in the cash price. In this sense, options provide protection against unfavorable events similar to that provided by insurance policies. To gain this protection, a hedger in an options contract must pay a premium, as one would pay for insurance. Options markets are closely tied to underlying futures markets.

Options that give the right to sell a futures contract are known as "put" options, while options that give the right to buy a futures contract are known as "call" options. The price at which the futures contract underlying the option may be bought (for a call option) or sold (for a put option) is called the "exercise" or "strike" price. As an example, suppose a wheat producer purchases a put option having a strike price of $8.00 per bushel. If futures prices move to $7.80, the option may be exercised for a net profit of $0.20 ($8.00 – $7.80), minus the premium paid for the option. If the harvest cash price is $7.70 per bushel, the farmer's return is $7.90 per bushel ($7.70 + $0.20), minus the premium. The effects on realized returns from hedging with futures and put options can be compared for a range of possible futures-price outcomes. For example, corn is stored in November and sold in May, output risk is absent, and the hedge ratio is 1.0. The May futures price is $7.80 per bushel at the beginning of the storage period, and the expected May basis is –$0.20. By hedging with futures, the farmer obtains an expected return for the corn in storage of $7.80 – 0.20, or $7.60. Alternatively, the farmer can buy an at-the-money put option with a $7.80 strike price for a $0.20 premium.

The put guarantees a price equal to the strike price, minus the premium, minus the basis, or $7.80 – 0.20 – 0.20 = $7.40, while allowing the farmer to gain if the futures price rises above $8.00 in May. By not hedging, the farmer gets the futures price minus the basis. The figure shows that the range of possible prices is greatest with the cash sale and least with the futures hedge. Unlike futures hedging, the put does not limit the potential profits associated with increasing prices, but the price must rise more than the premium cost before a profit is realized. The premium paid for an option typically consists of "intrinsic" value and "time" value. The intrinsic value reflects the difference between the underlying futures price and the strike price. If the price of the underlying futures contract is $7.90 per bushel, for example, and the strike price is $7.70, then the holder of a call option could gain $0.20 by exercising the option immediately. Consequently, the premium in this case must be at least $0.20 per bushel, and the option is "in the money." If the strike price is above the futures price, the intrinsic value of the call option is zero and the put is said to be "out of the money." When the strike price equals the futures price, the option is "at the money." The time value of an option, in contrast, depends on several factors, including the volatility of the underlying futures contract, the time until the option expires, the interest rate, the strike price, and the underlying futures price. Time value refers to the money that buyers are willing to pay for the possibility that the intrinsic value of an option will increase over time. An option on a futures contract with very low volatility, for example, will have a small time value because traders do not expect the intrinsic value to change to a great extent over time. If the futures price is volatile, in contrast, the probability is high that the option's intrinsic value would increase, and traders would be willing to pay more for the chance of such a gain. In addition, intrinsic value depends on the time until the option's expiration. The greater the time horizon, the greater the intrinsic value because price uncertainty is greater. Observed options prices can be used to provide information about anticipated price variability.[76]

The following example[77] illustrates the situation for a producer on March 15 who plans to produce 500 acres of corn and hedge with put options. The December futures price is $9.00 per bushel at planting time, and the premium for at-the-money puts is $0.50 per bushel. His expected yield is 150 bushels per acre, and his production costs are estimated at $450,000. Because the farmer expects his production to fall no lower than 50,000 bushels, he buys 10 put contracts (5,000 bushels per contract × 10 contracts = 50,000 bushels) and selects a strike price of $8.50. The cost associated with this purchase is $25,000 in premiums (at an assumed cost of $0.50 per bushel) and $900 in commissions (10 contracts at $90 per contract). As a simplifying assumption, suppose that the producer makes his decision on October 20 as to the sale of the option. If the futures price that day is $9.00 (equal to the strike price), the option has no intrinsic value (since the option is at-the-money) and an assumed time value of $0.14 per bushel (reflecting the probability that the futures price will decline before the option expires, raising the option's intrinsic value). Using this estimate, the producer's return to the purchase of the option is the time value on October 20 at $7,000 ($0.14 × 50,000 bushels), less the premium cost of $25,000 and the commission cost of $1,800 (10 contracts × $90 per trade × 2 trades), or –$19,800. If the producer had sold his crop at the harvest cash price, his return would have been $187,500, instead of the $167,700 ($187,500 – $19,800) earned in this hypothetical put-option situation.

The farmer's return to buying options depends largely on the futures price at harvest. With a high futures price in October (say, $9.75), the producer's loss associated with the option is even higher, while a low futures price (say, $8.50) would result in a higher gain than in the cash-sale-at-harvest-only case. In an efficient market, the producer's return from buying put options over a series of many years is expected to equal the return to either hedging with futures or simply selling the crop at harvest, except for commissions. Although returns are approximately the same in all three cases, hedging with either put options or futures reduces uncertainty about return.

Line item	Revenue, cost, and net returns	USD
1	Revenue from crop sale on October 20: (75,000 bushels × \$8.50/bushel)	637,500
2	Total production costs	450,000
3	Net return from crop sale	187,500
4	Premium for put option paid on March 15: (10 put options × 5,000 bushels × \$0.50 premium/bushel)	25,000
5	Return from put sale received on October 20: (10 put options × 5,000 bushels × \$0.14/bushel)	7,000
6	Commissions on put purchase and sale: (2 × 10 put options × \$90 commission)	1,800
7	Net return from put hedge (5 – 4 – 6)	–19,800
8	Net return from cash sale and put hedge (3 + 7)	167,700

The put option has a \$9.00 strike price and a \$0.14 time value on October 20. Although net returns from the hedge in this example are negative, the example could as easily have been constructed to show a positive net return.

5 Biofuels and Agriculture

Biofuels date back to the late 19th century, when ethanol was derived from corn and Rudolf Diesel's first engine ran on peanut oil. Until the 1940s, biofuels were seen as viable transport fuels, but falling fossil-fuel prices stopped their further development. Interest in commercial production of biofuels for transport rose again in the mid-1970s, when ethanol began to be produced from sugarcane in Brazil and since the 1980s from corn in the United States. During the 1990s, the industrialized economies of North America and Europe actively pursued policies in support of domestic biofuel industries to achieve energy security, develop a substitute for fossil fuels, and support rural economies. More countries have since launched biofuel programs, and over 50 countries have adopted blending targets or mandates and several more have announced biofuel quotas for future years.[1]

5.1 Biofuels: introduction and typology

Traditional biomass,[2] including fuelwood, charcoal, and animal dung, continues to provide important sources of energy in many parts of the world. Bioenergy is the dominant energy source for most of the world's population who live in extreme poverty and who use this energy mainly for cooking. More advanced and efficient conversion technologies now allow the extraction of biofuels – in solid, liquid, and gaseous forms – from materials such as wood, crops, and waste material.

5.1.1 Different types of biofuels

Biofuels are energy carriers that store the energy derived from biomass. A wide range of biomass sources can be used to produce bioenergy in a variety of forms. For example, food, fiber, and wood process residues from the industrial sector; energy crops, short-rotation crops, and agricultural wastes from the agriculture sector; and residues from the forestry sector can all be used to generate electricity, heat, combined heat and power, and other forms of bioenergy. Biofuels may be referred to as renewable energy because they are a form of transformed solar energy.

Biofuels can be classified according to source and type. They may be derived from forest, agricultural, or fishery products or municipal wastes, as well as from agro-industry, the food industry, and food-service by-products and wastes. They may be solid, such as fuelwood, charcoal, and wood pellets; liquid, such as ethanol, biodiesel, and pyrolysis oils; or gaseous, such as biogas.

A basic distinction is also made between primary (unprocessed) and secondary (processed) biofuels:

- Primary biofuels, such as firewood, wood chips, and pellets, are those where the organic material is used essentially in its natural form (as harvested). Such fuels are directly combusted, usually to supply cooking fuel, heating, or electricity production needs in small- and large-scale industrial applications.
- Secondary biofuels in the form of solids (e.g., charcoal), liquids (e.g., ethanol, biodiesel, and bio-oil), or gases (e.g., biogas, synthesis gas, and hydrogen) can be used for a wider range of applications, including transport and high-temperature industrial processes.[3]

In spite of their limited overall volume, the strongest growth in recent years has been in liquid biofuels for transport, mostly produced using agricultural and food commodities as feedstocks. The most significant are ethanol and biodiesel.

- Ethanol:[4] Any feedstock containing significant amounts of sugar, or materials that can be converted into sugar such as starch or cellulose, can be used to produce ethanol. Ethanol available in the biofuel market today is based on either sugar or starch. Common sugar crops used as feedstocks are sugarcane, sugar beet, and, to a lesser extent, sweet sorghum. Common starchy feedstocks include maize, wheat, and cassava. The use of biomass containing sugars that can be fermented directly to ethanol is the simplest way of producing ethanol.[5] In Brazil and other tropical countries currently producing ethanol, sugarcane is the most widely used feedstock. In OECD countries, most ethanol is produced from the starchy component of cereals (although sugar beet is also used), which can be converted fairly easily into sugar. Ethanol[6] can be blended with petrol or burned in its pure form in slightly modified spark-ignition engines.[7]
- Biodiesel: Biodiesel is produced by combining vegetable oil or animal fat with an alcohol and a catalyst through a chemical process known as transesterification. Oil for biodiesel production can be extracted from almost any oilseed crop; globally, the most popular sources are rapeseed in Europe and

soybean in Brazil and the United States. In tropical and subtropical countries, biodiesel is produced from palm, coconut, and jatropha oils. Small amounts of animal fat, from fish- and animal-processing operations, are also used for biodiesel production.[8]

- Straight vegetable oil: Straight vegetable oil (SVO) aka pure plant oil (PPO) is a potential fuel for diesel engines that can be produced from a variety of sources, including oilseed crops such as rapeseed, sunflower, soybean, and palm. Used cooking oil from restaurants and animal fat from meat-processing industries can also be used as fuel for diesel vehicles.

5.1.2 Biofuel feedstock

There are many supply sources of biomass for energy purposes, scattered across large and diverse geographical areas. Even today, most energy derived from biomass used as fuel originates from by-products or co-products of food, fodder, and fiber production. For instance, the main by-products of forest industries are used to produce fuelwood and charcoal, and black liquor (a by-product of pulp mills) is a major fuel source for bioelectricity generation in many countries. A considerable amount of heat and power is derived from recovered and/or recycled woody biomass, and increasing amounts of energy are recovered from biomass derived from cropland (straw and cotton stalks) and forestland (wood chips and pellets). In sugar- and coffee-producing countries, bagasse and coffee husks are used for direct combustion and to produce heat energy and steam. In terms of bioenergy, however, the big growth area in recent years has been in the production of liquid biofuels for transport using agricultural crops as feedstocks. The bulk of this has taken the form of ethanol, based on either sugar crops or starchy crops, or biodiesel based on oil crops.[9]

5.2 Biogas for heat, power,[10] and transport[11]

A range of biomass resources are used to generate electricity and heat through combustion. Sources include various forms of waste, such as residues from agro-industries, postharvest residues left on the fields, animal manure, wood wastes from forestry and industry, residues from food and paper industries, municipal solid wastes, sewage sludge, and biogas from the digestion of agricultural and other organic wastes. Dedicated energy crops, such as short-rotation perennials (eucalyptus, poplar, willow) and grasses (miscanthus and switchgrass), are also used.

Several processes can be used for power generation. Most biomass-derived electricity is produced using a steam-cycle process: biomass is burned in a boiler to generate high-pressure steam that flows over a series of aerodynamic blades causing a turbine to rotate, which in response turns a connected electric generator to produce electricity. Compacted forms of biomass such as wood pellets and briquettes can also be used for combustion, and biomass can also be burned with coal in the boiler of a conventional power plant to yield steam and electricity. The latter is currently the most cost-efficient method for incorporating renewable technology into conventional power production because much of the existing power-plant infrastructure can be used without major modifications.

- Anaerobic digestion

Biogas can be created through the anaerobic digestion of food or animal waste by bacteria in an oxygen-starved environment. The resulting biogas contains a high volume of methane along with carbon dioxide, which can be used for heating or for electricity generation in a modified internal-combustion engine. The conversion of animal wastes and manure to methane/biogas can bring significant environmental and health benefits. Methane is a greenhouse gas that has a global-warming potential that is 22 to 24 times more powerful than that of carbon dioxide. By trapping and utilizing the methane, its greenhouse-gas impacts are avoided. In addition, heat generated during the biodigestion process kills the pathogens present in manure, and the material left at the end of the process provides a valuable fertilizer.

- Gasification

Through the process of gasification, solid biomass can be converted into a fuel gas or biogas. Biomass gasifiers operate by heating biomass in a low-oxygen, high-temperature environment that breaks it down to release a flammable, energy-rich synthesis gas or "syngas." This gas can be burned in a conventional boiler, or used instead of natural gas in a gas turbine to turn electric generators. Biogas formed through gasification can be filtered to remove unwanted chemical compounds and can be used in efficient "combined-cycle" power-generation systems that combine steam and gas turbines to generate electricity.

- Biogas for transport

Untreated biogas is unsuitable as a transport fuel owing to its low methane content (60% to 70%) and high concentration of contaminants. However, it can be treated to remove carbon dioxide, water, and corrosive hydrogen sulphide and to enhance its methane content (to over 95%). When compressed, treated biogas has properties similar to those of compressed natural gas, making it suitable for use in transport.

5.3 Biofuel and agriculture

The current expansion and growth of energy markets, as a result of new energy and environment policies enacted over the past decade in most developed countries and in several developing countries, is reshaping the role of agriculture. Most significant is the sector's increasing role as a provider of feedstock for the production of liquid biofuels for transport – ethanol and biodiesel. Modern bioenergy represents a new source of demand for farmers' products. It thus holds promise for the creation of income and employment. At the same time, it generates increasing competition for natural resources, notably land and water,[12] especially in the short run, although yield increases may mitigate such competition in the longer run. Competition for land becomes an issue especially when some of the crops (e.g., maize, oil palm, and soybean) that are currently cultivated for food and feed are redirected toward the production of biofuels, or when food-oriented agricultural land is converted to biofuel production.[13]

5.3.1 The biofuel life cycle

Two of the main driving forces behind policies promoting biofuel development have been concerns over energy security and a desire to reduce greenhouse-gas emissions. Just as different crops have different yields in terms of biofuel per hectare, wide variations also occur in terms of energy balance and greenhouse-gas emission reductions across feedstocks, locations, and technologies. The contribution of a biofuel to the energy supply depends both on the energy content of the biofuel and on the energy going into its production. The latter includes the energy required to cultivate and harvest the feedstock, to process the feedstock into biofuel, and to transport the feedstock and the resulting biofuel at the various phases of its production and distribution. The fossil-energy balance expresses the ratio of energy contained in the biofuel relative to the fossil energy used in its production. A fossil-energy balance of 1.0 means that it requires as much energy to produce a liter of biofuel as it contains; in other words, the biofuel provides no net energy gain or loss. A fossil-fuel energy balance of 2.0 means that a liter of biofuel contains twice the amount of energy as that required in its production. Problems in assessing energy balances accurately derive from the difficulty of clearly defining the system boundary for the analysis.

Many studies[14] have been conducted on fossil-energy balances for different types of fuel, and progress on efficiency is being made every single day. The figures reveal wide "positive" variations in the estimated fossil-energy balances across feedstocks and fuels and, sometimes, for a feedstock/fuel combination, depending on factors such as feedstock productivity, agricultural practices, and conversion technologies.[15] The range of estimated fossil-fuel balances for cellulosic feedstocks is even wider, reflecting the uncertainty regarding this technology and the diversity of potential feedstocks and production systems. Similarly, the net effect of biofuels on greenhouse-gas emissions may differ widely.

Biofuels are produced from biomass;[16] in theory, therefore, they should be carbon neutral, as their combustion only returns to the atmosphere the carbon that was sequestered from the atmosphere by the plant during its growth – unlike fossil fuels, which release carbon that has been stored for millions of years under the surface of the earth. However, assessing the net effect of a biofuel on greenhouse gas emissions requires analysis of emissions throughout the life cycle of the biofuel: planting and harvesting the crop; processing the feedstock into biofuel; transporting the feedstock and the final fuel; and storing, distributing, and retailing the biofuel – including the impacts of fuelling a vehicle and the emissions caused by combustion. In addition, any possible co-products that may reduce emissions need to be considered. Clearly, therefore, fossil-energy balances are only one of several determinants of the emissions impact of biofuels. Critical factors related to the agricultural production process include fertilizing, pesticide use, irrigation technology and soil treatment. Land-use changes associated with expanded biofuel production can have a major impact. For example, converting forestland to the production of biofuel crops or agricultural crops displaced by biofuel feedstocks elsewhere can release large quantities of carbon that would take years to recover through the emission reductions achieved by substituting biofuels for fossil fuels.[17]

5.3.2 Second-generation biofuels

Current liquid biofuel production based on sugar and starch crops (for ethanol) and oilseed crops (for biodiesel) is generally referred to as first-generation biofuels. A second generation of technologies under development may also make it possible to use ligno-cellulosic biomass. Cellulosic biomass is more resistant to being broken down than starch, sugar, and oils. The difficulty of converting it into liquid fuels makes the conversion technology more expensive, although the cost of the cellulosic feedstock itself is lower than for current, first-generation feedstocks. Conversion of cellulose to ethanol involves two steps: the cellulose and hemicellulose components of the biomass are first broken down into sugars, which are then fermented to obtain ethanol.[18] As cellulosic biomass is the most abundant biological material on earth, the successful development of commercially viable second-generation cellulose-based biofuels significantly expanded the volume and variety of feedstocks that can be used for production. Cellulosic wastes, including waste products from agriculture (straw, stalks, leaves) and forestry, wastes generated from processing (nut shells, sugarcane bagasse, sawdust), and organic parts of municipal waste, could all be potential sources. Second-generation feedstocks and biofuels also offer advantages in terms of reducing greenhouse-gas emissions (see infra in the next chapter). Most of the second-generation biofuels have reached the commercially feasible stage by 2013. Biotechnology is now leading the way for third-generation biofuels to go mainstream (see infra in the next chapter for more on this).

6
Biofuels and the Sustainability Conundrum

The development of biofuels, which has emerged at the interface of agriculture and energy at the global level, has been one of the most significant agricultural developments in recent years. During the 1990s, the industrialized economies of North America and Europe actively pursued policies in support of domestic biofuel industries to achieve energy security, develop a substitute for fossil fuels, and support rural economies. In addition, the rising concern over climate change in the last decade propelled interest in biofuels as a possible means of mitigating greenhouse gas (GHG) emissions.

The need to address the growing challenge of climate change has led to closer scrutiny of biofuels to assess whether they can be produced, traded, and used sustainably. Criticism of biofuels centered around their perceived negative impacts on the environment through deforestation, spread of monocultures, loss of biodiversity and possible higher GHG emissions under uncontrolled land-use change. Moreover, the food crisis of 2007–2008 and the ensuing surge of commodity prices heightened the debate over food versus fuel and the possible consequences of biofuel production on food security. The potential of biofuels to contribute to a shift to more sustainable energy systems became contested, and scientists started to question the environmental superiority of biofuels. As a result of these concerns, sustainability has been promoted as an essential condition for biofuel long-term viability and for continued public support of renewable energy and climate-change mitigation. Consequently, a range of biofuel certification schemes emerged, all purporting to ensure sustainability. Yet these schemes also seem to be driven by the need to regulate the current and potentially huge future trade flows in feedstocks and biofuels between industrialized economies (with high potential excess demand for energy) and developing countries (with recognized comparative advantages in biomass production and huge potential excess supply).

The most recent and comprehensive report on this issue, which has been extensively drawn upon in the previous chapter, is an FAO 2013[1] report that addresses the central issue of biofuel sustainability using a global assessment of major commodities and feedstocks currently employed for bioethanol and biodiesel production. The approach taken was guided by two overriding considerations, being (1) the need to understand the basic dimensions of sustainability for biofuels (economic, environmental, and social), their linkages, and how they relate to the central challenges they address, namely land-use change, food security, and climate change, and (2) the need to critically evaluate the extent to which recent trends in biofuel certification schemes reflect true sustainability issues versus trade-flow regulation under the guise of sustainability – or put differently: are the initiatives essentially market driven or sustainability motivated, or potentially both? What follows is a reflection of the aggregate findings of the report and its major conclusions.

In countries where the biofuel industry is established, the first feedstocks utilized tend to be drawn from among the most important crops in the country (e.g., corn in the United States, rapeseed in the EU, sugarcane in Brazil, and oil palm in Malaysia/Indonesia). Biofuels tend to be led by a few dominant crops targeted through an active policy-support program that also accounts for domestic biofuel consumption patterns (e.g., ethanol in the United States and Brazil and biodiesel in the EU). Still, to meet expanding future demand in biofuels, there was and is growing interest in exploring other possible feedstocks (e.g., sugarcane, cassava, palm oil, sweet sorghum, jatropha) and dedicated energy crops (e.g., switchgrass, miscanthus, and short-rotation tree crops) for advanced ("cellulosic") biofuels.

6.1 Categorization of biofuels[2]

In essence the biofuel feedstocks can be divided into four broad categories: (1) high-efficiency feedstocks (e.g., palm oil, sugarcane); (2) moderate-efficiency feedstocks (e.g., corn, soybean, rapeseed, sugar beet); (3) feedstocks under development (e.g., sweet sorghum, jatropha); and (4) dedicated energy feedstocks (e.g., switchgrass, miscanthus, short-rotation crops, algae, waste). What follows is an overview of the categories.

6.1.1 Sustainability of efficient feedstock: a mixed bag

Sugarcane is an efficient crop (in terms of yield per unit of land), but its sustainability hinges largely on water availability, and the crop does better when there is ample rainfall and minimal need for irrigation (as in Brazil). Besides high biomass, sugarcane also produces a range of useful by-products, all contributing positively to its economic competitiveness. Sugarcane continues to be attractive even under second-generation technologies, as bagasse can be a prime feedstock source. Sugarcane also offers the possibility of using molasses (i.e., sugar production by-products) for biofuel in situations where sugar production has priority over biofuels (as in India). Sugarcane is also very demanding agronomically, requiring deep soils, high water use, and a full 12-month growing season; hence, sugarcane is less optimal in drier areas that require irrigation, especially if it has to compete with food crops for water use. Irrigated sugarcane is less of an option if water is obtained from depletable underwater sources or aquifers.

Another key concern with sugarcane with respect to sustainability is the potential undesirable impacts in terms of land-use change. This has been a particular issue in Brazil, the world's leader in sugarcane ethanol, where sugarcane expansion into grazing areas can push livestock systems into the forest zones. Brazil, being sensitive to these concerns, has placed restrictions on sugarcane expansion areas to minimize the negative impacts. Next to sugarcane, palm oil is by far the most efficient source for biodiesel (judged on yield per unit of land), far exceeding alternatives like rapeseed, soybeans, or sunflowers. The bulk of world palm-oil production is concentrated in Malaysia and Indonesia, but major investments in new plantations are also taking place in Africa and Latin America, driven by rising consumer demand, high potential for expanded trade, and opportunities for biodiesel production. In terms of environmental sustainability, oil palm presents a huge dilemma. On the one hand, this oil crop is highly efficient, and its GHG-emission potential and energy balance compares favorably with alternative feedstocks. However, oil palm plantations can also pose environmental problems when expansion takes place on sensitive lands (e.g., peat soils, forests). This is a particular concern in Malaysia and Indonesia where some oil palm is planted in drained peat lands, resulting in significant CO_2 emissions outweighing any carbon benefits arising from the new palm-oil plantations. A complicating factor is that investments in new palm-oil plantations are not driven by biodiesel alone, but rather (or more so) by increasing consumer demand in vegetable oil in many high-growth and populous developing countries. This in turn may minimize the impact of sustainability safeguards geared toward plantations focusing on biodiesel and not the underlying feedstock food crop.

6.1.2 Moderately efficient feedstocks: no guaranteed economic viability

Much of the burst in biofuel production in the United States and the EU depended on a few feedstocks that are only moderately efficient relative to alternatives. In the United States, maize is the predominant feedstock for ethanol, while rapeseed dominates biodiesel production in the EU. Maize has the advantage of high productivity per unit of land, although it also uses large amounts of fertilizers and pesticides, and hence consumes a lot of fossil energy. However, the increasing concern about climate change and GHG mitigation lessens the appeal of maize compared with sugarcane because its energy input-output balance, or carbon footprint, under current biofuel technology is relatively modest. In Canada and Europe, maize is traditionally used for feed, while in other countries (apart from China), white maize for food consumption predominates. Therefore, maize has not been favored outside of the United States as an ethanol feedstock because of concerns about competition with food.

The EU, on the other hand, centered its initial biodiesel development strategy around rapeseed, a domestically grown crop that can be promoted through subsidies. This strategy squared fully with the overriding objective of achieving energy security through the promotion of renewable energy, including biofuels. Although more rapeseed is grown in Canada, China, and India, only the EU (and to a lesser extent, Canada) has vigorously promoted rapeseed biodiesel production, largely through heavy subsidies and mandates. However, in terms of biodiesel yield per acre or GHG savings, rapeseed feedstock doesn't compare favorably with other alternatives (such as palm oil). Consequently, there is very little biodiesel production from rapeseed outside of the EU's direct support. Even within the EU, there has been some retreat from direct support to rapeseed biodiesel due to increased pressures on environmental grounds, seeing that rapeseed offers weaker benefits in terms of climate-change mitigation.

Soybean oil is the second-largest biodiesel feedstock after rapeseed oil. Biodiesel production from soy oil is concentrated in the United States and Latin America (e.g., Argentina, Brazil, Paraguay). China, a major soybean producer, does not produce biodiesel from this feedstock because of its ban on using food crops for biofuels and the fact that China is a net importer of soybeans. The largest expected expansion of soy oil for biodiesel is in Argentina and Brazil because of the availability of land and the relatively lower cost of production. However, soybeans in these countries, under the current market forces, tend to be grown under monoculture systems, which pose sustainability challenges. Moreover, expansion of corn for ethanol in the United States – which tends to reduce soybean acreage as corn-soybean rotation contracts – pushes up soybean acreage expansion in Latin America. This, in turn, raises concerns over potential undesirable land expansion and even encroachment into forested areas, with potentially negative environmental and GHG-emission consequences.

6.1.3 New feedstocks: opportunities and limitations[3]

The prospects of even greater expansion of biofuels in the future unleashed a search for alternative and highly productive feedstocks to meet future demand. Among these, sweet sorghum has been the object of sustained research and development programs in China, India, and the United States. Sweet sorghum is the closest competitor to sugarcane in terms of yield potential per unit of land. Sweet sorghum is an annual crop that is more versatile and can be grown in a variety of soil depths and water conditions. Sorghum is drought tolerant and can be grown in a shorter season with less labor requirements and is suitable in tropical areas too dry to grow sugarcane. The drawback to sweet sorghum is that it requires quick processing after harvest because its sugar content drops significantly after only three weeks. This presents a challenge for transportation and storage given the bulkiness of the crop (i.e., 70% water at harvest). This may limit the number of countries capable of developing the industrial infrastructure to produce, harvest, store, and process this bulky crop on a large scale. It further would lead to the need to concentrate production around processing facilities, limiting the options for more sustainable diversified production on that land (crop rotations). Another sustainability problem is the potential competition for food over land. A study on sweet sorghum in Mozambique showed that one solution to food competition is to plant sweet sorghum on fallow sugarcane land to be harvested and processed before the start of the sugarcane-harvesting season. As a biomass source, under this system, the sweet sorghum fibrous residues can be used in the same way as sugarcane bagasse to produce electricity, process heat, and power.

Another potential alternative feedstock for biodiesel is the nonedible crop jatropha. Jatropha is drought tolerant, has low input requirements, and is highly suitable for marginal lands. Jatropha can also improve the soil quality because of its deep root system; however, clearing jatropha land for conversion into cropland would be a considerable investment. Equally important is jatropha's suitability for small-scale production, as its seeds can be easily stored before processing. However, large-scale biodiesel production is capital intensive and thus requires tight supply arrangements such as out-grower schemes, in which producers deliver directly to local processing plants to ensure economic viability. India has been particularly keen on developing jatropha for biodiesel in line with its nonfood biofuel policy. Jatropha has also been tried in a few African countries (e.g., Ethiopia, Ghana, Mozambique). Still, the long-run economic viability of jatropha for biodiesel is untested. The key concern is that to ensure economic profitability, jatropha would require intensive crop management, which in turn would result in competition for top farmland unless explicit regulations on farmland use are in place. Consequently, the development of these feedstocks in more marginal areas by small-scale farmers is less likely without government incentives. In general, any feedstock will compete with food crops for land and water resources. In other words, economics will trump agronomy in terms of where the feedstocks can be grown. Cassava has also been targeted as a potential feedstock for ethanol because of its high starch content and high yield potential per hectare. However, cassava is a staple food crop in much of Africa and Asia and a critical food-security source for many poor rural communities. This raises concerns over its suitability as a biofuel feedstock, as the crop is for the most part grown by small-scale farmers for self-consumption. Moreover, cassava is a highly perishable crop, and cassava value chains, especially in Africa, are typically impeded by limited processing technologies and underdeveloped marketing channels. Given the agronomy of the crop, its central role for food security among the poor and rural households in many parts of Africa, and the largely underdeveloped cassava supply chains, there are serious doubts that such a crop can become a magnet for biofuel development at the local level, at least not on a large scale or when involving a significant share of small farmers.

6.1.4 Second-generation biofuels: the economics still don't work

Advanced biofuels (including "cellulosic" ethanol) are still under development and have yet to reach the commercial stage. Dedicated energy crops (e.g., alfalfa, switchgrass, miscanthus), fast-growing short-rotation trees (e.g., poplar, willows, eucalyptus), and agricultural and wood residues offer much greater potential for the biofuel industry. But the economics and high capital investments for the new supply chains remain serious obstacles for second-generation biofuels.

Assuming the commercialization stage has been reached, concern over land-use competition between food and fuel may not disappear simply because we can use agricultural residues or waste for feedstocks. The answer will turn essentially around economics and will depend on the relative costs of land-using feedstocks (e.g., dedicated energy crops) or non-land-using feedstocks (e.g., wood, municipal or other wastes). Even when agricultural residues (e.g., cereal straws) are targeted, this would alter the economics of traditional crops (i.e., pushing up their market value) and would increase competition for land and not lower it. Moreover, the advent of second-generation biofuels would create huge pressure for land to produce dedicated energy crops, hence worsening competition with food crops. The net effect on land competition will depend on whether the expansionary effect (resulting from a surge of investments in second-generation plants and the resulting high demands for feedstock) will dominate the substitution effect (i.e., away from traditional feedstock crops and into residues and waste). At any rate, second-generation biofuels, should they become commercially viable, would likely induce a fundamental shift in agricultural systems and would bring agriculture and energy markets much closer, with far-reaching consequences difficult to fully ascertain at this stage.

6.2 Biofuels and sustainability: a multidimensional issue[4]

The sustainability concept is complex and multidimensional, and its implementation on the ground requires an understanding of the specific local context. A sustainable biofuel-production system is one that is economically viable, conserves the natural resource base, and ensures social well-being. Moreover, the three core dimensions of sustainability (i.e., economic, environmental, and social) are interlinked and can best be approached holistically.

From a sustainability perspective, biofuels offer advantages as well as risks. On the upside, biofuels can contribute to increased energy security, help reduce GHG emissions, improve air quality in cities, and, in the process, spur growth in rural areas. On the downside, expansion of biofuels, especially under intensive production systems, could have negative impacts on biodiversity (e.g., replacement of natural forest with biofuel crops, spread of monocultures), water availability under scarcity, water quality, soil degradation, negative carbon and energy balances, and potential conflict with food production and food security, as well as worsening GHG-emission levels because of indirect land-use change.

Balancing the economic benefits with environmental and social impacts is a delicate act. Even when biofuels meet some environmental and social sustainability criteria, they need to first pass the economic sustainability (or viability) test. This means ensuring efficiency of production (through high yields and intensive management) and long-run profitability, access to productive resources (e.g., land, labor, technology), and reliable output markets. The challenge is achieving all this while ensuring economic viability and minimizing potential negative social or environmental impacts. Most of the initiatives on biofuel sustainability at the country or supranational levels come from industrialized economies where biofuel growth has been most dynamic and where there is broad scope for bioenergy demand and huge energy-substitution possibilities. Sustainability initiatives coming from Europe or North America largely mirror the industrial economies' priorities for biofuels (e.g., energy-security supply, protection of agriculture, and increasingly climate-change mitigation).

Because the EU (more than North America) depends relatively more on imported feedstocks for its biofuel needs, it took the lead in setting regulations and encouraging private-led schemes targeting biofuel sustainability. By contrast, because the United States' biofuel production is largely domestically oriented, there is no comparable push to require broad-based sustainability criteria for biofuels, apart from the requirement to regulate GHG emissions as required by existing legislation and Supreme Court rulings.

Outside Europe and North America, major feedstock exporters (such as Argentina, Brazil, Indonesia, and Malaysia) have responded to rising sustainability concerns, largely to protect their export markets. The other large developing economies, such as China and India, with huge populations to feed, have adopted a more cautionary strategy with biofuels, avoiding altogether any feedstock that could be used as food or feed.

6.2.1 Biofuel certification schemes: can they assure sustainability?

A number of sustainability initiatives defined through standards, principles, and criteria have been developed as a regulating instrument for biofuel and feedstock trade. These initiatives, both national and transnational, paved the way for several biofuel-specific certification schemes, either targeting all biofuels as a whole or tailored to specific biofuel feedstocks (sugarcane, soybeans, palm oil, etc.). Despite their diversity, most biofuel certification schemes followed a dominant type of governance: a voluntary, industry-led, multi-stakeholder forum with some input from civil society. It has advantages and disadvantages. On the upside, it allows the biofuel industry to self-regulate while preserving market efficiency. Specifically, these private-led certification schemes have the ability to (1) influence corporate social responsibility in biofuels, (2) influence businesses to improve efficiency within a supply chain, (3) decrease risk, and (4) raise awareness about problems in the supply chain. Also, the multiple forms of certification schemes (e.g., roundtables, consortia, private labels, industry-wide certificates) could generate positive pro-competitive effects, improving implementation and verification tools.

On the other hand, a commonly raised concern among exporting countries is that certification schemes are viewed as disguised trade barriers. Another limitation of the voluntary private-based certification schemes is that sustainability itself may take second place to efficiency, especially if some provision of public goods is required through a direct public intervention.

Economic sustainability (viability) requires long-term profitability, minimal competition with food production, and competitiveness with fossil fuels. The economics of biofuels have been in part driven by active policy-support measures (subsidies and mandates), which makes it difficult to assess the long-run economic viability of biofuels systems current or future. However, the protection of the domestic biofuel industry (sugarcane ethanol in Brazil from the 1970s, US corn ethanol, and EU rapeseed biodiesel) has managed to develop economies of scale and cut long-run costs through the introduction of technological improvements (diversification and market opportunities for by-products, efficient internal energy consumption, etc.).

The food crisis of 2007–2008 triggered the food-versus-fuel debate and raised concerns about out-of-control expansion of biofuels to meet ever larger energy needs. If left unchecked, biofuel expansion could well shift food production into more marginal lands, resulting in lower yields. Also, competition over resources such as water and fertilizers may also constrain food availability (depending on feedstock and location). Competition could also enhance yields as a result of higher rents (i.e., the market price of land) and the adoption of other productivity-boosting technologies (rotations, intermixed cropping). First-generation biofuels are also experiencing slow and progressive technological advances, including improved energy input-output ratios and increased market value and uses for by-products. However, these effects may vary depending on local market conditions and relevant policies or regulations in place. Increased demand from biofuels for feedstocks tends to push up agricultural commodity prices, and trade is a key determinant in a globalizing biofuel market. However, if trade barriers are lowered (including tariffs) and biofuels are more openly traded, market competition would moderate prices. Further, higher commodity prices, by making food more expensive, would draw resources from biofuels back into food production – a result of food-fuel competition over shared productive resources. The linkages between food and energy will likely grow stronger, affecting the relative competitiveness of biofuels and their long-term viability and sustainability. This is especially the case should second-generation biofuels become commercially available.[5] In that case, competition for shared resources will become even more intense, and it is unlikely that policies or regulations would not have to step in to balance between food and energy security.

Biofuels are bulk commodities with little scope for price premiums. Moreover, the quasi-mandatory requirements for certified biofuels (or biomass) entering the EU market also remove the conditions for price premiums. Yet despite the added certification costs, many producers in developing countries are still able to compete in the European market, as they can produce feedstocks more efficiently (at least the high-yielding ones such as sugarcane and oil palm). This partly explains the much concentrated focus of certification schemes on a few key traded biofuel feedstocks (sugarcane, oil palm, soybeans).

By contrast, commodities produced, transformed, and used domestically can fall largely outside the writ of these voluntary certification schemes, especially in the absence of strong and enforced domestic regulations (e.g., corn ethanol in the United States, sugarcane ethanol produced and used in Brazil, soybean biodiesel in Argentina, sugar in India, palm oil in Indonesia, and beef in Brazil).

One complicating factor in assessing the economic sustainability of biofuels is the multiple market outlets for feedstocks (e.g., food, feed, fiber, and now fuel). Yet sustainability requirements as articulated in current certification schemes appear to be limited to biofuels use only. A certification scheme established on the basis of a single final use (i.e., biofuel) may be ineffective in securing sustainability, resulting in indirect displacement effects. One remedy is to focus on sustainability at the biomass production side. However, the substitution possibilities among different final end uses of feedstock make it difficult to enforce sustainability compliance if tied only to biofuel supply chains.

6.2.2 Environmental sustainability: multiple indicators and the challenge of measuring impacts

Environmental sustainability encompasses a broad set of issues, some global (e.g., climate change, GHG mitigation, renewable energy), and others more location specific (e.g., water management, soil quality, erosion, and water and local air pollution). Environmental sustainability of biofuels has been largely defined in terms of reducing GHG (e.g., CO_2, methane, N_2O) emissions. For non-CO_2 GHGs, agricultural practices (e.g., soil tillage, irrigation practices, fertilizer use, pesticides, harvesting) are leading sources of emissions. Moreover, land use prior to biofuel conversion is also a critical factor in evaluating the environmental impact. A biofuel's GHG-reduction potential suffers markedly if grasslands or forests are used for biofuels. Definitive assessments of the GHG effects of biofuels continue to be hampered by a lack of reliable methodologies to measure indirect land-use change, soil carbon, and so forth. Life-cycle analyses are increasingly used to measure the sustainability of various biomass-biofuel systems, but the methodologies so far are not standardized and have yet to adequately account for indirect land-use change.

Another important motivation for biofuel is the promise of energy substitution to replace fossil fuels. Energy balance (i.e., the ratio between renewable-energy output and fossil-energy input) shows great variation among different biofuel feedstocks, with palm oil for biodiesel leading the pack with an energy balance up to 9.0 (i.e., nine times the energy required for its production). Sugarcane also has a relatively high but variable energy balance, ranging from 2.0 to 8.0. Most other feedstocks have energy balances that range from 1 to 4. Still, these calculations do not take into account the effect of indirect land-use change. Besides GHGs and energy, water-resource preservation may top other considerations in specific areas when evaluating environmental sustainability. In some cases, constraints regarding the quantity of water used and the impact on local water quality and future availability may be the most limiting factor against biofuels. Linked to water is the problem of fertilizer runoff – especially near streams and rivers.

Preserving biodiversity or avoiding biodiversity loss from biofuels is another critical criterion for sustainability. However, there are no standard ways to measure which systems to promote, except in general terms (such as use of rotations). Most current production systems do not indicate stability or even maintenance of biodiversity. Biomass production under intensive monoculture systems can have negative impacts on biodiversity, including habitat loss, the expansion of invasive species, and contamination from fertilizers and herbicides. However, the deployment of biomass in previously degraded land may benefit biodiversity, but this can only occur if there are strong enough incentives (including payments for environment services). A general problem with the environmental side of certification schemes is the difficulty of translating principles and criteria into effective sustainability indicators on the ground. This is partly due to inherent problems with identifying measurable, permanent impacts of certification schemes.

Another reason is the lack of available and meaningful data that enable proper comparison and assessment of compliance. Moreover, the principles and criteria themselves can be too broadly stated (with few exceptions) or, inversely, translated into indicators that are too narrowly specified, making it difficult to agree on broad values of sustainability. For example,the certification under the Round Table on Responsible good agricultural practices such as crop rotations or zero tillage are not mandatory, as they would reduce the market for soya qualifying under this certification.

6.2.3 Social sustainability: the weak link and inadequacy of current certification schemes

The social impacts of certification schemes are even less well documented. The key difficulty lies in the ability to translate social sustainability standards and criteria into measurable indicators. This is in part due to the wide range of social conditions, practices, and norms (e.g., labor structures, types of land ownership, local resource management). Another reason is the highly location-specific context of social impacts. For example, the indicator "all workers receive minimum wages" may mean little in countries where informal employment is widely practiced, particularly in the agricultural sector. If no formal contracts exist, compliance with this indicator might be difficult and costly to assess. While the enactment of certification schemes may have some positive impacts on workers and local communities, there is still limited evidence of direct poverty-related impacts, improved food security, or enhanced sustainable income opportunities through value addition, expanded market opportunities, and diversification. For most certification schemes and scorecards, the social aspects of sustainability are addressed only in terms of removing selected negative impacts (e.g., child labor, minimum wages), or calling for adherence to national laws or international conventions. However, critical social factors – such as participatory processes, common management of resources, health implications, and other aspects of poverty reduction or smallholder inclusiveness – are not typically addressed as primary concerns of existing certification schemes. This may seriously limit the scope of these schemes as designed in addressing social sustainability in an integrated way. In the end, the existing biofuel certification schemes are not properly structured to adequately address social sustainability. The private-led "voluntary" schemes are not the correct instrument to address social issues that are essentially public-good types. Rather, the appropriate sustainability measures require strong national supplemental policies and regulations that safeguard the potentially broad domestic social benefits as part of any biofuel development. More than the economic and environmental dimensions, social sustainability for biofuels and related feedstocks need a serious rethink of how to mainstream and implement sustainability. Essentially we need to move away from simply focusing on targeting selectively few of the most obvious negative impacts (child labor, minimum wages) and incorporate development goals where local communities share sustainably in the potential economic benefits from biofuels in comparison to current alternatives.

6.3 Biofuel and food security[6]

Another limitation of the prevailing biofuel certification schemes is the lack of inclusion of small-scale farmers. By design, certification schemes favor large-scale agribusinesses, as they require costly procedures with significant amounts of information and resources and also because big players have the means and incentives for scaling up production to absorb the certification costs. Moreover, larger companies typically already keep records needed for audits, but small-scale farmers often keep no written records on yields, fertilizers, and by-products – data that is needed for GHG estimations.

There are several ways to enhance smallholder inclusion, including enhancing the capacity and skills of small-scale producers to master compliance requirements (such as record keeping, facilitating farmer aggregation into producer organizations to reduce certification costs, and adopting more efficient and sustainable technologies that can facilitate certification). Though there are some incentives to address prohibitive certification costs for smallholders by some of the leading feedstock roundtables, a more sustainable solution is to ensure a more balanced representation in these roundtables, with active participation of smallholders' representatives in these multi-stakeholder certification schemes. Assessments of sustainable soy in Brazil and jatropha in India show that smallholders generally have good knowledge of on-farm conservation, but not the same options to extend native vegetation buffer zones. Similarly, field burning – an important emitter of GHGs – is mainly practiced on small farms, while many large plantations have already mechanized their production and can easily respond to this pollution issue.

6.3.1 Biofuels and the poorest developing nations

Much of the focus about the biofuel industry moving south has been on leveraging foreign direct investments (FDI) to bring large-scale capital-intensive biofuel plants closer to feedstock production sources, especially in developing countries presumed to have abundant land, water, and/or labor resources. The drive behind biofuel FDI in developing countries has been essentially driven by cost-cutting and efficiency-enhancing objectives. Yet despite the many touted advantages of biofuel investments for rural development, energy security, and employment, serious obstacles to biofuel growth remain in developing countries, including a lack of qualified labor, basic infrastructure, and the investment capital needed to develop feedstock supply chains.

These requirements are beyond the capacity of many developing countries, especially among the poorest. Even assuming FDI is forthcoming, this still requires an infusion of complementary and investment commitments from national governments to assure success and viability. Even under the best situations, one can expect smaller positive spillover impacts on the local economy because labor and capital are imported while biofuel is produced for export. There is also the issue of land acquisition for large-scale biofuel projects and the potential conflict with existing or traditional land rights, access, and use. This concern has become acute enough since the food crisis of 2007–2008 that FAO, along with other international organizations, developed a new set of guidelines for land access (Voluntary Guidelines on the Responsible Governance of Tenure of Land, Fisheries and Forests). An alternative model that can contribute both to food and energy security for many developing countries would be based on the promotion and development of small-scale biofuel or bioenergy systems that can be integrated into existing farm, household, or community-development activities. Such systems (e.g., biodiesel-fueled cooking stoves, solar lanterns, and biodiesel-fueled small power stations for electricity or small-scale irrigation) can be more effective in providing energy security for small-scale producers and local communities, especially in poor developing countries that traditionally rely on unsustainable exploitation of biomass which aggravates deforestation. An example of such a model is the integrated food-energy system (IFES) widely practiced in some Southeast Asian countries such as China and Vietnam with long traditions of closely linked livestock-fishery-crop intensive systems. Such biofuel-development models could boost agricultural production, raise land productivity, secure more rural employment opportunities, and offer greater positive economic impacts for local communities compared to large-scale biofuel-production systems that rely on imported capital and skilled labor and export the produced goods with fewer multiplier effects on the local economy. However, an IFES-like system would require an active policy support in line with national strategies that integrate energy needs with food security and sustainable long-term rural development. For poor countries, such a strategy could also be supported by Overseas development institutions (ODI), by international development agencies, and through bilateral aid funding, including funds for climate-change mitigation and adaptation.

Large-scale biofuel production (ethanol, biodiesel) could also be included as part of a national energy strategy, depending on the country's industrial capacity, energy needs, and comparative advantage (land abundance, established feedstock value chains – for example, palm-oil biodiesel in Malaysia or cassava ethanol in Thailand). The key criterion however is that the strategy must be dictated by domestic food- and energy-security needs, with trade playing a complementary role in case of excess supply.

6.4 Other remaining issues

Beyond the discussed topics, other remaining issue require tackling. For example, biofuel development is also facing, with increasing urgency, the rising challenges of climate change and the need to account for carbon footprints and reduce GHG emissions. How should the initial concern of biofuel certification be expanded to include carbon-footprint certification, or are these separate concerns? Are the social criteria of these existing certification schemes compatible with the recently endorsed Voluntary Guidelines on the Responsible Governance of Tenure of Land, Fisheries and Forests by the World Committee on Food Security? How could these voluntary guidelines be implemented within the existing biofuel certification schemes?

Overall, the increased awareness and pursuit of biofuel sustainability has gained momentum in recent years, which in itself is a positive development. However, the above assessment of the core biofuel sustainability issues and certification schemes revealed the limitations of the processes followed and the lack of an integrated approach. Moreover, there is a huge gap between the conceptual definitions of standards, principles, and criteria and actual testing and verification on the ground. Clearly the voluntary private-led certification schemes are not sufficient instruments to ensure balanced and integrated coverage of the essential elements of sustainability, be they economic, environmental, or social.

Strong complementary public policies including incentives, disincentives, and regulations are needed to ensure a more balanced treatment of the sustainability challenge, to safeguard the mobilized resources, and to enable inclusive value-chain development processes for smallholders. What is needed is a new approach that integrates sustainability with the pursuit of renewable-energy strategies and food security and is inclusive of marginal and small-scale stakeholders. What is required as a more coherent and integrated framework for sustainability that combines both private schemes and public regulations in such a way as to assure inclusive processes between large enterprises and small-scale producers, as well as between northern and southern countries' goals and interests.

Rethinking sustainability also requires incorporating full environmental costs in economic cost-benefit assessments and fostering business models that can reconcile sustainability with economic growth and integrate inclusive development with food security. Also required are policies, regulations, and incentives that broaden the biofuel development options to include small-scale, locally harnessed renewable-energy technologies and systems. Finally, biofuel sustainability will need to be mainstreamed into larger trends toward sustainable and climate-smart agriculture in line with the triple objectives of enhanced productivity, strengthened food security, and climate-change adaptation and mitigation.

The major conclusions[7] with respect to biofuels and the sustainability conundrum can therefore be summarized as follows:

- The initial surge of biofuels in industrial economies was driven by energy security and rising fossil fuel prices, but market forces alone were not sufficient to drive the process, which required heavy policy support (subsidies, mandates, and tariffs for imports) targeting few domestic-based feedstocks (corn, rapeseed, soybeans); meanwhile research and development of new feedstocks to support future biofuel expansions took off, including high-yielding (sweet sorghum) and more versatile crops (jatropha), as well as dedicated energy crops for second-generation biofuels. Yet the expected large gap between future demand and potential domestic supply in the North required expanding biofuel production in developing countries, which had the land and the climate needed to produce raw feedstocks on a large scale.
- However, rising concern about climate change and its necessary mitigation as well as the increasing awareness of the relationship between climate change and sustainability has altered views about biofuels, including a criticism of biofuels using feedstocks that are only moderately efficient but require direct subsidies. Moreover, the food crisis of 2007–2008 and the debate over food-versus-fuel competition has raised concerns about biofuels clashing with food security and ushered in a critical debate about the long-term sustainability of current biofuel systems.
- Measured in terms of efficiency and sustainability, feedstocks grown for biofuels are not alike. Crop feedstocks such as sugarcane or palm oil are relatively more efficient, in terms of biofuel yields per area, and can be economically viable without direct subsidies. However, their environmental sustainability comes into question when water irrigation is required (sugarcane) or when plantations are located in carbon-sensitive lands (palm oil). Sweet sorghum, still under development, offers high-efficiency potential and a wider scope for adaptability to soil and water conditions compared to sugarcane. However, sweet sorghum quickly loses sugar after harvest, therefore limiting its adaptability to those countries with well-developed infrastructure and supply-chain capabilities.
- Established feedstocks for ethanol (corn) and biodiesel (rapeseed, soybeans) have thrived largely under the protection of subsidies and mandates, but their long-term economic and environmental sustainability are not clear. The future prospects of these first-generation biofuels will depend on a range of factors, including the possible deployment of new and efficient feedstocks, improved economics of biofuels through continued innovations,

and future policy support, as well as the commercial deployment of second-generation biofuels and related feedstocks, including waste, residues, and other noncrop biomass.

- Alternative feedstocks with potential growth in developing countries, such as jatropha and cassava, may present attractive agronomic characteristics and good suitability in marginal lands with varying weather, water, and soil conditions, yet several obstacles may limit the scope of these crops as future feedstocks. Key among these is the economic need to ensure intensive management systems to maximize yields and efficiency, which may lead to direct competition for prime land, often with established infrastructure and where food production is already in place. In the end, the economics of production will likely trump the agronomy of the crops. Hence the prospects of seeing widespread use of these crops as biofuel feedstocks are not optimistic, especially for poor, less-developed countries that may have idle land but limited value-chain development capacity and the required infrastructure to support it.
- Assuming second-generation biofuels become commercially viable, we can expect a significant expansion of biomass use (a broader set of crop feedstocks, waste, and agricultural residues). Such a development will likely alter the demand and supply of biomass sources, and hence their economics, tightening even more the linkages between agriculture and energy, and potentially making even more intensive the competition for land between food and energy uses. This in turn will have uncertain implications for rural development opportunities, especially in poor, developing countries that continue to rely heavily on traditional uses of biomass that are neither sustainable nor climate smart. What is clear is that the economics of production will be the determining driver in sorting out how resources (land, labor, water, and other resources) are likely to shift between food and energy. If the past is any guide, market forces alone are unlikely to be the sole drivers of these processes, and the role of policy support (through incentives or disincentives) will also be critical in guiding the outcomes.
- From a sustainability perspective, biofuels offer both advantages (energy security, GHG reductions, reduced air pollution) and risks (intensive use of resources, monocultures, reduced biodiversity, and even higher GHGs through land-use change[8]), and measuring biofuel sustainability requires approaching economic, environmental, and social sustainability in an integrated way to maximize benefits and minimize risks. Yet the review of the biofuel sustainability initiatives taken as a whole does not show that such an integrated framework is being pursued or that the impacts of the core dimensions of sustainability are fully measured or understood.

- Biofuel certification schemes, despite their multiplicity, are dominated by a singular form of governance – namely voluntary, private-industry-led initiatives targeting sustainability assurances with input from nonindustry stakeholders. These schemes are driven as much by market-access and trade considerations as by the need to provide sustainability assurances. This may explain why the first schemes and initiatives have focused on those feedstocks and biofuels most involved in South-to-North trade (soybeans, sugarcane, and oil palm). This dual role of biofuel certification schemes also explains the tendency to target selected sustainability criteria and not others and hence the absence of full integration of the three core dimensions (economic, environmental, and social) into a coherent framework or strategy.
- National and supranational initiatives on biofuel sustainability have been led by Western Europe – a region that is most dependent on future imports of biofuels and feedstocks to meet projected domestic needs. Leading exporting countries such as Argentina, Brazil, Indonesia, and Malaysia have also responded with their own sustainability initiatives in part to protect their export markets and to meet importing countries' requirements. Transnational forums, such as the Global Bio-Energy Partnership, have been set up to harmonize sustainability initiatives across interested countries. Such forums emphasize consensus building around methodologies and other voluntary metastandards, but they are unlikely to agree to fully harmonize policies or approaches (outside voluntary guidelines) that may clash with their national biofuel or renewable-energy policies driven by domestic priorities.
- Competitiveness of biofuels in the long run will continue to depend on the economics of fossil energy, the policy-support environment, and the relative incentives and disincentives to encourage renewable versus fossil-energy sources. Economic competitiveness of biofuels will also depend on the substitution possibilities between food- and fuel-market uses, and the advances in technology and innovation in biofuel-production processes. Continued reliance on policy support through subsidies and mandates indicates the lack of market competitiveness of biofuels in the short and even the medium run.
- A full assessment of economic biofuel sustainability requires a complete internalization of the full cost of environmental effects (i.e., putting a market price on negative externalities). On the other hand, economic sustainability may clash with environmental considerations when the need to maximize returns on investments dictates pursuing intensive management practices that could clash with sustainable use of resources and exacerbate competition with food for productive resources such as land. Consequently,

large productivity gains are required to minimize such conflicts and bridge the gap between efficiency and long-run sustainability.

- Environmental sustainability assessments for biofuels are difficult owing to the complexity and the multiplicity of indicators, some of which are global (GHG, renewable energy), while others are local or regional (water management, soil and resource depletion, local pollution, etc.). Initiatives on sustainability via regulations, directives, or private-led certification schemes have had no clear and measurable impact, apart from increased awareness of their importance, and this despite the numerous initiatives and huge sustainability debates. A key problem continues to be a lack of consensus on measurement methodologies (such as life-cycle analyses and the way to tackle indirect land-use change). Moreover, certification schemes are of recent creation and continue to be impeded by inherent measurement and monitoring problems, which vary according to situation (location, feedstock, technology, alternative resource use, policy environment, and local capacity). Until progress is made on these obstacles, the approaches pursued so far will continue to be selective and haphazard, focusing on self-selected sustainability measures and ad hoc rules such as no-go zones for high-carbon stock or biodiversity-rich areas.
- The social impacts of biofuels certification schemes remain the weakest link in most sustainability initiatives thus far. Most certification schemes, scorecards, and regulations make mention of social impacts but only seek to mitigate a few of the obvious negative impacts (child labor, minimum wage, compensation for lost land and resources) or call for adherence to national laws or international conventions. However, evidence of how these measures are actually implemented, or their impacts on the ground, has been very limited, and successful cases are rare. Among the reasons are the complexity of social impacts and their inherently local context, often encompassing contrasting social norms, practices, capacity, community empowerment, and varied levels of political participation. Clearly, the social sustainability dimension requires a qualitative rethink that goes beyond mitigating a few negative impacts, but rather integrates participatory processes that ensure wider economic benefits to marginal stakeholders and local communities, and therefore guarantees broader acceptance and long-lasting stewardship of resources.
- From food security and rural employment perspectives, biofuel certification schemes, unless tied to specific public initiatives, are not structured to be inclusive of small-scale producers. Most certifications require costly, complex, and intensive information systems and management capacities that are easily absorbed by large-scale agribusinesses (with their advantage of economies of scale) but are largely out of reach for small-scale producers. This implies that private-led certification schemes may not be sufficient by

themselves to facilitate wider participation in promising feedstock–biofuel value chains or to offer small-scale producers opportunities for market diversification. Consequently, biofuel projects may have limited development, rural employment, and income-enhancing potential at a local or regional level. Filling the inclusiveness gap for small-scale producers, especially in poor, developing countries, requires active public interventions carefully tailored toward incentives to develop capacity, better organization, adoption of cost-cutting technologies, and new techniques to enable smallholders to better leverage the new market opportunities offered by any new possibilities for biofuel-led agricultural value addition and diversification.

- Linking biofuels to food security in developing countries also requires establishing closer links between food security and energy security. This requires choosing among different development-model paths, depending on the stage of industrial development of the country, the general state of food security, the extent of agro-industry development, and the capacities of producers and agribusinesses. No single model fits all situations. For poor countries with limited industrial capacity, emphasis should be placed on small-scale bioenergy systems that can integrate existing crop and livestock enterprises at farm, household, and community levels. Such schemes have larger developmental benefit potential in terms of local employment, productivity enhancement, and improved food security. FDI-induced larger-scale biofuel projects, on the other hand, may be suitable in those situations where countries have sufficient industrial capacity, besides land and biomass potential, and when these biofuel projects can be fully integrated into domestic energy strategies that do not conflict with food production potential and food security.

6.5 EU biofuels and agricultural commodity prices[9]

Biofuel mandates are being used in many parts of the world to encourage the substitution of fossil fuels in transport uses with renewable energy. Brazil has long had an extensive program of producing bioethanol from cane sugar, the United States has massively expanded production of ethanol from corn, and the EU's Renewable Energy Directive has led to a scale-up in EU biofuel use, most notably of rapeseed biodiesel. The 10% target for the use of renewable energy in transport set in the EU Renewable Energy Directive is anticipated to lead to a tripling of biofuel use in the EU in 2020[10] compared to 2008 levels, according to the National Renewable Energy Action Plans that all member states have been required to produce. The plans confirm that the current dominance of biodiesel over ethanol in the European biofuel market will continue. The EU's target predominantly will be met by first-generation biofuels produced from traditional food and feed crops, translating into a significant additional demand for these crops. It is clear that this additional demand, alongside the growing global demand for food and biofuel demand elsewhere, will increase agricultural commodity prices, but by how much is the subject of continued debate. The uncertainty about the extent of price increases arises from a range of factors, including the prevalent interactions between different crops and livestock markets and between world regions, and from the responses to price signals by consumers and producers. Yield increases stimulated by higher output prices are often suggested as a factor with the potential to mitigate price increases; another may be related to the rise of protein by-products as animal feed. The complexities of interacting factors can only be captured in reasonably sophisticated economic models.[11] Nonetheless, because of the wide variety in the context, scope, and methodology of models used to analyze the effects of biofuel mandates, the results of analyses of the price effects range substantially. Consequently, it is often extremely difficult to draw comparisons between models. In those modeling studies focusing on the impacts of EU (as opposed to global) biofuel policies, the most significant price increases are projected for oilseeds and vegetable oils, with increases in world prices by 2020 typically ranging between 8% and 20% and between 5% and 36%, respectively. Wheat prices are projected to increase by between 1% and 13%, and the majority of studies project increases of cereal/maize prices of up to 8% and of sugar prices of up to 2%. One model European Simulation model (ESIM)[12] projects these increases to be 22% and 21%, respectively. Many of the drivers of these differences in results are the same factors that have been under scrutiny in the indirect land-use change debate. Consequently, the studies at the forefront in that debate can be expected to deliver the most robust results with regard to agricultural market impacts.[13] Most notable is the International Food Policy Research Institute (IFPRI) study,[14] which projects increases in world rapeseed prices (anticipated to be the most significant feedstock for EU biofuel use in 2020) of around 11%. While looking at studies that model global biofuel policies does not allow singling out the EU policy impact, it is in some way a more complete scenario design, given that EU policies in the real world do not take place in isolation but other countries have in place biofuel policies as well. At least some of the global studies estimate substantially higher price effects for ethanol crops such as wheat, other cereals, and sugarcane.

In summary,[15] the price changes found in the studies performed are all positive, but not massive, especially in comparison to the recently experienced global-commodity price spikes. But they are not negligible either. Another rather robust result is that there is an important difference in impacts between commodity groups.

At the end of 2012 the EU Commission issued, as foreseen, her impact assessment on the social impact of biofuel policy in the EU.[16] In that report, besides the development of a number of strategies to achieve certain policy objectives in the field, a number of primary conclusions were drawn:

1. The estimated indirect land-use change (ILUC[17]) emissions are, despite the better understanding and recent improvements in the science, vulnerable to the modeling framework and the assumptions made.
2. The use of biofuels in the EU saves emissions, including when estimated indirect land-use change emissions are considered. In addition, the models indicate a hierarchy of biofuel types according to their indirect land-use change impacts, these being considerably higher for typical biodiesel feedstocks (oil crops) than for bioethanol feedstocks (cereals and sugar crops).
3. Given the strong reliance on conventional biodiesel, and to a lesser extent conventional bioethanol, in projected biofuel volumes to 2020, there is a high risk that the estimated indirect land-use change emissions will significantly reduce the expected savings from the policy if no action is taken to mitigate indirect land-use change emissions.
4. The development of advanced biofuels, using low-value resources such as straw, wood, and forestry residues, is slower than previously expected, as the costs associated with producing such fuels is higher than the alternative conventional biofuels.

There are reasonable grounds to believe that indirect land-use change emissions could partly undermine the greenhouse gas savings offered by using biofuels. Consideration has also been given to options for introducing additional sustainability requirements on certain categories of biofuels.

Therefore the last option analyzed, that is, limiting the amount of conventional biofuels counting toward the Renewable Energy Directive transport target to current production levels, would also be effective in reducing indirect land-use change (i.e., a reduction of indirect land-use change emissions of 55%, from 46 Mt of CO_2-eq./year to 21 Mt CO_2-eq./year in 2020). In addition, this option would require moderate industrial adjustment as it would only exclude vegetable oil biodiesel beyond current production levels in the run-up to 2020 and would not necessarily pose a technical challenge from a blending limit perspective, while providing a strong incentive for increasing the share of advanced biofuels. The incentives for producing advanced biofuels would be strong, as the amount of double-counted advanced biofuels would need to increase significantly.

The idea would be to reduce indirect land-use change emissions significantly while protecting existing investments. The impact assessment came with the draft legislation to limit food-based biofuels counting toward the EUs 10% target for renewable energy in transport at 5%.[18] Remarkably enough, in the whole analysis very little attention was paid to the role and impact of second- and third-generation biofuels, which, as discussed, avoid the claim on agricultural land. Or maybe the right assessment is that when there is a lower cap on first-generation biofuels, it leaves more room for the next generation of biofuels.

The EU impact assessment was followed by an avalanche of reports from a variety of institutions that variably promoted the land cap suggested combined with a continued criticism of the wrongheadedness of the EU biofuels policy. Often this included a refocus on reducing energy consumption and electrification of our transport needs[19] and on reducing GHGs rather than the objective of a land cap as a measure taken in isolation. Further arguments are the need for a waste hierarchy and the consideration of the complete life cycle of GHGs in the analysis.[20] That is not remarkable given the fact that in the year prior to the EU impact assessment report, a number of research papers convincingly questioned some of the data sets used by the EU to justify its current policy. Some of the researchers were also unable to back-test the EU results using the same data sets, or worse, yielded different (more negative) results, often pointing at a more limited reduction in GHGs,[21] thereby failing the sustainability test.

The EU report was then followed in early 2013 by the report assessing the impact of biofuels production on developing countries,[22] reiterating the interaction between weather conditions, the impact on developing economies of EU biofuel policies, and the like, as a reasonable portion of the biofuels consumed in the EU originated from outside the EU (broadly between 20% and 35%).[23] The Commission is also studying the sustainability of biofuels from a consumer perspective as a part of a broader study on the functioning of the fuels market.[24]

Further developments following these reports were as follows:

- On June 20, 2013, the EU Parliaments' energy committee voted to push the cap on food-crop fuels up slightly, to 6.5%. It also removed the stipulation that fuel suppliers report emissions using land-use change figures. Instead, the committee proposed gradually increasing mandates for use of advanced biofuels not made from food crops.
- On July 11, 2013, the European Parliament's environment committee voted to limit the amount of conventional biofuel that can be used to meet an EU target for renewable transport fuel. That limit was set at 5.5% (to be reached by 2020, as that was the target date in the original directive) as a compromise, being slightly higher than the 5% proposed by the EC in its proposal in 2012[25] and the 6.5% vote of the industry committee on the matter. They also voted to back the European Commission's proposal to require companies to measure the amount of 'indirect land-use change' (ILUC) their fuels cause, and this despite the argument raised that ILUC research is inconclusive[26] and the lack of clear evidence that biofuels contribute to

the volatility of agricultural commodity prices. This should be interpreted in conjunction with the postponement of the implementation of the Fuel Quality Directive to 2025. The environmental committee decided to raise the Fuel Quality Directive's requirement for fuel producers to reduce the carbon intensity of their fuels by 9% by the end of 2025, up from 6% by 2020, the current requirement. After the vote, it was echoed that the EU's line of thinking by defining a list of advanced biofuels[27] that are largely produced only in developed regions like Europe to replace the reduced cap would be in line with WTO legislation.

- On September 11, 2013, the plenary session of the EU parliament backed the[28] legislative proposals to limit the amount of food crops used to produce biofuel. In a close vote, MEPs voted, amid some intense horse trading, that "first-generation" biofuels – from crops like corn – should not exceed 6% of fuel used in transport by 2020, amending the target from 10%. The actual realized level in 2013 is 4.5% to 5% (geared toward EU transport demand). They also want "advanced biofuels" – sourced from seaweed or certain types of waste – to make up at least 2.5%. Before becoming law, the proposals still have to be agreed to by the 28 member states' governments represented in the EU Council. Lawmakers also backed an amendment that would force energy companies from 2020 to take account of the indirect emissions caused by crop-based biofuels, which increase overall demand for land and, as a result, encourage rainforest clearance or drainage of peat land. This would effectively ban the use of biodiesel from oil crops such as rapeseed, palm, and soy, which according to the EU's scientific models are more damaging than conventional diesel when their overall impact on the environment is taken into account. Lawmakers however demanded further talks about the rules before opening negotiations with EU countries to finalize them. With EU governments, yet to finalize their common position, talks on the proposals look likely to extend into next year. If they are not concluded by April, European Parliament elections scheduled for the following month could push back the law until 2015 or could take the whole thing back to square one. Overall, a mixed bag, and the question remains if the negotiated tweaking of percentages will do the trick of turning the situation of soaring global food prices around.

7
Financing the Agriculture Firm[1]

Within the context of this book, it would be (nearly) impossible to dedicate a full chapter to all the credit aspects of financing an organization and then to complement that with the specifics of financing the agricultural firm. Therefore it was decided to focus only on the specifics of the agricultural firm, thereby assuming that one is somewhat familiar with the principles of lending and the major fields of assessment – solvency, efficiency, coverage, leverage, liquidity, and profitability.[2]

Reality tells us that most lending institutions (financial institutions [FI] or nonbank financial institutions [NBFI] and microfinance institutions) are often confronted (more in developing economies than in advanced economies, however) with three main problems: (1) unfamiliarity with the agriculture sector, (2) lack of financial expertise tailored to the agriculture sector, and (3) limited access to medium- and long-term resources. Access to credit is a major constraint for actors in the agriculture sector and a complicated challenge. Challenges shared by banks and other financial institutions include (1) the need to open new market niches in the short or medium term and transition to the "universal bank" principle, (2) the need to expand credit activity to maintain or increase gross margin, and (3) the need to link the future profitably of the bank with growth of lending to the agricultural segment, that is, making agricultural lending an integral part of each institutions growth strategy. Banks are largely unaware of the potential of the agriculture sector and the problems and realities related to production, products, and the political and economic organization of the value chain. Entering the agriculture sector requires focus on the capacity of the borrower to repay rather than collateral or asset-based lending and a deeper understanding of how value-chain finance can provide additional guaranties and access to supplementary resources and technical competencies. Cash flow-based credit methodology should be combined with an understanding of the relationships and opportunities within a wide variety of value chains, carrying out mapping exercises to clarify specific opportunities to finance clients within each value chain. Credit officers need to better understand the specific agricultural crop, production cycle, yields, crop budgets, crop protocols,[3] and other issues unique to agriculture so that loan products and terms can be tailored to a farmer's needs.

The range of products required to meet credit needs includes long- and short-term loans for purchase of land, facilities, and breeding improvements; seasonal credit for purchase of seeds and other agricultural inputs; and lending, leasing, and factoring (loans based on accounts receivable due from sales contracts). Banks generally have a conservative, corporate approach to lending, only slightly adapting their products for small and medium enterprises (SMEs) and maintaining the same risk-analysis methodologies relying on a high percentage of collateral coverage and larger clients.[4] For example, in countries where one cannot buy or own the land but only lease it longer term, a critical piece of collateral to finance working capital for the agricultural firms falls apart. The economic countervalue in terms of economic value of the usufruct during the term of the lease is often too weak to be convincing for the traditional credit committee. What will follow is a bundling of experiences and best practices when it comes to lending to the agricultural sector across several levels of the value chain, including small-scale producers, processors, and distributors, and service companies and SMEs supporting them.

7.1 Scoping

Lending to the agricultural sector will normally include:

- Short- and medium-term financing for processors and service providers.
- Wholesale lending to smaller financial institutions, including smaller banks and microfinance institutions that, in turn, lend to market vendors and other entities serving the needs of subsistence farmers.

It will therefore cover the whole spectrum of the value chain to include:

- Input suppliers
- Producers
- Processors
- Exporters
- Wholesalers
- National and international retailers
- All providers feeding into that value chain

Understanding the market in which the borrower operates and where exactly he or she is positioned in the value chain is paramount in understanding the intrinsic risk in a specific lending opportunity. This will eventually lead to an evaluation of which is the most adequate credit product for the farming firm. From a bank's perspective, this will undeniably include a risk assessment associated with the product(s) considered or to be developed, approval criteria, policies germane to the products, product profitability, product costs, and management responsibility and accountability.[5] This analysis (from a bank's perspective) will include the "chapters" indicated in Table 7.1.[6]

Table 7.1 Financial analysis of an agricultural firm or project

Business care for a credit product	**Which risk and approval criteria need to be applied?**
Provide a description of the new type of credit product and the reason for proposing the product. Provide a business case structure that covers: • Brief introduction for agricultural products • Business impact of adding agricultural products • Methods and/or assumptions made for developing the products • Risks and contingencies associated with agricultural products • Conclusion and recommendations	Identify risk factors of associated with agricultural products and approval criteria. • Borrower capacity analysis • Borrower stability requirements • Minimum acceptable credit experience/history • Other borrower criteria • Age (minimum/maximum) • National citizenship or residence requirements • Maximum unsecured exposure (debt)/total exposure • Other exclusions: including less desirable employment or businesses • Borrower verification requirements • Collateral verification/appraisal process • Selection criteria for appraisers/evaluators • Key appraisal/valuation standards • Frequency of collateral reappraisals/revaluations • Maximum portfolio limits • Process exception limits
Policies and policy changes Describe the policy requirements or policy changes that will be necessary to implement and govern the agricultural products. Describe the processes that will be used to measure, monitor, and control risk associated with the products. • Minimum/maximum loan amounts • Collateral requirements • Other supporting security (promissory notes, guarantees, insurance)	**Product profitability**[7] Project the profitability impact of agricultural products in the portfolio as well as expenses, loss provisions, capital requirements, and other issues that will affect profitability and return. • Volume of exposure • Expected yields (including net interest income and fees and commissions) • Attributable direct costs

• Maximum loan to value for secured loans • Minimum/maximum tenor/renewal options • Pricing • Loan repayment method • Credit-risk exceptions • Product pricing exceptions • Target market exceptions • Loan application decline process • Nonaccrual and contractual write-off policy • Asset recovery • Fraud prevention • Fraud detection	• Expected default rates • Expected provisioning requirements • Capital requirements
Product development cost Prepare a projection of the cost of developing agricultural products, as well as the infrastructure to maintain it. Infrastructure requirements and costs IT requirements Human-resource requirements Marketing requirements	**Managerial accountability** Identify competent individuals and departments for introduction of the products, special procedures, and work-flow issues, and reporting, monitoring, and collection requirements. Procedures and process flows Approval requirements Monitoring Collection policy and strategy Reporting requirements

The bottom line is, however, or should be, the agricultural client and its financing needs. In the agriculture value chain, many participants are engaged in the process – from raw materials and inputs to the final product for sale and consumption. By this definition, potential borrowers include farmers, cooperatives, and other agricultural producers, input suppliers, processors, and other value-chain participants. These can include the following:

- Farmers and other agricultural producers: small and commercial farming, livestock, dairy, and aquaculture operations for local consumption; commodity farmers for export.
- Input suppliers: retailers (fertilizers, pesticides, etc.).
- Processors: millers, bakers, factories/plants, etc.
- Other participants in the sector: traders, storage facilities, transportation, retailers, market vendors.

This will undeniably lead to an adaptation of the product to fit the need of the agricultural borrower. Although loans to agriculture value-chain participants are similar to other small-business loans, agricultural loans often contain special pricing to make them compatible with crop cycles. These can include seasonal payments, allowing a borrower to make minimal payments during periods of planting and growing crops, and/or incremental payment systems with low payments early in the loan term to coincide with the farming process, followed by larger payments after harvest.

The types of agricultural loans include the following:

- Short-term loans (up to 12 months), including the following product types:[8]
  - Production or preharvest financing
  - Postharvest financing
  - Purchase-order financing
  - Inventory financing
  - Short-term equipment rentals
  - Hire-purchase
  - Credit sales
- Long-term loans (more than 12 months)

Applied to the market participants, this will provide coverage[9] as indicated in Table 7.2.

It is also very common for participants within the agricultural value chain to provide financing to each other. Eventually this will lead to product sheets

Table 7.2 Overview of the different types of financing needs of an agricultural firm

	Type of financing need		
Participant	**Short term**	**Medium term**	**Long term**
Farmers and other agricultural producers	Crop financing, postharvest financing, working capital	Equipment, vehicles	Land purchase
Input suppliers	Credit extension to suppliers, working capital	Equipment, vehicles	Expansion
Processors	Credit extension to suppliers of raw materials, working capital	Equipment, vehicles	Expansion
Other value-chain participants	Credit extension to customers and suppliers, working capital	Equipment, vehicles	Expansion

for the individual credit products.[10] Table 7.3 provides a summarized overview of the different agricultural financial product groups.

7.2 Putting the products in action

Building a strong and profitable agricultural loan portfolio is only possible if the bank's credit policies adequately recognize the borrowing needs of those clients so that a dynamic portfolio may be created. At the same time, policies must also be applied in a way that addresses the bank's need to control risk. This balance is best managed if the bank uses clear but flexible loan term requirements that are regularly reviewed and adjusted based on feedback from an active loan-rating system and problem loan–management process. The loan policy manual (LPM) should therefore include the following:

- Types of credit required for each segment of the value chain
- Pricing strategy
- Loan-rating system
- Conflicts of interest policy
- Loan approval levels of authority
- Problem loans–management policy
- Credit types for each segment of the value chain. This will include:

Table 7.3 Overview of agricultural financial product groups

Product name	Product description
Production/preharvesting finance	Short-term loan generally extended for purchase of essential farming inputs such as seed, fertilizer, feed, livestock, and fuel. The product is designed to accommodate start-up, development, and ongoing farm production activities.
Postharvest finance and/or working-capital financing	Short-term line of credit or medium-term revolving line of credit intended for financing of inventory, customer credit, and other time-sensitive purchases.
Warehouse receipts[11]	A warehouse receipt is a document that provides evidence of commodities storage. The receipt certifies the deposit of farm products of a particular quantity, quality, and grade. Warehouse receipts are negotiable documents and can be used as collateral to support loans.
Purchase-order finance	This is short-term financing that provides capital to businesses that can then pay suppliers up front for solid purchase orders they are ready to fill. Under purchase-order finance, the client does not take direct financing from the bank. Rather, the bank pays the customer's suppliers directly and collects directly from the purchaser on the purchase order. The purchase orders may come from commercial and/or government entities.
Real-estate mortgage financing	Finance long-term investment in real property
Equipment financing	Finance medium-term and long-term capital expenditures
Equipment lease	Rental arrangement for medium-term and long-term use of fixed (movable and nonmovable) assets
Securitization, exchanged controlled warehouses, and structured finance for farmers' production, "in the field" and physical inventory	Pre- and postharvest financing alternatives or complementors based on the following principles: Making equipment available, not liquidity Pledge of future production Loan guarantees by third parties
Credit sales and exchanged controlled warehouses based on future production	Capture the farmers' earnings through the buyer Group guarantees
Field warehousing, receivables-backed commercial paper, and collateral management for traders and processors	Shifting credit risk to traders and processors

1. *Preharvest financing.* Preharvest loans are made to farmers and cooperatives for purchasing inputs such as seed, fertilizer, feed, livestock, and fuel at the start of the growing season. The farmer receives an advance of funds for the cost of the inputs. These loans are short term and mature soon after harvesting, often less than 180 days from the time of advance. Interest is accrued on the borrower's account and paid in full at maturity.
2. *Livestock loans.* Livestock loans made to farmers to purchase feeding and breeding livestock when the moment is most advantageous for them. The loans are recommended to be short term (365 days or less). They may be structured as a demand loan, line of credit, or revolving line of credit. The important feature is that they are preapproved so that farmers may draw on them when the purchase of livestock is most opportune. The loans are structured to require repayment of principal and interest at the time the livestock is sold (bullet repayment). Real-property guarantees separate from the livestock would be required.
3. *Postharvest and working-capital loans.* Postharvest and working-capital loans are typically either a short-term line of credit or medium-term revolving line of credit intended for financing of inventory, customer credit, and other time-sensitive purchases.
4. Postharvest loans provide farmers and farming cooperatives sufficient liquidity to cover operating costs while holding an inventory of finished products until prices are advantageous for sales. Loans are made to farmers and cooperatives to support on-farm holding of products and to allow farmers to sell their nonperishable products at times of the year when prices are better. Such loans would pay off preharvest loans and allow farmers to avoid having to sell products immediately after harvest to pay their production loans. Product inventory would guarantee loans whose amount should not exceed 80% of the value of the product evaluated at harvest-season prices.
5. Working-capital loans provide other members of the agriculture chain with funds to support working-capital and operating expenses through the slower selling seasons. Preharvest and working-capital loans are structured to require repayment of principal and interest at the peak of the selling season.
    - A line of credit is a commitment from the bank to the borrower to allow the borrower to draw down and repay the loan throughout the year, with a maturity at the end of one year. Typically, the borrower repays the loan in full before the line of credit expires at the end of the year.
    - A revolving line of credit is a commitment from the bank to the borrower to allow the borrower to draw down and repay the loan for a period up to

three years. Frequently, at the end of the third year, the borrower is entitled to convert the remaining balance into a term loan with a maturity of three to five years. Experience with the line of credit would be required before such a conversion would be considered.

6. *Equipment and real-estate loans*. Equipment and real-estate loans are long-term loans for financing long-term assets and will invariably include the following types:
   - Equipment loans are term loans for fixed assets such as tractors, trucks, combines, dairy equipment, harvesters, planters, tillers, fertilizers, hay, and irrigation equipment. They are typically for five to seven years, with principal and interest payments made monthly or quarterly. The bank typically finances 70% to 80% of the equipment purchase price.
   - Real-estate loans are term loans for purchase of real property such as land, buildings, and other improvements. They come in many styles and packages, but generally they have terms of five to seven years in emerging markets. Loans are often structured with an adjustable-rate component and limited maturity (five to seven years), allowing the bank to decide whether to renew the loan with new terms and conditions at the end of the loan term. The bank typically finances up to 70% of the real property value as appraised by an independent appraiser. Principal and interest payments are typically monthly or quarterly. Such payments are adjusted in the case of agricultural loans to take into account the cash flow derived from the operation, and the loan payment is scheduled accordingly. Such transactions will require thorough knowledge of the regional regulatory environment and legal due diligence on any title presented.
7. *Equipment leasing*. Equipment leases are rental arrangements where the user (lessee) makes rental payments to the owner (lessor) for use of equipment, the title of which remains with the lessor. The lessee has the right to use the equipment unencumbered for a specified period, so long as the lessee makes rental payments according to the terms of the lease agreement. Typically, at the end of the lease the lessee returns the asset to the lessor. Interest rates and principal are implied in the rental payments. Equipment leases are a specialty area of finance and should be undertaken with caution. Equipment leasing requires special knowledge of the equipment and its residual value at the end of the lease term. That said, equipment leasing can be a profitable area of financing for a bank and can offer an alternative avenue for financing equipment for bank customers. For the customer, they can make expensive equipment available without incurring the cost of purchasing the equipment outright when there are many other demands on enterprise capital.

7.3 Loan pricing

Loan pricing is largely determined by market forces and often begins with a reference rate to which a risk margin is added. The common reference rates used by banks are (1) the London Interbank Offered Rate, referred to as LIBOR, for US dollar–denominated loans; (2) the Euro Interbank Offered Rate for euro-denominated loans; and (3) the South African Benchmark Overnight Rate for rand-denominated loans. The financial and capital markets areas of the bank are most familiar with the prevailing reference rate and can provide the most accurate, up-to-date information. Banks refer to interbank lending rates as a proxy for cost of funds. The risk margin is a rate of interest, often stated in basis points, that is added to the reference rate to compensate the bank for the credit risk of a particular borrower or type of credit exposure. To derive the risk margin, the bank must consider such factors as market conditions, competitor pricing, budgetary targets, and risk rating assigned to the borrower. For smaller agricultural loans, the bank may fix rates that are market sensitive but sufficiently high to compensate the bank for its cost of funding and operating expenses (direct and indirect), which may be higher because of a smaller loan size and the need for increased supervision by bank staff due to the vagaries of weather and the nature of agricultural activities. An advanced loan pricing strategy not only will compensate for the cost of funds and operating expenses, but will also incorporate expected loss from loans in the same risk category. A formula to measure income from the loan is as follows:[12]

Risk-adjusted income = (spread + fees – expected loss – operating costs) × (1 – statutory tax rate)

- Spread = difference between loan rate and bank's cost of funds
- Fees = commitment fees, loan origination fees, etc.
- Expected loss = expected default frequency × loss given default
- Expected default frequency = % expected default under rating system metric
- Loss given default = % of loss interest and principal resulting from default
- Operating costs = % loan officer time originating and monitoring the loan

Every loan must have a portion of bank capital assigned to it, which is described as the capital at risk. A standard measure of capital at risk is 8% of the loan amount. This process is and will be undergoing major changes due to the Basel III arrangements that are due to reach final implementation in 2019. In general the observation is that, with the exception of high-quality T-bonds, banks will have to hold more capital against loan portfolios, including loans provided to the agricultural sector. Once the bank calculates the risk-adjusted income for the loan, it can derive the return on capital. The bank then compares the return on capital with the required rate of return (also called the "hurdle rate") as specified by the bank's policy. The hurdle rate is the minimum return on invested capital that the bank will accept for any investment. A return on capital less than the bank's hurdle rate means that the bank is not generating a positive return for its shareholders.[13] Having calculated RAROC (risk-adjusted return on capital), the bank can now compare the results with the bank's statutory hurdle rate. If the RAROC exceeds the hurdle rate, the loan officer can go forward with the loan request. If it does not exceed the hurdle, the banker will need to negotiate further with the farmer to get the pricing up. The banker might pursue a higher loan rate, higher fee income, or more collateral to lower the percentage of expected loss in the event of default.

7.4 Rating loans: a key risk-management system

A sound loan-rating system (LRS) is the backbone of risk management. The LRS is the primary vehicle for classifying a borrower and identifying the risk of credit loss associated with a borrower. Through the LRS, risk management is able to identify the loan loss exposure of the bank and determine loan provisions. The LRS is also a key tool for managing exposure to any one borrower and managing the ongoing relationship with the borrower to prevent loss. However, the LRS must be sufficiently robust to allow for classification of credit exposure into appropriate categories. For example, an LRS with only three categories is too narrow to allow for sufficient classification of borrower types and credit risk. A more appropriate LRS will have at least five categories for classification of borrowers. In fact, it may serve the bank well to mimic rating categories of the major rating agencies, Standard and Poor's and Moody's. These agencies provide a broad range of rating categories and a wealth of information supporting their approach.

A robust LRS assigns a probability of loss to each risk rating category. An accurate measure of probable loss for each creditor allows for a better measure of risk-weighted capital to assign to the creditor exposure, and therefore a better measure for loan pricing. When applied to agriculture, the additional factors of weather (for rain-fed agricultural operations) and water availability for irrigated systems (which reduce risk, provided water is available at key times during the year), crop prices, import policy (dumping from Europe and elsewhere), and security considerations have to be factored into the system. The LRS must be reviewed on a regular basis to determine whether it is fulfilling expectations. Loan or obligor's ratings may change quarterly or semiannually, depending on the frequency of financial reporting or as stipulated in the loan agreement between the bank and client. In the case of agriculture, where production cycles are measured in months, a more frequent measuring may be required, and adverse conditions from various sources must be factored in at short notice. Without exception, all risk ratings must be reviewed annually to assure the integrity of the rating.[14]

7.5 Underwriting analysis for agricultural loans

Sound credit analysis and risk-mitigating factors are at the forefront of any financial institution, and it is no different for agricultural loans. The agricultural loan application often requires strong loan analysis skills to capture sufficient information about the business and its market to be able to understand and evaluate the risks of lending to that business. The following components should be expected as part of the loan underwriting process:

- Financial analysis[15]
- Proposed loan structuring
- Credit enhancement/risk-mitigation strategies
- Assessment of the business owner's management capacity and experience
- Market analysis for the product to establish risks to projections on sales
- Collateral assessment of condition, value, and potential selling price
- Reference checks for credit history and business experience

Financial analysis begins with a review of financial statements. These statements include a balance sheet, and income statement, and a statement of cash flow.

- The balance sheet provides a snapshot of the businesses financial position at a moment in time, often the last day of the year.
- The income statement provides a picture of the revenue, expenses, and profit during a given period (commonly one year). The income statement starts with revenue (sales) generated during the period and subtracts from revenue all expenses associated with generating revenue during the period.
- The cash-flow statement is a combination of the two statements, reflecting changes in the balance-sheet accounts (excluding cash and retained profit) and the addition of profit for the period.

Knowledge of crop dynamics is essential for understanding and interpreting the financial statements of a farming company. Increases in assets between the beginning and the end of the period reduce cash flow, while increases in liabilities increase the cash flow. Simply stated, an increase in accounts receivable means more cash is not collected from customers at year's end. An increase in accounts payable means more suppliers are not paid at year's end. Likewise, an increase in fixed assets means more assets were purchased during the year. An increase in long-term debt means a long-term loan was taken out during the year.

7.5.1 Credit analysis

Credit analysis usually refers to an evaluation of the five Cs of credit (see Table 7.4): character, capacity, capital, collateral, and conditions.

The client lending/investment risk breaks down into the categories displayed in Table 7.5.

Collateral is often the secondary or tertiary source of loan repayment. The bank will look for adequate collateral coverage for the loan. In the case of agricultural loans, the bank may find it difficult to realize the value of property or installed equipment located in remote rural areas. Location should be considered when deciding what value (if any) to assign to proposed collateral. The ability to repossess movable property (e.g., vehicles, tractors, harvesters, animals) also needs to be considered in assigning value (if any) to proposed collateral. If animals are taken as collateral, they need to be branded with the brand of the bank, and police/livestock-market authorities warned not to allow

Table 7.4 The five Cs of credit

Character	Integrity & temperament
Capacity	Management Technical Human & labor
Capital	Loan terms
Collateral	Production cycle
Conditions	Markets

Table 7.5 Client lending and investment risk breakdown

Client enterprise business risk	Client loan risk	Portfolio risk	Macroeconomic environment	Market risk
Enterprise ROI Capital rotation Profitability and reinvestment Capacity, skill, and technology Repayment capacity	Client and household	Market sector stability, competitiveness, and trends Systemic and unpredictable risks Institutional/investor profitability risk Transaction and operational cost risks	Currency risk Country/regional risk Land tenure risk	Competitiveness risk Market vulnerability risk Social and political risks

such animals to transit roads or to pass through markets without the brand having been appropriately canceled. For agricultural lending, two additional Cs are relevant: crop and chain. This is a critical aspect of risk mitigation.

Producers make up as much as 80% of the participants in the chain in some cases, so this layer is critical for understanding the value chain as a whole. Lending for agricultural production has specific inherent risks, such as the following:

- Long-loan terms with bullet repayments
- Production risk associated with weather and disease
- Price risk due to market volatility and weak markets
- Lack of regular monthly income for farmers
- "Side selling" by producers to cover short-term cash needs, in some cases at disadvantageous rates
- Over-indebtedness of borrowers, who turn to money lenders to meet short-term cash needs[16]

Financial institutions should familiarize themselves with other participants in a value chain as part of the credit-analysis process. For example, input suppliers who deal with producers in key local agricultural products may be a potential client segment for the bank. Their primary cash needs and selling seasons will align with the planting season. Suppliers of fertilizer, for example, may require a working line of credit to provide short-term supplier credit to farmers. Understanding the cash flows and seasonality of the underlying crops will provide the lender better insight into potential credit needs of the fertilizer supplier.

Each risk review normally contains eight chapters:[17]

- Management
- Operating margins and cash flow
- Balance sheet
- Competitive advantages
- Industry and market
- Credit history
- Collateral and secondary source of repayment
- Foreign exchange exposure

7.5.2 Structuring the loan

The structure of the loan can be critical to the success or failure of the borrower in meeting loan payments. Agricultural loans in particular must be structured to suit cash-flow needs of the borrower. Because agricultural production is by its nature seasonal, loans must be structured to allow repayment when cash flow is at its peak. For example,[18] the bank may offer a crop- or livestock-production loan. These loans generally provide funds at the beginning of the growing season to allow farmers to purchase inputs such as seed, fertilizer, feed, livestock, and fuel. The farmer receives an advance for the cost of inputs, usually in tranches of 40% at the time of purchase and 60% when the crop insurance report (if insurance is available) is issued. The bank may withhold fees from advances. These loans are short term and mature shortly after harvesting, often less than 180 days from the time of advance. Interest is accrued on the borrower's account and paid in full at maturity.

7.5.3 Credit-enhancement and risk-mitigation strategies

Simple enhancements may include taking collateral in the assets of the farm or business, requiring personal guarantees of key owners/management, and requiring that all operating accounts be maintained at the bank. As discussed in the relevant chapter, for farmers, crop and livestock insurance are important forms of risk mitigation, as are property and life insurance policies. If insurance is locally available, the bank should insist that the borrower have such insurance and that such policies are pledged to the bank. Another form of credit-enhancement includes third-party guarantees. Commonly, government-sponsored programs provide risk-management support to agricultural producers to encourage crop and livestock production and to insure against production losses for specified perils (weather, pests, and disease).

But there are more basic things that provide risk mitigation. Having full confidence in the capabilities of the owner/management is essential. The bank must have confidence that the owner/management can conduct business affairs successfully. Besides a full due diligence of the farm, farmer, his family, and coworkers, it is often recommended to rely on third parties, suppliers, agents, and other stakeholders and leverage on their knowledge base. This can include the following:

- University research, including market analysis for products to establish risks to projections on sales.
- Sector and country Strength, Weaknesses, Opportunities and Threats (SWOT)[19] analysis of the agricultural sector. With this analysis, together with the crop-dynamics analysis, the FI can assess whether the client's projections of sales are reasonable and achievable. The banker should never develop projections or impose projections on a client. Projections are the responsibility of the business, not the banker.
- Industry associations and development banks research.

7.5.4 Valuing collateral and execution value

Many farmers also have urban property and nonagricultural businesses. This property and assets of other businesses should also be considered for use as collateral. Collateral for farmers and other agricultural businesses may include assets such as livestock, trucks, tractors, combines, grain elevators, buildings, and land. The type and amount of collateral will largely depend on the type and size of the loan. For example, a preharvest loan might require land and livestock or warehouse receipts as collateral, while a real-estate loan will require security of the real property and plant.

There are several ways to value collateral: historical purchase price, cost to buy/reconstruct the asset today, prevailing market price for similar or comparable assets, and liquidation value of the asset if it were sold immediately. The values can come from book entries on financial statements, purchase and construction contracts, markets, appraisals, asset audits, and liquidation experts. The value applied to collateral is often more art than science.[20] Each approach is valid and probably will result in significant variation in price for the same asset. In the end, however, the value of collateral at the beginning of the loan process is less important than the value of the collateral in a worst-case situation (loan workout). It is important, however, that the collateral be properly secured and sufficient to repay the loan if the circumstance should arise. The value of rural property should be downgraded based on its location and to the extent that an easy sale may not be possible to realize the purported value of the asset. Land ownership and titles are problematic in some developing economies and present logical problems for the lender.

7.5.5 Reference checks

A reference check entails little more than contacting customers, input suppliers, agricultural extension agents, business associates, employees, and creditors and asking simple questions about the prospect. For example, the banker calls on the prospect's suppliers and asks about the prospect's payment history, timeliness of payment, number of orders, and payments during the year. The banker also calls on customers and asks questions about the prospect's reliability, quality of product, speed of delivery after purchase orders, and collection history.

7.5.6 Monitoring

Although agricultural lending is similar to other types of credit, it can be challenging due to its unique business cycles based on diverse crop requirements. Often the monitoring includes (1) individual loans and (2) loan portfolios, (3) nonperforming loans (NPLs), and (4) an action plan in case of overdues (different categories for 30, 60, 90, and 180 days, etc.) and execution guidelines on the collateral available. An organized and systemic loan-monitoring system will help in understanding possible issues in terms of loan repayments and whether those issues are client or industry specific.

7.5.7 Restructuring

Within agricultural lending, restructuring is not used as often as it could be to assist with delinquent loans. Delinquent clients must be divided between those who are "unwilling" to pay, those who are willing but "temporarily unable to pay," and those who are willing but have permanently lost the ability to generate profitable cash flows. Among these three types of clients, restructuring is appropriate only for the second type. A related question is whether the collateral held on the loan is maintaining its value in distress, or is it decreasing at a much quicker rate than the loan pay-down? With agricultural loans, this type of arrears is often due to a weak harvest, so the bank has to weigh a few additional factors:

- Can the client carry the interest throughout the next harvesting period?
- Can the client pay down the loan to a point that is manageable with a strong follow-up harvest?
- Does the client have additional sources of income that can offset the risks of continued reliance on the next harvest?

Loans are typically written off once they have reached 180 days in arrears.

7.6 Loan portfolio management

Careful monitoring of the agricultural loan portfolio will allow bank management to better control the portfolio's risks. For portfolio monitoring, the purpose is to continually gauge and analyze the overall portfolio's real level of risk and how this is related to the overall business strategy. For example, many portfolios can suffer from rising concentration risks due to recent portfolio growth within a sector or geographic region. Portfolio management in the broader sense deals with the integrity of the loan portfolio. From a portfolio point of view, the primary areas of risk are the following:

- Geographic risk
- Industry risk
- Product risk
- Collateral risk
- Loan personnel risk
- Overall economic risks, particularly within distinct sectors

Overall, the important question is whether resulting weaknesses are caused by personnel inadequacies, system inadequacies, or general economic weaknesses.

7.7 IT and MIS (management information systems)

Appropriate use of MIS systems to gather and analyze agricultural portfolio data is a core principle of financial-institution management, and is one of the primary success factors that can make a difference. MIS systems have moved from being a business enabler to being a business driver for financial institutions and play a vital role in achieving the strategic objectives of financial institutions entering the agricultural segment. However, there is often a disconnect between what MIS software provides and the actual management and operational needs of the agricultural business line. It is of critical importance to quantify and define reporting outputs required to operate a profitable agricultural business line.

7.8 Loan application and repayment capacity

Small-farm businesses have different characteristics from bigger companies. They are family based, they involve multiple economic activities and income, and expenditure is generally shared. There is often no clear distinction between the family and the business. The repayment capacity of a borrower depends on whether there is enough cash available in the "family pot" to service the loan. Loan installments normally do not represent "earmarked" funds but are simply taken out of the cash reserves of the household. Cash-flow analysis is the single most important analysis a lender has to do. The diversity of enterprise activity on a small farm makes it seem complicated, but it can and must be done. Some of the cash flows will be regular, while others will be irregular. For agricultural producers, most production-related cash flows are irregular, that is, seasonal in nature. Regular income may come from petty trade or the regular employment of some family members, although even trading activities may peak around certain dates. In order to make realistic cash-flow projections, the following items should be observed (besides the normal uncertainties and seasonality and price volatilities):

- Extraordinary reasons for payment delays.
- Understanding and underestimating household expenses.
- Listing unrecorded debts or commitments that are intrinsically debt commitments (lease, etc.).
- Rotation of crops.[21] Appropriate crop rotation is important to maintain soil fertility and structure, control diseases and pests, and facilitate weed control. If this is not done properly, soil quality deteriorates, resulting in decreasing yields and, hence, income.
- Erosion control. Erosion results in a loss of land for production. To prevent erosion and maintain the maximum area of fertile land possible, there are many

different techniques. These range from reducing grazing pressure to the introduction of cultivation practices like terracing or planting trees and hedges.
- Crop storage. Selling the harvest at the right moment is difficult for many farmers. Only few have sufficient and appropriate storage facilities to keep the harvest for a longer period and benefit from higher prices later. Most farmers are forced to immediately sell their harvest. Therefore it is important to know whether the loan applicant might face marketing constraints.

Once a cash-flow projection[22] has been prepared for all the economic activities of all household members, and all the family expenditure has been incorporated, it needs to be assessed in relation to the loan proposal. The most commonly used indicators for doing this are as follows:

- The accumulated repayment capacity.
- Net cash flow after loan repayment, or "free net cash flow." The accumulated repayment capacity indicator is calculated by adding up all the monthly balances during the envisaged loan term and comparing this figure to the total amount to be repaid (including both principal and interest). Since it is advisable to have a substantial security cushion for unforeseen events, it is recommended that the ratio be at least 2:1.

However, accumulated repayment capacity is more important than monthly free net cash flow. As many farm households have a highly variable income and expense structure, loan products that require equal repayment installments are not really appropriate. Ideally, a more flexible repayment schedule is required. A number of variations are possible:

- Monthly interest payments combined with a lump-sum repayment of the capital at the end.
- Various irregular payments of interest and capital.
- The entire loan amount plus interest paid at loan maturity.[23]

In these instances the monthly free net cash flow will not be of much help. However, the accumulated repayment-capacity ratio will be very critical to decide whether a loan should be approved or not.

In order to find out how cash flow might be affected by adverse factors, the loan assessment may include a sensitivity analysis. The objective is to know whether adverse circumstances would undermine repayment capacity to such a degree that loan repayment would be at risk.

Factors to be considered in the sensitivity analysis of cash-flow projections could include the following:

- Reduced yields due to bad weather conditions, diseases, or pests
- Delays in payments, for example, delays in payments for crops after harvest
- Lower-than-expected sale prices
- Higher input costs
- Additional labor costs, for example, replacing a sick family member with hired labor

When analyzing cash flow, the following risk-mitigating techniques must be taken into account as part of the risk profile of a farm household:

- Diversification of income sources. Prudent farmers tackle income insecurity by diversifying income sources.
- Liquidation of assets. The majority of farmers have savings in kind. One of the most popular forms of saving is buying livestock.

7.9 Microfinance and innovative finance for agriculture

Agricultural finance has been one of the most prominent elements of the rural development strategies used by development agencies and national governments. But this financing has long been characterized by poor loan repayment rates and unsustainable subsidies.[24] Accordingly, agricultural credit from some donors and multilateral development banks has dropped dramatically in recent decades and is now often considered too risky. The renewed international emphasis on poverty reduction has put rural populations, particularly agricultural households, back in the spotlight of development efforts. Agricultural development programs often include credits for agricultural production, which has renewed the debate about how to provide finance in rural areas. Traditional providers of agricultural finance insist that it is time to recognize their role in specialized lending to meet the crop-based cash-flow cycles of small farmers, now that microfinance institutions have successfully expanded into rural areas with their one-size-fits-all techniques. Microfinance institutions have generally managed default risk very well, while traditional agriculture lenders have developed products that respond well to the cash-flow cycles and marketing relationships of farming communities. But it is important to remember that, for many small farmers, the main source of credit is not a bank or even a microfinance institution, but agribusiness actors such as input suppliers (for example, sellers of seed or fertilizer), traders, and processors. Moreover, self-finance continues to play a vital role in agricultural production. Agriculture is widely considered more risky than industry or trade.[25] Thus it is not surprising that agricultural lending projects have had poor repayment performance. Such risks are higher for farmers engaged in monoculture of crops that are particularly sensitive to the correct use of high-quality inputs or the timing of harvesting. Risk in agriculture can also be traced to farmers seeking to increase their incomes through higher-risk, higher-return cropping strategies. Complicating the scenario is the fact that decision making in agriculture is not an exact science; it depends on many variables that change from year to year and are beyond the farmer's control. With the entry of new players, growing competition in international markets can fundamentally change the competitiveness of a local industry. The precision of crop schedules generates specific risk for agricultural finance. Loan disbursements need to be tailored to irregular cash flows; yet the timing of final crop income may vary based on when farmers choose to sell. (They may delay selling until market conditions are favorable.) These characteristics of agricultural production require lenders to be quite efficient and physically close to their farmer clients. Thus, for banks and other financial institutions, agricultural lending involves a risk of causing default due to their own inefficiency. The production of most improved cash crops is relatively complex, involving careful timing of numerous steps, from preparing land through planting, fertilizing, and harvesting. Mistakes or delays at any step can substantially reduce returns or eliminate them altogether. An agricultural finance model has been developed that combines the most relevant and promising features of traditional microfinance, traditional agricultural finance, and other approaches, including leasing, area-based insurance, use of technology and existing infrastructure, and contracts with processors, traders, and agribusinesses, into a hybrid defined by ten main features. Most of the features address issues specific to financing agriculture, some respond to the general challenges of operating in rural areas, and some reflect good practice in delivering small unsecured loans:[26]

- Repayments are not linked to loan use. Lenders assess borrower repayment capacity by looking at all of a household's income sources, not just the income (e.g., crop sales) produced by the investment of the loan proceeds. Borrowers understand that they are obliged to repay whether or not their particular use of the loan is successful.
- Character-based lending techniques are combined with technical criteria in selecting borrowers, setting loan terms, and enforcing repayment. To decrease credit risk, successful agricultural microlenders have developed lending models that combine reliance on character-based mechanisms, such as group guarantees or close follow-up on late payments, with knowledge of four-crop production techniques and markets for farm goods.
- Savings mechanisms are provided. When rural financial institutions have offered deposit accounts to farming households, which helps them to save

funds for lean times before harvests, the number of such accounts has quickly exceeded the number of loans.

- Portfolio risk is highly diversified. Microfinance institutions that have successfully expanded into agricultural lending have tended to lend to a wide variety of farming households, including clients engaged in more than one crop or livestock activity. In doing so, they have ensured that their loan portfolios and the portfolios of their clients are better protected against agricultural and natural risks beyond their control.
- Loan terms and conditions are adjusted to accommodate cyclical cash flows and bulky investments. Cash flows are highly cyclical in farming communities. Successful agricultural microlenders have modified loan terms and conditions to track these cash-flow cycles more closely, without abandoning the essential principle that repayment is expected, regardless of the success or failure of any individual productive activity, even the one for which the loan was used.
- Contractual arrangements reduce price risk, enhance production quality, and help guarantee repayment. When the final quality or quantity of a particular crop is a core concern, for example, for agricultural traders and processors, contractual arrangements that combine technical assistance and the provision of specified inputs on credit have worked to the advantage of both the farmer and the market intermediary.
- Financial-service delivery piggybacks on existing institutional infrastructure or is extended using technology. Attaching delivery of financial services to infrastructure already in place in rural areas, often for nonfinancial purposes, reduces transaction costs for lenders and borrowers alike and creates potential for sustainable rural finance even in remote communities. Various technologies show enormous promise for lowering the costs of financial services in rural areas, including automated teller machines (ATMs), point-of-sale devices linked to "smart cards," and loan officers using personal digital assistants.
- Membership-based organizations can facilitate rural access to financial services and be viable in remote areas. Lenders generally face much lower transaction costs when dealing with an association of farmers as opposed to numerous individual, dispersed farmers if the association can administer loans effectively. Membership-based organizations can also be viable financial-service providers themselves.
- Area-based index insurance can protect against the risks of agricultural lending. Although government-sponsored agricultural insurance schemes have a poor record, area-based index insurance, which provides payouts linked to regional levels of rainfall, commodity prices, and the like, holds more promise for protecting lenders against the risks involved in agricultural lending.

- To succeed, agricultural microfinance must be insulated from political interference. Agricultural microfinance cannot survive in the long term unless it is protected from political interference. Even the best-designed and best-executed programs wither in the face of government moratoriums on loan repayment or other such meddling in well-functioning systems of rural finance.

8
Farmland I: An Investible Asset Class?

Farmland has gained significant attention in recent years by investors.[1] Not surprisingly, since the start of the financial crisis, investors have been faced with economic uncertainty and poor investment visibility, periods of significant price volatility in many (mainstream) asset classes, inflation-related concerns, and dismal returns on fixed-income products (with only a few notable exceptions), and overall the fact that the monetary strategies deployed by many central banks in the world have perverted the proper pricing of risk of pretty much every asset around as pricing of assets ultimately feeds into the central banking rate. No wonder most investors were or are looking for hard tangible real assets that have a core capital value embedded, preferably with a pricing mechanism that shows no correlation with securities on the mainstream financial markets, where capital growth is driven by solid fundamentals, and where direct ownership (versus paper ownership) is a distinct possibility.[2] Productive agricultural land ticks all those boxes, explaining the frenzy around agricultural land worldwide and the snapping up of land in often developing or even frontier nations by large overseas investors with a view toward benefiting from capital appreciation of farmland.

8.1 Qualifying the opportunities

Investors may either decide to acquire productive land that is leased to a tenant farmer, or they may go down the route of engaging in a more participatory role, sharing in harvest revenues under a contract farming agreement. Another option could be the acquisition of land with agricultural potential and adding value through conversion into productive agricultural assets, including investing in on-farm infrastructure such as roads, storage, and irrigation. These types of deals often occur in emerging economies where infrastructure needs catch-up or a technological upgrade to enhance efficiency.

In their quest for superior risk- and inflation-adjusted returns, farmland, in contrast to other hard assets like gold, has the benefit that it has investment fundamentals that can be judged and evaluated. These fundamentals are driven by annual income, capital appreciated based on increased demand for arable land, and enhanced crop yields. From that perspective, farmland often is a way for investors to diversify their portfolio away from the traditional asset classes and their enhancing levels of correlation, in particular in downward-moving markets where correlation can spike to 80% or more under severe distress.

During the last decade, the best financial deals where those that involved the development of previously unused land into productive agricultural properties through the establishment of transport infrastructure, irrigation, and soil-quality controls. This way substantial capital value can be added to land assets, and it also generates an annual income stream which can be reinvested or utilized to cover the cost of development financing.

8.1.1 Driving value through investment fundamentals

Farmland has, in contrast to some other hard assets, investment fundamentals that can be judged. The drivers behind those fundamentals are mentioned on various occasions and coincide with the investment fundamentals of other areas of the agricultural investment spectrum. In short they come down to any or a combination of the following dynamics: (1) global population growth, (2) changing dietary habits, (3) agricultural productivity (which is both an issue and an opportunity), (4) changing climate patterns, (5) the rise of biofuels production and consumption, and (6) the increasingly limited availability of farmland.

Many of the elements of these investment drivers are discussed across the different chapters of this book, so in this context I will limit the discussion to a number of specific features that relate directly to farmland fundamentals and summarize some of the underlying data.

1. **Population growth**: the human population has increased by nearly 400% during the past 100, years with a net increase of 225,000 new people per day during the last decade. Currently, more than 5% of the total number of people that have ever lived are alive today, and over 1 billion people have been added to the global population in the last 12 years. On the current trend, the global population is projected to be roughly 40% higher than today by 2050.
2. **Changing dietary habits**: a rising middle class, higher income levels, and urbanization (particularly in emerging economies accounting for 80% of the world's population) have led to a higher-protein diet, a more resource-intensive diet, and millions of new meat eaters. The latter is particularly a problem. Let's provide some context: While livestock production has

historically been supported by grazing and crop/food waste, an increasing demand for meat has led the global livestock industry to become increasingly reliant on grain as a primary livestock feed. In modern intensive livestock farming where the majority of feed is grain based, seven kilograms of grain is required to produce one kilogram of beef. On a global average basis, given that part of the production is based on other sources of feed, such as grazing land and organic waste, three kilograms of grain is required to produce one kilogram of meat.[3] As meat production now depends on grain as a key input, any increase in demand for meat results in an acceleration of demand for arable and grazing land area. At least 35% to 40% of all cereal produced in 2008 was used as feed for livestock. This leaves an estimated 43% of cereal production available for human consumption after losses from harvest, postharvest, and distribution are taken into account.

In percentage terms, the effect of increased income on diets is greatest among lower- and middle-income populations, which currently consume the lowest percentage of animal products.[4] This indicates great potential for increased meat demand on a global basis given that low-income countries, which account for 5.1 billion of the world's population, consume less than half as much meat (as a percentage of dietary energy intake) as high-income countries, which account for only 1.3 billion of the world's population.[5]

3. **Agricultural productivity**: With more than 20% of the global population undernourished,[6] the options are limited. Only three possible solutions exist: bring more land under production, increase yield from existing land, or a combination of both. The latter is more a necessity than an option. Higher demand than supply dynamics allow for productivity gains to be achieved either through developing more arable land or enhancing crop yield per unit of arable-land cultivated. Now in 2013 there is very little suitable land left to be developed, and the application of fertilizers, herbicides, and pesticides results in annual yield increases of less than 1%. In the 1960s, agricultural yields increased by 3.5% per year, but by the 1990s annual increases had fallen to only 0.5% per year, despite the fact that global use of nitrogen fertilizers during the period increased by approximately 700% and water use doubled. Put differently, with regard to agricultural productivity, the world reached a tipping point sometime in the early part of the last century and the 1980s respectively, from a state of increasing supply of food and farmland availability, to a state of decreasing supply relative to the population. We have entered an era where, for the second time in recent human history, per capita food supply is in decline (the first time being prior to the Green Revolution when there was widespread starvation in Asia and Africa), and this is all happening at a time when climate change threatens to constrain both further expansion of agricultural land as well as yields on existing lands.

4. **Changing climate patterns:** Climate change, whether man-induced or not, impacts agricultural productivity through soil degradation, water scarcity, rising global temperatures, and as a result of climate-driven weather events. Rising surface temperatures have had the effect of distorting normal rainfall patterns and creating a greater frequency of extreme-weather events. Based on the FAO, one can conclude that the total number of natural disasters (excluding earthquakes) has increased by approximately 1,000% between 1960 and the present time, with the number of food emergencies roughly doubling since 1980. It is estimated that on average, yields of the dominant regional crops may fall by 15% to 35% in Africa and Western Asia once temperatures rise by three or four degrees.[7]
5. **Biofuels:** As agricultural production both consumes and produces energy, agricultural and energy markets have always adjusted to each other, with output and consumption rising or falling in response to changing relative prices. Rapidly increasing demand for biofuels is now tying agriculture and energy together more closely than ever. There are two reasons for this correlation: First, higher oil prices drive demand for alternative fuels, which require feedstock such as corn or maize produced on agricultural land. Second, oil and natural gas are key agricultural inputs, and as such, fluctuations in price and availability of fossil fuels could have a major impact on agricultural economics and productivity. The United States is the largest producer and consumer of bioethanol, followed by Brazil, which now uses 2.7 million ha of land for biofuels production, equivalent to 4.5% of its cropland area, mainly planted to sugarcane. Globally, biofuels, including bioethanol (mainly from sugarcane and corn) and biodiesel (mainly from soybean, palm oil, and other oilseed crops), accounted for roughly 1% of total fuel consumption for road transport in 2005, and it may reach 25% by 2050. The higher the oil price, the more economically viable biofuel production becomes (even without subsidies or climate-change-mitigation incentives) and the greater will be the competition for cropland. At oil prices above the 50 US dollar mark, biofuel production becomes capable of producing significant profits, thus creating stiff competition and forcing up the price of maize, wheat, and other feedstock crops.
6. **Available farmland:** The expected shortage is caused by a combination of two factors: urbanization, which I expect to rise from 50% to 80% globally over the next 30 years, and soil degradation. What makes matters worse is the fact that urbanization tends to occur at the expense of cropland, as cities tend to develop around the most agriculturally productive locations. A total of 40 million hectares of cropland are expected to be lost by 2030, and another 27 million by 2050. Deforestation and inappropriate agricultural practices have resulted in approximately 2 billion hectares of agricultural land becoming degraded.[8] Some estimates suggest that at current

Table 8.1 Global overview of erosion and degradation dynamics

Major types and causes of soil degradation			
Types (%)		**Causes (%)**	
Water erosion	56	Overgrazing	35
Physical degradation	4	Agriculture activity	27
Chemical degradation	12	Industrial activity	1
Wind erosion	28	Deforestation	30
		Overuse of vegetation (e.g., fuelwood)	7

Source: FAO/UNEP.

Table 8.2a Major types and causes of soil degradation

Causes/ million hectares	**Europe 219**	**Asia 7 47**	**Africa 494**	**South America 243**	**North and Central America 158**	**Southwest Pacific 103**
Water erosion	115	440	227	123	106	83
Wind erosion	42	222	187	42	39	17
Chemical degradation	26	73	61	70	7	1
Physical degradation	36	12	19	8	6	2

Source: FAO/UNEP.

rates, up to 30% of all agricultural land will be unusable by 2020. Annually, the global rate of land degradation, which is due chiefly to soil erosion, is estimated to be between 20,000 and 50,000 km^2. Also, due to overgrazing, compaction, and erosion from livestock, some 70% of all grazing land in dry areas is considered degraded. Some of the data sets are reflected in Tables 8.1 and 8.2a.

8.1.2 Evaluation of investment dynamics

a. Preservation of capital

Although positions of investors change over time, one of the fundamental drivers of long-term investors is the "preservation of capital." The recent financial crisis has also driven many investors to prefer preservation (or minimal volatility) over return generation. Studies have shown that even taking into account the transaction costs associated with property transactions, farmland still constitutes a substantial portion of the optimal portfolio across a wide range of scenarios.[9] Farmland tends to hold its value over time, although questions are being asked about the alpha that is still in the purchase price at the levels of farmland prices anno 2013.[10] Undoubtedly, there will be upward dynamics on those land prices, but they often don't happen in a monotonic upward-sloping function.[11] Most speculative bubbles are characterized by a few key features common to most speculative bubbles. The fundamental conditions that give rise to bubble formation are as follows:

- Bubbles result from human behavior. Traditional market participants stop behaving as they have traditionally.
- Buying decisions are no longer supported by traditionally sound underlying economic fundamentals.
- Uninformed market participants begin bidding up the price of an asset.

The fundamentals (see Table 8.2b) most important in spotting a potential bubble are economic conditions (supply and demand), financial soundness (profitability, leverage); and market drivers (presence of speculation). The best proxy for farmland values is the price of corn. It should be noted here that corn is used as a proxy for farmland values because its price is the underlying force driving profitability. Corn is, in essence, the driver of all crop prices – as the price of corn goes, so goes the price of all other crops.[12]

Some recent research indicates that (1) sector fundamentals appear very strong,[13] (2) farm-sector profitability is at an all-time high, (3) the farm sector has very little leverage, (4) income-capitalization rates are improving, and (5) farmland is very widely held, with institutional ownership representing less than 1% of US farmland.[14]

b. Protection against inflation

Agricultural assets and farming land in particular show a positive correlation with inflation. It therefore constitutes an effective hedge against the effect

Table 8.2b Evolution of arable land per capita (hectare per 1000 people)

Region/year	**1975**	**1985**	**1995**	**2005**	**Delta over period**
Australia	3,050	2,995	2,200	2,600	–17%
Canada	1,850	1,790	1,500	1,480	–25%
Russia	990	930	870	860	–17%
United States	900	850	750	680	–32%
Brazil	450	400	420	390	–16%
India	300	250	230	210	–45%

Source: FAO and World Bank.

of inflation of the value of investment portfolios. Between 1941 and 2002, average farmland values increased by almost 2% more than the average rate of inflation over that time period.[15]

c. Income-generating capacity

There are two different channels through which farmland generates income: either through lease payments collected from tenant farmers, or as a percentage of harvest revenues under a joint venture between landowner and farming operator, known as a contract farming agreement (CFA). Farmland enjoys a near 100% tenant occupancy rate, as demand for quality farmland is always high, regardless of the economic environment, ensuring that farmland incomes remain stable throughout all market conditions, thus further dissipating income volatility within a diversified investment portfolio. CFA arrangements carry more risk than simple lease agreements with fixed annual payments, as the investor is exposed to commodity prices and agricultural yields, therefore benefiting from good times and sacrificing returns when conditions are less attractive. Yields vary depending on location, business strategy, and a host of other endogenous factors that affect agricultural economics and farmland profitability. Annual yields vary from 1.6% to 8% depending on the location and type of cultivation. On the other hand CFAs can generate annual yields of up to 20% in many regions, including Australia, Latin America, Africa, and Asia, while in less than perfect conditions it is possible for investors to turn a negative yield if harvest revenues are insufficient to cover the cost of inputs and labor.

d. Superior risk-adjusted total return

Investing in agricultural land captures both operating profits and capital growth through a combination of rental income and appreciation in the value of the asset. When taking into account total returns, the investment performance of farmland has repeatedly outperformed mainstream assets including stocks, bonds, and commercial real estate across a wide range of markets and timescales, despite relatively low levels of risk (measured in terms of the standard deviation of annual rates of return) – see infra.

e. Hedging tool in recessionary times

Both income streams and capital appreciation have stayed relatively stable over longer periods of time. It performs comparatively well during times of market uncertainty, thus acting as an ideal recessionary hedge. A comprehensive study on US farmland conducted in 2002 compared the effects on portfolio efficiency of including farmland in a mixed-asset portfolio under market conditions of certainty and uncertainty. It concluded that, in both certain and uncertain world models, farmland can be shown to improve portfolio efficiency.[16]

f. Portfolio diversification

A number of studies have shown that farmland returns have a low or negative correlation with traditional asset classes such as stocks and bonds and only a modest positive correlation with commercial real estate. A study in the United States, using data over a period of 33 years up to the 1980s, considered six asset classes, including farm real estate, large and small capitalization stocks, long-term corporate bonds, and treasury bills. The study concluded that inclusion of farmland in the portfolio had highly attractive characteristics, particularly in view of the low correlation with other assets in the portfolio, especially large capitalization stocks (see infra). These characteristics have the potential to make farmland an attractive diversification tool that can help reduce the impact of broader market volatility on a diversified portfolio. The farmland component can be further diversified by varying crop types, management styles, and geographic distribution within the portfolio. In a direct-ownership structure, investors can acquire farmland across a range of farms in different countries and/or climate zones and under different asset managers.

A diversified portfolio will have exposure in key farming regions, creating geographic and commodity diversification as well as a globally competitive portfolio of crops. Farmland investing is broadly categorized by row (annual)[17] and permanent[18] crops/cropland.

This can lead to the portfolio-development matrix depicted in Table 8.3.

Table 8.3 Portfolio-development matrix

	Risk tolerance		
Portfolio factors	**Minimal**	**Moderate**	**High**
Asset mix	100% row, 0% permanent	50% row, 50% permanent	0% row, 100% permanent
Management style	100% leased, 0% operated	50% leased, 50% operated	0% leased, 100% operated
Geographic/ commodity diversification	Broad diversification	Moderate diversification	Limited diversification

g. Tax-efficient investment

In many regions of the world, in both developed and developing economies, there are a range of tax-related incentives associated with agricultural properties. These may result in favorable treatment across one or all of the standard taxes (such as income taxes and capital gains taxes) that would normally have an adverse effect on returns in other asset classes. In some instances, there are also special exemptions with respect to inheritance taxes that may make farmland particularly attractive for estate-planning purposes. Some countries have additional incentives for forestry-related usage of farmland.[19]

8.1.3 Investment performance

Farmland investments capture both capital growth and income. Lower but more stable income is achieved through rentals, and higher, but potentially more volatile returns are achieved through exposure to harvest revenues as part of a joint venture. Farmland returns in emerging markets also tend to outperform more mature markets due to lower asset prices, lower labor costs, and in many cases more fertile land and better growing conditions.

What follows is an attempt to summarize the wide portfolio of research available with respect to the performance of farmland in different parts of the world. I will reference the studies in the endnotes to enable further reading and analysis. Having said that, there has been massive investment in agricultural assets including farmland, handling and trading, technology, fertilizer, and others. Studies about investing in farmland have been extensive but have limited their focus to investing in nonfarmland agricultural assets. Recent research,[20] however, analyzes the role of farmland and other agricultural investments in class-specific portfolios. The Copula-VaR and Copula-VaR with restrictions methods have been used to find, compare, and contrast the optimal portfolio compositions among US farmlands, classified agricultural equities, and grain futures. The results illustrate that farmland is attractive as an investment. However, as risk tolerance is increased, a shift to other agricultural assets would potentially bring higher returns.

These studies conclude the following:

Farmland is one of the more limiting agricultural inputs. The greatest prospect for future returns in farmland are likely greatest in regions/countries going through new technology adoption whereby input values become capitalized in land from past technology and new technology increases value of inputs (land). Investment opportunities are similar in other agricultural inputs that reduce production and marketing costs, and improve technical efficiency of growers. These include fertilizer, farm machinery and equipment, biotechnology, seeds/germplasm, storage as well as investments in facilitating functions including food safety, GPS and farmmanagement...

The purpose was to analyze the role of farmland and other agricultural investments in class-specific portfolios. We derived optimal portfolios comprised of farmland and other agricultural investments to determine the extent that these are included in efficiently derived portfolios of agricultural assets...

The results indicate that land dominates the portfolios, particularly for lower risk tolerances. As risk tolerance increases, returns increase (substantially), and the composition of assets changes...

In conclusion, these results again support the robustness of agriculture, broadly defined to be inclusive of land, technology, seeds and traits, logistics, etc., is performing well as an asset class.[21]

What follows is a review of the investment performance per region of the world,[22] accompanied by Figure 8.1 reflecting the historical average price evolution per continent.

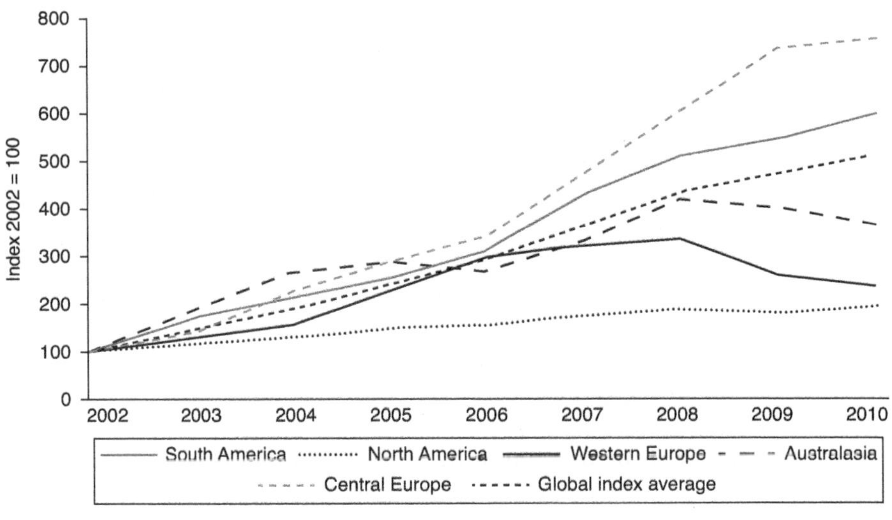

Figure 8.1 Evolution of farmland prices per continent
Source: USDA, Eurostat and various data estimates.

a. North America

The primary measure of farmland-investment performance in the United States is the National Council of Real Estate Investment Fiduciaries (NCREIF) Farmland Returns Index. The index provides investors with a measure of the investment performance of a large pool of individual agricultural properties acquired in the private market for investment purposes.

According to the index, US farmland returned 8.6% in 2010, 14.55% in 2011, and 17.59% in 2011. Also 2013 started off in sixth gear with an increase of 5.44%. Regional US farmland growth figures vary from state to state. A new report by the Federal Reserve Bank of Kansas City showed a 12.6% increase in mountain states farmland values over 2011. The Minneapolis Federal Reserve Bank District reported farmland values as of the second quarter of 2011 up 17% from the same period a year before, while the Kansas City District reported farmland prices up 20%.[23] Nebraska has seen one of the largest increases, with nonirrigated land up 30%. Oklahoma ranch land suffered from a prolonged drought, and values were up just 6.4%. What little value increase there was, was driven by oil and gas exploration. For 2012 and 2013, reports or estimations show the same upward momentum as the NVREIF.

However, there are some ownership restrictions in the United States; for example, there are some areas of the Midwest where companies and foreigners cannot buy land, which is one reason, in addition to the lower population density, why US land values look relatively good compared to those in Western Europe.

In Canada, average values have continued to rise steadily (infra). In western Canada, states such as Saskatchewan do not permit foreign ownership of agricultural land, while in some other states foreigners can own up to 49% of the shares of a company that owns land. In eastern Canada, however, there are no restrictions, and this is reflected in the higher land values.

As previously mentioned, there has been (and is) some concern among the agricultural community in the United States (and beyond) that land values have spiraled out of control, with demand for assets fueled almost entirely by investors seeking to diversify out of the stock market and into tangible assets. About 60% of all farmland is being purchased by active operators, with 15% purchased by nonlocal investors, 13% by local-area investors, 7% by institutions and investment groups, and 5% by other entities. In short, the same concern exists with respect to the presence of financial investors in the future commodities sphere.[24]

In Canada, Farm Credit Canada (FCC) monitors the value of a basket of 245 benchmark farm properties every six months.[25] On average, Canadian farmland increased 9.5% for the year ending June 2011. The average value of

Canadian farmland increased 10.0% during the second half of 2012. Farmland values remained stable or increased in all provinces.[26] Canadian farmland values have continued to rise over the last decade and continued to do so in 2013 at record rates. The average national increase of 10.0% in 2013 was the highest since FCC began reporting on farmland values in 1985. The second highest increased was 8.6 % in H1 of 2012.

b. Europe

Farmland values across emerging economies in Europe continued to perform well during 2011–2013 as increased investor appetite for productive agricultural assets, as well as active agriculture investment policies from China and Arab states who are interested in bolstering food security, has provided stable short-term pricing support.

Throughout Europe there is little data available to effectively measure the performance of farmland investments. In Central and Eastern Europe, prices tend to be only a fraction of the prices in Western Europe but are moving consistently upward (at rates between 5% and 15%) as is the case in Western Europe.[27] Across the Central and Eastern Europe (CEE), there are restrictions in place on the foreign ownership of farmland for set derogation periods post EU membership. For example, in Poland 12 years and Hungary 7 years, with a possible extension of 3 years. However, in most cases, local CEE companies owned 100% by foreigners can buy land, with some exceptions, including Hungary. Values in Hungary have been slower to grow compared to some other CEE markets due to the general heavier ownership restrictions imposed on this market. Overall, they are large countries with rich agricultural soils. With the correct investment in resources and infrastructure (roads, ports, and processors), they are ideally positioned. The trend in Western Europe was upward, but the recession has taken the steam out of that uptrend, with the Netherlands, Ireland, and Denmark taking the biggest hit, but they had experienced the highest increases as well. Other Western countries have been experiencing stable prices or even upward prices for many years.[28]

Russian[29] land prices are among the cheapest of major agricultural nations, depressed by a switch to selling among farm operators, with even the most expensive sites costing less than one-fifth of French or German values. The cheapest land, at $250 a hectare, is worth some 2% of those in the major EU producing states; the maximum price of Russian farmland is at about $2,100 per hectare, in the southern region of Krasnodar, whose proximity to a major Black Sea port attracts a premium. However, most Russian farmland sells for a lot less, with the Krasnodar average at $1,711, and land selling for $548 per hectare in Voronezh, in the heart of the black-earth region. In Saratov, in the Volga Valley, land averages $331 per hectare, with some going for less than

$250 per hectare. The prices compare with an average of $4,500 per hectares in eastern European Union countries, and $11,250 per hectare further west in France and Germany, the bloc's main agricultural producing nations. Prices in Brazil and the United States average $9,000 per hectare, and in Argentina $7,000 per hectare.

Although there was significant interest from international buyers in the period 2005–2009, from 2010 onward that interest faded, a phenomenon that was accelerated by export curbs, which deny growers the ability to tap high international prices. Admittedly, there are not many places (left) that have large amounts of land available with water and infrastructure – and most importantly, with land at low prices. Also productivity in Russia and the Commonwealth of Independent States (CIS) is much higher than expected and higher than other parts of the world judged over the last decade.[30] Prices are also dependent on the completeness of registration and the accompanying infrastructure. In Russia, 49-year leases are common. Foreigners can own companies that own land through cross-share ownership, but land ownership can be a difficult process, often taking more than a year to achieve land registration. However, this can more than double the value of the investment. In the Ukraine only leases are available for farmland.[31] These are typically 15 years. Romanian and Ukraine legislators were, in late 2013, considering draft laws at curbing farmland investments. The Romanian draft law, in its current form, limits individuals' ownership of farmland to 100 hectares and will make it harder for Romanian farmers to purchase land rather than discourage foreigners. It also prefers buyers under 40. Moreover, the creation of a state agency that will be granted preemption rights to buy land that is up for sale will lead to market distortions. In Ukraine, the maximum size of privately owned land is proposed to be installed at a total of 100 ha, and the minimum period of the land use is to be 15 years. The right to purchase agricultural land belongs to citizens of Ukraine, the state, the Land Bank, and territorial communities.

c. United Kingdom

According to the latest Knight Frank UK index report, average farmland values in England rose by almost 3% in 2012 to £6,214/acre. In 60 years, prices have risen almost 11,500% from just £54/acre. Land prices are predicted to increase by around 5% in 2013, and the growth curve is expected to flatten from then onward. In fact, according to research undertaken by Savills, leased agricultural land generated an average return of 14% per year over 15 years with volatility indicators of only 8%, whereas equities generated a return of only 8% per year over the same period with an associated volatility of over 18%. According to the same research, farmland investments in the UK have delivered returns of 8.2% per year on average for 30 years; 12% per year over 5 years; and 6.3% per year over the past 3 years. In the most recent reporting, they indicate that they expect the growth pattern to flatten somewhat, but with a clear upward-sloping trend after a stellar year in 2012. They forecast growth of 40% over the next five years for average farmland values across Great Britain.[32] They further indicate that average farmland values will continue to grow in the short to medium term, driven by competition for top-quality farmland. However, we expect weaker demand and slower growth in values for smaller farms with a significant residential weighting, at least until the mainstream residential market revives.[33] Or put differently, prime farmland will continue to outperform while the rest had, to a large degree, its best run. The rest includes Grade 3 arable and poor grassland. With the supply of farmland in general decreasing,[34] rural assets continue to outperform alternative assets despite economic uncertainty remaining a real issue. However, it is this uncertainty that has a positive effect on values and enhances farmland's reputation as a safe haven from economic storms. The principle motives for buying were (1) relocation, 4%; (2) expansion, 44%; (3) investment, 19%; and residential/sporting purposes, 26%.[35] Noteworthy is the fact that in most deals cash remains the largest component of the financing package.

To put things in perspective, Table 8.4 indicates the 3-, 10-, 20-, and 30-year performance of UK agri-farmland.[36]

d. Latin America[37]

Emerging markets in Latin America offer farmland investors one of the best opportunities to capture capital growth and income from farmland investments, due in no small part to the ideal growing climate found in Latin America, as well as improving infrastructure and its proximity to high-demand growth markets in emerging economies in the region.

Table 8.4 UK farmland performance

Type/return	3-year	10-year	20-year	30-year
Wetland	8.4	13.8	13.2	8.9
Farming land, top quartile	15.7	14	11.8	10.2
Forestry	18.7	13.2		
Let residential	5.8	9.7	11.6	13.8
Commercial, all	8.2	7.0	8.2	9
Equities	12.3	4.9	8.1	11.8
Gilts	6.4	6.4	8.2	10.3

Argentina offers one of the largest stocks of underutilized quality land in the world, and opportunities exist for greenfield development projects involving the development of unused land with on-site infrastructure, storage facilities, roads, and irrigation. Substantial capital value can be added, and double harvesting combined with low labor costs increases operational profitability, with yields of 10% to 15% possible with the right land. Although aggregate country-by-country reviews of the evolution of land prices in Latin America are few and far between, some wrapping together of individual pieces of data sets evidence an average annual 10% to 20% rise of farmland values in Argentina. First-quality agricultural land values in Argentina have increased by two and a half times since 2003, and almost four times for second-quality agricultural land. Argentina is an attractive area for agricultural investment, as there is good-value quality land, the potential to farm large areas with good legal title, sophisticated farming techniques, and probably the lowest production costs in the region despite the imposition of export taxes. Generally, markets are well established and water availability is good. Low costs and the potential for high productivity offer good investment opportunities, but location is critical. For Brazil, that range is about 10% to 15% annually during the last decade. This increase in return could be attributable to the enhanced use of farmland for sugarcane development within the context of biofuel production, which the country has decided to bet big on. Agricultural land and commodities are major investment topics in Brazil. Brazil has many natural advantages, including fertile soil, an unlimited supply of water, and an abundance of land. With 19% of the world's arable land, Brazil is one of the world's top eight producers of 28 different agricultural commodities. The government, however, has introduced laws making land ownership by foreigners more restrictive. In other parts of Latin America, investors and farmers from Argentina and Brazil are looking to expand operations and diversify risk. The perception of lower political risk in Uruguay has meant that investors have seen prices rising to the point that returns are now less than in Argentina. Land values in Uruguay for good-quality land typically range from US$3,750 to US$9,000 per hectare, but farmland is not generally as good as in Argentina, with typically more mixed farming operations and lower crop yields. Most farms would have a mix of perhaps arable, cattle, and forestry enterprises, as well as larger areas of unusable land.

e. Asia

It is impossible to provide a complete review of the investment performance of farmland across the entirety of the Asian continent within the context of this book. Reporting in many countries is unreliable, and the majority of transactions are conducted at either the subsistence-farmer or governmental level, where the flow of information is poor if not nonexistent. No complete data sets are available and/or truly reliable. In fact it is often true to say that recent interest from institutional investors is forming the basis of a previously non-existent agricultural property market in many of the countries surveyed. It seems that the capital value of farmland assets that form part of larger institutional investment are assessed using an income-capitalization model, taking into account the cost of establishing a modern agricultural operation on the land, weighed against the projected revenues from harvesting.

It can be observed, however that large institutional investors are working their way into the Asian farmland environment. For example, the Kuwait Investment Authority (KIA) has been targeting farmland investments in Asia, notably in Vietnam, Cambodia, and Laos, via a specially established investment vehicle – the Kuwait China Investment Company (KCIC). The investing environment in Asia can be obstructive in certain countries, however. When KCIC sought to acquire rice-farming land in Cambodia directly through the central government, it provoked widespread resistance among the population. Other examples are investments by the Quatar investment authority and Abu Dhabi's Al Dahra Agricultural Company.

f. Africa[38]

African farmland is very high on the agenda both of sovereign investors seeking to shore up domestic food supplies and of financial investors seeking superior returns. While risk certainly features heavily in the African land market, especially concerning productive agricultural land, it is true to say that there is substantial profit potential for those prepared to take on board some extra risk. Ample opportunities exist for greenfield developments in Africa, with large tracts of suitable land yet to be cultivated, while improved infrastructure and modern agricultural practices enhance on-farm profitability and reduce losses due to poor storage and transport facilities. The introduction of modern irrigation systems can dramatically improve the capital value of suitable land assets as well as productivity and yield. In South Africa, irrigated land traded for R$11,000 per hectare in 2012, while equivalent dry land achieved only R$3,000. The same trend can be seen in every African nation that allows direct farmland ownership, creating opportunities for investors capable of delivering infrastructure improvements on the ground. Ample opportunities also exist to improve agricultural practices, including soil management and improvements that take advantage of excellent fertile soils and improve profitability. Some deals, such as the Daewoo deal in Madagascar, not only reflect poor corporate governance on the side of some African governments managing these precious public goods but also demonstrate the sensitivity of foreign investors buying into African farmland despite demand for food by the the local population in Africa, given the expected 100% rise in population growth over the next 40 years. Similar land grabbing emerged when South Sudan became independent, involving US hedge funds and even US officials.[39] Tectonic shifts in politics are often used to unlock value in hard assets, with reference to a similar process that unfolded when the USSR disintegrated more than 20 years ago. Many of the countries in Sub-Saharan Africa do not allow land ownership, but land is generally available on long-term leases. Countries where ownership is permitted include Mozambique, South Africa, Botswana, Namibia, Tanzania, and Zambia. Irrigated land comes at a significant premium, although there are no clear boundaries between either irrigated or dry land. This is region dependent and varies with the amount of water available. It is "ideal" to invest in farms of 1,000 to 5,000 hectares, big enough to allow economies of scale while offering a lower risk of touching land-ownership sensitivities. Larger acquisitions may attract negative media attention. Issues to be tackled are technical expertise at the farm level, availability of local credit, infrastructure, and as mentioned, issues related to land tenure.[40]

African agriculture is dominated by smallholder farming for subsistence and local markets. The average holding is in the region of half a hectare. Most of the land is either state owned in one form or another, or held by tribal or other local communities. The land-tenure system and regulations vary for each country but are generally based on a long leasehold interest of between 50 and 99 years, often with a renewable clause written in to allow effective ownership in perpetuity. It is anticipated that the market will become more transparent and structured/regulated as it develops, and therefore common terms and transactional evidence are likely to emerge.

There are fundamentally two options available to an investor wanting to access this market, and in both cases security and title are essential:

1. Invest in (early-stage) projects and develop a greenfield asset (primary market).
2. Source and acquire one of the limited "developed" farms (secondary market). Although the core market is still in its primary phase, there is evidence that a secondary market is starting to develop.

The secondary market is a consequence of the original pioneers of commercial agriculture who invested four or five years ago and have successfully developed their farms from greenfield to fully operational aggregated agribusinesses. In some cases these successful investors and pioneers are now looking for a full or partial exit to realize the true value of their asset. We anticipate that some of these developed commercial farms will come to the market in the next year or two offering significant opportunities to investors.

In addition there are smaller titled farms available for sale, which are often a result of postindependence family-farm development. More evidence of asset values will emerge as the market develops and there is a transaction history for these more developed assets.

g. Oceania

Both Australia and New Zealand are considered safe investment destinations in farmland terms. New Zealand has the advantage over many areas in that water is abundant due to reliable rainfall and good irrigation infrastructure. In both countries we have seen an upturn in prices (average +/–7%) for farmland after a downbeat during the early years of the financial crisis.

Opportunities therefore exist in various places around the world, and drivers are pretty common. But farmland values vary significantly, and although having outperformed other asset classes on an inflation-adjusted basis over the last decade, only prime farmland will protect against overall potential bubble formation in some of these farmland markets.[41]

8.2 Investment strategies

Whether deciding to invest in farmland directly, or indirectly via a managed investment vehicle such as a real-estate investment trust (REIT) or farmland-investment fund, there are two underlying investment strategies to consider when approaching productive agricultural land as an alternative investment asset class. Investors that are looking for capital preservation, income, and an inflation hedge may best consider acquiring suitable productive land and leasing to a farming tenant, and those with a little more appetite for risk may decide to engage in profit-sharing arrangements.

For those investors seeking to generate superior investment returns, green-field developments involving the development and cultivation of previously unutilized land may seem more appropriate, as substantial capital value can be added through on-site soil preparation and infrastructure investments, while long-term revenue is captured through ongoing cultivation of a range of crops. In both cases, investors must also consider the operational strategy, including the type of crops grown, whether food crops for human consumption or feedstock for biofuel production, or even permanent crops such as vines, or oil-producing trees for biodiesel production. In most cases the most profitable crops with acceptable risk parameters will ultimately win the competition for production on arable land. Investors are often drawn to farmland based on its cash-flow potential (rents or crop yields) in addition to its capital appreciation. Additionally, many investors who are seeking less exposure to paper assets often find that farmland offers the type of diversification that they desire.

8.2.1 Generating rental income from farmland

The most straightforward and lowest-risk approach to farmland investing is the identification and acquisition of agricultural land and leasing the asset to a tenant farmer for a fixed annual income, often linked to inflation or periodically reviewed. Rental rates[42] vary from region to region and ultimately depend on the productivity and profitability of that specific holding. Farmers are willing to pay higher rents for more productive land, or land that is in close proximity to a guaranteed water supply and/or storage and transport infrastructure. Net lease returns can vary from as small as 1% (UK) to 4% (US on average) to 6% to 8% in Australia.

8.2.2 Active engagement in (new) developments

Investors seeking to generate superior capital growth and income opportunities might consider that developing land with agricultural potential can generate appreciation in the capital value of the asset, while also generating an ongoing operational or lease income (average return 10%). Such developments, especially in emerging markets where there is still virgin land to be exploited and favorable growing conditions, provide investors with an opportunity to "decorrelate" a portion of their investment returns from the performance of traditional investment assets during periods of extreme market volatility. Often, the value of developed agricultural land can increase by up to 500%, although taking into account the cost of conversion, net gains amount to circa 100% depending heavily on the specific nature of the project, its location, and a host of other endogenous variables. Also, geopolitical risk must be considered a factor in most emerging markets including Latin America, Asia, and Africa.

8.3 Risk exposure and management

While the investment performance of farmland has historically been less volatile than traditional assets like equities, it is important for investors to recognize that farmland investments, like all investments, carry risks that are specific to the asset class. Risks involved in owning farmland properties can be separated into two broad categories:

- Endogenous risks: those risks having an internal origin relating to specific crops, on-site management, and site location.
- Exogenous risks: those risks having an external origin such as commodity prices, extreme climatic events, and international trade policies.

While risks cannot be eliminated, both endogenous and exogenous risks can be managed and mitigated through proper planning and execution.

The risk categories are as follows:

1. General investment risk

The value of agricultural land can and will fall and rise within a short-term framework as commodity prices, local availability, and agricultural yield all impact current market values. While strong demand fundamentals underpin future farmland value growth, should land values fall significantly, then the performance of farmland investments may suffer. Farmland could best be viewed as a mid- to long-term hold, affording investors the opportunity to ride out any potential short-term price volatility.

2. Commodity price risk

Commodity markets are known for their volatility, and the farmland market is exposed to that, the price of crops being the key driver. But the cost of fuel, labor, and other commodity-based inputs such as fertilizers may also have a bearing on the cost of generating income from productive land.

3. General agricultural risk

Weather conditions, such as floods, droughts, undesirable rainfall, hail, frost, or uncharacteristic cold spells; weeds, pests, and diseases; fire; and the possibility of generally worsening conditions associated with climate change – any of these factors individually or in combination may have adverse consequences on incomes and/or values. Against some of these risks, insurance can be taken out.

4. Geographic risk

As most of the opportunities are emerging in developing economies, investors are exposed to the potential risk of unstable regulatory and logistical environments complemented with a poorly functioning judiciary that is overcrowded, potentially corrupt, or does not possess specialist knowledge regarding the industry.

5. Liquidity risk

Direct farmland investments are illiquid in nature, and only certain buyers qualify. Often investors prefer liquid/listed investment vehicles to get exposure.

6. Regulatory risk

New legislation may adversely affect income through price controls, export restrictions, or the imposition of duties. Governments may also impose more stringent environmental regulations upon the agriculture sector, thus increasing compliance costs, with possible negative consequences for farmland profitability. There is also the possibility of governments imposing limitations on the volume of agricultural land that can be owned by overseas investors. A very good example of what might move the needle on US farmland prices coming in via the regulatory cycle is the farm bill (1,198 pages) discussed throughout 2013, which is a copy-paste story of a lot of measures coming together. It includes an initiative to guarantee prices for sushi rice. Also included is insurance for alfalfa and a marketing plan for Christmas trees, and profit-margin insurance for catfish farmers, etc. In short it is a potpourri of measures that come down to either subsidized insurance, promotional programs, or protections against imports. As it currently stands, the draft[43] would reduce payments to growers of corn, wheat, and other crops by eliminating a $5-billion-a-year program of direct subsidies while expanding subsidized crop insurance (which should facilitate better bank lending to farmers). Their different price tags mainly result from variations in food-stamp spending.

7. Inflation/deflation risk

Farmland investments have been shown to share a strong correlation with inflation, which makes the asset class particularly attractive to those individuals who expect to see extended periods of high inflation.

8. Currency risk

This occurs when the investment is in a different currency than that of the capital base or when trade is conducted and the international price is quoted in US dollars.

9. Counterparty risk

Agricultural investments require detailed knowledge of farmland topography and operational knowledge to avoid landslide- and erosion-prone areas. Insufficient infrastructure, expensive access, and long distances may also determine the feasibility of a farmland investment.

10. Asset-specific risk

Issues that may affect the profitability of a farmland investment include soil quality, topography, water availability, distance from infrastructure, and security of tenure. These may all have an impact on the risk patterns and price or revenue-generating capacity of the agricultural asset.

11. Others risks
    - Tenure risk/concession risk: includes a set of problems related to control over a given parcel of land.[44]
    - Infrastructure is of paramount importance for access to farms and to transport outputs to markets economically. In addition, an appraisal of the availability and provision of utilities and services including electricity and telecoms is essential.
    - Subsidization of agriculture has, in several parts of the world, become an integral part of farm incomes. The current and long-term agricultural policy is hugely influential.
    - Research and development activity is a useful area to consider as part of the longer-term investment strategy. This is to have a grasp of the willingness of governments and industry to embrace new technology.
    - Agronomic potential of any investment is probably the most important area to understand, as the success of the whole investment will hinge on the land being managed by a successful farm business. However, it is a complex subject and requires local agribusiness expertise and experience. In many cases, the highest agronomic potential is in areas where the risk score is highest. Assessing the relative opportunities between countries and regions can be done based on the following factors: (1) scalability, the opportunity to farm large-scale holdings, maximizing economies of scale; (2) soil quality, but care should be taken in the final due diligence, as this varies across countries; (3) potential for yield/output increases, but this will require top-quality management with the correct resources to maximize; (4) water availability, either natural sources or the potential for irrigation; and (5) sustainability – this is high on the agenda of consumers and retailers across the world, who will ultimately be the purchasers of farm outputs.

8.4 What are the opportunities left in agricultural farmland?

The total investible global agricultural farmland market is estimated at about USD 8.3 trillion. However, what sets developing markets apart from the developed world is the fact that in the developed world institutional investors are commonplace in farmland investments, whereas in developing nations (and some developed nations outside the United States), most farmland is owned by family farmers who own and operate their farms. One of the major challenges in effectively analyzing the sector is to obtain accurate and timely market data and provide insight into the size, scope, and ownership structures that shape the opportunity to invest in agricultural land globally. In analyzing the major agricultural regions and the countries being targeted by agricultural investors, the data demonstrates that while areas such as the United States have higher levels of institutional investor ownership, farmland ownership globally is heavily dominated by family farmers who both own and operate their farms. Given the small scale of the average farm globally and the challenges for such businesses accessing capital, the scope and need for institutional capital to be deployed in agriculture in order to improve efficiencies and generate higher returns is significant. To get an idea of the total market size (which includes arable land, pastureland, forestland, and permanent cropland), the regional volumes are as follows (million ha):[45]

- Canada 378
- USA 707
- Chile 32
- Russia 1,024
- New Zealand 20
- Australia 559
- Argentina 170
- Brazil 786
- Uruguay 17
- Ukraine 51
- Kazakhstan 213

On a per capita basis, Canada and Australia top the ranks, with 1.7 and 2.4 ha per capita. The FAO manages a database that collects these data. The two major classifications of farmland it makes are *arable land* – the land used for row crops or broad acre – and *permanent crops*: the land planted with trees or vines.[46] The availability of arable land is of interest to investors for the following key reasons:

- It faces the greatest demand of all agricultural land and is decreasing in availability on a per capita basis. Since 1960, the amount of arable land per capita available for agriculture globally more than halved.
- For the land that is agronomically suitable, crops are predominantly the highest-value and best use of the land.
- The land is mostly used for producing grains (e.g., wheat, corn, barley), oilseeds (e.g., soybean, canola/rapeseed, sunflower), and fiber (e.g., cotton). These commodities are experiencing increasing demand for use as food, as animal feed, and especially in biofuel and industrial production.

One complication is the structure of land ownership globally. Of the abovementioned 8.3 trillion USD, an absolute majority is owned by the operator (fragmented industry). Only 30 to 40 billion is owned by institutional investors,[47] and about 70 to 100 billion USD is owned by noninstitutional investors.[48] Therefore, one of the attractive factors from an investment perspective is the opportunity for consolidation given the importance of scale in driving returns from agriculture. Only Brazil, Argentina, and Australia have over 5 million hectares of suitable land yet to be brought into crop production. Australia has the most arable land per capita, with over two hectares per person. This is more than four times that of the United States. The higher the proportion of arable land per capita a country has, the greater its opportunity to export its agricultural produce. While there are significant areas of arable land in Russia, Kazakhstan, and the Ukraine, utilization rates are much lower, and crop yields are lower mainly due to poor farming practices limiting productivity. Low productivity is symptomatic of challenges to agriculture such as the lack of infrastructure, government regulation stifling activity, or a lack of human resources. Average crop yields are low in Australia due to the large proportion of low-density cropping that constitutes the overall arable land area. Due to all these reasons, the investible universe is considerably smaller. Table 8.5 compares (estimates) agricultural and timberland space.[49] The total land value is reduced to the investible universe due to a range of factors that significantly restrict or prevent institutional investment, including a lack of basic infrastructure, adverse governmental policies, and unacceptable social and environmental risks.

Despite the low correlations, relative stability of returns, and inflation protection shared by both agriculture and timber land that are driving institutional interest, agriculture is far behind timber in terms of institutional investment relative to industry capitalization. One factor explaining this is the fragmented nature of farming and the absence of large operations with appropriate diversification and institutional-grade management in place. Compare this to the origins of timber as an asset class, where forests owned by timber mills and state land authorities were divested, as those organizations sought to run more "efficient" balance sheets. This meant that large-scale assets, which in some cases were run by existing management teams,

Table 8.5 Comparison of agricultural and timberland space and valuation

In USD	Agricultural land	Timberland
Total land value	8.3 trillion	425 billion
Investible land value (estimate)	1 trillion	300 billion
Owned by institutional investors	35 billion	60 billion

became available for investment. Another factor is the liquidity in the sector, albeit there are varying degrees to which this is evident from country to country.[50]

In conclusion,[51]

- Arable-land scarcity is the result of a range of human and climatic factors, including degradation, climate change, soil constraints, urban encroachment, and unequal land distribution.
- There currently remains some 2.7 billion hectares of land with potential for crop production in the world, concentrated in South and Central America and Sub-Saharan Africa.
- The solutions to addressing the availability of arable land are threefold: the production of more arable land, an increase in the productive capacity of existing arable land, and the conservation of arable land in order to prevent degradation.
- Despite more than an adequate supply of arable land to meet future demand, land availability will continue to be a major factor in meeting future food security because of the need to find a balance between competing interests and uses and finite resources.

8.5 Achieving alpha in farmland

But let's take it a few steps further and look behind the curtain, especially if you are not convinced by the remaining potential in land-price appreciation or feel that this "land-grabbing" type of investment is not yours and you want some more direct fundamentals you can judge before investing.

Farmland-investment advisors frequently refer to postinvestment improvements in farm-level management as the primary mechanism for achieving alpha in farmland investing. By contrast, the actual returns achieved by the majority of corporate farming entities and farmland funds show that this rarely if ever delivers the promised outperformance. The claim made is that farm income can be improved by installing better farm managers using more advanced technologies. While this concept may have some logical basis in emerging or frontier markets where outdated management practices are more common, it is less likely to be a source of alpha in more advanced agricultural economies. In these markets the majority of land assets are already controlled by the most competent farmers with decent economies of scale and the latest technologies. What's more, many of the factors affecting farm incomes are beyond the control of the manager (e.g., weather, commodity prices, input prices, and exchange rates). Among farm managers of a similar agronomic skill level, the difference in farm profits is marginal, deriving primarily from a particular proficiency at managing internal costs, the only financial element over which the manager has any significant control. Outperformance is, by definition, only possible if a truly exceptional farm manager can be recruited. Such farm managers are either already running their own farms or are able to lease additional land at rates that enable them to earn a more attractive return from their expertise than would be possible working as an employee for a farmland fund. The reality is that it is only possible to attract and retain exceptional farm managers by offering exceptionally high levels of remuneration. Thus it is likely that agency costs reduce net returns to the extent that alpha is not in fact achieved.

This is not to say that farm-level management and tenant selection are not important from the perspective of safeguarding the value of the underlying land asset and securing an income stream in line with the market benchmark, merely that, taken alone, they are unlikely to be a source of alpha. This leads to two conclusions:

1. "Active" farm management is unlikely to deliver higher risk-adjusted returns than "passive" farm management (i.e., leasing).
2. In order to achieve outperformance, investors need to look to other mechanisms.

Other mechanisms can include and/or will benefit from the following considerations:[52]

- The land-value discounts that arise due to the relatively illiquid nature of farmland markets[53] provide the most promising opportunities to enhance returns. As such, the key to achieving outperformance lies in developing an analytical model that allows land-value discounts to be recognized and exploited as and when they arise.
- The illiquid nature of the farmland market favors investors who take a long-term view. Such investors wait patiently for pricing opportunities to arise and are able to deploy capital decisively when they do.
- Valuing farms accurately enough to capitalize on discounting opportunities is difficult due to a general lack of detailed and reliable historical financial and production data. Investment decision making is further hampered by the fact that farm-to-farm variations in agronomic/climate conditions and enterprise type undermine attempts to make apples-for-apples comparisons between farms.
- The comparability quandary can be solved by developing an analytical model with simulated production forecasting and financial modeling at its core, thus allowing differentiation by all the key variables required to accurately assess the comparability of individual farms. Such a model should be

capable of producing a single consistent set of risk-return metrics by which all farms can be judged according to the same set of rules.

- Income returns are higher for lower-rainfall farms. This is due to the fact that, despite having lower crop yields, land prices are substantially lower in low-rainfall areas, so income returns as a percentage of farm asset values are higher. Income returns are also higher the more variable the income stream, with owner-managed farms showing the highest average rate of income and the fixed-lease model showing the lowest average rate of income. However, the return premium associated with the owner-managed model is significantly counterbalanced by the higher level of volatility and downside risk associated with the income stream.
- In a geographically diversified portfolio of farms, crop-share and fixed-lease investment models produce positive income in all years, even in lower-rainfall zones. As such, these investment models show higher risk-adjusted returns compared to the owner-managed and share-farm models, especially when considered from the perspective of downside risk.
- Total returns for crop-share and fixed-lease investment models can be further enhanced due to the fact that farms can often be acquired under more favorable terms where the buyer is prepared to offer the seller a leaseback as a condition of the purchase.
- Given that arable farms expend 50% of annual farm overheads in the form of crop inputs at the beginning of the growing season before knowing if seasonal conditions will result in a profitable crop, the reliability, rather than the magnitude, of crop yields is the more important determinant of long-term farm profitability.
- Due to the fact that the impacts of weather variability on yields are not geospatially consistent, geographical diversification in a farmland-investment portfolio is very beneficial with respect to reducing income volatility (particularly on the downside) and increasing risk-adjusted returns.
- Understanding the effect of crop-yield volatility on investment returns is a key step toward achieving outperformance.

Beyond that, there will be opportunities due to a wave of consolidation under way. Partly due to the need for economies of scale, and partly due to family-related positions,

- the average age of agribusiness owners continues to increase,
- many business owners do not have descendents that are willing or capable of managing an increasingly more complex agribusiness,
- the capital availability of descendents may be limited, and
- problems with divergent and extended shareholders may arise, sometimes three or four generations removed from the business operations.

8.6 Sustainability and legal issues with respect to foreign ownership of agricultural land[54]

Investing in farmland has been on the radar of (international) investors since agricultural commodities pricing went structurally up and biofuels became economically feasible, as well as the reduced availability of certain areas of the world Gulf Cooperation Council –Gulf Region (GCC[55]) or certain parts of a country (the United States).[56] A snapshot of recent international farmland deals would include the following:

- 2008: Egypt buys from Uganda 800,000 hectares of land to grow corn and wheat.
- 2008: The UAE buys from Pakistan 324,000 hectares of land to grow a variety of crops.
- 2009: South Korea buys from Sudan 690,000 hectares of land to grow wheat.
- 2010: Egypt buys from Russia 427,000 hectares of land to grow a variety of crops.
- 2010: China[57] buys from Sudan 400,000 hectares of land to grow a variety of crops.
- 2011: China buys from Tajikistan 110 hectares of land to grow cotton and rice.
- 2013: China (allegedly) buys (leases) from Ukraine 3 million hectares of land to grow a variety of crops.
- 2013: Indonesia buys from Australia 1 million hectares of land to raise beef/cattle.
- A variety of deals over the last few deals with a total amount of 27,500 hectares in Indonesia to grow palm oil to produce biofuels.

However, it hasn't always been this way. Since the 1980s, foreign ownership of land for agriculture significantly declined as other types of foreign investment, particularly contract farming[58] and investment in other aspects of agricultural production – including seeds, fertilizers, machinery, processing, manufacturing, and retail – have been preferred by investors and transnational organizations.[59]

A number of observations can be made about the prevailing trends in agricultural land deals. The lease period is typically 50 to 99 years, and the land acquisitions are often in excess of 10,000 hectares, with some reports of deals of up to 1 million hectares. The main actors are states, sovereign wealth funds, and the private sector, including agribusinesses, investment banks, commodity traders, and mining companies. The current investment flows fall into different categories. The private sector in Europe, the United States, and

Japan are looking for land all over the place, from Russia to Sudan to Australia. They are likely to produce crops for food, feed, and fuel – whichever takes the highest price or supplies other integrated production chains. Gulf states are looking for land in countries with which they share cultural or historical ties or geographical proximity. The primary goal is food production for home-state consumption. China, Japan, and South Korea are looking for land predominantly in Africa and Asia. They are interested in producing food, animal feed, and energy crops. While the data appear to indicate that this is not exclusively for home consumption, it is widely understood that it is primarily for this purpose.

Smaller and Mann indicate in their reporting an anomaly compared to previous periods of above-normal levels of interest in foreign farmland:

> It is this feature perhaps more than any other that helps redefine the nature of the current land purchase and lease arrangements: they are not market-driven in the classic sense of seeking comparative advantage for production for global markets. Rather, they are about shifting land and water uses from local farming to essentially long-distance farming to meet home state food and energy needs. It is, in practice, purchasing food, with the risks that local food needs, land users and water rights will be displaced. This is not completely new. However, the growing scale of this practice today, combined with the increasing economic and environmental concerns that are motivating this surge, are creating a new dynamic of global importance. It is no longer just the crops that are commodities: rather, it is the land and water for agriculture themselves that are increasingly becoming commodified, increasingly subject to globalized rights of access.[60]

The government is often key in negotiating and closing the deal. They, in their turn, incentivize the private sector to invest.[61] When the private-sector investment is backed by the government, the agreement is more likely to contain export-home-only provisions. Other private-sector investments – although a minority of deals – are more strongly motivated by profits and do not place restrictions on the destination of exports. Some of the land acquisitions are reported to include trade-offs or other incentives for the host government. A lease can be given in exchange for oil contracts or investments in infrastructure projects in the host country, including roads, ports, and bridges. Tax incentives, including tax exemptions on the import or purchase of goods and machinery, or income-tax relief for investors are also included in some of the deals.[62]

In many of the host states, there is either no, insufficient, or unclear domestic law concerning land rights, water rights, pollution controls for intensive agriculture, human health, worker protection, and the like. This has the potential to shift rights from local to foreign actors. The two primary sources of international law that relate to this issue are international contracts, which are commercial in nature, and international treaty law on investment. There is critical concern that the economic interests of the foreign investor are prioritized, potentially at the expense of the social and economic interests of the local users.

What follows is a brief overview of the three sources of law most immediately relevant to land-related investments: (1) domestic law, (2) international investment contracts, and (3) international investment agreements.[63]

1. The primary source of law that should be used to regulate all investments in agricultural land and water is domestic law in the host state. Given the range of issues involved, this might typically include laws relating to the admission of foreign investors, laws and regulations on incentives for foreign direct investment (FDI), taxation, property law, water rights and rates, and an array of laws relating to the potential impacts of the investment on the local community: environmental, human health and safety, worker safety, labor rights, and possibly others.[64]
2. In most cases of agricultural land and water investment, the role of the contract between the host state and the investor is critical, especially for the investor. The contract will set out not just the price, quantity, and duration for the purchase or lease of land, but it will in most cases also address a range of other issues. These include taxation and investment incentives for the investors, rights to export production, rights to import equipment and personnel, any associated infrastructure or logistics requirements (e.g., road, rail, shipping), and other operational matters. The contracts will often address environmental and development issues either expressly or through general provisions.[65] Investment contracts increasingly have a propensity to become a legal code for the investment. In several developing countries, investment contracts become enshrined in legislation. In some instances, they even have a quasi-constitutional status. When such contracts depart from the application of otherwise generally applicable national or subnational laws, they attain a lawmaking status for that investment that goes outside normal commercial contracts. This lawmaking function can come in the direct terms of the contract and/or through the application of what are known as "stabilization" clauses.

   These are clauses in the contracts between investors and host states that either preclude the application of, or require compensation for, new or changed regulatory measures that affect the investments. They may apply for a certain period, or for the full life of the contract and any renewal period. It can be noted that such clauses can vary in scope of coverage (from narrowly defined fiscal issues to all laws and regulations applying to the investment) and degree of impact (the extent to which they preclude the application of new or changed laws or require compensation for them). In addition to the lawmaking function, investment contracts also determine which law applies to interpret the contract in the event of a dispute and which court or tribunal will be responsible for resolving a dispute arising under the contract. The three main choices are: (1) the courts of the home state, (2) arbitration in the home state under domestic law, or (3) an international arbitration process. In many instances, investors choose international arbitration to avoid possible biases in the domestic court.

3. International investment agreements (IIAs) are treaties between states that are intended to provide investors from one state investing into the territory of the other state with special protections under international law. IIAs can come in several forms. The most common is bilateral investment treaties (BITs) between two states. Any rights included, in legal terms, are layered over the domestic law, which must comply with the terms of the treaty. IIAs have been designed to protect foreign investors from a range of governmental measures. Key provisions are the following:
   - *The requirement of national treatment.* This is a commitment by the host state to treat foreign investors that are "in like circumstances" to domestic investors in a manner no less favorable than the domestic investors.
   - *The requirement of most-favored nation treatment.* This is a commitment to extend the treatment provided to the most-favored foreign investor to all other foreign investors. This has been interpreted to include not just the treatment under domestic law and policy, but also under other investment treaties.
   - *Fair and equitable treatment,* also known as the minimum international standards of treatment. This has become a broadly conceived standard that includes elements of government transparency in decision making, and a prohibition against arbitrary or discriminatory acts.
   - *A prohibition against expropriation without compensation.*
   - *Rights to export the products produced.* It is commonplace in investment agreements to provide investors with the capacity to operate their investment in accordance with their own needs (to export what is produced).
   - *Safeguards.* Many IIAs include provisions that allow the host state to limit the export of capital by the investor in the event of economic or fiscal crisis.
   - *National security.* Almost all IIAs have provisions that allow the obligations in the treaty to be superseded by national security concerns. The scope of this exception clause has generally been thought to mean national security in a policing and military sense; however, many of these provisions are self-judging by the invoking state.

- *Dispute settlement.* Virtually all of the IIAs concluded since 1990, which is the great majority of the existing treaties, have special dispute-resolution processes known as investor-state arbitration.
- In addition to the above forms of investor rights and protections, an increasingly common additional feature of IIAs is the inclusion of what are often called *preestablishment rights*. This is the right of an investor from one state to make investments into the other state on the same footing as a domestic investor. In the agricultural land and water sector, this would mean that an investor could purchase land and access water rights on the same basis as domestic landowners and water users. Often accompanying these preestablishment rights is a prohibition on imposing *performance requirements*. Performance requirements are legal or policy measures that are used to require the investor to have specified levels of business linkages with the local community or state businesses, to employ a given number of employees, to share revenues, or other such constraints on business operations.

The different legal instruments and different levels of legal interaction does provide a challenging read and does hold quite a few ramifications. We will need to engage in a in-depth legal review of these but limit ourselves to a review of the pertinent questions investors would like to see reviewed and answered before considering an investment in overseas agricultural land.

Those key questions[66] will most likely include the following:

- **Do foreign investors have a right to buy land and water rights?** International law generally does not give foreign investors rights to invest in land and water in another state. Normally, whether a foreign investor can acquire land, and in particular agricultural land, is a matter of domestic law within each state unless a commitment is made in an international treaty to allow foreign investment in any given sector, or the economy at large. Such commitments provide "preestablishment rights" for foreign investors. It is not clear under existing case law whether an investment liberalization commitment made in one IIA will be available for investors from third states that benefit from a treaty with a most-favored nation (MFN) treatment clause.
- **What rights do foreign investors acquire if they do invest?** Absent any international contracts or treaties, foreign investors would be treated the same as a domestic investor under the applicable domestic law. However, when a contract between the state and investor is in place, the investor may acquire, depending on the terms of the contract, additional rights not set out in domestic law relating to water use, land-tenure rights, the right to

export all products of the investment, and so forth. In addition, a foreign investor may obtain favorable taxation terms and other economic incentives available under domestic law to foreign investors. The contractual rights may also replace parts of the generally applicable domestic law. Finally, an investor that includes a stabilization clause in its contract may also be able to avoid complying with domestic laws that come into force after the date the contract was entered into.

- **What happens to the rights of previous users of the land or water?** The rights of prior users of purchased or leased land or water is a critical issue in the debate over "land grabs." Under domestic law, where these rights are clear and vested in local owners or users, they will be entitled to be the vendor of the property or water rights, and thus to participate in the contracting process. If the government determined that an investment should take place despite the opposition of a land or right holder, expropriation might be possible, subject to the relevant compensation requirements. The problem is that in most states where such contracts are being completed, land and water rights are vague, based on local traditions, or nonexistent. Title is often vested in the government, or in local chiefs or other community structures. This is particularly problematic as legal negotiations often happen with central governments, and only when the deal goes operational do issues emerge when confronting local communities or previous local exploiters of the land bought.
- **Can obligations to promote sustainable development be imposed?** An often-identified approach to improve the equity and development impacts of land-investment contracts is to include certain requirements of investors to contribute to the local community in economic terms. Normally domestic law will allow such conditions to be imposed. A contract could include such provisions as well. Once signed, they would become binding and enforceable under the law of the contract, though it is unclear whether communities could seek to enforce these provisions or only the contracting state. However, if an investment treaty is in place that includes preestablishment rights, it may well include prohibitions on the imposition of such performance requirements as well.
- **What happens if land or water rights are expropriated or redistributed, especially if there is a food or water shortage?** The first recourse, in the event of an expropriation of land rights or water use rights by the government, is usually under domestic law. Most legal systems provide some form of recourse to either challenge the expropriation or seek compensation. This is so even when the property or rights are taken for a legitimate public purpose. All IIAs impose a requirement for compensation to be paid if investor rights are expropriated. The matter becomes less clear

where critical rights for operating the enterprise are reduced but not fully taken away. If water resources drop to a level below the requirements of the investment, the host state will not be able to do much, and no compensation could be foreseeable – this would ultimately be part of your risk assessment with respect to the investment. Obviously, the situation will be different in case the reduction in water supply is artificial to support other purposes (drinking water, other agricultural land, etc.).

- **What happens if other applicable laws relevant to the investment change, for example pollution control laws?** Smaller and Mann indicate some possible situations:
  - New environmental measures to protect against runoff of pesticides and fertilizers
  - The banning of certain chemicals to protect the health of workers or the local community
  - Mandatory increases in the minimum wage
  - Promoting the right of workers to unionize.

This is one of the most difficult and untransparent situations from a legal point of view. It is one that clearly links the state of domestic law, the contract, and treaty rights into an interlinked whole, but with no definitive answer. Generally speaking, and absent unjustified discrimination between domestic and foreign investors, existing laws at the time an investment is made will not create a problem. Changes in the laws, however, can lead to assertions of breaches of the contract or of treaty protections for foreign investors. Due to ideological differences, there are two contradictory directions in the case law on whether regulatory measures enacted for legitimate public objectives, including public health, safety, and the environment, can be considered an indirect expropriation and therefore subject to compensation, or a breach of the fair and equitable treatment obligation. As a result, it is impossible to determine in advance if a regulation will be classified as an expropriation if it is challenged under a treaty. Governments are increasingly including provisions in IIAs that specify that public health, safety, and environmental measures do not, or do not generally, constitute an indirect expropriation. The question remains whether the nature of the financial impact of the changes on the investor will be a key driver of whether or not compensation is granted – although economic impact will never be a decisive element on its own. Also, the linkages to contracts have become critical. Most critically, if a contract includes a stabilization clause that limits the ability of the government to make new laws applicable to the investment, tribunals will take this into account in determining whether there has been a breach of either the expropriation provision of a treaty or the fair and equitable treatment provision.

- **What happens if a host government implements a trade measure?** This might occur as a consequence of a food shortage in the host state. Although they are permissible under trade law, they can create problems for host governments if they negatively affect investor rights. This is particularly pertinent for contracts where agricultural production is for export to the home country only. Policy measures consistent with international trade law may now find that they are in breach of international investment law if they impact rights granted to foreign investors and can therefore most likely be upheld under dispute settlement systems (arbitration), although it is unclear if a "national security" exception often written into contracts would be helpful to the host states in justifying temporary export bans.

As one can observe, the legal and sustainability dynamics of investing in foreign farmland is a mixed bag, and some of the sustainability issues can pose a significant risk for the buyers and the countries alike.[67]

8.7 Evidence with respect to foreign investments in agricultural land and farmland in particular

The available data confirms that there has been a marked rise in FDI into the agri-food sector of developing countries since 2007. Although agricultural FDI flows contracted after their peak of 2009, their level in the 2010–2011 period was still higher than the average for 2003–2007. The flows are characterized by regional patterns whereby intraregional flows are greater than interregional flows except for Africa. The share of FDI that goes to the agri-food sector almost doubled between the periods 2000–2005 and 2006–2008 but is still low compared to other economic sectors, accounting for less than 5% over the period 2006–2008. The bulk of agricultural FDI flows is directed to the food-manufacturing sector, while primary agricultural production accounted for less than 10% over the period 2006–2008.

Studies[68] have been conducted analyzing various types of impacts (economic, social, and environmental) of foreign agricultural investment at national and local levels in the host country. The observed impacts were very diverse, and the studies shed light on the various factors that condition the success or failure of agricultural investments. The lands acquired by foreign investors tend to be among the best ones, including good soil quality, high production potential, irrigation, and proximity to infrastructure and markets. As a majority of foreign investment projects aim at export markets or the production of biofuels, they may pose a threat to food security in low-income, food-deficit countries, especially if they replace food crops that were destined for the local market. The net effect on food security will also depend on the additional income generated by the project, its sustainability, and how it is distributed in the local economy. Large-scale acquisition of agricultural land can have other adverse impacts, especially in countries where there is a lack of good governance, rule of law, transparency, and clear land-tenure rights. These negative effects include the displacement of smallholder farmers, the loss of grazing land for pastoralists, the loss of incomes and livelihoods for rural people, the depletion of productive resources, and, in general, negative impacts on local livelihoods due to reduced access to resources, which may lead to social fragmentation. There is also evidence of adverse environmental impacts, in particular the degradation of natural resources such as land, water, forests, and biodiversity.

The main type of benefits, which show up very inconsistently across the projects analyzed, appears to be the generation of employment, but there are questions as to the sustainability of the created jobs. There is also the issue of the quality of the created employment and who benefits, as managerial positions tend to be occupied by expatriates or persons originating from other areas than that where the project is located. Another expected advantage of FDI in developing countries is the transfer of technology. In the case of large-scale land acquisition, the evidence is mixed. There is obviously a time dimension in the assessment of this effect, as for other outcomes of investment. In conclusion, the studies suggest that for investment involving large-scale land acquisitions in countries where land rights are unclear and insecure, the disadvantages often outweigh the few benefits to the local community, especially in the short run. This outcome is even more likely when the acquired land was previously utilized by local people, whether in a formal or informal manner. Consequently, acquisition of already-utilized land to establish new large farms should be avoided. The studies suggest that investment projects that give local farmers an active role and leave them in control of their land tend to have positive effects on local economic and social development. Successful projects combine the strengths of the investor (capital, management expertise, and technology) with those of local farmers (labor, land, traditional know-how, and knowledge of the local conditions) – inclusive business models. Business models that leave farmers in control of their land give them incentive to invest in the improvement of the land. Since the bulk of agricultural investment comes from farmers themselves, these models are more likely to raise the level of agricultural investment in developing countries. Also, inclusive business models empower farmers by giving them a say in the implementation of the project or even its management. However, their benefits do not arise immediately. The time factor is essential. By their nature, inclusive models involve more stakeholders; hence building consensus on the project requires time, and decision making is slower. Transaction costs are high, especially in the initial phase. They should be viewed as a necessary investment that will enable higher returns in the longer term. However, most companies need relatively rapid returns on their investment, and their time frame is not compatible with that of local economic development. There is a need for "patient capital" provided by investors with a longer time horizon initially to ensure that the expected benefits materialize. Such investors are usually from the public sector or the nonprofit sector, but some private companies such as "impact investors" and "social investors" also have longer time frames. The high transaction costs inherent in inclusive models and their heterogeneous nature makes them very fragile in the beginning. There is a high risk that faced with high initial costs, slow progress, and the absence of tangible benefits in the starting phase, both the investor and local players might become discouraged and abandon the project. Consequently, inclusive investment models require substantial support from an independent and competent third party that can play the role of honest broker (NGOs, government agency, development banks, etc.) and facilitate collaboration between the investor and the local community. Among the many factors that condition the impacts of foreign investment on the local economy, the domestic laws and institutions governing agricultural investment and land tenure are the most critical.

However, although the findings presented are based on case studies and do indicate useful directions, one should refrain from drawing general conclusions for several reasons. First of all, the case-study approach has inherent limitations and cannot fully capture the wide variety of situations. Some observed changes may be due to other factors than the considered investment. Another reason is the issue of time frame. Most studies analyze recent investments, while the full effects may materialize many years after the investment has taken place. Over the long run, the outcomes of a project may change drastically. Finally, it is difficult to compare the results due to differences in local context, and drawing general conclusions is problematic because the studies use different analytical frameworks. There is a need for normalizing the approaches of the various research activities on the impacts of agricultural investment.

As a conclusion on this topic, I would like to refer to a statement demonstrating the relative awareness that we are exposed to when it comes to land grabbing and international deals. D'Odorico indicated after the release of the report he co-authored,[69] "There's still an enormous amount that researchers don't know about land-grabbing. It took years of painstaking work just to assemble data on how much land and freshwater is actually being bought up abroad. And there are many basic things that researchers can't yet detail, including where the crops grown on grabbed land actually go, or how much yields improve when foreign investors come in." Fred Pearce[70] completed that thought by turning to the practical aspects of land grabbing, a phenomenon that is likely to increase in incidence in the years to come, when he reported in an interview, "The net result is that poor farmers and cattle herders across the world are being thrown off their land. Land grabbing is having more of an impact on the lives of poor people than climate change."[71]

9
Farmland II: Capita Selecta

9.1 Introduction

Population growth, resource scarcity, and climate change are the three defining economic trends of our modern times. Any sector positioned at the nexus of their convergence will offer investors the best midterm opportunity and the potential for stellar returns over the long term. Agriculture is one such sector. The consequent increasing scarcity of farmland has resulted in rapidly rising farmland prices across almost all regions of the world. Farmland values are determined by the relationship between demand on the one side, driven primarily by the profitability of agricultural enterprise, and the supply of productive farmland on the other. More specifically, rising demand for agricultural commodities will exert demand-side pressure on farmland values, while restrictions on cropland expansion will exert supply-side pressure. The interrelationships between supply and demand for farmland are complex. Demand for land increases when commodity prices rise. In response, supply increases if further land is brought under cultivation. However, there are many other factors at play. For example, efficiency increases or yield-enhancing technologies might mean that less land is required to produce the greater supply of commodities required in the future. On the other hand, losses in productivity from climate change and land degradation could have the opposite effect. The core objective of this chapter is to assess prevailing and emerging trends in supply and demand in the agricultural sector.[1] The intention is to provide a clear understanding of the interrelationships between these forces and what this might mean for the investment prospects of farmland as an asset class in the mid- to long term. Farmland investment should be considered a long-term strategy. The objective is to gain a clear understanding of the longer-term trends required to assess the best opportunities for reliable, long-term investment performance.

Let's begin with a brief look at the dynamics of supply and demand for farmland and how this affects farmland values. It highlights the fact that both farmland incomes and values are rising in step with agricultural commodity prices despite rising input costs. Farmland values are dictated, over the long term, by the relationship between supply (measured by the per capita availability of farmland) and demand (for the limited supply available). On the demand side of the equation, farmland values are driven primarily by the profitability of agricultural enterprise. The greater the income yield a farmer is able to earn from farming activities, the higher the price they will be prepared to pay for the land upon which that yield is derived. This relationship applies for both farmer landowners and investor landowners. Tenant farmers will be prepared to pay higher rents on land if the financial incentive to do so exists, and investors will be prepared to pay a higher price for land as yield ratios improve.

The profitability of an agricultural enterprise is dictated by the relationship between input prices such as fertilizers, pesticides, herbicides, and fuel (all of which influence yield per unit of land area) and the output value of the crops these inputs produce. Thus, agricultural commodity prices play a crucial role in dictating land value. It is this relationship between agricultural commodities and farmland values that goes a long way to explaining the marked gains in farmland prices in recent years. On the supply side, if there is a high level of availability of farmland in a particular market, then prices will likely be lower compared to a market with more limited availability. In any given market, the most productive land is taken into production first, as this land provides agricultural enterprises with a higher level of income. For two different regions in which land characteristics are broadly similar, the availability of farmland explains much of the variation in farmland prices. As the most productive land is developed first, in a mature farming region, land currently under cultivation generally commands higher prices than land yet to be brought into production. In order for land at the less productive extremity of the land-supply curve to be brought into production, commodity prices and agricultural enterprise profits would need to be high enough to justify the greenfield development costs while taking into account the diminished returns from lower productivity. There is also proportionality between rental rates paid by tenant farmers and land values. Land values can trend away from rental rates in densely populated regions such as Japan. This is because farmland prices can also experience upward price pressure from development speculators, thus driving prices beyond pure agricultural value. This price differential can encourage the market to pay prices over and above the level dictated by agricultural earnings in more densely populated regions. This direct proportionality between cropland expansion and rising population has collapsed in recent years. This makes perfect sense when considered in the context of Ricardo's theory of rent. Less productive land is brought into cultivation only when farm profitability rises. As this occurs, previously developed more productive land will increase in value relative to newly developed less productive land. This means that in a future where less productive land is brought into production, investors who had the foresight to acquire more productive land in the past will benefit more than new investors. Despite this apparently simple relationship between farm profits (or rents), farmland availability, and farmland values, predicting the future is a lot more complex than it seems because agricultural commodity prices are also set by supply and demand. Agricultural commodities are renowned for their cyclical behavior. During a good harvest year, global grain stocks are high, creating a supply surplus relative to demand, and commodity prices fall. This reduces the financial incentive to farmers to invest in planting more of that particular crop during the next planting season. This in turn results in lower harvests in the next phase of the cycle, thus reducing global grain stocks and causing commodity prices to rise once more. These higher prices incentivize further investment in production, and the cycle begins again.

For this reason, predicting land values in the short-term is challenging, but as farmland investment is a mid- to long-term strategy, longer-term trends in supply and demand for commodities are of more interest to the farmland investor. Returns, particularly the capital-growth component, are linked to long-term agricultural commodity trends rather than short-term price volatility. It is the long-term fundamentals of food demand growth and food supply constraints that are likely to result in a continuation, and perhaps an acceleration, of the historical upward trend in farmland asset values. On a very basic level, the mere fact that global population continues to increase while the carrying capacity of the planet remains intrinsically finite gives an instinctive sense that prices must rise over time. Although this simple generalized view appears logical, answering more specific questions about the implications for future capital values and rental yields on farmland is more challenging.[2]

If one looks at agricultural productivity and commodity prices in a historical context. The impression is of an agricultural economy that has displayed a long-term cyclical trend throughout recent history. It has alternated between periods of supply tension characterized by rising commodity prices and periods of supply expansion characterized by falling commodity prices.[3]

The implications for commodity prices in the foreseeable future are noteworthy. The surge in 2008–2009 in commodity prices was unique in many respects. Despite increased speculation in commodities, the unprecedented rises were supported by real fundamentals of rapidly rising demand against a backdrop of comparatively fixed supply, leading to the historic lows seen in global food inventories in 2008. The consensus is that commodity prices will stay at or close to all-time highs, combined with the cyclical movements up and down as the seasons go by (see the outlook chapter for more on this).[4] Despite the recent fallback, mid-2013 prices for most agricultural commodities were close to all-time highs. There is no direct indication of that changing dramatically anytime soon.

Going forward, the big questions are, how much further expansion in agricultural production can be achieved (both in terms of yield and total cropland area), at what rate can this expansion take place, and for how long can that expansion continue? If the expansion can take place as easily, cheaply, and rapidly for the next half century as it has done for the last half century, perhaps the prospects for agricultural investment aren't as bullish as the demand-side fundamentals suggest. Alternatively, it may be that, as far as expansion is concerned, the low-hanging fruit has already been taken. In fact, the numbers suggest just that. The evidence indicates that the law of diminishing returns is asserting its inevitable effect on world food production. Indeed, this has become increasing apparent in recent years. The law of diminishing returns refers to how the marginal contribution of a factor of production usually decreases as more of that factor is used. A consequence of diminishing marginal returns is that as total investment increases, the total return on investment as a proportion of the total investment also decreases. The same logic applies to finite resources such as land, fertilizers, pesticides, and any other variable external input, such that each additional unit of the variable input yields smaller and smaller increases in output. The return curve is first linear and then begins to decline toward zero. For example, for each additional kg of fertilizer applied (above 0 kg), yield increases linearly up to the point where diminishing returns begin. After that point, the line curves toward the horizontal, and may even become negative if applications reach toxic levels, the result being that both unit and total returns decrease. The reality is that data supporting the increasing emergence of diminishing returns is already being observed, and this despite the fact that the global use of nitrogen fertilizers during the period increased significantly, as did water use. To achieve each unit of production gain, it took a much higher increase in the use of variable inputs, roughly fourfold in the case of nitrogen fertilizer. Data implies that future gains in productivity due to increased application of fertilizers will be lower than the rates of increase seen in the past.[5] These trends have taken place in the presence of another greater trend, that of population increase. This explains the fact that the per capita production of grain has been in decline since the early 1980s, roughly the same point in time as productivity gains due to nitrogen fertilizer bottomed out. One of the primary reasons for this is the "fixed input" component of the agricultural productivity equation: farmland.[6]

What potential is there to expand irrigated land? The trend for the expansion of irrigated land has flattened off in recent years despite the rise in demand for food and the accompanying financial incentive. During the last decade (1998 to 2007), the average annual rate of expansion in the area of agricultural land equipped for irrigation was 0.6%, whereas during the 1960s the average annual rate of increase was 2.1%, over three times higher. This is due in part to the fact that the areas most suitable for irrigation have already been brought into cultivation.[7] Like the best rain-fed lands, the best irrigated lands are developed first. There is a development cost associated with putting in place irrigation infrastructure. The law of diminishing returns has been making itself felt here as well, meaning that the average cost per unit of production of developing irrigation infrastructure has increased over time (as has the unit cost per hectare). Due to higher irrigation development costs and already stretched water supplies in many regions, it is assumed that the downward trend in the new development of irrigated cropland will continue in the future.[8] The reality of diminishing returns has implications for many of the medium-term agricultural commodity price forecasts that are based on historical productivity increases. The implication is that using more recent data would produce lower future productivity estimates.

It is therefore clear that commodity prices have implications for farmland values. An intertwining factor is the already often discussed population growth, with China being the 800-pound gorilla in the room when it comes to this issue.[9] As a nation, China is a growing net importer of food, placing an increasing strain on global food supply. It is earmarked by the FAO as a country suffering from "severe localized food insecurity."[10] Other correlating factors are constraints and noneconomics regarding livestock, urbanization, biofuels, and changing dietary habits; increasing levels of calories per human due to increased global wealth; and the impact of climate change[11] and related issues such as soil degradation, urbanization, and so forth, as are discussed in the dedicated chapters.[12]

9.2 Demand/supply and future farmland values

Standard economic theory suggests that high prices are their own worst enemy. A high price spurs producers to find new means of raising output. In other words, price increases lead to supply-and-demand responses that result in due course to lower prices. One such response is the propensity of consumers to choose alternative products as prices rise. Trying to predict prices at a given point in the future is extremely challenging, but directional price evolutions should be able to be generated by looking at fundamentals. As food demand as such is pretty inelastic, it has always been seen as a defensive industry. Food expenditure as a share of total expenditure is lower in more developed countries with a higher per capita income. This is described by Engel's law, which displays an inverse relationship between the percentage share of food expenditure and income. This relationship has strong implications for agricultural commodity prices when considered in the light of changing dietary habits in countries such as China and India where per capita GDP is rising (between 13%

and 20 % versus 10% in Europe and the United States on average). The price inelasticity of food demand is most clearly demonstrated by recent data from developing countries where incomes are much lower and the share of food expenditure is much higher.[13] The price spikes of 2007–2008 only had a minor impact on the level of dietary energy intake in even the poorest countries of the world, with only a marginal difference between the richest and poorest sectors of the population in terms of the extent to which food purchases dropped as a result of higher prices. This lends further support to the notion that even in lower-income sectors of the market, price elasticity of demand for food remains low.[14] The effects of price inelasticity of demand for food are further amplified in more developed countries because the commodity component forms a smaller proportion of the final food bill. This is due to the fact that developed economies have more complex, costly, and wasteful food distribution, processing, packaging, and marketing systems (urbanized systems). This means that the extent to which changes in agricultural commodity prices feed through to food prices is lower than in developing countries. Farm-level commodity prices account for 20% of the total consumer cost in the West versus 25% to 35% in the developing world.[15] The world as a whole now spends less than a quarter (measured as a proportion of total merchandize trade) of the money on food that it did in the early 1960s. In addition, despite recent price spikes, the actual price of agricultural commodities in real terms actually remains low by historical standards. Commodity prices relative to incomes illustrate this point even more convincingly. Consumers are now spending a significantly lower proportion of their income on food than they have in the past. The complex and dynamic interrelationships between supply and demand, the state of the broader economy, environment, geopolitics, energy prices, and other factors make predicting actual future prices difficult. What is certainly clear is that there is a lot of room for the price of agricultural commodities and the farmland upon which they are produced to rise in the decades to come, judging based on a historical comparative analysis.[16] A final driver has been the upcoming peak oil[17] and the overall shortage of fossil fuels. Although discussed in the biofuels chapter, there a few pointers in relation to farmland.

The extent to which oil prices will affect demand for biofuels is dictated by future price levels of crude oil because the economic incentive to produce biofuels increases in proportion with oil prices. This affects food commodity prices in two ways. First, new market demand for feedstock crops is created for "first-generation biofuels" (biofuels made from edible plants containing sugar, starch, or oil or from animal fats). Second, competition for arable land for "second-generation biofuel" feedstocks (e.g., the use of nonedible plant matter such as wood and cellulose as feedstocks) will reduce the availability of arable land for food production.[18] The higher the oil price, the more economically viable biofuel production becomes (even without subsidies or climate-change-mitigation incentives), and the greater will be the competition for cropland. At oil prices above the $50 mark, biofuel production becomes capable of producing significant profits, thus creating stiff competition and forcing up the price of maize, wheat, and other feedstock crops.

The single biggest culprit in the consumption of fossil fuels in industrial farming is the use of agrochemicals (pesticides, fertilizers, growth agents, etc). These chemicals are absolutely critical to the supply side of the equation. As much as 40% of energy used in the food system goes toward the production of artificial fertilizers and pesticides. Nitrogenous fertilizers are synthesized from atmospheric nitrogen and natural gas, a process that takes a significant amount of energy. Producing and distributing them requires an average of 62 liters of fossil fuels per hectare.[19] During the 2008 price spikes, fertilizer prices reached historic highs, due to a combination of rising global demand and high oil prices, and farming input costs increased dramatically. Farmers in low-income countries who could not afford fertilizers reduced their usage, relying on the residual nutrient content in the soil from natural stocks and previously applied fertilizers, a process known as "nutrient mining."

An inevitable consequence of nutrient mining is the lowering of yields over time. This in turn resulted in a drop in worldwide grain production, which fed through to higher agricultural commodity prices. This would place upward pressure on agricultural commodity prices, causing farmland profits in high-income countries more capable of absorbing rising fertilizer prices to rise.

The commonly accepted viewpoint among agricultural investors is that rapid economic growth in emerging markets has resulted in increased demand for and consumption of world commodities, and that this increased consumption is the primary driver behind rising commodity prices. However, this statement can be questioned, and when it is questioned, the analysis is built around China as the 800-pound gorilla in the room.[20]

9.3 Farmland and portfolio considerations

Traditional arguments with respect to including farmland in an investor's portfolio are typically the following:

- Steady, stable return – greater return than a traditional fixed-income inflation hedge
- Noncorrelation with other asset classes (although it is possible to take a position through an equity position in a farming company)
- Qualification as a "hard asset" – protection of value and stable outlook
- Superior performance in an inflationary environment, to mitigate market risks due to global monetary policy
- Decent visibility with respect to the valuation of farming assets

- Underowned asset class
- Good long-term demand/supply fundamentals

When investing directly in a farming asset, an investor might expect to earn a return in three possible ways:

- Income from agricultural tenancies
- Capital growth from rises in agricultural land values
- The potential for windfall returns if land is developed for alternative uses at some point in the future (e.g., infrastructure, industrial, commercial, or residential development)

It was mentioned earlier in previous chapter that a number of studies have shown that, historically, farmland returns have a low or negative correlation with traditional asset classes such as stocks and bonds and only a modest positive correlation with commercial real estate.[21] These characteristics make farmland an attractive diversification tool that can help reduce the impact of broader market volatility on a diversified portfolio. The farmland component can be further diversified by varying crop types, management styles, and geographic distribution within the portfolio. Farmland returns have been shown historically to have a positive correlation with inflation, making farmland an effective inflation hedge and capital preservation vehicle.[22] When seeking capital preservation, data show that farmland has exhibited strong capital protection characteristics over prolonged periods of time. For the risks of farmland investing, see supra in this chapter- and how they can be avoided by ensuring effective due diligence and title of the property and normal efficient management of the farm. Studies have shown that even taking into account the transaction costs associated with less liquid assets, farmland still constituted a substantial portion of the optimal portfolio across a wide range of scenarios.[23] The risk-return paradigm also seems to be positive. The Sharpe ratio for US farmland, for example, was higher than large-capitalization US equities, T-bills, and commercial real estate for the period 1991–2007. In terms of total returns, farmland was outperformed only by stocks (by 0.3% annually). Stocks, however, had more than double the risk (in terms of standard deviation) compared to farmland.[24] The performance of farmland against other asset classes also shows positive dynamics. In a study conducted in the United States, where farmland was specifically analyzed from the perspective of capital growth and income from rents, the findings were consistent with other studies in terms of its benefits compared to other asset classes.[25] Historically, land (and agriculture in general) has repeatedly benefited from "flight-to-quality" investment behavior. It performs comparatively well during times of market uncertainty, thus acting as an ideal recessionary hedge.[26] Although farmland is qualified as real estate, it does have a number of unique features that shelter it from the typical real-estate market swings:

- Limited supply.
- P/E and yield tend to move in check. Agricultural land values are, in general, supported by earnings (although some agricultural areas may experience upward price pressure from development speculators, thus driving farmland values up beyond pure agricultural value – see for that the current US farmland value discussion supra in this chapter). Farmland has therefore shown itself to be a reliable vehicle yielding, over a prolonged period of time, returns that include both the capital gains on principal as well as income. However, purchasing farmland at above-market prices in regions where earnings from farm operations do not support current real-estate values is essentially a bet on future real-estate development, and as such is inherently higher risk
- Both the debt-to-equity and debt-to-asset ratios are low in the farming sector.
- There are a range of tax-related incentives associated with farm real estate.

9.4 Getting exposure

There are a number of different ways of investing in the agriculture sector.[27] The choice of method will be determined by the investor's objectives: short, medium, or long investment term; liquidity and income requirements; risk appetite; and so forth. Listed below are the principal methods used by the investing community to gain exposure to the agriculture sector followed by a brief assessment of their relative merits and demerits.

1. Agricultural commodities: Investors can get direct exposure to individual crop prices through investing in agricultural derivatives such as futures or options, or buy into a mixed basket of agricultural commodities through exchange-traded funds that track commodity indexes.
2. Direct investment in agricultural equities: There are a number of agricultural equity plays available to investors. It is possible to invest in large-scale commercial farming enterprises involved directly in crop production, or in other industries that supply the agriculture sector, such as fertilizer, pesticide, and seed producers and agricultural machinery manufacturers.
3. Collective investment funds: Investing in farmland through a securitized or unitized fund that invests over a portfolio of different farms or agricultural equities. This can also include ETF[28]/ETN[29]/ETPs or real-estate investment trusts (REITs), etc.

4. Direct farmland investment: Direct investment in agricultural land with a fixed rent, possibly with an additional variable element linked to farm income, profitability, or commodity prices. Alternatives are through an agriculture- focused private equity (PE) fund.

While a direct comparison is somewhat challenging given the different risk profile of each method, there are a number of benefits to direct farmland investment, both in terms of asset characteristics and mid- to long-term supply-and-demand fundamentals. As already indicated, farmland has performed well against the majority of other assets in a number of different markets over a wide range of different time frames. There is also evidence to suggest that this holds true not just across different asset classes but also across different instruments in the agricultural sector.[30] By owning farmland through agricultural companies, or by owning companies that depend on the sector for revenue, the investor obviously assumes additional risks such as enterprise risk and other risks generally associated with investing in companies. This additional level of risk has not shown itself to be sufficiently rewarded by proportionately higher historical returns.[31] Indeed, whatever the resource sector, you always want to invest as upstream as possible – closer to the source. It was already indicated that there is less volatility in direct farmland investment compared with other asset classes (including agricultural equities and commodities). Farm-enterprise profits are more vulnerable to fluctuations in commodity prices and input costs. It is this perceived volatility within the farming sector that has caused the nonagricultural investor to believe that farmland ownership is riskier than it actually is. Diversification is frequently cited as one of the advantages of investing in farmland through a collective investment-fund structure. For institutional investors (i.e., those with a sufficient capital base), similar diversification is possible through direct farmland investment by investing in a portfolio of farmland assets.[32] When doing so, a long-term investment horizon is recommended. The argument that there is a lack of clear and consistent market data to make effective investment decisions is only valid short term.[33]

9.5 Portfolio weighting of farmland[34]

Despite all of the advantages, agricultural assets have not been high on the agenda in the mainstream investment community.[35] The modern asset manager, adhering to modern portfolio theory, seeks to increase return, or maintain return at stable levels, while reducing overall volatility (risk) by combining assets with low or negative return correlations. The question of appropriate portfolio weightings for farmland needs to be considered not only in the context of the current rather unique and unpredictable market conditions but also in terms of the investor's future expectations of market activity. While farmland investing is expected to provide a diversification benefit, an inflation hedge, and a greater return than fixed income, it entails challenges and risks (see supra in this chapter[36]), especially given the relatively low level of expertise in the investment community compared to the traditional asset classes. Due diligence is therefore key.[37] Nevertheless, farmland seems to be delivering on its promise both before and after the 2008 financial crisis.[38]

Modern farming technology, especially the size and capacity of farm machinery, has driven a strong trend of consolidation in the agricultural sector as the financial benefits of economies of scale become more pronounced. This trend is apparent in many parts of the world. A recent report takes an interesting look at the evolving relationship between farm size and farm economics (from a US perspective), providing useful insights into what the future may hold for investors in this and other regions.[39] The next question then becomes to see if the cash rents follow farmland values to determine whether cash rents have changed at the same pace as land values.[40]

Buying and selling farmland, like other types of real estate, carries higher transaction costs than many other assets. Additionally, transactions generally take longer to execute than more heavily traded unitized assets. This limited liquidity leaves options for those who are able and willing to look further than next month or quarter.[41] Farmland prices and farming profits vary depending on local conditions in the area being considered. The downside of investing in low-rainfall areas, however, is that farm income is less reliable and more volatile than in high-rainfall areas. This may adversely affect a low-rainfall tenant's ability to pay rents. Besides the already-mentioned climate-change impact on farmland values, the geopolitics cannot be stressed enough. Honoring of the legal title or long-term lease is essential, as with the long-term investment horizon the possibility of negative impacts on returns accumulates. Legal stability and a properly functioning judiciary are crucial.[42]

9.6 Financialization of farmland[43]

What remains is the question that relates to the financialization of farmland as an asset class. The current farmland-investment boom could be seen to indicate a deviation from the process of financialization. Farmland investors often draw from discourses that stress the profitability of long-term, productive investments and frequently choose an "own-operate" approach that involves investment in agricultural production as well as just land. In many ways, however, this trend represents a continuation of financialization into new territories. Financial returns linked to productivity reveal that this is the newest area of reality subject to financialization rather than a shift away from financial to real assets. Investments depend on both the use- and exchange-value aspects of land. Fairbarn concludes that "perhaps what we are seeing is the emergence of a new type of financialization for an era of growing resource scarcity – one in which farmland's role as a quasi-financial asset will be even more prominent."[44] As we have seen since the 1970s with real estate, the securitization of real assets (REITs, etc.) takes away its potential to protect against inflation and increases the correlation with other asset classes and increases volatility.[45] The risk of that occurring short term is unlikely, but longer term it cannot be ignored. Since farmland investments are longer-term oriented, it cannot be ignored, even today. The use of land as a financial asset is not restricted to professional investors. "Instead the sector is characterized by crossover; financiers are using land as a productive asset, while operators are using land as a financial asset. Rather than a situation in which land is increasingly treated as a purely financial asset, ... land's financial qualities are increasingly valued but not necessarily divorced from its productive qualities."[46] However, short-termism is around the corner, and if capital gains are to be realized, rather than just serving the purpose of value storage, then the land (or the company that owns the land) must eventually be sold. Timing becomes important in purchasing and disposal decisions.[47]

10

Farmland III: Land-Expectation Value and Timberland Valuation[1]

Wood and commodities made of it have worldwide importance. The global forest area consists of 3.9 billion hectares of forestland, which is about a third of the total global land area. In 2000, the amount of round wood felled was 3.4 billion cubic meters, half of it firewood. The global production of forestry commodities was about one billion cubic meters in 2000. The export value of these commodities in world trade was USD 134,000 million and the import value USD 141,000 million.

In order to maintain free and unchallenged capital transfers globally, it is more and more important for decision makers to have trustworthy and comparable facts at their disposal. This applies to enterprises and accounting systems (IFRS) as well as different kinds of property valuations (IVS).

Some argue that present valuation standards pay too little consideration to the valuation of forest properties. They suggest that completely independent and specified valuation guidance should be created for the valuation of forest properties.[2] In this chapter we are going to explore the different viewpoints and meaningful valuation analysis for farmland, forests, and timberland, and we will do so taking into account the different angles of valuation from which this can be looked at.

10.1 Valuation models[3]

The three main concepts that are being used to establish the "value" of land[4] are: economic value, noneconomic value, and market value.

10.1.1 Economic value

The economic value of farmland is comprised of productive value and capital-gain value.

A. Productive value

The productive value of land is determined by the land's ability to generate a financial return. To estimate the productive value, or the return to land, all income and costs (cash and noncash) must be accounted for.

There are four steps to determine land's productive value:

1. Estimate gross income from the land to be purchased.
2. Estimate all costs of production, *except* for interest on the land.
3. Determine the return to land (Step 1 minus Step 2).
4. Divide the return to land by the capitalization rate.

1. **Estimating gross income**

   Gross income on the land to be valued should be based on the crop rotation to be followed using long-term average yields for the management level that applies and the long-term market price of the grain.

2. **Costs of production**

   Industry or government cost-of-production estimates do not necessarily represent an individual's situation. As there is great variation in farming practices, crop rotations, soil type, and climate, producers must calculate their own expected costs. The cost of production should include an estimate of all costs associated with the land to be valued, except for an interest charge on the land investment. This would include all cash costs, such as seed, fertilizer, machinery repairs, taxes, and so forth, and noncash costs, such as depreciation on buildings and machinery, interest on building and machinery investments, and an allowance for labor and management.

   Where available, a farmer should use estimates based on his or her own past records and trends. An estimate for cash expenses such as seed, fertilizer, and chemicals will need to be based on the condition of the land being valued and the crop rotation being planned. Other cash expenses, such as machinery, fuel and repairs, insurance,[5] etc., can be based on the costs per acre on the existing land, with the assumption that the costs on the new land will be similar.

   Noncash costs (interest and depreciation on buildings and machinery) should be determined for the whole farm and then prorated to the additional land to be purchased. This procedure accounts for the noncash costs on the existing buildings and machinery, plus the noncash costs on additional buildings and machinery that are acquired as a result of the land purchase.

3. **Return to land**

   The return to land is the gross income minus operating costs, depreciation, and interest costs for buildings and equipment, and an allowance for labor and management. Interest paid on term loans should not be included in the operating costs.

Small variations in crop yields and prices can have a dramatic effect on the return to land. Furthermore, it is difficult to predict prices accurately, and yields can vary significantly in the short run. Therefore, several calculations should be made using different yield and price assumptions. The range of results provides a basis for assessing the risk involved in buying land. This is often called sensitivity analysis.

4. **Capitalization** is the conversion of future profits or earnings from the land into a current economic value.

   By using the appropriate capitalization rate, farmers can estimate the value of land. The capitalization rate should be the rate of return that could be earned on other investments. A minimum rate is the earnings on a savings account, while the maximum rate would be the bank interest rate on loans.

   The capitalization procedure converts future returns into today's value. For example, if the goal is to earn a 6% annual return on an investment in land with an expected net income of $15 per acre, the value of the land would be $250 per acre. The capitalization formula is as follows:

   Annual return to land/capitalization rate = productive value of land

   Productive value of land: $15 per acre/0.06 = $250 per acre.

B. Capital-gain value

Productive value is not the only factor to consider when determining the economic value of land. Another important consideration is the amount of capital gain that might be realized when the land is sold.

To incorporate expected capital gain into the economic value formula, reduce the capitalization rate used by the expected annual rate of capital gain. For example, if the capitalization rate is estimated to be 6% (as in the example) and the annual capital gain is expected to be 1%, then the adjusted capitalization rate would be 5% (6% – 1%).

The formula to determine the economic value of land (with the productive value adjusted by a potential capital gain) is

Annual return to land/(capitalization rate – annual capital gain) = Economic value of land

Using the data of the above example, the calculation is

$15.00/(.06 – .01) or $15.00/(.05) = $300 per acre

The economic value of land is equal to the productive value plus its capital-gain value.

10.1.2 Noneconomic value

Noneconomic factors affect the value of land. For example, a buyer may be willing to pay more to live in a particular community. A farmer may place more value on land that is adjacent to land already owned. Land may be of interest to some buyers because of its aesthetic value – it may be located near a river or a picturesque creek. In some cases, nonfarmers may be competing with farmers for land.

10.1.3 Market value[6]

Land's "market value" refers to the price of transactions between informed buyers and sellers. It is always recommended that you hire an accredited appraiser in cases where an accurate valuation needs to be done. You can, however, get a good idea of market value by doing some analysis yourself.

The technique to determine market value is to use "comparable sales" of similar property in the same area. When doing so, it is extremely important to compare properties that are very similar in every way, and to examine the conditions and terms under which these properties were sold. It is important to obtain as many comparable sales as possible, as the more information, the more accurate the average becomes. It then becomes a simple process of calculating the average selling price, usually per cultivated acre, or per forage acre, or per grazing acre.

Here are some factors to consider when looking at comparable sales:

1. Proximity is important. How close are the comparable properties to the one that is being sold? If the properties are too far away, it may not reflect your "local" market for land.
2. Comparable properties should, as much as possible, have the same soil types, similar topography, and similar stones and sloughs – those things that change how much someone views a property to be worth.
3. Determine the motives of the purchasers of the comparable sales. A purchaser buying land adjacent to his/her farm may be prepared to pay more.
4. Improvements to property, such as buildings, make comparisons of land more difficult. It is often better to use bare-land sales only.
5. Sales between relatives or close friends should not be used for comparative purposes because they may not reflect market conditions.
6. Adjustments should be made if there are differences in the number of cultivated acres per quarter section. In fact, as stated above, the comparisons are usually done on a "per cultivated acre" or "per forage acre" or "per grazing acre" basis.
7. If the property has not been farmed properly, the estimated value should be reduced by the cost of bringing the property back to normal condition.

Another quick method for making a more general comparison is "multiples of assessment." By using these same comparable sales, make a ratio of selling price divided by municipal/regional or country assessment(s). This is the multiple of assessment. Take the assessment and multiply it by the ratio to get an estimate of market value.

In summary, land used for agricultural purpose shall be valued using solely the income approach to value without any allowance for urban or market influences. The income of the property shall be determined using the capitalized average annual net cash rental of the property.

1. The average of the annual net cash rental, excluding real-estate and sales taxes, is determined through an analysis of typical arm's-length rental agreements collected for a five-year period before the year for which the valuation is being determined for comparable agricultural land used for agricultural purposes and located in the vicinity, if practicable, of the property being valued.
2. The outcome of (1) shall be capitalized at a rate 1.5 percentage[7] points higher than the average long-term annual effective interest rate for all new farm-credit services loans for the five-year period before the year for which the valuation is being determined.

The income approach is used to value commercial or industrial properties, or properties which are bought and sold by investors primarily because of their income-producing potential. This approach to value depends on reliable and detailed information on the income and the costs of doing business for a particular business or enterprise. This is referred to as the "income stream" of the property. The income approach defines value as "the present worth of future benefits of owning a property." These are composed of the annual income for an estimated number of years (called the economic life of the property) plus a capital amount representing land value or land value plus some remaining worth of the improvements. This approach emphasizes investment components rather than physical components of a property.

The steps in the income approach are as follows:

1. Estimate potential gross income (PGI).
2. Deduct vacancy and collection losses.
3. Add miscellaneous income to derive effective gross income (EGI).
4. Deduct operating expenses to derive net operating income (NOI).
5. Select appropriate capitalization rate and method.
6. Develop an estimated value.

10.2 Land-expectation value and timberland valuation[8]

Appraisers and other professionals often use discounted cash flow (DCF) techniques to value timber and timberland. Land-expectation value (LEV) is a standard DCF technique applied to many timberland situations. LEV calculates the value of bare land in perpetual timber production and is often used to value even aged-pine plantations. However, it is also useful in the valuation of immature timber stands and uneven-aged timber stands cut periodically. These models have wide applicability in timberland-appraisal situations.

Discounted cash flow analysis derives the net present value (NPV) of the net income stream produced by a property. It is a relatively simple calculation, applicable to many timberland-appraisal situations. Forestry and timberland-investment analysts commonly use a specialized DCF technique to calculate the value of bare land in timber production. Land-expectation value (LEV) is simply the value of a tract of land used for growing timber. It is the NPV of all revenues and costs associated with growing timber on the land in perpetuity (not just those associated with one "rotation of timber" or other time period). LEV is thus a special case of DCF where a perpetual stream of revenues and costs are considered. LEV can be interpreted as the maximum price possible for a tract of timberland if a rate of return equal to the discount rate used to calculate LEV is expected. If the NPV of all cash flows expected from growing timber on a specific tract of land is estimated, the expected value of the land has been estimated (hence the name "land-expectation value"). The LEV criterion is also called "soil-expectation value" and "bare-land value," because many applications assume the cash-flow stream begins with bare land. LEV also is sometimes called the Faustmann formula. The technique was first published in 1849 by Martin Faustmann, a German appraiser who developed the formula to place values on bare forestland for tax purposes.

While the calculation is not complex, it is not commonly used by appraisers. The formula does require judgments with respect to stumpage prices, reforestation costs, and forest yield. Evaluating a site with respect to these items may be beyond the expertise of many appraisers and will frequently require the assistance of a forester. Current stumpage-price information and cost-of-forest-practices data are available for most regions. A forester's expertise might be most necessary in establishing forest yield. Forest yield depends on site index (i.e., a measure of the land's ability to grow timber). The calculation is no more precise than the quality of the data used as input.

10.2.1 The LEV calculation

LEV simply calculates the value of bare land in perpetual timber production. It is based on the standard discounting formula for the present value of a perpetual periodic annuity:

$$PV = \frac{a}{(1+i)t-1} \tag{a}$$

where
PV = present value of a perpetual periodic annuity
a = value received every t years in perpetuity
t = years between annuity payments
i = interest rate, expressed as a decimal

This is actually a standard DCF calculation, but with several critical assumptions:

1. The values of all costs and revenues are identical for all rotations. All costs and revenues are compounded to the end of the rotation to get the future value of one rotation. This value will be the amount received every t years.
2. The land will be forested in perpetuity.
3. The land requires regeneration costs at the beginning of the rotation.
4. Land value does not enter into the calculation. Land value is what you are calculating. The value calculated is the present value (PV) for a perpetual series of rotations. Many timber companies and pension funds do not buy timberland with the intention of holding it to perpetuity. The LEV does give the value of bare land in permanent forest production, however, and is the standard forestry DCF calculation. Because it is a standard DCF calculation, it can be applied to single or multiple rotations on a consistent basis. For example, if the future value at the end of a single rotation is \$916.76 (see Table 10.1), LEV is \$408.65 at a 4% interest rate. The PV of the first rotation is \$916.76/(1.04)30 = \$282.65. This means that the PV of the remaining perpetual rotations must be \$408.65 – \$282.65 = \$126.00. The LEV at the end of year 30 must also be \$408.65, and if that value is discounted to today, we obtain \$408.65/(1.04)30 = \$126.00. Thus, LEV does have practical applications to any situation where a forested tract will be in permanent timber production but might change owners at different times.

The calculation is quite easy and involves two steps. First, each cost and revenue is compounded to the end of the first rotation. The net value at rotation represents the dollar amount available at the end of each rotation in perpetuity. Second, the PV of the dollar amount is calculated on a perpetual periodic basis using Equation a. To calculate LEV for even-aged management (e.g., a pine plantation with all trees equally aged) on bare land, a simple three-step process is used:

Table 10.1 Revenues and costs of a typical forestry investment and calculation of net future value (i = 4%)

Year	Item	Amount per acre in USD	Compounding formula	Future value (year 30) in USD
0	Site preparation	(40.00)	$(1.04)^{30}$	(129.74)
0	Tree planting	(40.00)	$(1.04)^{30}$	(129.74)
18	Thinning income	160.00	$(1.04)^{12}$	153.70
25	Thinning income	912.00	$(1.04)^{5}$	194.66
30	Final harvest income	912		912.00
1–30	Annual property tax	(1.50)	$\frac{(1.04)^{30}-1}{0.04}$	(84.12)

Note: Net future value: 916.76.

1. Determine all of the costs and revenues associated with the first rotation. These values should include initial costs of planting, site preparation, and so on, as well as all subsequent costs and revenues.
2. Place the costs and revenues on a timeline and compound all of them to the end of the rotation. Subtract the costs from the revenues.
3. Use the PV of a perpetual periodic series formula to calculate the PV of an infinite series of identical rotations. (Divide by $(1 + i)t - i$ where t is the rotation length.)

Thus, the formula for LEV is simply

$$LEV + \frac{NFV}{(1+i)t-1} \qquad \text{(b)}$$

where
LEV = land-expectation value
NFV = net future value of one timber rotation
t = length of timber rotation
i = interest rate expressed as a decimal

Note that the LEV formula uses constant dollars and a real interest rate. The LEV calculation can include prices or costs adjusted for real price increases by using the formula for a geometric series of cash flows (cash flows that increase or decrease by a fixed % from one time period to the next). Of course, the annual percentage increase must be less than the discount rate, or the LEV will tend toward infinity.

LEV is the theoretically correct criterion for valuing bare land in timber production, for evaluating the value of various forest-management alternatives, or even for determining the age of final timber harvest (rotation age). It is so widely recognized as the standard criterion that appraisers certainly ought to include it in their "menu" of valuation techniques.

10.2.2 Sample calculation for even-aged management

Much timber is grown in plantations, or in stands of same-aged trees. This is called even-aged management. This is the ideal situation for the use of Equation b. It should be remembered that we are dealing with bare land. In a typical appraisal situation such as this, there may be a cutover tract of land, with remaining logging debris and a few scattered unmerchantable trees. A forester can "prescribe" the proper management regime (i.e., timing of management events over one rotation of trees). Assume in this case that site preparation and tree planting are required. The forester prescribes the proper timing of thinnings and the final harvest. The forester can specify the timber yields expected and should be knowledgeable on current prices of the expected timber yields. The appraiser will be required to project future timber prices, or will probably do a constant-dollar analysis and use a real interest rate as the discount rate. Timber prices have easily kept up with inflation since World War II, so a constant-dollar analysis would be appropriate.

Assume the revenues and harvests exist as described in the first three columns of Table 10.1. A 30-year rotation is described for southern loblolly pine, and the real cost of capital is 4%. Site preparation and regeneration will occur in year 0 at a cost of $80 per acre. Annual management costs and property taxes will be $1.50 per acre. Thinnings will occur at ages 18 and 25 and will yield 6 and 10 cords per acre, respectively. Final harvest will yield 57 cords per acre. Pulpwood is worth $16 per cord. If a buyer intends to follow this management sequence and wants to earn at least 4% on the investment, how much can the buyer afford to pay for the bare land?

All revenues and costs must be compounded to the end of rotation (year 30 in our example). The calculation for the net future value (NFV) of one rotation is as follows:

$$NFV = \sum_{n=0}^{t} R_n(1+i)t-n - \sum_{n=0}^{t} C_n(1+i)t-n \quad \text{(c)}$$

where

NFV = net future value of one rotation at year t

Rn = revenue received in year n

Cn = cost incurred in year n

t = rotation length in years

n = year of a particular revenue or cost

i = real discount rate, expressed as a decimal

Table 10.1 illustrates the use of Equation c in determining the NFV of the example rotation. The NFV as calculated in Equation c is substituted into equation b to determine LEV:

$$LEV = \frac{916.76}{(1+i)t-1} = \$408.65 \tag{d}$$

LEV represents the maximum amount that could be paid for a tract of land and still earn the required interest rate. A buyer could pay $408.65 per acre for the tract and earn 4% on the investment, assuming that the land is used to grow timber according to the management schedule outlined.

This simple example does not include some common costs and revenues. For example, there is no provision for revenue from hunting leases. In the Southeast, income from hunting leases could be significant. These types of costs and revenues could easily be added to the calculations in Table 10.1 (e.g., hunting-lease revenue could be netted with the annual property tax). Also the LEV calculation applies to a forest with a predictable periodic timber yield. As a practical matter, unproductive land may have to be averaged into the expected yields, or its value calculated on a separated basis.

10.2.3 Valuing immature even-aged stands using the LEV criterion

Precommercial timber holdings pose a difficult valuation question. The stands of trees have value, but, by definition, they have no current potential for conversion to timber products. The value is intrinsic and is equal to the DCF expected from future timber harvests. Precommercial timber's value changes with its temporal progression toward mature commercial timber. This value is affected by the sunk cost of stand establishment and the opportunity cost of holding land to grow trees.

Comparable sale information often does not reflect the value of immature timber. To value a parcel of land and immature timber at near bare-land value, however, clearly does not make economic sense. Fortunately, a second method using LEV can clearly establish the value of immature timber. Consider the same forestry investment described in Table 10.1. Assume the timber stand is 15 years old. A simple calculation can be used to estimate the value of this immature stand:

$$V_m = \frac{NVt + LEV}{(1+i)t-m} - LEV \tag{e}$$

where

V_m = value of m-aged timber stand

m = age of the immature stand

NVV = net value of the income and costs associated with the immature stand between year m and rotation age t

Table 10.2 Calculation of the value of an immature even-aged timber stand (age = 15 years)

Year	Item	Amount per acre in USD	Compounding formula	Future value (year 30) in USD
18	Thinning income	96.00	$(1.04)^{12}$	153.70
25	Thinning income	160.00	$(1.04)^{5}$	194.66
30	Final harvest income	912.00		912.00
15–30	Annual property tax	(1.50)	$\frac{(1.04)^{15}-1}{0.04}$	(30.04)

Note: Net value at age 30 (NVt): $1,230.32

$$V_m = \frac{1.230.32 + 408.65}{(1.04)30-15} - 408.65 = 910.06 - 408.65 = 501.41$$

The value of this immature stand is calculated in Table 10.2. The value of the immature timber is $501.41. Note that the value of the immature timber and the bare land is $910.06. The bottom of Table 10.2 shows how the $910.06 was derived.

Why does this calculation work? If the interest rate and future management decisions are as originally assumed in the LEV calculation, the value of an immature stand has two components (see Figure 1).

Figure 1 Calculation of the value of an immature timber stand

NVt We add the net value for the period between year *m* and year *t* (*NV*)

+

LEV to *LEV,* which represents the future net value of the income after year *t*,

$V_m = ?$

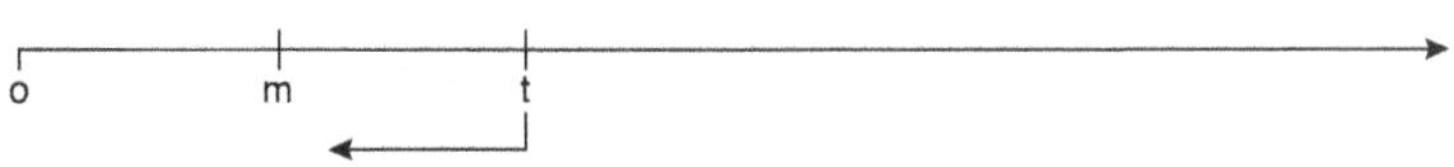

This sum is then discounted to year *m* (the number of years is therefore *t – in*).

$$V_m = \frac{NVt + LEV}{(1+i)t - m} - LEV$$

Why do we then subtract LEV to obtain Vm?

With LEV included, we have the value of the land and timer. When we subtract LEV we have the value of the immature stand of timber only.

1. The discounted net value of the income and costs associated directly with the existing, immature stand (NV).
2. The discounted LEV. LEV is also discounted for years because of the delay in harvesting subsequent stands. The LEV of all subsequent stands isn't realized until the existing stand is harvested in year *t*.

10.2.4 Valuing uneven-aged timber stands using the LEV criterion

Uneven-aged timber stands contain trees of various ages. Usually "mature" trees are selectively harvested on a cycle of some sort. The tract may be harvested annually, removing a small timber volume each year off each acre; or perhaps timber volume is removed every *c* years. In this case, the value of the land and timber must be estimated concurrently, and one cannot be separated from the other. Unless all the trees are cut, bare land never exists under uneven-aged management. In effect, DCF is used to value a perpetual timber-production "factory." The simplest case is when an annual income stream is produced. This is a perpetual annual annuity situation. The standard equation is as follows:

$$LEV = \frac{a}{i} \tag{f}$$

where

LEV = land-expectation value

a = net annual income generated

i = interest rate, expressed as a decimal

Consider a 1,000-acre tract of timber that produces 1,500 cords of pulpwood annually. Pulpwood is worth $35 per cord, so annual revenue is $52,500. The tract costs $3,000 annually to manage, and property taxes are $1.50 per acre per year. Annual costs, then, are $4,500, and net annual revenue is $48,000. Using a 4% discount rate, the LEV of this tract is

$$\text{LEV} = \frac{\$48{,}000}{0.04} = 1{,}200{,}000.00\text{, or } 1{,}200 \text{ USD per acre} \tag{g}$$

The other situation takes place when net timber revenue occurs on a periodic basis – say, every other year, or every five years. The standard LEV calculation is appropriate in this case. Such a forest is said to have "cutting cycles," where a "reserve growing stock" is permanently maintained, and growth from this constant reserve is cut periodically. This is analogous to maintaining the principal in a savings account and periodically withdrawing interest. Note that annual management and property-tax costs are subtracted from net timber revenue using the future value of a terminating annuity formula. LEV is calculated as follows:

$$LEV = \frac{NTR - \left[ax\frac{(1+i)c-1}{i}\right]}{(1+i)c-1} \tag{h}$$

where
LEV = land-expectation value
NTR = net timber revenue received every c years
a = annual management and property-tax costs
c = Number of years in the cutting cycle
i = Interest rate, expressed as a decimal

Consider a 1,000-acre forest that produces \$200 of net timber revenue per acre per year every five years, beginning in five years. Annual management and property taxes are \$2.50 per acre per year.

LEV at a 4% interest rate is

$$LEV = \frac{\$200 - \left[2.50x\frac{(1.04)5-1}{0.04}\right]}{(1.04)5-1} = \$860.64 \text{ per acre} \tag{i}$$

Note that land and timber cannot be separated in this case. Again, we are dealing with a timber-production entity. In addition, the value of an off-cycle uneven-aged timber stand can be calculated. The same concepts used to develop Equation e are necessary. The value of an uneven-aged stand between cutting cycles is given as follows:

$$V_u = \frac{NR_c + : EV}{(1+i)c-m} = \frac{NTR - a\left[\frac{(1+i)(c-k)-1}{0.04}\right] + LEV}{(1+i)(c-k)} \tag{j}$$

where
V_u = value of an uneven-aged stand between cutting cycles
NRc = net timber revenue at beginning of cutting cycle
LEV = land-expectation value, uneven-aged stand
c = cutting cycle in years
k = number of years since last harvest
i = interest rate, expressed as a decimal

For example, the value of the 1,000-acre forest described previously three years after a harvest is given by

$$V_u = \frac{\$200 - \$2.50\left[\frac{(1.04)2 - 1}{0.04}\right] + \$860.64}{(1.04)^2} = \frac{\$200 - \$5.10 + \$860.64}{1.0816} = \frac{1,055.54}{1.0816} \quad \text{or 975.91 USD per acre} \qquad \text{(K)}$$

Notice that Equation j calculates the value of the timber stand immediately after a harvest ($k = 0$), which should equate to the results of Equation h.

When $k = 0$, Equation *j* gives a value of

$$V_u = \frac{\$200 - \$2.50\left[\frac{(1.04)5 - 0 - 1}{0.04}\right] + \$860.64}{(1.04)5 - 0} = \$860.64 \qquad \text{(l)}$$

When $k = 5$, just prior to a harvest, the value of the timber stand is at a maximum.

To reiterate a key point, the value of land and timber cannot be separated in the uneven-aged stand case. The two are intertwined to produce a single value.

In conclusion, a standard forestry DCF calculation is appropriate for most timberland valuation situations. In many cases the value of land and timber can be separated. The LEV model described assumes a cycle or rotation of growing trees. The standard assumption is that valuation takes place at the beginning of a cycle or rotation. The rigid assumptions of the model may make the formula inappropriate in some valuation situations, but it represents the standard forestry DCF valuation model.

The model can easily be adapted, however, to immature even-aged stands, uneven-aged stands, and uneven-aged stands between cutting cycles. Appraisers should find these techniques very useful in timberland valuation.

10.3 Forest valuation guidelines

Forestry enterprises are increasingly attracting interest both due to the increasing demand for forest products as well as from investors looking for long-term stable investments. In 2012, the International Valuation Standards Council (IVSC) produced their draft "Forest Valuation Guidelines."[9] Another consideration indicating the need for international valuation guidance has been the adoption of the International Financial Reporting Standards (IFRS) around the world. An ever-increasing number of entities involved in forestry are required to account for their interest under IAS 41 Agriculture, which requires the "fair value" of the "biological asset," represented by the tree crop, to be estimated. Because the trees cannot exist without the land on which they are growing, this can create some conceptual difficulties in allocating the value of the complete forest to its different components. The IVSC distinguishes three main valuation techniques: (1) the cost approach, (2) the income approach, and (3) the market approach.

Below, an extensive summary with some running commentary is provided of the main aspects of the discussed guidelines:

10.3.1 Introduction and scope

The essence of the guidelines is to provide guidance on the main recognized methods that are used for valuing forests. In the context of the valuation standards, forests are defined as "land and trees held for the growing and harvesting of forest products for commercial purposes." Out of scope are therefore forests that have no commercial value and that are held purely for social, amenity, or private enjoyment.

Valuations of forests, as is the case for all valuation exercises, happen for a purpose, and often different purposes yield different techniques and a different weighting of aspects of what is considered important in that specific case. Reasons for valuing forests can include the following:

- Acquisitions, mergers, and sales of businesses or parts of businesses
- Purchases, sales, and leases of forest assets
- Reporting to tax authorities
- Litigation and insolvency proceedings
- Financial reporting

10.3.2 Identifying the asset

From planting until harvesting, the land and the tree crop that comprise a forest are physically a single, indivisible item. However, for valuation purposes the asset is the interest, or rights, that a party holds in the forest.

There are different types of interest related to the same forest. The first step of any valuation is to identify the interest that needs to be valued and the assets or activities included in that interest. For example,

- An entity operating a forest may own the land and the tree crop in perpetuity or under a lease for a fixed period, and therefore will have a real property interest.
- Otherwise the entity may have no interest in the real property at all, simply a right to cultivate and harvest the trees within a defined area.

The relevant interest may be subject to contractual or statutory restrictions, for example, the requirement to replant/restore the land after the trees are harvested or conditions imposed that limit the operations or require the operator to provide access to third parties or the general public.

10.3.3 Further aspects to consider

Depending on whether the valuation includes a single stand or an estate consisting of multiple stands, the practical applications may differ, although the basic principles will be similar. Some of the following aspects will require attention:

- Site, size, and location
- Supply and demand for products
- Historical performance of standing timber
- Projected growth rates and production-related risks
- Rotation length

10.3.4 Site and location

Site- and location-related issues can include the proximity and means of access to the market for the timber that can be produced. The nature of the terrain can also have an impact on the microclimate of the site and its suitability for the crop, as well as affecting the cost of cultivation and harvesting and soil quality.

10.3.5 Forest inventory

This stage essentially tries to estimate the production potential of the site and the quantity, quality, and growth rate of standing timber. For larger properties, aerial photography and remote sensing are used to determine the area and density for stands of different maturity, often supported by sampling strategies on the ground. There are a wide number of other considerations, including the following:

- Year of establishment or regeneration treatment
- Initial stocking
- Fertilizer and weedicide treatments
- History of pruning or thinning
- Insect and disease attacks and treatments

Different practices can lead to differences in the log yields or qualities of particular sizes of what otherwise could be considered identical stands.

10.3.6 Time to maturity

The date on which a crop of trees reaches maturity and is ready for harvesting (optimum financial maturity) is often not equal to the anticipated biological maturity.

10.3.7 Estimating wood flows

To forecast future wood flows, aspects such as expected silviculture, growth rates, and mortality need to be taken into account. For the most common species, biometricians have developed specialized growth models. An understanding and investigation of the demand patterns for the timber products that will be produced is crucial for both the analysis of data used in the valuation and in developing the valuation opinion. The demand for, and price of, these products will significantly impact the value of forests, and a forest-specific product price analysis is an important input to the valuation process. Price comparisons are either based on the "stumpage price" or the "mill door price." Stumpage is the price that a buyer would be willing to pay for the standing timber. Mill door is the price that a processor of timber products will pay for logs delivered to the mill.

10.3.8 Alternative uses

Market value reflects the highest price the buyer is willing to pay that can best use the asset. Land within a forest may for example have a higher value for an alternative use.

Typical examples include mining, agriculture, or building development. The condition is that the alternative use is physically possible (e.g., no technical constraints), legally permissible, and financially feasible.

10.3.9 Three Valuation approaches and methods

The IVSC provides, as indicated, three valuation approaches. The choice of the most appropriate approach, or approaches, will depend on the purpose for which the valuation is required.

- **Market approach**

Due to a variety of elements, including the heterogeneous nature of forests, direct comparison with sales of other forest interests is rarely possible. Nevertheless, sales analysis and adjustments can still be important in the valuation process. Ratio analysis can be applied for indirect sales-comparison purposes, and market analysis often isolates relevant elements of value. However, the usefulness of any analysis will depend on the quality of the information and the factors that influenced the price paid, such as the relative maturity of the tree crop. Previous transactions involving the same forest may provide useful information if the conditions of the transaction were materially similar to those in the case under review or if adjustments can be made to reflect those material differences and it can be accurately measured which impact those differences have/had in order to properly reflect price changes.

The market approach works best in the case of an absolute ownership interest in the land and the tree crop combined, although it can also be used to value partial interests in a forest. However, its reliability in that situation will depend on the extent of the information that is available about the rights and obligations of the interest involved in potentially comparable transactions.

- **Income approach**

The timing of income from forests is different most of the time from the timing of costs incurred in the cultivation, maintenance, and protection of the crop. Therefore the discounted cash flow method (DCF) is often used. Essential in that respect are the "projected cash flows" and the "discount rate" used in the DCF valuation model. The DCF tries to intrinsically estimate the market value of the forest; consequently, inputs need to reflect market conditions as accurately as possible. Therefore any costs or revenues predicted by the current owner should be consistent with market expectations, and the discount rate used should reflect a market rate based on the intrinsic risk and inflation expectations in the project.

The DCF method requires, as is the case in a normal DCF model, a period for which future cash flows are explicitly forecast (the explicit forecast period) and an estimated terminal value (reflecting the period after the growth rate). The length of the explicit forecast period is always somewhat problematic but will invariably depend on the extent of the fluctuations in the expected cash flows and the certainty with which their timing can be predicted. The discounting rate will cause the effect of an explicit forecast on the present value to diminish over time. As indicated, the value of cash flows beyond the end of the explicit forecast period is reflected in the terminal value. The most significant chunk of income from forests arises at the maturity of the crop, which may be many years in the future. The explicit forecast period is often longer for forests than for a normal company-valuation model where it reflects the (expected) growth period.

Materially, cash outflows might typically include the costs of establishment, cultivation, protection, and maintenance over the period until each stand is harvested. The cost of infrastructure such as fire protection and roads may be apportioned where needed and appropriate. The cash inflows might typically include the calculation of the stumpage price for either the whole forest or for each individual stand, based on the estimated date of harvesting. If the market price of timber products vis-à-vis customers is used as a basis for revenue forecasts, the associated costs of harvest, overhead, delivery, administration, and time should also be reflected in the cash outflows. Further, the cash flows should take into account not only the tree crop at the valuation date but also replanting (rotations) scheduled within the forecast period. If a DCF method is being used to value only the tree crop, the cash flows will need to include an appropriate outflow for the contribution of the land and other assets.

Adjustments may be needed to represent the contribution of infrastructure improvements to the land, for example access roads and irrigation systems. In cases where the land is leased to the forestry operator, the amount or basis of computation of the rent will normally be in the lease contract. However, in other cases it may be necessary to estimate a rental value for the land by using the market approach to analyze rents agreed for other land used for comparable forestry activities or by the return that a market participant would require on the capital cost of acquiring such land. Where there is comparable market data for land that has comparable infrastructure in place, it may be possible to calculate a notional rent for the improved land.

One approach is to measure the value of infrastructure by reference to the cost savings it provides for future forestry operations. Other adjustments needed would represent the contribution of infrastructure improvements to the land, for example, access roads and irrigation systems. The terminal value will reflect the value of the precommercial and commercial stock estimated at the end of the forecast period, as well as the residual value in the land. Adjustments to the terminal value may include the cost of any (contractual) restoration obligations. The discount rate traditionally used is either the cost of equity calculated using one of the variations of the capital asset pricing model

(CAPM) or the weighted average cost of capital (WACC). Both are intended to reflect the return that would be required by a market participant in a case with a similar risk profile.

The same theory indicates that for investment property it is the convention to use a rate derived from analysis of market transactions, or when there is insufficient market data to reliably determine a discount rate, a rate may be estimated using a "buildup" method (infra). Given that the market for forests is not as deep or as liquid as for some other asset classes, there will typically be fewer sources of information on transactions or investor expectations. However, market transactions may be analyzed in order to derive the discount rate implied by the price paid. The implied discount rate is the rate at which the transaction price matches the attributed cash flows. If an appropriate discount rate cannot be directly observed in the market, a buildup method may be used. It needs to be observed at all times that the different components of risk associated with the forest investment are identified and measured so that meaningful comparisons can be made.

Risks associated with forests include environmental, biological, and economic risks, some of which will fluctuate depending on the stage of the rotation on the valuation date. The risk associated with a newly planted forest is often highest in the early years up until the first commercial thinning. Each subsequent thinning ameliorates the risk as the number of trees per hectare reduces over the rotation and as the tree size increases.

- **Cost approach**

The cost approach attempts to value forests by calculating the current replacement cost of an asset and by making deductions for physical deterioration and all other relevant forms of obsolescence. The underlying principle is substitution, that is, unless undue time, inconvenience, risk, or other factors are involved, the price that a buyer in the market would pay for the asset being valued would not be more than the cost to assemble or construct an equivalent asset. The cost approach is best suited to value recently planted forests, where the cost of creating an equivalent asset may be judged with a reasonable degree of certainty. In the case of young trees, more weight is likely given to the current cost of planting on the valuation date and the opportunity cost of the time required for a new plant to grow to the age of plants under consideration than to the expected cash flow on harvest. Relevant costs mentioned by the guidelines are the following:

- the cost of acquiring suitable land for planting (assuming the interest being valued includes land),
- the cost of infrastructure,

- the cost of cultivation and preparation,
- the cost of buying and planting and establishing the young trees, and
- any unrecoverable taxes that would be incurred in creating the above.

The cost approach is more difficult to apply for established forests because not only is it more difficult to establish the cost of an equivalent, but it may not even be possible to create an equivalent because of the time required for the tree crop to reach the same stage of maturity. The value attributable to the land can be a significant component of the overall value of the forest, especially where the plants or trees are young.

10.3.10 Combination of methods

Given the heterogeneous nature of forests, all three principal valuation approaches may be used so that the results can be compared and any major differences considered and reconciled. Reconciliation of differing results from different approaches enables the valuer to better understand the critical inputs of different methods and to ensure that these are as realistic as possible, given the consistency of the inputs.

It is often observed that the preferred valuation approach changes as a forest matures due to physical and economic changes.

10.3.11 Valuation for financial reporting

Tree crops are accounted for under IAS 41 Agriculture, which provides that "biological assets" shall be measured on initial recognition and at the end of each reporting period at fair value less costs to sell. Fair value under IFRS has to be determined in accordance with the provisions of IFRS 13 Fair Value Measurements. The biological asset is defined in IAS 41 as the living plant or animal. In the case of forests, a value therefore needs to be ascribed to the tree crop. This requirement can create difficulties in practice because of the need to exclude any element of value attributable to the land. The approach suggested by IAS 41 is to take the value of the entire forest and deduct from this the value of bare land suitable for planting and any other improvements, such as roads, structures, irrigation systems, and so forth, in order to arrive at a residual value for the tree crop. The land element of a forest is accounted for under IAS 16 Property Plant and Equipment. An entity initially accounts for land at its cost but after initial recognition may use either the cost model or the revaluation model.

- Under the cost model, land is carried at its cost less any accumulated depreciation and any accumulated impairment losses.
- Under the revaluation model, land is carried at a revalued amount, being its fair value at the date of the revaluation less any subsequent accumulated depreciation and subsequent accumulated impairment losses.

Needless to say, under IAS 16 land is only depreciated if it has a limited useful life to the entity. In the case of forestry this might be the case if there is no right to replant following the harvesting of the current tree crop and the owning entity has an obligation to restore the land to an alternative less-valuable use.

The IVSC guidelines further include a number of "illustrative examples" in the appendix.

10.4 Farm and stock valuation[10]

It happens often that for reasons other than an outright sale of land or a farm, the farm or its individual components need to be valued. The most frequent reason for doing so is for annual (business or corporate) tax purposes. Although guidance is provided by local tax authorities, other methods of valuation may also be acceptable to use in particular cases, provided they are recognized by the accountancy profession as giving a true and fair view of the results for the period concerned and do not violate the taxing statutes as interpreted by the courts. A valuation that pays insufficient attention to the facts, although in form it is made on a recognized basis, will often not be accepted.

The reason for valuing stock at the end of an accounting period is to identify and carry forward those costs incurred before that date but which will not give rise to income until a later period. By carrying forward those costs, they can be matched with the income when it arises. Profit will be understated if stock is not brought in. However, if there is no reasonable expectation that the proceeds from the sale of the stock in a future period will be enough to cover the costs, then relief for the expected loss may be obtained in the period for which the accounts are being prepared by valuing the stock at what it is expected to realize when sold in the normal course of trade.

For tax purposes the objective often is to arrive at a figure that represents the cost or, if lower, the net realizable value of the stock.

In some circumstances there may be more than one acceptable method of calculating the value of stock, but the basis of valuation in a particular case should be consistent. Occasionally it is observed that the stock figure in the accounts is the net of a provision (reserve), such as for dilapidations. If the creation of such a provision is considered appropriate, we should be made fully aware of it. Provisions are only allowable for tax purposes if profits would not be properly stated in their absence and the amount referable to the year can be quantified with reasonable accuracy. Even if these conditions are met, tax law provides that some provisions are not allowable for tax purposes (for example, for repairs to premises which are not allowable unless expended).

Although livestock should be valued on an animal-by-animal basis, it is acceptable for farmers to value animals of a similar type and quality together on a global or average basis, classified according to age. If deemed cost is used, home-bred animals should be distinguished from animals that have been bought in.

10.4.1 Livestock, growing, and harvesting crops

Production cost

- Production cost is the actual cost of getting the stock into its condition and location at the balance-sheet date. Farm stock valuations should include the costs directly attributable to producing or rearing the stock in question. From an accountancy point of view, it is preferable, but not mandatory, except in the case of certain limited companies, to also include a reasonable proportion of the costs only indirectly attributable to the production of the stock to the extent that those costs relate to the period of production. This will result in a more accurate matching of costs with related sales income. Either method, if applied consistently, is acceptable to us.

Direct costs

- Costs that are directly attributable to buying, producing, and growing the livestock or crops should be included. Such costs will consist not only of the expenses of acquiring the "raw materials", such as seeds, but also of any expenses that directly relate to producing or rearing the stock in question. There can be no definitive list, but the following are examples of direct costs:

1. Livestock
   - purchase costs, or
   - insemination costs, plus additional maternal feed costs in excess of maintenance, plus
   - costs of rearing to the valuation date, or maturity if earlier, including
     - feed costs, including forage
     - vets' fees, including drugs
     - drenches and other medicines
     - ringing, cutting, and dehorning
     - supervisory employee or contract labor costs

2. Growing and harvesting crops
   - seeds
   - fertilizers

- "beneficial sprays" (beneficial sprays include preventive sprays and any sprays that are not applied to remedy a particular infestation or crop deficiency)
- seasonal license payments (for example, short-term hire of land to grow a particular crop) but not normal farm rents
- drying
- storage
- employee (including director) or contract labor and direct machinery costs (for example, fuel, servicing, rental, spares, and the reduction in value due to wear and tear caused by actual usage for the activity concerned) incurred on cultivation-crop working and harvesting

Indirect costs

Examples of such costs are the following:

- depreciation and maintenance of farm buildings
- rent and rates (excluding license payments added)
- general employee (including director) or contract labor and machinery costs

Costs are to be based on expenditures incurred, except where the deemed-cost method is used, and must include the actual costs incurred by the particular farmer on producing the stock, as established from the farmer's own personal records. Larger and specialized businesses, such as intensive pig-rearing units, will usually have adequate records to calculate the cost.[11]

Labor costs should not include anything for the notional cost of own labor for sole proprietors or partners.

10.4.2 Deemed cost acceptable in some circumstances

If it is not possible to work out actual costs from the farmer's records, often a deemed-cost valuation will be accepted (see infra).

Net realizable value

If there is no reasonable expectation that the net realizable value of stock will cover costs incurred, then the stock should be stated at net realizable value. Net realizable value consists of the following:

- The sale proceeds that it is anticipated will be received from the eventual disposal of the stock in the condition in which, at the balance-sheet date, the farmer intended subsequently to market it. It is important that the valuation should be made on a normal commercial basis. For instance, it is not acceptable to value stock on the basis that it would have been sold in a forced sale on the balance-sheet date in its then, possibly immature, state.

- Grants and subsidies intended to augment the sale prices of stocks.
- For breeding/production animals, the ancillary stream of income from the sale of their progeny and produce.
- Minus the further costs to be incurred in getting the stock into marketable condition and then marketing, selling, and distributing that stock. Where the proceeds from the sale of progeny/produce are brought in, then the costs relating to their production and marketing should also be deducted.

It is not acceptable to treat cull value as the only future revenue from production animals, as this does not recognize the value of the future income stream from the produce and/or progeny.

It is recognized, however, that farmers may not have the extensive records necessary to calculate net realizable value with reasonable accuracy; therefore,

- for production animals, such as laying hens and breeding sows, that are not usually sold except for slaughter at the end of their productive lives, it is accepted that, at the balance-sheet date, a reasonable approximation of the net realizable value is the value arrived at by consistently writing off the cost, down to anticipated cull value, on a straight line or other appropriate basis over the animal's expected productive life.
- For other production animals, we will accept the use of the open-market value of animals of the same kind, quality, and condition, based on the assumption that, at the balance-sheet date, there is a willing buyer and a willing seller of the particular animal as a production animal. Where net realizable value is used as being less than the cost, we may want to establish the basis of valuation.

a. **Cooperatives**

In the same way as any other stock held by a farmer, stock marketed through cooperatives, acting as agent for the farmer, must be included in the valuation unless it has been sold. Stock held off the farm which is identifiable as belonging to the farmer must also be included. Where stock held off the farm has been pooled and cannot be identified as belonging to a particular farmer, the unsold proportion must be included. This may be computed by taking

$$A \times B/C$$

where A is the amount in the pool that came from the farmer, B is the amount in the pool not sold at the valuation date, and C is the amount in the pool not sold at the valuation date plus the amount sold from the pool up to that date.

Where a cooperative acts as agent for the farmer but the relevant stock can be identified as not being part of a pool, no apportionment is necessary. It should be included in the valuation. Stock that has been sold to a cooperative that does not act as agent should not be included in the valuation.

b. **Grants and subsidies – effect on stock valuations**

Grants and subsidies toward specific expenses should be regarded as reducing those expenses. If those expenses are included in the cost for stock valuation, then the figure used should be the net cost after deducting the related grants. Grants and subsidies intended to augment the sale prices of stocks should be taken into account in calculating their net realizable values.

c. **Consumables**

Consumables include spares for plant and equipment, oil, diesel, sprays, fertilizers, feedstuffs, and bags. For any stock of unused but usable consumables held at the balance-sheet date, the valuation should normally be made at cost. If, however, the consumables have deteriorated or become obsolete, then use their net realizable value if it is lower than cost.

d. **Deemed-cost valuation**

When deemed cost is acceptable
Valuations should only be based on deemed cost where it is not possible to ascertain actual costs from the farmer's records. Deemed cost should not be used for purchased animals if it is less than the original purchase price plus (if the animal was immature when purchased) the costs of rearing from the date of purchase to the valuation date or, if earlier, to maturity. In such situations, it will be accepted that a reasonable estimate of "deemed cost" is given by a specific percentage of open-market value. It may be necessary, from time to time, to review the percentages if the relationship between costs and market-value changes. Current observed percentages are set out below. For production animals, open-market value should be based on the assumption that there is a willing buyer and a willing seller of the animal as a production animal free from, for example, movement restrictions. It is not acceptable to treat cull value as the open-market value of production animals because this does not recognize the value of the future income stream from produce and/or progeny.

Livestock
The percentages in the case of livestock are as follows:

- cattle – 60% of open-market value
- sheep and pigs – 75% of open-market value

Please note that deemed-cost valuations are only valid for home-bred or home-reared stock or stock acquired some time before maturity and matured on the farm. It is preferable for deemed cost to be fixed at maturity, but we will accept valuations at deemed costs based on open-market value at the balance-sheet date if that method has been used consistently. Farmers should be aware that using deemed cost at each balance-sheet date may result in profits coming into tax earlier. The valuation of immature and unweaned animals using deemed-cost methods based on the open-market value of animals of a similar age and type is acceptable to us except in the situation described in the next paragraph. If it is appropriate to value mother and progeny together because that is the market unit, this should be done. The method described in the previous paragraph is not appropriate where the mother is on the herd basis and where there is no market or a very limited market in unweaned progeny (for example, unweaned lambs at foot). In this situation, failure to recognize the young stock at all in the valuation is not acceptable. The costs of producing the progeny should be carried forward to be set against the eventual sale price.

Deadstock (harvested crops)
Deemed cost based on 75% of open-market value at the balance-sheet date will normally be acceptable.

10.5 Farmland valuation and the carbon cost

As more and more countries around the world have and are introducing carbon-emission (trading) schemes and accompanying legislation, the question can be asked to what degree the costs and benefits of carbon emissions need to be taken into account when valuing agricultural land and forests. This as more and more schemes have introduced the approach of "all sectors, all gases." It would be beyond the scope of this book[12] to go into detail on the problem, but since the topic is increasingly important, we are not prepared to ignore it. The most important features in the debate can be summarized as follows:

1. Valuation methodology to determine impact

   a. Comparable sales analysis: Using transaction data to assess the market evaluation of carbon impact on land prices. Limited and heterogeneous data (at this stage) make it a difficult method to use.
   b. Discounted cash-flow analysis (DCF): involves as discussed discounting expected cash flows at an appropriate discount rate. A common mistake is that it leaves little room for changing conditions relative to assumptions made when performing the valuation. Sensitivity analysis is used to partly offset this problem (e.g., when forests are managed differently over time due to the arrival of new information).

c. Real-options analysis (ROA): extends the DCF method by explicitly accounting for the value of real options, reflecting the owners' ability and flexibility to defer time irreversible investment decisions (harvesting, etc.) in the face of uncertain decisions (log, land conversion prices, etc.). Applying the ROA is often complex even with a limited number of uncertainties, etc. A more sophisticated model is needed to reflect the complexities involved. A bootstrapping real-options analysis (BROA) is suggested incorporating the Monte Carlo simulation technique in the ROA. This implies that simulation is used to determine the optimal solution by trial and error. It allows one to improve the situation based on the learning curve of the previous experience (level in the bootstrap). The resulting valuations are approximate only, meaning that in general they will not be the "true" valuations but present a lower bound on value in a normal ROA application. Key uncertainties are implementation period for an ETS and later regulatory changes (and timing) to the system, the evolution of carbon prices and their pricing mechanism, the outlook post-Kyoto, and the implementation phase (the Cancun and Copenhagen rounds were not particularly transparent and clarifying on a detail level).

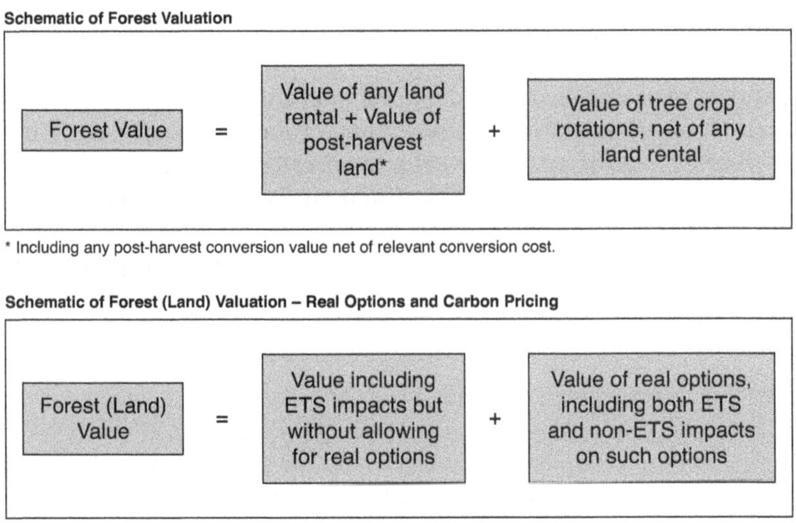

Figure 10.1 Forest-valuation components including and excluding carbon pricing
Note: ETS = emission trading scheme.

The evolution in forest-valuation approach can be depicted as shown in Figure 10.1 in a pre- and post-ETS fashion.

What could potentially complicate the forest-valuation analysis even further is the fact that the land can be owned separately from the trees. In a wider context, similar questions can be asked taking into account the wider impact of climate change on forests.[13]

11

Advanced Technologies and Agriculture: Genomics, Nanotechnology, and Beyond

Advances in genomics, the study of all the genetic material (i.e., the genome) of an organism, have been remarkable in recent years.[1] Publication of the first draft of the human genome in 2001 was a milestone, quickly followed by that of the first crop (rice) in 2002 and the first farm animal (chicken) in 2004. Huge technological advancements have meant that sequencing has become dramatically quicker and cheaper over time, so the genomes of many of the important crops, livestock, forest trees, aquatic animals, and agricultural pests are now already sequenced or soon will be.[2]

All living things are made up of cells that contain genetic material called DNA, a molecule made up of a long chain of nitrogen[3]-containing bases (of four kinds: A, C, G, and T). DNA is organized as a double helix, where two DNA chains are held together through bonding of the bases, where A bonds with T and C bonds with G.

Only a small fraction of the genome typically codes for proteins, that is, where the DNA is first "transcribed" to a molecule called messenger RNA (mRNA) that is then "translated" to a protein. The remaining and major share of the DNA represents sequences that do not code for proteins and whose role is not yet clearly understood.

The field of genomics is still relatively young. The first genome sequence of a virus (called phi X 174) was published in 1977, and the first genome of a bacterium (a strain of *Haemophilus influenzae*) was sequenced nearly 20 years later. The landmark Human Genome Project began in 1990, and the draft sequence (covering about 90% of the entire 3 Gb genome) was published in 2001, and the full sequence in 2003. This project, in particular, stimulated major technological advancements in DNA-sequencing and bio-informatics.

These dramatic improvements have led to a plethora of genome-sequencing projects. An overview of these can be seen from the National Center for

Biotechnology Information (NCBI), which hosts an important database containing, inter alia, information about genomes.[4] The huge investments that have been directed toward genomics are driven by the prospects of using the knowledge in several areas, particularly toward human disease diagnosis and prevention and drug development. In addition, genomics is being used to assist wildlife conservation.

11.1 Genomics in food and agriculture

Since the first plant genome was sequenced in 2000 a lot of progress has been made.[5] What interests us most here is the areas of application, which can be summarized as follows:[6]

11.1.1 Genetic improvements of populations

The potential to genetically improve crop, livestock, forest tree, aquatic animal, and microbial populations for specific purposes is probably the major driving force behind the use of genomics and the other "-omics" in food and agriculture. They allow researchers and breeders to gain direct access to knowledge about the makeup and functioning of the genetic blueprint of the population or species of interest and to use that knowledge for their genetic improvement. Genome sequencing leads to the identification of unprecedented numbers of molecular markers (such as SNPs) spread throughout the entire genome. Their association with genes of interest can be harnessed for genetic improvement, an approach called marker-assisted selection. A number of strategies are available to use the large number of markers for genetic improvement. One strategy is the use of association mapping, or genome-wide association studies (GWAS), where genome-wide marker alleles associated with the trait of interest (e.g., yield) are first identified and then, in a second step, used to predict breeding values. Another strategy is genomic selection, where, unlike GWAS, all of the genome-wide markers are used to predict breeding values.

11.1.2 Characterization and management of genetic resources for food and agriculture

Domestication of plants and animals began about 10,000 years ago, and studies of the timing, location, and selection pressures behind the different domestication events provide valuable information about the current genetic resources for food and agriculture. Genomics has provided a whole range of new tools to explore such issues. Genomics and genomics-derived molecular markers are also playing an important role in the characterization, study, and preservation of wild populations, such as capture-fish and forest-tree populations.

11.1.3 Food and agricultural product authentication

Genomics and molecular markers derived from genome sequencing can be used to confirm the authenticity of commercially available food and agricultural products.

11.1.4 Pathogen detection

Knowledge of the genome sequence of specific strains of mircoorganisms makes it possible to accurately identify the agents causing food contamination as well as plant and animal diseases. The specificity involved also makes it possible to trace the source of the pathogen and to monitor its geographical and temporal spread.

11.1.5 Vaccine development

Genomics is also used to develop vaccines to manage diseases in livestock and fish. One of the key steps in vaccine development is the identification of potential antigen candidates that may be effective in vaccines (an antigen is a molecule, usually a protein foreign to the animal, that elicits an immune response on first exposure to the immune system by stimulating the production of antibodies specific to its various antigenic determinants. During subsequent exposures, the antigen is bound and inactivated by these antibodies). Study of the host genome can also assist in vaccine development.

11.2 Nanotechnology

The application of nanotechnology to the agricultural and food industries was first addressed by a US Department of Agriculture road map published in September 2003.[7]

11.2.1 What is nanotechnology?[8]

Nanotechnology is the manipulation or self-assembly of individual atoms, molecules, or molecular clusters into structures to create materials or devices with new or vastly different properties. Nanotechnology can work from the top down (which means reducing the size of the smallest structures to the nanoscale, e.g. photonics applications in nanoelectronics and nanoengineering) or from the bottom up (which involves manipulating individual atoms and molecules into nanostructures and more closely resembles chemistry or biology). The definition of nanotechnology is based on the prefix "nano-," which is from the Greek word meaning "dwarf." In more technical terms, the word "nano-" means 10^{-9}, or one-billionth of something. For comparison, a virus is roughly 100 nanometers (nm) in size. The word nanotechnology is generally used when referring to materials with the size of 0.1 to 100 nanometers; however it is also inherent that these materials should display different properties from bulk (or micrometric and larger) materials as a result of their size. These differences include physical strength, chemical reactivity, electrical conductance, magnetism, and optical effects.[9]

11.2.2 Nanotechnology and agriculture

Nanotechnology has the potential to revolutionize the agricultural and food industry with new tools for the molecular treatment of diseases, rapid disease detection, enhancing the ability of plants to absorb nutrients, and so forth. Smart sensors and smart delivery systems will help the agricultural industry combat viruses and other crop pathogens. Nanostructured catalysts are available that increase the efficiency of pesticides and herbicides, allowing lower doses to be used. Nanotechnology also protects the environment indirectly through the use of alternative (renewable) energy supplies, and through filters or catalysts to reduce pollution and clean up existing pollutants. So three main areas are covered by nanotechnology in relation to the agricultural sector: (1) food safety and biosecurity,[10] (2) material science,[11] and (3) food processing and product development.[12] Examples are the identification of contamination, less fatty ice cream, smart packaging, fresher food, nanofertilizers and insecticides, and healthier french fries.

Another area where nanotechnology plays a role is precision farming.[13] One of the major roles for nanotechnology-enabled devices will be the increased use of autonomous sensors linked into a GPS system for real-time monitoring. These nanosensors could be distributed throughout the field where they can monitor soil conditions and crop growth. Nanosensors are also being used to monitor grocery stores. This technology enables shopkeepers to identify food items that have passed their expiry date and also reminds them to issue a new purchase order. Equipment of increased sensitivity has been created, allowing an earlier response to environmental changes. Following are some examples:

- Nanosensors utilizing carbon nanotubes or noncantilevers are small enough to trap and measure individual proteins or even small molecules.
- Nanoparticles or nanosurfaces can be engineered to trigger an electrical or chemical signal in the presence of a contaminant such as bacteria.
- Other nanosensors work by triggering an enzymatic reaction or by using nanoengineered branching molecules called dendrimers as probes to bind to target chemicals and proteins.

11.2.3 Smart delivery systems

Nanoscale devices with novel properties could be used to make agricultural systems "smart." For example, devices are being used to identify plant health issues before these become visible to the farmer. Such devices are capable of responding to different situations by taking appropriate remedial action. If not able to take such action, they will alert the farmer to the problem. In this way, smart devices will act as both a preventive and an early-warning system. Such devices could be used to deliver chemicals in a controlled and targeted manner in the same way as nanomedicine has implications for drug delivery in humans.

Technologies such as encapsulation and controlled-release methods have revolutionized the use of pesticides and herbicides. Many companies make formulations that contain nanoparticles within the 100–250 nm size range that are able to dissolve in water more effectively than existing ones (thus increasing their activity). Other companies employ suspensions of nanoscale particles (nanoemulsions), which can be either water or oil based and contain uniform suspensions of pesticidal or herbicidal nanoparticles in the range of 200–400 nm. These can be easily incorporated in various media such as gels, creams, and liquids and have multiple applications for preventative measures, treatment, or preservation of the harvested product. In other areas, scientists have delivered various technologies to make fertilizer- and pesticide-delivery systems that can respond to environmental changes. The ultimate aim is to tailor these products in such a way that they will release their cargo in a controlled manner (slowly or quickly) in response to different signals, such as magnetic fields, heat, ultrasound, or moisture. New research also aims to make plants use water, pesticides, and fertilizers more efficiently, to reduce pollution and make agriculture more environmentally friendly.

Nanotechnology can improve our understanding of the biology of different crops and thus potentially enhance yields or nutritional values. In addition, it can offer routes to added-value crops or environmental remediation. Particle farming is one such example, which yields nanoparticles for industrial use by growing plants in defined soils. Nanotechnology can also be used to clean groundwater using aluminum oxide nanofibers two nanometers in diameter (NanoCeram) as a water purifier. Filters made from these fibers can remove viruses, bacteria, and protozoan cysts from water. Alternatively, water purification is the way to go: lanthanum nanoparticles absorb phosphates from aqueous environments. Applying these in ponds and swimming pools effectively removes available phosphates and as a result prevents the growth of algae. Ultrafine, nanoscale powder made from iron can be used as an effective tool for cleaning up contaminated soil and groundwater. The iron nanoparticles catalyze the oxidation and breakdown of organic contaminants such as trichloroethene, carbon tetrachloride, dioxins, and PCBs to simpler carbon compounds that are much less toxic. Further, oxide particles are extremely effective at binding and removing arsenic from groundwater.

11.2.4 Nanotechnology in the food industry

The definition of nanofood is that nanotechnology techniques or tools are used during the cultivation, production, processing, or packaging of the food. It does not mean atomically modified food or food produced by nanomachines.

A. Packaging and food safety

Developing smart packaging to optimize product shelf life has been the goal of many food companies. Such packaging systems would be able to repair small holes or tears, respond to environmental conditions (e.g., temperature and moisture changes), and alert the customer if the food is contaminated. Nanotechnology can provide solutions for these, for example, modifying the permeation behavior of foils, increasing barrier properties (mechanical, thermal, chemical, and microbial), improving mechanical and heat-resistance properties, developing active antimicrobic and antifungal surfaces, and sensing as well as signaling microbiological and biochemical changes.[14] Other organizations are looking at ways in which nanotechnology can offer improvements in sensitivity or the ease by which contamination of food is detected.

This is a challenge for both developed as well as developing nations. However, it is more for developing nations. The food-packaging sector of developing countries derives its strength from the large volume of agricultural production, the steady growth in food commodities, and the continually increasing food demand fueled by rising incomes. Industry limitations that beset the packaging sector across developing regions include limited packaging solutions to meet international market requirements, and the small size of the domestic demand for packaging materials that consequently leads to low investment by the packaging industry. This in turn limits developing countries' ability to enhance product quality to meet standards of increasingly discriminating consumers, both in domestic and international markets. As many products exported to developed countries are already processed at the point of origin, the demand for these to be packaged in retail-friendly form is on the rise, and this is an opportunity for developing economies to take advantage of. Of the three subsectors of the packaging industry (manufacturing, machinery, and service), the greatest potential appears to be to develop package manufacturing in developing countries, given the tendency to process packaging nearer to the production source and the availability of paper and paperboard packaging materials. As to packaging machinery, creativity in coming up with locally designed equipment suited to local needs and conditions will help propel packaging-industry development. While this strategy is in its infancy, developing countries will mainly have to rely on secondhand machines, with their lower cost offsetting the cost of maintenance. This is because of the prohibitive cost of brand new machinery. Developing the packaging-service provision subsector, specifically the prepacking of food products for export, and outsourcing the packaging of food products to specialized companies, will provide a much-needed boost to the packaging industry in developing countries. It will not only lend affordability of packages and packing but will likewise enhance the efficiency of handling and distribution to exporters, specifically to SMEs. Initiatives toward the growth of packaging industries in developing countries may come either from improving traditional technologies or importing new ones. However, more relevant than the appropriateness and balance of the selected mix of strategies, the success of these initiatives will be contingent on a country's enabling environment, both economic and political. As packaging materials are in short supply in developing countries, relaxing packaging regulations without sacrificing food safety considerations may help the industry to develop. An example is allowing the use of recycled packaging materials when they pose no problem of contamination (i.e., for use with dried foods). Such relaxation of packaging standards will help increase the supply of packaging materials while at the same time addressing food-loss concerns. While prospects are bright for the growth of the packaging-industry sector, it will be for the good of the industry, the stakeholders comprising it, and society at large if focused packaging solutions, taking into account sustainability concerns, are provided to meet the sector's needs. This will not only bring about cost savings but will likewise be good for the environment.[15]

B. Food processing

In addition to packaging, nanotechnology is already making an impact on the development of functional or interactive foods that respond to the body's requirements and can deliver nutrients more efficiently. Various research groups are also working to develop new "on-demand" foods, which will remain dormant in the body and deliver nutrients to cells when needed. A key element in this sector is the development of nanocapsules that can be incorporated into food to deliver nutrients. Other developments in food processing include the addition of nanoparticles to existing foods to enable increased absorption of nutrients.[16] Nanotechnology has further already been used in the cosmetics industry to produce transparent creams. Food and cosmetic companies are working together to develop new mechanisms to deliver vitamins directly to the skin.[17] Nanotechnology and the agricultural and food sector need to be aligned with intense risk assessment and management systems.[18]

C. Impact on agricultural supply chains

The development of nanotechnology creates an excellent opportunity to address complex technical issues of agricultural supply chains and heralds revolutionary changes. The potential application of nanotechnology to the agricultural and food-supply chain is wide and has many variations. Although there is evidence that nanotechnology could enhance agricultural supply chains,[19] further research is necessary to determine whether awareness of the technical advances and benefits alone will be sufficient to overcome resistance to the implementation of these new technologies. Failure to embrace nanotechnology will deny the sector an opportunity to capitalize on improved product visibility, food safety, quality, and security and associated economic benefits. Agricultural supply chains have unique characteristics that distinguish them from generic supply chains. First, the agricultural chain from production to consumption is highly fragmented. As a result, information about market supply and demand and competitors and partners is widely dispersed and difficult to analyze. Second, the market price of agricultural production is subject to seasonal variation. And finally, the perishable nature of fresh produce limits the level of adjustment that can be made to accommodate variations across regions and seasons, in particular managing demands for logistics, warehousing, transportation, and distribution-level activities.[20]

11.3 Genetically modified organisms (GMOs)

Agricultural biotechnologies[21] encompass a wide range of tools and methodologies that are being applied to an increasing extent in crops, livestock, forestry, fisheries and aquaculture, and agro-industries to help alleviate hunger and poverty, assist in adaptation to climate change, and maintain the natural resource base, in both developing and developed countries.[22]

One of these biotechnologies is genetic modification, and it is used to produce genetically modified organisms (GMOs), which are organisms in which one or more genes (called transgenes) have been introduced into their genetic material from another organism using recombinant DNA technology, a set of techniques for manipulating DNA, including the identification and cloning of genes, the study of the expression of cloned genes, and the production of large quantities of gene product.[23] The genes may be from a different kingdom (e.g., a bacterial gene introduced into plant genetic material), a different species within the same kingdom, or even the same species.

While there has been little controversy about any of the other biotechnologies, there has been considerable debate about the current and potential implications that genetic modification and GMOs have for food security, the environment, biodiversity, human health, farmers' income, the global food system, and other issues. This often-polarized debate began in the 1990s, and it continues today without showing significant signs of abating. It is also an area of major R&D investment, and it is predicted that the range of modified traits and species that will be commercially available to farmers in developing countries in the future will be far wider than it is today.

11.3.1 GMOs in food and agriculture[24]

The up-and-running GMOs in the food and agricultural sphere are as follows:

- GM crops: GM crops were first grown commercially in the mid-1990s. While the vast majority was grown in developed countries in the past, this is changing, and almost half of the global GM crop area is now estimated to be in developing countries.[25] Almost all GM crops grown commercially worldwide are genetically modified for one or both of two main traits: herbicide tolerance and insect resistance, for example, Bacillus thuringiensis (Bt) crops.[26]
- GM trees: Commercial release of GM forest trees has been reported in one country, China (since 2002).
- GM livestock: No GM livestock have been commercially released for agricultural purposes. Outside the field of agriculture, GM animals have been approved to produce a small number of pharmaceutical proteins that are commercially available.[27]
- GM fish: No GM fish have been commercially released for food purposes. A number of ornamental GM fluorescent fish have been commercialized in some countries.[28]
- GM microorganisms: Microorganisms (or microbes) are living organisms that are microscopic in size and include bacteria, fungi, and viruses. Although documentation is generally quite poor, use of genetically modified microorganisms (GMMs) in food processing and the animal feed sector is routine. In the agro-industry sector, enzymes (i.e., proteins that catalyze specific chemical reactions) are commonly used in food processing and in the production of food ingredients, and many of them are commonly produced using GMMs. For example, since the early 1990s, preparations containing chymosin (an enzyme used to curdle milk in the preliminary steps of cheese manufacture) derived from GM bacteria have been available commercially. In animal nutrition, feed additives such as amino acids and enzymes are widely used. The greatest use is in pig and poultry production, where, over the last decade, intensification has increased, further accelerating the demand for feed additives. For example, most grain-based livestock feeds are deficient in essential amino acids such as lysine, methionine, and tryptophan, and for high-producing monogastric animals (pigs and poultry), these amino acids are added to diets to increase productivity. Metabolic modifiers are a group of compounds that alter the physiology and metabolism of animals to improve efficiency of meat and milk production. One of these is somatotropin (also known as growth hormone), and GM bacteria are used to produce recombinant bovine somatotropin (rBST) to increase feed conversion efficiency and milk yield and to decrease milk fat in dairy cows,

and recombinant porcine somatotropin (rpST) to increase muscle growth, reduce body fat, and improve carcass composition in pigs.[29]

Poor animal health is a major factor that impacts negatively on productivity. Animal disease control can be improved through vaccination, where a host organism is exposed (usually by injection) to biological material (antigen) that allows it to mount a specific immune reaction, giving it a better capability to fight subsequent infections of a specific pathogen. Recombinant DNA technology is now used to develop different kinds of vaccines to manage diseases in livestock and fish. Some of these vaccines offer advantages over conventional vaccines, such as safe and cheaper production, earlier administration and immunity of the animal, more protective immunity, and the possibility to differentiate vaccinated from infected animals. These include gene-deleted vaccines, where pathogens (bacteria or viruses) with deletions in genes associated with virulence or involved in key metabolic pathways are used as live vaccines; recombinant vaccines based on vectors, where avirulent viruses or bacteria containing foreign genes coding for antigens are used for delivery of these antigens to the host animal; subunit vaccines, composed of semipure or purified pathogen proteins produced by recombinant DNA technology; or DNA vaccines, where bacterial plasmids (i.e., self-replicating nonchromosomal DNA molecules found in many bacteria) encode protein(s) of an infectious agent.[30]

Several of these new kinds of vaccines are commercially available, such as a gene-deleted bovine herpesvirus 1 (BoHV-1) for cattle or viral vector vaccines against poultry diseases such as Marek's disease, fowlpox, and gumboro, or against West Nile virus (WNV) in horses. Several influenza vaccines for poultry (and humans) are produced from reverse genetic systems (which make it possible to introduce designed mutations, insertions, and deletions into the viral genome of live viruses) and are produced and applied in developing countries against H5N1 (the highly pathogenic avian influenza virus currently circulating in a number of countries) in poultry. For DNA vaccines, the main commercialized product is a WNV vaccine for horses, which contains genes for two WNV proteins and does not contain any whole WNV, live or killed. DNA vaccines are also commercially available against infectious haematopoietic necrosis virus for salmon.[31]

11.3.2 What GMOs are in the making?

The pathway leading from research to the eventual commercial release of GMOs is typically long and complex. For example, in crops this process has been recently described,[32] broken into seven "activity stages," some of which overlap in time. The first stage is "early discovery," involving preliminary screening and identification of genetic sequences with the potential to deliver the trait of interest. The second is "late discovery," where the candidate genetic sequences are evaluated in model plant systems (such as the much-studied *Arabidopsis thaliana*). The third is "construct optimization," where the candidate genetic sequences are combined with different promoter sequences to develop the most suitable genetic construct. The fourth is "commercial event production and selection," where the optimized genetic constructs are introduced into the target crop species for subsequent evaluation under greenhouse and/or field conditions. The fifth is "introgression, breeding, and wide-area testing," where a number of genetic events are identified or selected on the basis of their biological activity for introgression into the most elite germplasm. The sixth is "regulatory science," which involves conducting regulatory science studies and generating data in the field, greenhouse, growth chambers, and laboratories. The seventh and final stage is "registration and regulatory affairs," where results of the regulatory studies are submitted to the relevant regulatory bodies to seek approval for cultivation or for import. If approval is granted, commercialization can then begin. The experiences of six major multinational companies were analyzed and reported that, for a single GM event, it took them on average 11.7, 12.0, 12.7, and 16.3 years to go from early discovery of the new trait to commercial sale in canola, maize, cotton, and soybeans respectively.

Apart from the particular target species used, the length of time required for commercialization is also affected by the sector involved (e.g., field trials of GM trees will take longer than those involving annual crops) and by the strictness and requirements of the national regulatory framework.[33]

In discussing the accuracy of the commercial pipeline of GM crops, experts consistently observed big discrepancies in the past between the announced and actual dates of release for certain GM events and concluded that commercial pipelines are not very reliable when dealing with the longer term and should instead be focused on events in the later stages of R&D.

Researchers attempted to predict what GM crops might be commercially available by 2015.[34] Compared to 2008, they predicted that the total number of commercial GM varieties cultivated worldwide by 2015 would be far higher for those species where GMOs were already approved (such as soybeans, maize, and cotton) and would also include those from species that were not yet approved, such as potatoes and rice. They also predicted that a wider range of traits would be commercially available by 2015, with herbicide tolerance and insect resistance continuing to dominate, but with new traits such as crop composition (mostly type/proportion of oil and starch content), virus resistance, abiotic stress tolerance, and disease resistance featuring prominently among the new approvals. They also predicted that substantial numbers of the new GM crops commercialized by 2015 would be developed in Asia (mostly India and China).

Comparable comprehensive and detailed studies of the GMO pipeline are not available for the other food and agricultural sectors. Nevertheless, we know that, compared to crops, investments in development of GMOs are modest in livestock and fish[35] and in trees,[36] but there is substantial research ongoing worldwide in these areas. For example, for forestry, Kanowski reported that over 700 field trials with GM trees of 30 genera have been carried out, most of them in the United States, of which over 70% involved *Populus*, *Pinus*, and *Eucalyptus* species; 84 field trials have been approved in China, most with *Populus* and *Robinia*; and 18 trials of GM *Eucalyptus* have been approved in Brazil.

In livestock, it was reported[37] that a small number of countries, particularly Argentina and China, have invested heavily in GM animals for food production, while more have focused on GM animals for medical purposes. In China, it is reported that nearly 800 million US dollars were invested in GM pigs, cattle, sheep, and crops between 2008 and 2012, and that over 20 GM food animals are being developed, including a fast-growing carp.[38] Another fast-growing GM fish, the AquAdvantage Atlantic salmon, which has been modified by the addition of a Chinook salmon growth hormone gene under an ocean pout antifreeze protein promoter, is currently awaiting commercial approval in the United States.[39]

For microorganisms, there is active ongoing research involving GMMs in all the areas described above as well as in a wide range of other areas where microorganisms are useful in food and agriculture, such as the use of GMMs to modify rumen function in livestock[40] or to convert biomass to biofuels.[41] As probably noticed, it is quite a challenge to provide indications about what GM organisms will reach commercialization in the years to come. The FAO biotechnology and GMO department hosts an annual conference to try and get a grip on the ever-changing landscape of GMOs.[42]

11.4 Agri-mechanization and production-enhancing techniques

But not everything is on the high end of the spectrum. In many parts of the world, a big leap forward can be made using existing techniques and mechanization procedures, particularly in the developing world.[43]

11.5 Directing research and development[44]

The rate of growth in agricultural research and development (R&D) investment has been declining globally, while a large number of developing countries have experienced negative growth rates over the past decade. For example, stagnating investment in Sub-Saharan African agricultural research[45] is particularly worrisome. General underinvestment is evidenced by (1) the continuing high rates of return to research demonstrated in studies at the commodity level, and (2) by macroeconomic studies showing that the relevant Millennium Development Goals (MDGs) cannot be reached without a doubling or even tripling of research investment given estimated growth-poverty-reduction elasticities. Also of concern is new evidence that a change in the composition of research away from productivity enhancement at the farm level is statistically related to a decline in the growth of agricultural productivity in advanced economies below historical levels. Capacity in agricultural research[46] is increasingly concentrated in a few leading countries in each region. One of the main indicators to compare relative R&D investment levels is the ratio of agricultural research investment over agricultural output, the so-called agricultural research intensity ratio (ARI). An ARI of 1% has been seen by many as a target that low-income, agriculturally based countries should strive for. However, the ARI by itself is influenced by several factors that need to be studied in depth at the country level. The ARI can be decomposed into four components: (1) priority of research within agricultural expenditure, (2) priority of agriculture in total public expenditure, (3) fiscal capacity measured as the ratio of public expenditure to gross domestic product (GDP), and (4) the (inverse of the) share of agriculture in the GDP. Emerging challenges, such as adaptation to climate change and increasing variability of weather, water scarcity, and increased price volatility in global markets, will be faced by many countries that are least able to adapt to existing stresses. This lends increasing importance to developing the human and institutional capacity in agricultural research at the national level to interact with regional and global efforts under way.

In summary, technology-based solutions have been and are proving their results in three major areas. Greater investment in and broader adoption of science and technology can enable the world to meet the growing demand for food as the population increases by (1) improving the livelihoods of farmers and their families by producing more and higher-quality crops for a growing population, (2) enhancing the nutritional value and safety of food to improve the health and well-being of people around the world, and (3) contributing to agriculture sustainability through reduced resource use. Those techniques include, as discussed,

- *Plant breeding*: Plant breeding, the science of optimizing a plant's genetic makeup to produce desired characteristics, can be accomplished through a number of techniques, including hybridization and more complex molecular techniques. Through plant-breeding techniques, we can produce higher-yielding crops that are better in quality, tolerant to environmental pressures, resistant to pests and diseases, and tolerant to insecticides and herbicides.

- Hybridization: Hybridization involves crossing two or more crop lines to produce hybrid crops with more favorable traits, resulting from combining genes from the selected parents.
- Molecular market-assisted selection: Molecular markers are small sequence differences between various lines in a plant-breeding population that can be used, when physically linked to traits, as a surrogate for the presence or absence of a desired trait without having to field test for the attributes of that trait. Molecular markers are detected through DNA-sequencing methods using DNA derived from plant samples. The practice of molecular marker-assisted selection enables plant breeders to combine desirable plant traits rapidly and in large numbers. Through this technique, breeders can reduce the time it takes to develop new crop varieties. Additionally, it increases the efficiency of plant breeding by enabling breeders to genetically prescreen multitudes of potential varieties with high precision prior to selecting lines or hybrids with the highest genetic potential for costly field evaluation. Consequently, this technique is an increasingly common breeding technique in crops where marker systems have been developed and marker–trait associations have been established. Genetic markers are also being used to monitor and increase genetic diversity in breeding programs. Diversified crop varieties protect farmers, including smallholders in food-insecure countries, from being vulnerable to widespread disease and environmental stresses that impact certain varieties.
- Plant breeders use agricultural biotechnology as another source of genetic variation to produce superior crops with improved yields while requiring fewer inputs. The products of this technology have been widely used by farmers for over a decade in varieties of corn, cotton, soybeans, and canola. Biotechnology expands the genes available for crop improvement beyond those present in the breeding populations and uses the tools of genetic transformation to bring specific genes into the genetic makeup of the plant. To date, this method has been used to enable crops to tolerate insects, viral diseases, and certain herbicides; to produce grain with improved nutritional quality; and to resist stresses caused by extreme weather.

- *Crop-protection solutions*: Advances in crop protection have been a powerful tool in combating the pests, diseases, and weeds that can be devastating to crop yields. Through the use of crop-protection products, which include chemical (e.g., insecticides, fungicides, and herbicides) and nonchemical tools (e.g., biological pest control and barrier-based approaches), farmers have significantly curbed these losses and increased their productivity yields.
- *Other technologies*: such as modern irrigation practices, mobile technology, fertilizer, and mechanization

Table 11.1 Anchors facilitating change within the context of the Green Revolution

Type of change	Biochemical	Mechanical	Social
Innovations	Hybrid seed selection; use of fertilizers, herbicides, and pesticides	Diesel- and electric-pump-powered irrigation, mechanization, transport improvement	Land reforms, loans, changes in distributive system
Consequences	Increased yields, weeds and pests controlled, increased cost for farmer, possible environmental degradation	Water supply controlled, less labor needed, increase in arable land, increase in access to market	Farm consolidation, better seeds and other inputs available to poor farmers

The Green Revolution has three anchors embedded, as shown in Table 11.1.

11.6 GHGs and seed technology[47]

Agriculture's significant global contribution to greenhouse gas (GHG) emissions has spurred consumer and retailer interest in GHG mitigation and may lead to incentive programs for producers to lessen GHG emissions. Along those lines, one choice for producers is the use of embedded-seed technology designed to enhance the marketable portion of yield through improved disease, weed, and pest management with the same or lower use of inputs. Commonalities and differences have been examined across three recent studies on rice, sweet corn, and cotton that addressed the impacts of embedded-seed technology on yield, input use, and GHG emissions. Embedded-seed technology can be any method of improving the physical or genetic characteristics of a seed. These seed enhancements can include physiological quality, vigor, and synchronicity (consistency across seedlings in time of emergence and size) through traditional breeding, hybrid breeding, or biotechnology.[48]

12
Challenges in Agricultural Production and Natural-Resources Management[1]

It has been mentioned already on a number of occasions that the expected growth rates in agricultural output are slowing down while population growth powers on. Regional or global studies do unavoidably mask intercountry differences. This chapter brings together the most prevailing challenges and supports them with data, albeit within the indicated parameters.

12.1 Productivity growth in agriculture

It is expected that by 2050 the world's average daily calorie availability could rise to 3,070 kcal per person (about a 10% increase from current levels). For these projections to materialize, annual world agricultural production would need to increase by some 60% from 2005–2007 levels to 2050, consisting of a 77% increase in developing countries and a 24% increase in developed countries. Over the same period, world population is projected to rise by some 39%, meaning that per capita production would rise by some 15%. The fact that this would translate into a smaller (11%) increase of per capita calorie availability is mainly[2] due to expected changes in diet, that is, a shift to higher-value foods of often lower calorie content (e.g., vegetables and fruits) and to livestock products, which imply a less efficient conversion of calories of the crops used in livestock feeds.

At the world level, agricultural production increases are about equal to increases in demand for agricultural products, although some projections show that demand outstrips supply over extensive periods in the next couple of decades. Simple growth accounting shows that the increase in global demand comes from about 70% on account of population growth, 22% on account of the increased availability of calories per person, and 8% on account of other factors, mainly changes in commodity composition driven by shifts in diets. The latter two factors are predominantly determined by the increase in per capita income, so that overall one could state that at the world level increases in demand and production are 70% determined by population growth and 30% by per capita income growth.[3] The annual growth of world agricultural production is projected to fall from 2.2% over the last decade to 1.3% over the period to 2030, and 0.8% from 2030 to 2050. Nearly 90% of the increase in (annual) production would take place in developing countries, which would raise their share in world agricultural production from 67% in 2005–2007 to 74% in 2050. This increase would be particularly strong for livestock production, going from 55% in 2005–2007 to 68% in 2050.

With a view to analyzing natural-resource use in agricultural production, one should bear in mind that the bulk of the foods consumed are produced locally. On average at present, only 19%[4] of world production enters international trade (corresponding to 17% for cereals and 14% for meats), with of course wide variation among individual countries and commodities.

12.2 Crop production and sources of growth

Crop-production growth is seen to decelerate in all regions, in particular in developed countries and East Asia. Naturally, growth prospects differ among countries and crop sectors, with in general slow growth foreseen for cereals like rice and more vigorous growth for coarse grains and some oilseeds used for feeding purposes in the livestock sector. Growth in crop production comes on account of growth in crop yields and expansion in the physical area (arable land) allocated to crops, which, together with increases in cropping intensities (i.e., by increasing multiple cropping and/or shortening of fallow periods), leads to an expansion in the actually harvested area. The FAO conducted a study to (roughly) measure the upside potential of present and future land/yield combinations for 34 crops under rain-fed and irrigated cultivation conditions for 105 countries and country groups. Some selected findings regarding the sources of crop production (%) are summarized in Table 12.1.[5]

Table 12.1 Assessment of upside potential of present and future land/yield

	Arable-land expansion		Increase in cropping intensity		Yield increases	
	1961–2007	2007–2050	1961–2007	2007–2050	1961–2007	2007–2050
All developing countries	23	21	8	6	70	73
Sub-Saharan Africa	31	20	31	6	38	74
MENA	17	0	22	20	62	80
South Asia	6	6	12	2	82	92
Latin America	40	40	7	7	53	53
East Asia	28	0	−6	15	77	85
World	14	10	9	10	77	80

Some 80% of the projected growth in crop production in developing countries would come from intensification in the form of yield increases (73%) and higher cropping intensities (6%). These summary results, as indicated, mask a wide variation among countries. The actual combination of the factors used in crop production (such as land, labor, and capital) in the different countries will be determined by their relative prices. Taking the physical availability of land as a proxy for its relative scarcity and hence price, one would expect land to play a greater role in crop production the less scarce it is. In developed countries, the area of arable land in crop production peaked in the late 1960s, then remained stagnant for some time and has been declining since the mid-1980s. Hence growth in crop yields accounted for all of their growth in crop production and in addition compensated for declines in their arable-land area. This trend is foreseen to continue during the next decades. As a result, intensification (higher yields and more intensive use of land) is at the world level seen to contribute 90% to the growth in crop production over the projection period. An attempt was made by the researchers to unfold crop production into rain-fed and irrigated production, which offers an opportunity to estimate the contribution of irrigated crop production to total crop production. At present, irrigated agriculture, covering some 16% of the arable land in use, accounts for 44% of all crop production and some 42% of cereal production in the world. Similar estimates for developing countries are somewhat higher, with 21% of arable land, accounting for 49% of all crop production and 60% of cereal production. The aggregate result of individual country projections is that these shares would change little over the projection period. It should be emphasized that except for some major crops in some countries, there is only limited data on irrigated land and production by crop, and the results presented here are in good measure based on expert judgment. Nevertheless, the results suggest a continuing importance of irrigated agriculture.

12.3 Land with crop production potential

About 12% (more than 1.5 billion ha) of the globe's land surface (13 billion ha, excluding "inland water") is used for crop production (arable land and land under permanent crops). Arable land at present takes up some 28% of the prime (very suitable) and good (suitable and moderately suitable) land. This leaves a gross balance of unused prime and good land of some 3.2 billion ha and a net balance (i.e., excluding forests, strictly protected land, and built-up areas) of some 1.4 billion ha. These balances of land with crop-production potential suggest that there is still scope for further expansion of agricultural land. However, there is also a perception in some quarters that no more, or very little, additional land could be brought under cultivation. Notwithstanding the predominance of yield increases in the growth of agricultural production, land expansion will continue to be a significant factor in those developing countries and regions where the potential for expansion exists and the prevailing farming systems and more general demographic and socioeconomic conditions are favorable. But how much land is there exactly (left) that could be used to produce food to meet the needs of a growing population?

The most recent version of the Global Agro-Ecological Zones (GAEZ v. 3.0) analysis[6] combines soil, terrain, and climate characteristics with crop-production requirements, estimating the suitability (in terms of land extents and attainable yield levels) for crop production of each land grid cell at the five-arc-minute-level at four technology and management levels (low, intermediate, high, and mixed). The suitability assessments provide extents for a range of suitability classes as follows: very suitable, suitable, moderately suitable, marginally suitable, very marginally suitable, and not suitable. Fischer et al. (2010)[7] condense these six classes into three classes, (1) prime land, (2) good land, and (3) marginal and not suitable land. Prime land is characterized as very suitable land with attainable yields of over 80% of maximum constraint-free yields. Good land represents suitable and moderately suitable land with attainable yield levels of 40% to 80% of maximum constraint-free yields, and marginal and not suitable land includes all land with estimated attainable yields that are less than 40% of maximum constraint-free yields. Summing[8] over all the crops covered in GAEZ and the technology levels considered ("mixed level of inputs"), about one-third (34%) of the world's land surface, or 4.5 billion ha, is estimated to be of prime (very suitable) or good (suitable and moderately suitable) quality for rain-fed agriculture. Of this area, some 1.6 billion ha is already under cultivation. It is interesting to note that of this 1.6 billion ha, some 300 million ha (or 19%) of agricultural land is on areas the GAEZ deems only marginally suitable or even not suitable, at least for rain-fed agriculture. Such areas might have been made productive by applying irrigation (e.g., 32 million ha in desert areas). Excluding irrigated areas, still some 220 million ha (or 17%) of rain-fed agriculture is apparently taking place on marginally and not suitable areas. An explanation could be that farmers might have no choice or maybe are prepared to accept (relatively) low yields, or that not suitable land (43 million ha) has been made suitable through man-made interventions. The remainder of the discussion below is limited to areas of prime and good quality.

Likewise, developing countries as a whole have some 2.9 billion ha of prime and good-quality land of which a quarter (700 million ha, or 24%) was in 1999–2001 in use in agriculture. The gross land balance at the world level of 3.2 billion ha (2.2 billion ha in developing countries) would therefore seem to provide significant scope for further expansion of agriculture. However,[9] this favorable impression needs to be qualified by a number of considerations and constraints.

First, the gross balance ignores land uses other than for growing crops, so forest cover, protected areas, and land used for human settlements and economic infrastructure are not taken into account. Excluding prime and good land currently under forests or built-up areas or on strictly protected land (grasslands, scrub and woodland, and nonvegetated land not yet in agricultural use), the remaining net balance amounts to 1.4 billion ha (960 million ha in developing countries), still a considerable amount compared with the current arable area.

Second, the net land balance is very unevenly distributed among regions and countries. For example, some 85% of the remaining 960 million ha in developing countries is to be found in Sub-Saharan Africa (450 million ha) and Latin America (360 million ha), with very little or no land remaining in the other regions. In addition, about half of the remaining land is concentrated in just seven countries (Brazil, Argentina, Sudan, China, the Democratic Republic of the Congo, Angola, and Mozambique). At the other extreme, there is virtually no prime or good land left in many countries in the Near East and North Africa, South Asia, or in Central America and the Caribbean. Even within the relatively land-abundant regions, there is great diversity of land availability, in terms of both quantity and quality, among countries and subregions.

Third, and probably more important than allowing for nonagricultural uses of land with crop-production potential, is the method used to derive the estimates: it is enough for a piece of land to support a single crop at a minimum yield level (40% of the maximum constraint-free yield) for it to be classified as suitable (prime or good) land. For example, large tracts of land in North Africa that permit the cultivation of only olive trees (and a few other minor crops) are counted as suitable, even though there may be little use for them in practice. The notion of overall land suitability is therefore of limited meaning, and it is often more appropriate to discuss suitability for individual crops.

Fourth, much of the remaining land suffers from constraints such as ecological fragility, low fertility, toxicity, high incidence of disease, or lack of infrastructure. These factors reduce its productivity and require high input use and management skills to permit its sustainable use or prohibitively high investments to make it accessible or disease free. Fischer et al. (2002)[10] show that more than 70% of the land with rain-fed crop-production potential in Sub-Saharan Africa and Latin America suffers from one or more soil and terrain constraints. Natural causes and human intervention can also lead to deterioration of the land's productive potential, for example through soil nutrient mining, soil erosion, or salinization of irrigated areas. Hence the evaluation of suitability may contain elements of overestimation, and much of the land balance cannot be considered as a resource that is readily useable for food production on demand.

These considerations underline the need to interpret estimates of land balances with caution when assessing land availability for agricultural use. Cohen (1995)[11] summarizes and evaluates all the estimates of available cultivable land, together with their underlying methods, and shows their extremely wide range. Young (1999)[12] offers a critique of the estimates of available cultivable land, including those given in Alexandratos (1995),[13] stating that they often represent gross overestimates.

12.4 Possibilities of land expansion for crop production

The question has been raised to what extent agriculture will be able in the future to sustain a growing world population and growing levels of consumption. The continuing decline of arable land (in use) per person is often cited as an indicator of impending problems. The underlying cause for such problems is perceived to be an ever-increasing demand for agricultural products facing finite natural resources such as land, water, and genetic potential. Scarcity of these resources would be compounded by competing demands for them originating in urbanization, industrial uses, and use in biofuel production, by forces that would change their availability such as climate change and the need to preserve resources for future generations through environmentally responsible and sustainable use.

Naturally, one could interpret the declining arable land per person in parallel with the observed increasing average food consumption per person as a sign of ever-increasing agricultural productivity (crop yields). In practice, changes in arable land (in use) per person will be the result of these countervailing forces (population/demand growth and increasing crop yields), with the exact outcome differing between countries. The perception that there is no more, or very little, new land to bring under cultivation might be well grounded in the specific situations of land-scarce countries and regions such as South Asia and the Near East/North Africa but may not apply, or may apply with much less force, to other parts of the world. As discussed above, there are still large, as-yet-unused tracts of land with varying degrees of agricultural potential in several countries, most of them in Sub-Saharan Africa and Latin America, with some in East Asia. However, as noted, this land may lack infrastructure, it may be partly under forest cover or in wetlands which should be protected for environmental reasons, or the people who would exploit it for agriculture lack access to appropriate technological packages or the economic incentives to adopt them.

In reality, expansion of land in agricultural use continues to take place. It does so mainly in countries that combine growing needs for food and employment with limited access to technology packages that could increase intensification of cultivation on land already in agricultural use. It also has been expanding in countries with abundant land resources that could profit from the growth of demand for their exports, such as Brazil.[14] The data shows that expansion of arable land continued to be an important source of agricultural growth in Sub-Saharan Africa, Latin America, and East Asia.[15] This includes countries with ample land resources with potential for crops facing fast demand growth, particularly for exports and for nonfood uses.

12.5 Irrigation-water requirements and water-resources management

A key question is whether there will be sufficient freshwater to satisfy the growing needs of agricultural and nonagricultural users. Agriculture already accounts for approximately 70% of the freshwater withdrawals in the world and is usually seen as one of the main factors behind the increasing global scarcity of freshwater.

The estimates of the expansion of land under irrigation presented in the preceding section provide a partial answer to this question since the assessment of irrigation potential already takes into account water limitations. The renewable water resources available to irrigation and other uses are commonly defined as that part of precipitation that is not evaporated or transpired by plants, including grass and trees, and which flows into rivers and lakes or infiltrates into aquifers. The annual water balance for a given area in natural conditions, that is, without irrigation, can be defined as the sum of the annual precipitation and net incoming flows (transfers through rivers from one area to another) minus evapo-transpiration, runoff, and groundwater recharge. Average annual precipitation at the global level is about 800 mm per year, but it varies from a low 160 mm in the most arid region (Near East/North Africa) to a high precipitation of about 1,530 mm per year in Latin America. These figures give an impression of the extreme variability of climatic conditions facing developing countries, and the ensuing differences observed in terms of water scarcity: those countries suffering from low precipitation and therefore most in need of irrigation are also those where water resources are naturally scarce. In addition, the water balance presented is expressed in yearly averages and cannot adequately reflect seasonal and intra-annual variations. Unfortunately, such variations tend to be more pronounced in arid than in humid climates.[16]

Irrigation efficiency (below termed "water-use efficiency") can be defined as the ratio between crop water requirements, estimated as consumptive water use in irrigation plus water needed for land preparation and weed control in the case of paddy rice, and irrigation-water withdrawal. To make projections going forward about water withdrawal for irrigation purposes, very little evidence is available to support potential hypothesis. Two factors, however, will have an impact on the development of water-use efficiency: the estimated levels of water-use efficiency in the base year and water scarcity. Water scarcity, or "stress," is measured as consumptive water use in irrigation as a percentage of renewable water resources. Researchers at the FAO designed a function to capture the influence of these two parameters, bearing in mind that improving water-use efficiency is a very slow and difficult process. The overall result is that efficiency could increase marginally at the global level and in water-rich regions, but more so in water-scarce regions and countries (e.g., a 9% increase in the Near East/North Africa region). Indeed, it is expected that, under pressure from limited water resources and competition from other uses, demand management will play an important role in improving water-use efficiency in water-scarce regions. In contrast, in humid areas, the issue of water-use efficiency is much less relevant and is likely to receive little attention.[17] At the global level, irrigation-water withdrawal is expected to grow by about 6%. The 6% increase in irrigation-water withdrawal should be seen against the projected 12% increase in the harvested irrigated area. This difference is in part explained by the expected improvement in water-use efficiency leading to a reduction in irrigation-water withdrawal per irrigated hectare.

12.6 Crop-yield growth

It is expected that growth in crop yields will continue to be (and even more so than in the past) the mainstay of crop-production growth, accounting for nearly 80% of the latter (well over 70% in developing countries and all of it in developed countries). Although the marked deceleration in crop-production growth foreseen for the future could point to a similar deceleration in growth of crop yields, such growth will continue to be needed. Questions often asked are, will yield increases continue to be possible, and what is the potential for a continuation of such growth? Looking back over the last 25 years, it indeed is so that crop-yield growth is coming down for most agricultural commodities. The reasons for such slowdown, however, are more likely to be found in the observed slowdown in world production than in certain resource constraints (including the genetic potential) becoming binding.

Also on this front, FAO experts have been wrapping their heads around what they think are plausible land-yield combinations of agro-ecological rain-fed and irrigated environments, taking into account whatever knowledge is available. A major input into this evaluation are estimates regarding the availability of land suitable for growing crops and of yields attainable in each country and in each agro-ecological environment as originated in agro-ecological zones work.[18] Such estimates are introduced as constraints to land and yield expansion, but they also act as a guide to what can be grown where. The resulting land and yield projections, although partly based on past performance, are not mere extrapolations of historical trends, since they take into account present-day knowledge about changes expected in the future. The overall result for yields of all the crops covered in this study is, at the global level, a more than halving of the average annual rate of growth over the projection period as compared to the historical period: 0.8% p.a. during 2007–2050 against 1.7% p.a. during 1961–2007 (for developing countries, the annual growth rates for these periods are 2.1% and 0.9% respectively). This slowdown in yield growth is a gradual process that has been under way for some time and is expected to continue into the future. It reflects the deceleration in crop-production growth explained earlier.[19] Despite increases in land under cultivation in land-abundant countries, much of the agricultural production growth has been based on the growth of yields, and will increasingly need to do so. What is the potential for a continuation of yield growth? In countries and localities where the potential of existing technology is being exploited fully, subject to the agro-ecological constraints specific to each locality, further growth, or even maintenance, of current yield levels will depend crucially on further progress in agricultural research.

The indicated FAO research suggests that there has been and still is considerable slack in the crop yields of the different countries, which could be exploited if economic incentives so dictate. However, the fact that yield differences among the major cereal-producing countries are very wide does not necessarily imply that the lagging countries have scope for yield increases equal to intercountry yield gaps. Part of these differences of course simply reflects differing agro-ecological conditions. However, not all, or perhaps not even the major part, of yield differences can be ascribed to such conditions, as wide yield differences are present even among countries with fairly similar agro-ecological environments. In such cases, differences in the socio-economic and policy environments probably play a major role. The literature on yield gaps distinguishes two components of yield gaps, one due to agri-environmental and other nontransferable factors (these gaps cannot be narrowed), and another component due to differences in crop-management practices such as suboptimal use of inputs and other cultural practices. This second component can be narrowed provided that it makes economic sense to do so, and it is therefore termed the "exploitable yield gap" or "bridgeable gap."

In order to draw conclusions on the scope for narrowing the yield gap, one needs to separate its "nontransferable" part from the "exploitable" part. One way to do so is to compare yields obtained from the same crop varieties grown on different locations of land that are fairly homogeneous with respect to their physical characteristics (climate, soil, terrain), which would eliminate the "nontransferable" part in the comparison. One can go some way in that direction by examining the data on the suitability of land in the different countries for producing any given crop under specified technology packages. These data would then make it possible to derive a "national maximum obtainable yield" by weighting the yield obtainable in each of the suitability classes with the estimated land area in each suitability class. The derived national obtainable yield can then be compared with data on the actual national average yields. This type of analysis has been undertaken frequently over the last two decades or so, and by now there exists a fairly extensive literature clearly showing and confirming that a good part of the yield gap is of the second, exploitable type.[20] This implies that crop production could increase through the adoption of improved technologies and practices to bridge some of the gap that separates actual yields from obtainable yields. The broad lesson of experience seems to be that if scarcities develop and prices rise, farmers quickly respond by adopting such technologies and increasing production, at least those living in an environment of not-too-difficult access to improved technology, transport infrastructure, and supportive policies. However, in countries with land-expansion possibilities, the quickest response comes from increasing land under cultivation, including shifting land among crops toward the most profitable ones.[21] The preceding discussion may create the impression that all is well from the standpoint of potential for further production growth based on the use of existing varieties and technologies to increase yields. This statement should, however, be heavily qualified since the exploitation of bridgeable yield gaps means further spread of high external-input technologies, which might aggravate related environmental problems, and perhaps more important from the standpoint of meeting future demand, ready potential for yield growth does not necessarily exist in the countries where the additional demand will be. When the potential demand is in countries with limited import capacity, as is the case in many developing countries, such potential can be expressed as effective demand only if it can be predominantly matched by local production.

12.7 Fertilizer consumption

As discussed, the bulk of the projected increases in crop production will probably come from higher yields, with the remaining part coming from an expansion in harvested area. Both higher yields, which normally demand higher fertilizer-application rates, and land expansion will lead to an increase in fertilizer use. Increases in biomass require additional uptake of nutrients, which may come from both organic and mineral sources. Unfortunately, for most crops there are not enough data to estimate the relation between mineral fertilizer consumption and biomass increases.[22] Increased use of fertilizer is becoming even more crucial in view of other factors, such as the impact on soil fertility of more intensive cultivation practices and the shortening of fallow periods.[23] Research demonstrates that increases in food production can be obtained with a less-than-proportional increase in fertilizer nutrient use.[24] Other management techniques such as precision agriculture offer abundant opportunities to substitute information for fertilizer.[25] It is expected that this trend of increasing efficiency of nutrient use through better nutrient management, by improving the efficiency of nutrient balances and the timing and placement of fertilizers, will continue and accelerate in the future. Projections evidence that fertilizer consumption could increase from 166 million tons in 2005–2007 to 263 million tons in 2050. This would imply a continuing slowdown in the overall growth of fertilizer consumption. Developing countries account at present for almost 70% of world fertilizer consumption, and this share could increase further to over three-quarters of world consumption in 2050. However, the growth in fertilizer consumption is slowing in the United States and Western Europe, the two largest and most mature regions from a fertilizer-use point of view. They account for much of the projected slowdown in fertilizer-consumption growth. Also, changes in agricultural policies, in particular reductions in support measures, contributed to a slowdown or even decline in fertilizer use in this group of countries. Increasing awareness of and concern about the environmental impacts of fertilizer use are also likely to hold back future growth in fertilizer use.

Cereals, in particular wheat, rice, and maize, account at present for some 60% of global fertilizer use, and they are expected to still account for just over half of fertilizer consumption by 2050.[26] Since the early 1960s, the use of mineral fertilizers has been growing rapidly in developing countries, admittedly starting from a very low base. This has been particularly so in East and South Asia following the introduction of high-yielding varieties. By now, high application rates have been reached in East Asia, and growth of fertilizer consumption in East Asia is expected to slow down drastically; and eventually fertilizer consumption is expected to decline.[27] For Sub-Saharan Africa, above-average growth rates are foreseen, starting from a very low base, but fertilizer consumption per hectare is expected to remain at a relatively low level. The latter probably reflects large areas with no fertilizer use at all, combined with small areas of commercial farming with high levels of fertilizer use, and could be seen as a sign of nutrient mining.[28]

Average fertilizer productivity, as measured by kilogram of product obtained per kilogram of nutrient, shows considerable variation across countries. This reflects a host of factors such as differences in agro-ecological resources (soil, terrain, and climate), in management practices and skills, and in economic incentives. Fertilizer productivity is also strongly related to soil moisture availability. Furthermore, a high yield/fertilizer ratio may also indicate that fertilizer use is not widespread among farmers, or that high yields are obtained with nutrients other than mineral fertilizer (e.g., manure is estimated to provide almost half of all external nutrient inputs in the EU). Notwithstanding this variability, in many cases the scope for raising fertilizer productivity is substantial, but more so for N fertilizer than for P and K fertilizers.[29] The degree to which such productivity gains will be pursued depends to a great extent on economic incentives.

The projected slowdown in the growth of fertilizer consumption is in the first place due to the expected slowdown in crop-production growth. As mentioned, fertilizer is most productive in the absence of moisture constraints, that is, when applied to irrigated crops. For this reason, the expected slowdown in irrigation expansion will also slow the growth of fertilizer consumption. Another factor is the continuing improvement in fertilizer-use efficiency, partly driven by new techniques such as biotechnology and precision agriculture, which will continue to reduce mineral fertilizer needs per unit of crop output.[30] Then there is an increasing concern about the negative environmental impact of high rates of mineral fertilizer use. Finally there is the spread of organic agriculture and the increasing availability of nonmineral nutrient sources such as manure; recycled human, industrial, and agricultural waste, and crop by-products. All these factors will tend to reduce growth in fertilizer consumption[31] (albeit with large fluctuations across shorter periods of time).

12.8 Livestock production

Like crop production, growth in livestock-production mirrors growth in total agricultural production, although the observed deceleration in growth is slightly less than for crop production, as the consumption of livestock products continues to increase its share in total food consumption. Sub-Saharan Africa is the only region where livestock-production growth will continue to be fairly strong, while only slow growth is foreseen for the other regions.[32] Livestock production is the world's largest user of land, either directly through grazing or indirectly through consumption of fodder and feed grains. Globally, livestock production currently accounts for some 36% of the gross value of agricultural production. In developed countries this share amounts to half of total production, and in developing countries for almost one-third.[33] Increased production can be achieved by a combination of expansion in animal numbers and increased productivity. Higher productivity is a compound of higher offtake rates (shorter production cycles by, for example, faster fattening) and higher carcass weight and milk or egg yields. The projections show that the increase in livestock numbers will remain significant, but less so than in the past. Higher carcass weights will play a more important role in beef and mutton production, while higher off-take rates (shorter production cycles) will be more important in pig and poultry meat production.[34]

12.9 The future of agriculture: a perfect storm in the making

Based on the above, one can derive different views about the long-term impact of several constraints in the agricultural sphere. The different sections provide a balanced view of the dynamics of expected future constraints and suggest that whatever beliefs one might have, the future of agriculture is not one straight road to hell. Nevertheless, the different constraints and limitations combined with a number of the discussed exogenous factors (the "unknown unknowns") does lead some authors to believe that the glass is more half empty than half full.[35]

Quite often their conclusions are built around the following framework:

1. Impacts on commodity (food) prices caused by exogenous shocks to commodity demand (e.g., due to biofuels mandates), yields (e.g., due to climate change), or land supply (e.g., due to urbanization) are mediated by three elasticities: the intensive margin of supply response, the extensive margin, and the price elasticity of demand. The smaller the elasticities (the denominator), the larger the price change.
2. Changes in agricultural land use caused by exogenous shocks to commodity demand (e.g., due to biofuels mandates), yields (e.g., due to climate change), or land supply (e.g., due to urbanization) is also mediated by three elasticities: the intensive margin of supply response, the extensive margin, and the price elasticity of demand (however with a larger denominator than the first equation above, suggesting less land-use sensitivity to economic response than for prices).

The resilience of the global agricultural system can therefore be improved by focusing on raising the following factors:

- **Elasticity of substitution between variable inputs and land**

Some operational implications would be the following:

- Improve the performance of factor markets.
- Market infrastructure (roads, information), organization (farmer organizations), and policy (competition, input trade, price).
- Invest in agriculture R&D.
- Ensure long-run responsiveness of public research systems to focus on land-saving technologies.

- **Price elasticity of demand for aggregate agricultural output**

Factors that may lower global own-price elasticities of demand for aggregate agricultural products include trade barriers, mandated demand (e.g., biofuels mandates), and share of institutional investor demand in commodities/land (which is driven by relative volatility and diversification rather than own-price levels).

Some operational implications would be the following:

- Improve agriculture trade to ensure that food moves from (low-demand) surplus areas to (high-demand) deficit areas.
- Promote responsible agriculture investment to ensure local benefit from foreign investments (to offset the potential impact on raising long-term global food prices through lowering global demand elasticities).

- **Elasticity of land supply with respect to agricultural output prices**

Factors that may lower the elasticity of land supply with respect to commodity prices include restrictions to land sales and rental markets. One operational implication of this would be to improve performance of land sales and rental markets.

The higher price volatility may lead to a decline in the magnitude of the (above mentioned) elasticities (or at least a reduction in their summed total), thus contributing to long-term commodity-price escalation. This would suggest a continued focus on ways to dampen and better manage price volatility at global, national, and local levels.

13
Sustainability of Agricultural Productivity Growth[1]

Global agriculture will face multiple challenges over the coming decades. It must produce more food to feed an increasingly affluent and growing world population that will demand a more diverse diet, contribute to overall development and poverty alleviation in many developing countries, confront increased competition for alternative uses of finite land and water resources, adapt to climate change, and contribute to preserving biodiversity and restoring fragile ecosystems. Climate change will bring higher average temperatures, changes in rainfall patterns, and more frequent extreme events, multiplying the threats to sustainable food security. Addressing these challenges requires coordinated responses from the public and private sectors and civil society that will need to be adapted to the specific circumstances of different types of farmers in countries at all levels of development.

Improving agricultural productivity, while conserving and enhancing natural resources, is an essential requirement for farmers to increase global food supplies on a sustainable basis.[2] The role of smallholder farmers and their families in increasing agricultural productivity growth sustainably will be crucial. Half a billion small family farms produce most of the food consumed in developing countries and farm over 80% of the land in Asia and Africa, but their productivity is generally lagging. The success of developing countries in increasing agricultural productivity will have global implications in strengthening the resilience of food markets, enhancing food security, improving well-being, and promoting sustainability. The two following sections focus on four broad areas that require attention: providing an enabling environment conducive to investment and innovation in agriculture; investing in agricultural innovation, broadly defined; improving national and international research collaboration; and closing the gap between actual and potential productivity levels of agriculture in developing countries.

13.1 Sustainable productivity

The growing global demand for food, feed, and biofuel is well established. The growing global population and more affluent citizens require a significant increase in global agricultural output, as mentioned earlier in this book. Throughout history, productivity in agriculture has shown high growth rates. Some recent estimates suggest that total factor productivity (TFP), the most comprehensive measure of productivity reflecting the efficiency to turn all inputs into outputs, grew at an average rate of around 2% per year since 2000 across major world regions.[3] The picture, however, is more complex when looking at individual countries or subregions. Some large countries like Brazil, China, Indonesia, Russia, and Ukraine have achieved much higher TFP growth rates than the corresponding regional average. The most popular indicator of land productivity is crop yield. The average global rates of growth in yield of most of the major cereals are declining. Since the 1980s, growth in wheat and rice yields fell from 2.5% to 3% to around 1%.[4] Efforts to increase food production will take place within an environment characterized by a scarcity of natural resources. In many regions, there is little room for expansion of arable land, with virtually no additional land available in South Asia, the Near East, and North Africa. Where land is available, in Sub-Saharan Africa and Latin America, more than 70% of the land suffers from soil and terrain constraints.[5] Unsustainable land-use practices, such as overuse, poor land management, and nutrient mining, result in global net losses of land productivity of an average 0.2% per year.[6] Land degradation makes the topsoil vulnerable to water and wind erosion and reduces the productivity of inputs such as fertilizer and irrigation, which in turn leads to production and income losses.

At the global level, agriculture is the largest water user worldwide, representing about 70% of total withdrawal. In some countries, over 90% is withdrawn for agricultural purposes. Cities and industries are competing intensely with agriculture for the use of water, and an increasing number of countries, or regions within countries, are reaching alarming levels of water stress and pollution. Global freshwater resources will be further strained in the future in many regions, with over 40% of the world's population projected to be living in river basins experiencing severe water stress by 2050.[7]

Agriculture is also a major source of water pollution, from nutrients, pesticides, soils, and other contaminants, leading to significant social, economic, and environmental costs. It also damages the wider environment through the emission of greenhouse gasses. In some intensive farming systems, up to 50% of available inorganic and organic nutrient inputs are not always utilized by crops or pastures, leading to significant pollution from nutrient run-off.[8] The opposite is the case in large parts of the developing world, where crop farming leads to a net extraction of nutrients from the soil. In large parts of Sub-Saharan Africa, soil productivity has been on a declining trend.

Biodiversity underpins agriculture and food security through the provision of genetic material needed for crop and livestock breeding. The past century has seen a great loss of biodiversity through habitat destruction, mainly due to deforestation.[9] Maintenance of biodiversity is crucial for sustainability and resilience of farming systems as it builds the capacity to absorb shocks and continue to function within a changing set of circumstances. The challenge is to maximize agriculture's positive contributions to biodiversity while minimizing its negative impacts.

Global agriculture will need to adapt to climate change. There is growing evidence that climate change has had negative effects on agriculture and widespread agreement that agriculture, particularly in developing countries, will be for the most part negatively affected by climate change.[10] In the near term, climate variability and extreme-weather shocks are projected to increase, affecting all regions with negative impacts on yield growth and food security, particularly in Sub-Saharan Africa and South Asia in the period up to 2030.[11] Agriculture (including deforestation) accounts for about one-third of greenhouse gas emissions; for this reason, it contributes significantly to climate-change mitigation. While crops can be adapted to changing environments, the need to reduce emissions will increasingly challenge conventional, resource-intensive agricultural systems.[12] Productivity growth needs to increase to keep up with demand growth, but also to increase resilience of the sector to supply shocks, whether due to climate change or to resource limits more generally.

13.2 Agriculture innovation systems

The productivity of farms can be improved through economies of scale and the adoption of more technically efficient production systems. However, long-run productivity growth for the sector as a whole requires continuous technological progress, as well as social innovations and new business models. For agriculture to respond to future challenges, innovation will not only need to improve the efficiency with which inputs are turned into outputs, but also conserve scarce natural resources and reduce waste.[13]

Estimates of the rates of return to agricultural R&D suggest a very high social value of such research. Annual internal rates of return of investments on agricultural R&D estimated in the literature range between 20% and 80%.[14] In developing countries, the dollar-for-dollar impact of R&D investments on the value of agricultural production is generally within the range of 6% to 12% across countries.[15] Those countries that have heavily invested in R&D while simultaneously investing in extension have had the strongest productivity growth.[16]

Driven by policy incentives, recent productivity improvements in developed countries have occurred with lower levels of variable input use, and thus more sustainably. Innovation systems have responded to the demand articulated by users, policy makers, and international development partners and developed innovations that allow for more sustainable use of resources, such as no-till farming, insect-resistant crops, more efficient irrigation, water-management systems, sensors for nutrient status in crops, remote sensing and geographic information systems (GIS) to improve and monitor land use, and SMS messaging for enhancing advisory services to farmers.

The predominant model for innovation has been mostly supply driven: scientists in the public sector create new technologies which are then disseminated by extension officers to farmers who are asked to adopt them. Many countries have reviewed their agricultural innovation systems (AIS) in recent years in response to concerns about lack of adoption of innovation and the need to increase performance to respond to emerging and pressing challenges. Indeed, sustainable production intensification requires a major shift from the supply-driven innovation model to knowledge-specific and often location-specific farming systems which conserve and enhance natural resources. Nontechnological innovation such as marketing or organizational innovations also receive more attention today.

Unlike the experience of the Green Revolution, which relied on improved and genetically uniform high-yield varieties complemented by high levels of inputs, increasing agricultural productivity in today's context will require gains among a large number of smallholders in very different agro-ecological regions. Traditional technologies and practices have proved their relevance to increasing productivity and ensuring environmental sustainability. Recognizing the need to move away from supply-driven to demand-driven innovation, some countries have taken steps toward a more inclusive, interactive, and participatory approach.[17]

Innovation is increasingly taking place in a network-based setting that fosters interaction and learning. AIS display a large diversity, reflecting different country contexts and different degrees of transition from the traditional top-down approach. While R&D remains an important component of agricultural innovation systems, there is a growing recognition of the role of other actors: farmers, extension services, upstream and downstream industries, consumers, civil society, and information brokers. Working with farmers to validate and adapt technologies in an integrated way lies at the heart of AIS. The participation of more diverse actors along with institutional reforms have improved the responsiveness to AIS to specific needs, with resulting innovations better suited to resource-poor farmers.[18] However, most developing countries, particularly in Sub-Saharan Africa, face challenges in implementing an AIS approach due to weak resources and institutions.

Upstream and downstream industries have increased their role in the diffusion of innovation to farmers, and the private sector is increasingly involved in R&D activities with high potential market returns, such as biotechnology. Partnerships between public research and the private sector are being developed, including with local industries. Governments have encouraged public research institutions to engage in public–private partnerships with producer organizations and the agri-food industry.

Another notable trend in AIS is less government involvement in the delivery of extension services. This has permitted the emergence of other intermediaries, such as innovation brokers, who can articulate farmers' demand for research and help them access technology and knowledge, or who are associated with creating linkages in value chains.[19] However, public services still dominate extension in developing countries, and they face widespread problems of limited funding, insufficient technology and knowledge, poorly trained staff, weak links with research, and limited farmer participation.

Public institutional changes have generally aimed to increase coordination at the national level both within agricultural systems and between agriculture and other sectors. Some countries have merged or strengthened the links between agricultural R&D and higher-education institutions. Others have reformed their agricultural universities. Mechanisms and networks to set priorities for agricultural research have been strengthened and made more inclusive. In two recent OECD reports,[20] the role of governments was once more stressed with respect to AIS and the impact it has on economic performance of agriculture across countries.[21] In parallel, the World Bank issued a report[22] reviewing and synthesizing the literature on trends and fluctuations in market distortions and the political-economy explanations that have been advanced.

13.3 R&D funding

In all low- and middle-income country regions as a whole, public expenditure on agricultural R&D increased from the 1980s, but there are important variations across countries within regions. Several countries have well-managed and funded systems producing world-class research; others, some of which are highly dependent on agriculture, have experienced significant reductions in their R&D expenditure and capacity levels since the early 1990s. Overall, the average share of agriculture in total R&D expenditures is decreasing in both high-income and low- and middle-income countries.

Public R&D expenditure on agriculture accounts for above 1% of agricultural GDP in most OECD countries, and 4% in the United States. However, public expenditure on agricultural R&D in OECD countries grows slowly (e.g., by 0.2% per year in the United States and 0.5% in Japan in the 2000s) or even decreases. In some high-income countries, the slowing or negative growth rates of public R&D expenditure on agriculture in the last decade are raising concerns about the ability of the sector to maintain and increase productivity growth rates because agricultural research has to spread its efforts across an increasing number of topics.[23] Public R&D expenditures on agriculture in low- and middle-income countries are generally lower as a percentage of agricultural GDP than in OECD countries, and there is wide diversity across countries. In East Asia and the Pacific, China accounted for about two-thirds of total public agricultural R&D spending in the low- and middle-income countries in 2002 and onward. Following a period of stagnation in the 1990s, China's agricultural research spending doubled during 2001–2008.[24] In Sub-Saharan Africa, after a decade of stagnation in the 1990s, investment in agricultural research rose more than 20% between 2001 and 2008. However, most of this growth occurred in only a handful of countries.[25]

While public expenditure is the main source of funding for agricultural R&D, private-sector investment has increased but is generally focused on high-value and market-oriented production systems. Greater protection of intellectual property, rapid progress in molecular biology, and the integration of global output and input markets have generated strong incentives for the private sector to invest in R&D. At the same time, the record of private research in natural-resource management and in maintaining biodiversity is limited, with the exception of a few public–private partnership initiatives.

Investments by the private sector in the developing world remain small, and agricultural research continues to be mostly funded by governments.[26] The evidence suggests that, on average, government allocations have accounted for 81% of funding since 2000, and only 7% of funding was derived through donor contributions. These latter contributions have been in the form of both loans and grants, and mostly attributed to countries in Sub-Saharan Africa and a few low-income countries in Asia and Latin America.[27]

International R&D, in particular by CGIAR, has in many instances successfully led to the development of technologies well suited to smallholder production systems. In the 1990s, more centers were added to the CGIAR, and although total funding continued to grow, average spending levels per center declined. Since 2000, overall funding to the 15 centers of the CGIAR has increased, but a larger portion of this funding is support for specific projects and programs of research involving different centers and non-CGIAR research organizations.[28]

13.4 (Additional) investments in agriculture

Investments in agriculture encompass both public and private spending on natural capital (such as land, water, and biodiversity); physical capital (such as animals, machinery, irrigation systems, storage, processing and marketing facilities, roads, ports, and other "hard" infrastructure on- or off-farm); human capital (such as health, education, training, and advisory services); and knowledge capital (such as research, technology development, and organizational and other innovations). These various types of investments play complementary roles in the production process and contribute to increased productivity. Investment needs will vary depending on the stage of economic and agricultural development.

Investment can be financed by both public and private sources, including domestic savings of households and private companies, government savings, external borrowing, and foreign investment. Public investments often focus on the provision of public goods, while private investments tend to focus on areas generating private returns. Private investors have been particularly involved in technology generation as IPRs have been strengthened. But these lines are becoming increasingly blurred, and public–private partnerships are emerging where mutual benefits are anticipated.

In terms of physical capital, increased mechanization by investment in tractors and power threshers spurred crop-yield growth in many industrialized countries a few decades ago and continues to do so in developing countries. While this type of investment can be characterized as private investments in agriculture, such investments in agriculture play an important enabling role. They have public-good (or quasi-public-good) characteristics and are consequently predominantly financed from public sources. Investment in infrastructure in rural areas, in particular transportation (ports and rural roads), soil and water conservation, irrigation systems, electrification, and information and communication technologies, is an effective way to stimulate productivity growth.[29] It allows smallholders to connect to markets and thereby provides higher incentives to increase productivity. Irrigation systems allow for increased land productivity, particularly in countries that depend on rain-fed agriculture and face water shortages. Often, infrastructure and road development are ranked among the top two sources of overall agricultural growth, second to R&D investments. Especially in Africa, irrigation and feeder roads are shown to have large output-increasing and poverty-reducing effects. The importance of transport infrastructure is demonstrated in Africa where, for certain landlocked countries, transport costs can be as high as 77% of the value of their exports. The establishment of development corridors linked to major ports can be an effective way to stimulate local economies.[30] Improving market facilities such as warehouses, storage facilities, and market-information systems is important in creating an enabling environment and facilitating the integration of farmers into markets as well as providing incentives to increase investment and hence productivity.

Expanding the capital base of agriculture is intrinsic to the process of increasing productivity and generating higher and more stable returns to those whose livelihoods depend on this sector. In developing countries, smallholders provide the bulk of domestic private investment, notably on-farm. Better investment of their own labor and part of their income to improve land, acquire new equipment, expand livestock herds and farms, and invest in storage and the postproduction chain will increase their efficiency to produce food and minimize losses.

Mechanization and the use of animal traction to replace human labor has dramatic potential to increase productivity. Land improvements, such as land leveling, terracing, and bunding for runoff and erosion control, can contribute significantly to the management of natural resources. In sum, other forms of investment, such as public investment or foreign direct investment, will have limited or no impact if they are not accompanied by increased on-farm investments.

Globally, on-farm investments, reflected by the volume of agricultural capital stock, have increased over time, with the increase concentrated in low- and middle-income countries. On-farm investments in high-income countries have remained relatively stable. However, the rate of growth of agricultural capital stock has been uneven over time, probably reflecting the impact of both market and policy incentives to farmers. Average annual growth in the global agricultural capital stock of about 1% during the 1980s was followed by significantly lower rates during the 1990s.

On-farm investment growth has recovered since 2000. This was partly due to the increasing trend in agricultural commodity prices, although these remain at a level significantly lower level than that which characterized the period 1980–1990.

Foreign direct investment (FDI) is often referred to as a major and growing source of finance for agricultural development. Higher agricultural prices and better incentives to invest in agriculture are likely to stimulate further international investment flows to this sector.

Flows of FDI to agriculture (agriculture, hunting, forestry, and fisheries) as well as the food, beverages, and tobacco sectors have increased significantly over the past decade. In many cases, this new momentum has translated into large-scale acquisitions of farmland in lower- and middle-income countries. Less attention has been paid to a wide range of collaborative arrangements between large-scale investors and local smallholders, such as various types of contract-farming schemes, joint ventures, management contracts, and new supply-chain relationships.

Data issues hamper the analysis of FDI, as the increasing flows are partly explained by wider country coverage of the statistics. Reporting in current USD also tends to overestimate the real increase. Nevertheless, in spite of the increase in recent years, current flows of FDI to agriculture remain limited in size. In 2008, out of total recorded flows of FDI of USD 1.2 trillion, flows to the food, beverages, and tobacco sectors amounted to only USD 87 billion, less than 1% of the total. Flows to primary agriculture amounted to USD 5 billion, a far less significant sum.[31]

13.5 Policy development and institutional capacity building

The incentives to innovate and adopt better technologies, as well as to invest in agriculture, depend on the overall policy environment, including macroeconomic and sectoral policies and regulations. In developing economies, policies have long worked against investments in agriculture. More recent estimations reveal that over the past four to five decades the nominal rate of protection afforded to nonagriculture steadily declined for developing countries as a whole, from 45% in the 1960s to less than 10% in the 1990s.[32] The impact of specific agricultural support policies on farm productivity depends on how and why it is delivered. Commodity-based support has the largest impact on production, but protecting farmers from competition does not encourage them to increase productivity. Market interventions often treat the symptoms of market failure and underdevelopment rather than the cause. For example, food-price stabilization can provide a more stable investment climate but can also impose very high costs on consumers, thwart the development of private risk management, and can export instability onto world markets. It also thwarts the development of the private sector, which is crucial for the long-term development of the food sector.[33] Like price support, input subsidies also distort production. However, they can redress, at least temporarily, market failures such as the underdevelopment of infrastructure, missing markets for credit and inputs, and a lack of knowledge of the benefits of using improved seeds, animal breeds, and fertilizer. To this extent, they can help farmers acquire improved technology and thus foster productivity, but over time they can also impede the development of private markets and do not tackle the problem of market failure directly. More generally, if support is targeted to a specific input, it can encourage an input mix that will not necessarily be economically or environmentally sustainable. For example, irrigation subsidies can affect sustainable water use and may not encourage the adoption of water-saving irrigation systems if appropriate regulations are not in place. Providing producers the tools they need for risk management is important for the adoption of innovation, but too much government support in risk-management schemes may prevent the emergence of market solutions. As seen earlier, public expenditures on agricultural R&D have positive and large impacts on agricultural productivity, but public expenditures on extension and advisory services are also important and complementary as they promote the adoption of new production systems that enable productivity growth on a sustainable basis.

13.6 Regulation affecting agriculture

Plant and animal diseases are a cost to producers, reduce productivity, and pose health risks to consumers. Food-borne diseases have an enormous impact on health and livelihoods and are of great concern to consumers, producers, and policy makers. Policy interventions are necessary to tackle such problems, and governments have a number of options available to intervene. Some of these policies can also have implications for international trade. Import conditions for food products defined by public regulation and private standards continue to differ significantly across countries.

Policy recommendations have been formulated at regular intervals to support the G20 and others to support meaningful action.[34]

13.7 Improving productivity and seed quality[35]

There are essentially five components to every productivity-enhancing program in agriculture:

- Soil and water management and crop nutrition
- Plant breeding and genetics
- Animal production and health
- Insect and pest control
- Food quality and safety[36]

According to FAO estimates, as much as 40% of the world's food is grown using irrigation, but large amounts of this water are lost to leakage in the irrigation system itself. Improper irrigation is also a major cause of soil salinity. Roughly one-tenth of the world's irrigated land has been damaged by salt. With the threat of climate changes, more and more regions of the world are also at risk of drought and desertification. Improved irrigation practices will help conserve water and protect vulnerable land. Stable and radioactive isotopes can be used as "tags" at the molecular level to monitor how plants use essential nutrients. Such techniques allow scientists to determine the exact nutrient and water requirements of a crop in particular conditions, making it possible to find sustainable alternatives for that area. The International Atomic Energy Agency (IAEA) is using nuclear science to improve crop-production techniques, particularly in regions where water is scarce. Through the FAO's Soil and Water Management and Crop Nutrition program, the IAEA develops and transfers techniques that use stable and radioactive isotopes to study crop growth and nutrient requirements to increase yields, while conserving water and preventing further degradation of marginal lands. Microirrigation systems and fertigation – the application of fertilizers through an irrigation system – help to control water and nutrients in the root zone, saving labor and equipment costs. In an IAEA regional project over a five-year period, isotope techniques were used to identify improved water- and fertilizer-management practices. Tremendous savings of water and nitrogen fertilizer were observed when these methods were compared with the traditional method of applying water and nitrogen fertilizer to the field. Insect pests are another serious threat to productivity. They can devastate crop yields and transmit disease to both crops and livestock. Conservative estimates put losses of food production in the range of 25% to 35%, even with the use of pesticides. Moreover, there are concerns that reliance on pesticides to maintain yields not only has negative impacts on the environment, but may also lead to the insects developing resistance to the pesticides themselves. Through its Insect and Pest Control program, the FAO is using nuclear science to develop environmentally friendly alternatives for pest control. One of the most successful techniques developed to date is the sterile insect technique (SIT). SIT involves mass rearing insects and then sterilizing them using gamma radiation. These sterile insects are identical to those in the wild except that they cannot reproduce. The sterile insects are then released to compete with wild males for mating. Successive releases of sterile insects over time will gradually reduce, and eventually eradicate or control, the targeted insect population. This technique has been successfully used to eradicate several insect pests of agricultural significance. Two of the most significant are the Mediterranean fruit fly (Medfly) and the tsetse fly.

Another productivity challenge deals with improving the performance of dryland crops in dry, mainly semiarid environments. Although such crops are often limited by water, the development of the notion of water-limited potential yield has shown that their yields are often limited strongly by other factors. These factors are explored by dissecting the water-limited potential yield into a framework involving four largely independent components, namely, the potential water supply for the crop, the fraction of that water supply that is transpired, the efficiency with which the crop exchanges water for CO_2 in producing biomass (transpiration efficiency), and the fraction of the biomass that ends up in the grain (harvest index). This framework can then be used to explore a wide range of agronomic possibilities for managing crops so as to get close to the water-limited potential, including managing previous crops, forages, and fallows to increase soil water at sowing, reducing evaporative losses from the soil surface, ensuring that sowing and flowering occur at the right times, maximizing soil water extraction by the crop, and ensuring that there is adequate water available during late floral development and grain filling. Such operations often involve trade-offs and risks that must be managed.[37]

14 Commodities I: Derivatives Markets

14.1 Functioning of the commodity-derivatives markets

In the last ten years, commodity-derivatives markets underwent major changes in at least two areas: (1) the amount of money invested, and (2) the types of investors. An important source of information is provided by the CFTC (Commodity Futures Trading Commission), which releases weekly data on the investment positions held by the different types of traders on the US commodity-derivatives markets.[1]

The size of the commodity-derivatives markets has grown dramatically during the last decade, with a significant acceleration at the beginning of 2006 (open positions grew from 2000 to 2008 by a multiple of six, of which two-thirds was realized during the 2006–2008 period), when the number of outstanding positions in regulated commodity futures markets almost doubled in only six months (Bank of International Settlements, 2009). This is part of a broader pattern, common to other derivatives markets (interest rates, equity indices, exchange rates); nevertheless, the relative weight of commodity contracts on total financial derivatives has also increased from 1.5% to 2.3% between 2004 and 2010. Moreover, the increase in commodity-derivatives trading accompanied a widespread rise in prices, which contributed to bringing the value of outstanding positions in the main commodity markets from about \$100 billion in 2002 to almost \$700 billion in mid-2008. The evolution ever since 2010 until mid-2012 is somewhat stable as reported by the Bank for International Settlements (BIS).[2]

Another way to gauge the evolution of these investments is to relate them to developments in the physical commodity markets. Since 2005 wheat consumption has grown by 7.7%; instead, the total number of contracts in US-regulated markets has more than doubled during the same period. Investments in commodity derivatives are often made via over-the-counter (OTC) financial instruments, which have developed remarkably in the past few years. According to data provided by the BIS, the notional amounts of outstanding forward and swap contracts on OTC markets reached about $7 trillion in June 2008 before dropping during the crisis. Afterward, the number of contracts on regulated markets resumed its upward trend, exceeding the previous mid-2008 peak by 50% in the first quarter of 2011. On the contrary, the notional amounts of OTC contracts have remained subdued. This may reflect an enduring exacerbation of counterparty risks, which favor a shift of investors away from OTC instruments toward the regulated markets, and it is also due to the development of new instruments – such as exchange-traded products (ETPs) – which allow more flexible investment strategies. This evolution is confirmed by the estimates made by Barclays Capital of the overall assets under management by financial institutions in commodity-related instruments.

Since mid-2010, invested resources have increased even more rapidly than before the financial crisis; commodity-related financial assets recorded a historic peak in July 2011 ($431 billion, 65% above the precrisis level of 2008). Before falling abruptly as a result of the crisis, investments linked to commodity indices gained weight, rising from $75 billion in 2006 to $170 billion in June 2008.

From mid-2009 they resumed growth at a slower pace than other instruments, never exceeding the previous peak. The postcrisis dynamic of investments in commodity indices is another possible cause of the slackness of the OCT markets, as they are usually carried out through OTC swap agreements. Although the proportion of contracts traded in organized exchanges varies over time, all the commodity-related financial products are interconnected via arbitrage opportunities (for instance, financial intermediaries active in the OTC markets normally hedge their net exposures on the regulated exchanges).

In principle, three broad types of participants[3] can be identified on futures commodity markets depending on their investment scope and time horizon – that is, "hedgers," "speculators," and "commodity-index investors." While the first two types of investors were present from the beginning of the derivatives markets, index investors have come to play an important role only more recently.

1. Hedgers use derivatives markets to hedge business risks. Supposedly, they have an exposure on the physical commodity market, for example, in relation to mining companies, agricultural producers, refiners of oil and metals, or airline companies (whose costs are heavily affected by fuel prices).
2. Speculators enter the commodity derivatives market to make profits, taking positions according to their expectations of future price movements. Their investment horizon is usually relatively short (from minutes up to a few weeks or months), and they are supposed to revert to their positions before the delivery date. For instance, "scalpers," who often trade in and out of a position within a few seconds, exploit small differentials to earn money. They guarantee the immediacy of execution for a trade. In general, they

should play a stabilizing role by injecting liquidity, acting also as counterparts for hedging transactions and by improving market efficiency, aiding price discovery through their efforts to gather information on fundamental price drivers. However, in a world with asymmetric information, they may adopt procedures to predict price movements that are not necessarily based on market fundamentals, such as trend extraction techniques (in this case they are named "trend followers"), and thus their actions may amplify misalignments and feed speculative bubbles.

3. Commodity-index investors (also named "CIT investors") use commodity derivatives as alternative investment assets as part of a portfolio-diversification strategy and are less concerned with the evolution of fundamentals. Most of these investments are made through OTC intermediaries (the swap dealers) by institutional investors, such as pension funds or sovereign wealth funds. Commodity yields have historically shown a positive correlation with inflation and low correlation with equity returns (although since 2009 the latter has increased noticeably), so that they are a natural choice in a long-term portfolio-optimization strategy. Exposures by these investors tend to reproduce indices that compound different commodities.

The two indices most tracked are the Dow Jones-UBS and the S&P-GSCI. In 2011, around a fifth of investments on commodity indices involved agricultural products (wheat, corn, sugar, and live cattle). Based on weights in dollars, 18.3% of all positions are held in agricultural commodities: 3.8% in wheat, 3.3% in corn, 2.5% in sugar, 2.5% in live cattle, and 2.4% in soybeans. Commodity-index investor strategies are characterized by a relatively long time horizon, and investors would always acquire long positions in futures markets (directly or through intermediaries or other financial instruments); thus commodity-index investors may represent a natural counterpart to commercial hedgers, who more often hold net short positions.

The role of CIT investors on futures markets is quite controversial. Masters[4] likens the entry of CIT investors to a demand shock, as their primary objective is to allocate a given amount of money to commodities, the demand for which is considered rather price inelastic. There is a crucial distinction between traditional speculators and index speculators: traditional speculators provide liquidity by both buying and selling futures. Index speculators buy futures and then roll their positions by buying calendar spreads. They never sell. Therefore they consume liquidity and provide zero benefit to the futures markets.

In general, there are concerns that commodity-index investors may affect price quotations through investment strategies that ignore expectations for fundamentals. Soros[5] has pointed out that during the last commodity-price boom, CIT investors in search of higher returns intensified the trend generated by market fundamentals. Their distorting influence was thus similar to a speculative bubble, with quite large and long-lasting consequences. Moreover, one could also argue that CIT investors, having set a trend, may induce "traditional speculators" to follow it, in the belief that it will be long-lasting. Thus, even informed speculators may be induced to delink themselves from market fundamentals, exacerbating a bubble spiral or increasing price volatility. This view, however, is challenged by other experts. According to Radetzki[6] and Greely and Currie,[7] the "passive" investment strategy of index investors prevents them from having an effect on price quotations; since they are instead seen as a natural counterpart to commercial hedgers, they may actually improve market liquidity and reduce price volatility.

As mentioned, the CFTC publishes weekly data on the positions held by each type of investor on US futures markets. Until 2007 it used to classify them into two broad groups "commercials" and "noncommercials" according to their main economic activity. Nevertheless, the association between "commercials" and hedging behavior has indeed weakened over time. By now, this category includes nontraditional hedgers, such as "swap dealers," who operate as counterparts of various types of clients (both commercial and noncommercials, including CIT investors) in the OTC markets and hedge the net exposure on the regulated markets. For this reason, nontraditional hedgers were associated with commercial traders and granted a special exemption from the position limits imposed on other kinds of noncommercial traders. However this privilege has been challenged by the most recent regulation (stemming from the Dodd-Frank Act), which maintained it only for investors operating in the physical market.

As early as 2007, the CFTC responded to the need for more transparency by publishing (for a limited number of agricultural products) a supplement to the standard weekly report on futures positions, with a categorization of investors as either "commodity index traders," "non-CIT commercials," "non-CIT noncommercials" (also called "other speculators"), and lastly "nonreportables."[8]

Subsequently, in October 2009, the CFTC released a new Commitment of Traders report aimed at reconciling investors' trading activity with the filing classification. It distinguishes major categories of investors into "producers, merchants, processors, and users;" "swap dealers;" "money managers" (encompassing commodity trading advisors, commodity pool operators, and hedge or pension funds); and "other reportables."[9]

The data shows the evolution of long and short positions by type of investor in the wheat market (although true for other agricultural commodities as well). CITs are the most important buyers, while they rarely hold any short positions; commercials are constantly net short, while noncommercials are basically net long; nonreportables (small investors not subject to position reporting) are a nonnegligible fraction of the market.

The patterns of CIT and swap dealers' positions resemble each other quite a lot, indicating a large degree of overlapping; the same is true as regards positions of commercials (not CITs) and the group of producers/merchants/users. Similar evidence also emerges for the other main agricultural commodity markets.

14.2 Leading exchanges[10] for agricultural commodity derivatives[11]

The most important exchanges for agricultural commodities are summarized in Table 14.1.

Table 14.1 Overview of the most relevant agricultural commodity exchanges

Exchange	Relevant derivatives	Relative importance
Chicago Board of Trade (CBOT) – part of CME Group	Maize, soft red winter wheat – futures, options wheat–maize intercommodity spread options	Leading exchange for soft red winter wheat and maize
Dalian Commodity Exchange (DCE, China)	Maize – futures	Most important exchange for maize in Asia
Intercontinental Exchange (ICE)	United States: cocoa, raw sugar (no. 11) – futures and options Europe: Brent, WTI – futures and options Canada: barley – futures and options OTC: crude oil (various) – swaps	Leading exchange for raw sugar and cocoa futures (ICE Futures United States) and Brent crude oil futures (ICE Futures Europe)
Kansas City Board of Trade (KCBT)	Hard red winter wheat – futures and options	Specialized exchange for wheat
Minneapolis Grain Exchange (MGEX)	Hard Red Spring Wheat Index (HRSI), Hard Red Winter Wheat Index (HRWI), Soft Red Winter Wheat Index (SRWI), National Corn Index (NCI) – futures and options	Leading exchange for hard red spring wheat
Multi-Commodity Exchange of India (MCX)	Brent crude oil, crude oil, barley, wheat, feed maize, white sugar	Among leading exchanges for crude oil
New York Mercantile Exchange (NYMEX) – part of CME Group	Cocoa, raw sugar (No. 11) – futures (settlement: financial) WTI, Brent, others – futures and options	Leading exchange for light, sweet crude oil futures; among leading exchanges for other commodities
NYSE LIFFE	London: white sugar, cocoa, feed wheat – futures and options Paris: milling wheat, malting barley, maize – futures and options	European exchange for agricultural commodities
Zhengzhou Commodity Exchange (ZCE, China)	Hard white wheat, strong gluten wheat, white sugar – futures	Largest number of contracts for white sugar, but contract size is 20% of that at NYSE LIFFE

There are additionally a massive number of regional exchanges (with an overweight for Asia) on all continents that trade some or all agricultural commodities and biofuels.

Overall, it is difficult to obtain a reliable quantitative ranking of exchanges by volume, as rankings are usually based on the number of contracts traded (e.g., by the Futures Industry Association, FIA). This may be misleading, because futures contracts for the same commodity at different exchanges may differ substantially in size. For instance, the white sugar contract at the London International Financial Futures Exchange (LIFFE) refers to 50 tons, whereas the respective contract at the Zhengzhou commodity exchange refers to 10 tons. The FIA rankings can provide a very rough idea as to the relative importance of different exchanges and contracts in the global trade of commodity derivatives, but not enough to allow any reliable quantification.

14.3 The role of commodity markets: The role of information

It has been highlighted already that since the early 2000s, we have seen commodity prices steadily on the rise (the first index tracker was put on the market in 1998!).

These developments coincide with major shifts in commodity-market fundamentals, particularly in emerging economies which are experiencing fast growth, increasing urbanization, and a growing middle class with changing dietary habits, including an increasing appetite for meat and dairy products. In addition, in an attempt to reduce the use of fossil fuels in energy consumption, a range of food crops are now being used in the production of biofuels. The related conversion of land use from crops for food to crops for biofuel production has also affected the prices of food crops. At the same time, a decline in the growth rates of production and productivity, partly due to the adverse effects of climate change, has adversely affected the supply of agricultural commodities.

However, these factors alone are not sufficient to explain recent commodity-price developments; another major factor is the financialization of commodity markets (see infra).

The role of information flows is crucial for price developments in commodity-derivatives markets. Traditionally, the so-called efficient market hypothesis (EMH) is assumed to hold in financial markets, including in commodity-derivatives markets and especially in futures markets. The EMH postulates that all publicly available information is immediately reflected in prices. In its strong form, the EMH contends that even private information – available only to individual market participants – is reflected in the price through the effects of the transactions of the persons in possession of the information. If the EMH were to apply, commodity-price developments would reflect nothing but information on fundamentals.

A number of studies[12] show that the EMH does not apply to the present commodity futures markets. Market participants also make trading decisions based on factors that are totally unrelated to the respective commodity, such as portfolio considerations, or they may be following a trend. Therefore, it is difficult for other agents in the market to discern whether or not their transactions are based on information about fundamentals, which in any case is sometimes difficult to obtain and not always reliable. Trading decisions are thus taken in an environment of considerable uncertainty. In such a situation, it is rational to follow other participants' trading decisions. A wide range of motivations leads traders to engage in this so-called intentional herding on a perfectly rational basis, the most important one being imitation in situations where traders believe that they can glean market information by observing the behavior of other agents.

In an environment of herd behavior, there are limits to arbitrage. Acting against the majority, even if justified by fundamentals, may result in large losses, often of borrowed money. It may therefore be rational for market participants to ignore their own information and follow the trend. This is what many financial players do by default, basing their trading decisions purely on the behavior of price series (algorithmic trading), which can lead to a commodity-price bubble.

Contrary studies, including those that evidence there is still a demonstrably close relationship between futures price and an underlying demand/supply relationship, will be discussed later in this chapter as well as Chapter 16. In situations where multiple variables play a key role and data sets can differ significantly across time, and where (advanced) statistical tools are used to model results, nothing more can be wished for that sufficient intellectual honesty to reflect the different outcomes, viewpoints, and conclusions. After years of analysis, it can only be concluded that a holistic answer to some of the multifaceted questions in the agri-commodities space is not available, or at least not without framing the conditionalities that relate to the data set and methodologies used. Nevertheless, there is both academic (understanding the level playing field and its limits) and professional (variability in inputs) that allow, the person who applies the models, to make decisions based on those often conflicting data sets and outcomes rather than to let data sets guide (i.e., force) certain investment decisions. More effort, however, will need to be put in this direction.

There is considerable empirical evidence that points to financial investors' impact on commodity prices:

- A number of studies (see infra) find evidence of commodity-price bubbles. Analyses show that position taking by index investors, who passively replicate the price movements of an index based on a basket of commodities, has an impact on price developments, particularly of crude oil and maize. The fact that these effects are persistent – especially in the case of

crude oil – points to the presence of herd behavior. Whereas index investors were identified as significant price drivers prior to the financial crisis, the importance of money managers (e.g., hedge funds), which follow more active trading strategies and take positions on both sides of the market, has increased since then. This is reflected in the very close correlation between price changes and position changes of money managers since 2009.
- Cross-market correlations between currency and commodity markets have increased in recent years and point to factors other than fundamentals that are driving commodity prices. Information flows in other financial markets increasingly influence the dynamics of commodity futures. In addition, an analysis of the reactions of commodity prices to announcements of economic indicators shows that, within minutes of an announcement, commodity prices react in a similar manner across different commodity markets that do not have much in common.
- The behavior of commodity prices over the business cycle has changed fundamentally. In earlier business cycles, commodity prices and equity prices evolved differently. Increases in commodity prices did not occur until well after the trough. In the most recent business cycle, on the other hand, prices surged immediately after the trough, even before share prices started to rise. This surge was based simply on the expectation, not the actual occurrence, of an upswing.

Due to their financial strength, financial investors can move prices in the short term. This leads to increased volatility, which may harm markets and drive hedgers with an interest in physical commodities away from commodity-derivatives markets. The increased volatility results in more margin calls and thus higher financing requirements. Although most if not all research stresses the role of fundamentals in medium- to long-term commodity-price formation, it concedes that substantial price distortions and herding effects could occur in the short term due to the participation of financial investors.

The analysis clearly shows that information flows play a vital role in commodity-price developments. The market distortions described above are closely related to the fact that market participants make decisions under conditions of substantial uncertainty.

As we will leave the analysis of the wide body of research regarding the impact of financial investors on commodity prices and levels of volatility for later, we can now focus on the basics of the commodity-derivatives markets, that is, its price-discovery function.

14.3.1 The role of futures exchanges and commodity markets in the price-discovery (formation) function

Commodity derivatives are traded either on organized exchanges or bilaterally "over the counter" (OTC), usually with a financial institution, depending on the concrete requirements of a trader. However, a number of exchanges (such as IntercontinentalExchange – ICE) also offer OTC transactions and clearing services.

As the choice of exchange-traded standardized contracts is limited, there may not be a futures contract that exactly tracks the price developments of the underlying asset. Differences may be due, for example, to the delivery point or the quality. Due to this so-called basis risk, the standardized contracts do not always provide a perfect hedge.

This is also why traders tend to choose tailor-made, nonstandardized OTC contracts to hedge their risks, usually in the form of swaps. This means that substantial counterparty risk is an issue. Currently, OTC markets in all parts of the world are still rather opaque, both with respect to the concrete positions taken and the way prices are formed.

Price discovery in OTC markets relies heavily on the services of price-reporting agencies (such as Platts or Argus), which provide thousands of cash reference prices per day. These benchmarks are commonly used to determine the floating price component for the settlement of swaps, though there may be some doubts about the reliability of these prices.[13]

In contrast to the OTC market, futures exchanges trade standardized products with clear definitions of the quality and quantity of the respective commodity, and predefined delivery points. Qualities deviating from these standards or with different delivery points are partly accepted but traded at a discount. Futures exchanges thus offer high liquidity, price transparency, and reduced counterparty risk. Counterparty risk is limited by the requirement to deposit an initial margin and settle the account on a daily basis. If the balance on the account falls below a predefined threshold (the so-called maintenance margin), a margin call is triggered. The respective market participants then have to provide additional funds or close their position.

The initial margin is generally only a fraction of the value of the contract, which means that a trader can take a position that is several times the value of the initial margin. Due to the high degree of standardization of contracts, exchanges attract a large volume of trade (i.e., there is high liquidity).

Price developments at the exchanges are immediately reported to news agencies, such as Reuters or Bloomberg, via the exchanges' price-reporting systems. There is a high degree of price transparency, but the positions of various types of traders are only reported in the United States in an aggregated way and on a weekly basis (infra). On the whole, prices on futures exchanges are much more transparent than those in spot markets, which are comparatively opaque. This is also emphasized in the International Organization of Securities Commissions (IOSCO) report,[14] which states, "The transparency and functioning of cash markets for commodities remains a prominent concern." It is therefore not surprising that futures markets play such a vital role in commodity price discovery.

A recent study by the International Food Policy Research Institute analyzes the dynamic relationship between spot and futures prices of selected agricultural commodities (Hernandez and Torero, infra).

Using data on weekly returns and weekly volatility for maize, hard wheat, soft wheat, and soybeans, the study shows that for these commodities, changes in futures prices lead changes in spot prices more often than the reverse. The study thus supports the findings of several earlier ones that reached similar conclusions. According to the findings of Hernandez and Torero, "the information flow from futures to spot markets has intensified in the past 15 years, probably due to the increase in the relative importance of electronic trading of futures contracts over open auction trading, which results in more transparent and widely accessible prices." In liquid standardized markets, such as commodity exchanges, any substantial price differentials would not normally be expected to persist for an extended period, as arbitrage is expected to eliminate such differentials quickly. The extent to which price differentials of similar qualities of commodities can persist also depends on concrete contract specifications. The most important specification is whether the settlement is financial or physical, and in the latter case, the relevant delivery points. In the case of physical delivery, transaction costs, such as transport costs between delivery points, need to be taken into account.[15]

14.3.2 Price evolution and fundamentals

Grain prices, as with most agri-commodities including live cattle, have been very volatile in recent years. Having peaked in 2008, they declined sharply, but started rising again in 2010. In February 2011 maize prices exceeded the level of June 2008. Even all the way into 2013, most agricultural future prices are either stable or stubbornly climbing with little downside risk, although some weaknesses have come in during the Q2/Q3 of 2013 due to excellent harvests. That could be easily reversed later on in 2013 as the "flash drought" in the United States will take its toll.

A number of supply-and-demand factors contribute to rising food-commodity prices. Supply growth is slowing, because agricultural land is limited and productivity growth has slowed. Supply constraints are exacerbated by the effects of climate change (such as extreme-weather events), which are already felt in many regions of the world but are expected to grow dramatically over the next decades.

On the demand side, the rising world population and changes in emerging economies toward more protein-rich diets are major long-term factors. As incomes in emerging economies have risen sharply with accelerated economic growth, consumption patterns of the population have also changed. Between 1995 and 2005, world meat consumption rose by 15%, East and Southeast Asia being the region with the highest increase at almost 50%.[16]

Taking into account that the production of one kilogram of meat requires about seven kilograms of grains, the impact on grain demand is substantial. Biofuel production (see separate chapter) is another decisive demand factor. The decision by some governments to introduce blending requirements and subsidies for biofuel production is considered to play a significant role in the recent price hikes of grains. Biofuel production also affects price movements of agricultural products that are not used in the production of biofuels because agricultural land is diverted to producing crops needed for biofuel production. As biofuels partly replace petroleum products, they strengthen the link between the oil market and markets of agricultural products used in the production of biofuels (i.e., maize, sugar, oilseeds, and palm oil). High oil prices also affect agricultural commodity prices via higher production costs, especially for energy and fertilizers. This may also explain the co-movement of oil prices and some agricultural commodity prices.

In the short run, weather effects have a strong impact on price developments. Often, these are exacerbated by policy measures such as export bans or taxes. Thus, wheat prices were driven up in August 2011 by the drought in the Russian Federation and an export ban.

In contrast to grains markets, high and rising prices are not a new phenomenon in the sugar and cocoa markets, judged by historical standards. These two soft commodities already experienced extreme price spikes in the 1970s and 1980s. Recently, the cocoa price has come under pressure due to political tensions in Côte d'Ivoire, the world's largest cocoa producer.

Unlike in earlier periods, the recent price hikes have occurred in an environment of general price increases across a wide range of commodities, from energy to agricultural commodities. Most of the factors that are often cited as price drivers, such as population growth or changing consumption patterns, have been at work for an extended period, often coinciding with low commodity prices. Their role in explaining recent price hikes is therefore doubtful. Experiences with the weak forecasting performance of econometric models for oil prices based on fundamentals (e.g., Kaufmann, 2011)[17] also suggest that physical supply and demand are not the only factors that drive oil prices.

14.3.3 Does investing by a financial investor in the agricultural sphere serve a purpose?[18]

A common driver for a financial investor when investing in the agricultural sphere is to achieve a certain level of diversification they cannot find elsewhere. In their search for uncorrelated assets, they can turn to either agricultural stocks (private or publicly listed, grouped or on an aggregate basis) or buy long-only agricultural futures and other derivatives.

With fluctuations in the stock market and low interest rates, investors are looking for alternative places to put their money. The first question then becomes, how good of an investment are commodities for the typical investor? The analysis of Zapata, Detre, and Hanabuchi[19] suggests that commodity prices are negatively correlated with stock indices. Their second analysis shows that during a time period when stocks did poorly and commodity prices rose, investment in a commodity index would have been part of an optimal portfolio.

Much of the first analysis is spent on measuring long-term cycles in the ratio of commodity and stock prices. Even with a 140-year data period, a 31-year cycle will only be observed four and a half times. This makes for a small number of observations. No strong theoretical explanation is offered for the cycle, and structural change over this time period has been substantial. While this is a fun thing to do, I am skeptical of attempting to trade based on this analysis. Their second analysis shows that stocks have gone up at a rate ten times that of commodities over the last 100 years. Holding commodities would typically have storage costs that are not included in the analysis. So, commodities are not a good investment for an investor with a long time horizon. Their second analysis looks at a very short time period when stocks have done poorly and commodity prices have risen. It is not surprising that commodities are part of the optimal portfolio. Thus, commodities are a place to park money when stock returns are low. Note that this time period is not representative, and a quarterly planning horizon is much shorter than that of most investors.

On the other hand, Schnitkey and Kramer[20] examined the returns of agricultural stocks versus returns from the S&P 500 stock index during 2000–2011. Agricultural prices increased substantially over this time period, and incomes of agricultural producers have gone up accordingly. They seek to determine if stocks of agricultural companies have also done well. They find that agricultural companies have performed better than the S&P 500. The better performance of agricultural stocks occurred before the main rise in agricultural prices. So this suggests that either stock investors foresaw the rise in commodity prices or that the correlation between the prices of agricultural stocks and agricultural prices is even weaker than it first appears. They use time-tested methods, so there is little to complain about in their procedures. With an increase in demand for agricultural products such as corn for ethanol, we would expect returns to initially increase for all producing sectors. In a competitive market, the excess profits should be competed away, and the returns should eventually go to holders of resources such as land or to specialized labor and management. To the extent that they control stocks of phosphorus and potassium, fertilizer firms are resource holders – recall the 2010 $40 billion (hostile) bid by BHP Billiton for Potash Corporation of Saskatchewan Inc., which controls about one-fifth of the global potash inventory in the world. Equipment and seed producers also hold some limited resources in the form of patents and seed varieties.

Otherwise it is hard to see how agricultural firms would benefit that much unless markets are not competitive. That is what is found. Stocks of agricultural firms did well in this period, but probably not quite as well as investment in some other parts of the agricultural sector such as agricultural land.

Irwin and Sanders (2012)[21] documented some of the changes that have occurred in futures markets during the last few years. There has been an explosion in volume and open interest. Most trading has switched to electronic platforms rather than open outcry. The portion of small traders has shrunk. A major focus of their study is index funds. Index funds include a variety of investment vehicles that use a buy and hold strategy in commodity futures markets. Zapata, Detre, and Hanabuchi argued that if investors have a short planning horizon, then an investment in commodities can belong in an optimal investment portfolio. Firms have been successful in marketing these investments. A major policy question is how has the growth of index funds influenced the commodity markets?

Since 2006, as Irwin and Sanders demonstrate, index-fund investment has been a relatively constant % of the market. It is indeed difficult to see how index funds could have been responsible for a price bubble in commodity markets. Their position in earlier research, such as Irwin and Sanders (2011),[22] was that index funds had no effect at all. It can be observed that they have moderated their position slightly. They now mention that index funds could have reduced risk premiums paid by short hedgers and in some cases could even cause risk premiums to be paid to short hedgers. They also discuss the possibility of index funds increasing the demand for storage. Most would consider these two effects to be positive (although an increase in demand for storage could increase price levels in the short run). Thus the policy implication is that there does not seem to be a reason to limit investment in index funds. Irwin and Sanders (2012) do not discuss managed funds, which take both long and short speculative positions. Managed funds often use trend-following trading systems, and therefore managed funds could help cause a price bubble.

14.3.4 What can be concluded?

The three pieces of research discussed are related in that they all address the general issue of investors in financial markets using the agricultural sector to diversify their portfolios and thus reduce risk. In conclusion, Schnitkey and Kramer (2012) find that investing in agricultural stocks has modest diversification potential. Zapata, Detre, and Hanabuchi (2012) argue that investment in commodities themselves has strong diversification potential. While the risk reduction appears to be present, expected returns from commodities are low, and thus investment in commodities would appeal more to a shortsighted investor. Investment vehicles that provide the ability to invest in commodities by taking buy and hold positions in commodity futures markets have been created, and investment in them is substantial. Irwin and Sanders (2012) review the literature on possible effects of this investment. Their review and analysis suggest that this investment should not be a cause of concern for the agricultural sector.

14.4 What is the relationship between agribusiness and nonagribusiness stocks?

But what about the fact that during extensive periods of time, the correlation between and within asset classes is very high, especially when the downward trend in the market is negative? During the financial downturn of 2008, asset classes that investors traditionally found to have low correlation with US stocks became more highly correlated at the most inopportune time. In fact, that statement was true for investors around the world.

Post-downturn, investors increasingly looked for alternative assets that offer diversification benefits, one of which being farmland. One of the challenges of investing in farmland is that the asset is not a securitized, low-cost investment. D'Antonia and Detre (2013)[23] investigated whether exposure to farmland via an index of agribusiness stocks provides significant diversification benefits. They estimated the dependence between daily returns of the S&P 500 and an index of agribusiness stocks from 1970 through 2008 using copulas. They found significant evidence that agribusiness stocks have strong lower-tail dependence with large US stocks and are actually less correlated in the upper tail of the distribution. This means that the agribusiness index moves in near lockstep with US stocks in downturns and more independently in large upswings. This provides little evidence to support the investment strategy of purchasing agribusiness stocks broadly to gain exposure to farmland.

14.5 Does the presence of financial investors invalidate the hedging position of commercial agricultural firms?[24]

The instability of prices and the hypothesis that speculative behavior was one of its sources has brought renewed interest in the futures markets. As pointed out by Lence (2009),[25] vulnerability to risks is among the most important problems faced by commodity producers in developing and developed countries. The use of futures and options contracts has been extended to the area of food security, as they have been proposed as a way in which importing countries could manage price volatility. After all, futures markets perform several functions: they provide the instruments to transfer price risk, they facilitate price discovery, and they offer commodities as an asset class for financial investors, such as fund and money managers who had not previously been present in these markets.

As discussed, commercial participants use futures contracts to hedge their crops or inventories against the risk of fluctuating prices; for example, processors of agricultural commodities, who need to obtain raw materials, would buy futures contracts to guard against future price rises. If prices rise (i.e., both cash and futures prices), then they use the increased value of the futures contract to offset the higher cost of the physical quantities they need to purchase. However, hedgers are not the only agents operating in futures markets, as one can also find noncommercial participants, who do not have any involvement in the physical commodity trade in contrast to commercial participants, such as farmers, traders, and processors.

The question arises regarding the usefulness of futures prices for hedging against price risk now that the market is predominantly crowded by financial investors.

As is well known, what is important when hedging is not the absolute movements of the futures and the cash/spot prices but the relationship between them, that is, the basis. Most of the studies on volatility focus on the behavior of commodity prices; however, there seems not to be interest in the behavior of the basis, particularly on analyzing the basis risk (i.e., when futures and cash prices do not evolve in a similar way during the period before contract expiration). Therefore, the actual question is whether the basis has become more erratic in particular futures markets.

From the chapter on speculation in commodities markets (see infra), it can be learned that any estimation of the effects of speculation on futures prices has to be done indirectly, or at least indirectly through statistical tools. Every analysis will, as such, be an indirect test of whether the increasing presence of speculation in futures markets has made them divorced from the physical markets and therefore not useful for price hedging.

Although a minority of farmers still use hedging to reduce risk, the question remains as to whether hedging in futures markets has become an even less attractive operation or whether it is still a valid mechanism to reduce vulnerability to price risk given the presence of financial speculators,[26] or put differently, whether those markets still allow substitution price risk for basis risk. This implicitly would be a test of whether the increasing presence of speculation in futures markets have made them divorced from the physical markets and therefore not useful for commercial entities. Two aspects are relevant: efficiency and hedging effectiveness.

Revoredo-Giha and Zuppiroli (2013)[27] based their methodology on the applied methodology of Carter (1984)[28] and Castelino (1989).[29] It consisted of analyzing two issues in the selected markets: first, the efficiency of the futures markets, and second, the effectiveness of hedging before and after 2006. The year 2006 is a watershed between periods of very different volatility levels. For the purpose, several subanalyses were performed, such as the correlation between spot prices and futures prices by contract, an analysis of the basis volatility, the value of the hedging ratios, and estimates of the effectiveness of hedging.

14.5.1 Efficiency analysis

The continuous flow of information, public or not, forces prices to fluctuate. A futures market is considered to be efficient when

1. it demonstrates that prices adjust to the information available (i.e., "price efficiency"), and
2. it does not persistently favor one side of the market, no matter if they are the "long" or the "short" positions (i.e., "market bias").

As regards price efficiency, were the analysis due to data availability was limited only to historical prices, only the notion of weak price efficiency was tested (Fama, 1970).[30] Other notions of efficiency (i.e., semi-strong or strong) would have required availability of public information on the fundamentals (supply-demand sheets, ending stocks, stock-to-use ratio, and so on) or private information, respectively. From the price-efficiency point of view, one would expect that the increasing role of hedge funds and commodity-index traders in futures markets would have reduced the price efficiency of the market.

The second efficiency test refers to the theory of "normal backwardation," which assumes that an inefficient market should give a structural advantage to the long positions taken by speculators with respect to the short positions taken by hedgers. According to Carter (1984),[31] this characteristic, called a thin market, is usual of markets that are less active and where futures contracts lack interest among speculators. Hedgers, interested in transferring risk to other agents, would accept market returns in the long run favoring the buyers of the contracts. The existence of this bias in favor of speculators has been tested using the implication that a trade routine such as the long position taken by speculators in futures markets should have earned them positive profits over time (in contrast, hedgers are supposed to be continuously net short and the losses they made are a "payment" for the price insurance they receive).

Gray's trading routine[32] assumes that the speculator takes a net long position all year round. If the annual harvest is immediately hedged, the price at harvesttime must be low enough to induce speculators to invest on the long side of the hedge. Futures prices must rise continuously over the postharvest life of the contracts in order to ensure profits for speculators as a whole. Gray's hypothetical trading routine involves purchasing the futures contract closest to maturity, buying it on the first trading day in the delivery month of the preceding futures contract. Then every contract is sold on the first trading day of its own delivery month. Gray's assumption is that if hedging is always net short, then speculators as a group must be net long. Cootner[33] instead noted that hedging is not always net short: when commitments to deliver at fixed prices are larger than commitments to buy, hedging may be net long. During the period of declining short hedging, interest prices must fall: under this condition, a rational behavior of speculators is to be long not for all the months but only for a part of the year; otherwise they were short.

It should be noted that in order to apply Cootner's routine to current data in a consistent way, one needs to explore the existence of seasonality and what the seasonality pattern was. This then allows us to adapt the trading routine to the actual price dynamics, determining the months which are better for a long position and for a short one.[34]

14.5.2 Hedging effectiveness analysis

According to Sanders and Manfredo (2004),[35] minimum-variance measures of hedging effectiveness have not changed dramatically since Ederington's (1979) initial use of the correlation coefficient to measure the relationship between changes in cash and futures prices. In fact, they point out that minimum-variance hedging effectiveness is most commonly evaluated through an ordinary least squares (OLS) regression of the change in cash price as a linear function of the change in the futures price, where the resulting R-square is the measure of hedging effectiveness.

While the economic theory behind hedging is still the minimum-variance portfolio approach, the econometrics when estimating hedging ratios has evolved with the progress in time-series econometrics. Lien (2002)[36] provides an overview of relatively recent econometric methods to compute the hedging ratio.

It is important to note that the type of hedging varies depending on the type of the operator. All the operators working along the wheat supply chain have a potential interest for hedging, but for everyone hedging has its own meaning. Thus, in order to evaluate hedging effectiveness for farmers, it is necessary to define planting interval and postharvest period. The season for growing wheat is a lengthy one, generally 10 to 11 months, beginning and ending in different periods according to the country and the type of wheat marketed (spring or winter). The farmer lifts the hedge after 10 or 1 months and starts in a fixed period: the kind of hedging suitable for farmers is a long-term hedge, "seasonally specified."

Hedging is also of use for merchants and for processors in the supply chain. The length of the hedge suitable for these categories is shorter than for farmers ("short-term" hedge) and is not seasonally specified. Merchants and processors usually hedge their physical (spot) positions all over the year, holding position in the futures market for less than 10 or 11 months: the lengths assumed here are 30, 60, and 90 trading days.

These intervals imply approximately a month and a half, three-month, and four-month periods respectively. It follows that the evaluation of the effectiveness for the hedges in question needs a separate computation comparing the dynamics of spot and futures prices for all of the 30, 60, and 90 trading-day intervals available. Finally, in comparison with Carter (1984), hedges at very close range (seven trading days, which imply approximately 10 calendar days) were also calculated. Every test at the same time gives the result not only for short hedging but also for a long-hedging procedure. The last type of hedge is common for processors; however, both types are used by merchants and traders depending on their counterpart in the transaction.

In general, Revoredo-Giha and Zuppiroli's (2013) results indicate that the increasing role of non-related-with-the-physical investors, as hedge funds and commodity-index traders, did not reduce the price efficiency and usefulness for hedging of the three selected exchanges (Liffe and Matif, two European wheat markets, and the US CBOT for comparison purposes). The several weak-form tests such as serial autocorrelation analyses (cross correlation of a variable with itself) did not reject the hypothesis that all the studied wheat futures markets were efficient.

As regards whether holding a speculative position, structured in a rational way, would bring consistent profits, the results show that in the last 20 years, these profits are not statistically different from zero. Based on this evidence, the European exchange futures contracts perform as well as the US CBOT. With respect to the effectiveness of hedging, the results indicate that this is still a viable alternative for commercial entities as spot and futures prices evolve closely. In other terms, particularly as short-term hedges, the basis has not been affected by the instability observed in commodity markets.

The results of the hedging effectiveness can provide an assurance (as an implicit test) that the increasing presence of financial speculation has not made futures markets divorced from the physical markets. It is only for lengthy-period hedges (such as farmers' hedges) that there appear to be some concerns of their effectiveness. The results imply that futures markets are not only still an efficient tool in risk management but may also be a useful tool for food-security purposes; however, it is important to stress that the analysis carried out is only valid for the regions where the exchanges are located, and the use of these exchanges or others for food security would require computing the basis using appropriate spot prices and the most adequate futures contracts. So in conclusion, there is still a good connection between physical and futures markets, and therefore hedging can still play an important role protecting commodity handlers against price volatility.

It can be added that Hernandez, Ibarra, and Trupkin (2012),[37] who applied their analysis across a variety of markets worldwide, found that agricultural markets are highly interrelated, and there are both own- and cross-volatility spillovers and dependence among most of the exchanges. Chicago particularly plays a major role in terms of spillover effects over other markets (compared to both Europe and Asia), particularly for corn and wheat.[38] Additionally, the level of interdependence between exchanges has only increased in recent years for some commodities[39] but definitely not across the board.

In parallel to the previous study, the question was raised to what extent (rising) energy prices stimulate food-price volatility. Gardebroek and Hernandez (2012)[40] examined volatility transmission in oil, ethanol, and corn prices in the United States between 1997 and 2011. They followed a multivariate GARCH approach to evaluate the level of interdependence and the dynamics of volatility across these markets. The estimation results indicate a higher interaction between ethanol and corn markets in recent years, particularly after 2006 when ethanol became the sole alternative oxygenate for gasoline. They further observed significant volatility spillovers from corn to ethanol prices but not the converse. They did not find major cross-volatility effects from oil to corn markets. The results do not provide evidence for volatility in energy markets stimulating price volatility in grain markets.

15
Commodities II: Trading Agricultural Commodities[1]

Many economic factors, including temperature, precipitation, changing customer needs, substitute products, and new market participants, are far beyond our control. However, they are key to determining the supply and demand for vital commodities such as corn, wheat, soybeans, soybean meal, soybean oil, rice, and oats.[2] As a result of continuously changing global supply and demand for grains and oilseeds, commodity prices can vary substantially from day to day.

It is possible, however, for food prices to remain relatively stable and predictable when the prices of commodity ingredients used in food production are volatile. To a large extent, the financial effects of futures markets are responsible for this price stability. Agricultural option platforms are essential for managing price swings and for providing the global benchmark price for grains and oilseeds.

One note up front: Although the initial ambition was to be fully comprehensive on all aspects of trading agricultural commodities, including the different trading strategies analyzed for each of the agricultural commodities separately, very quickly reality forced me to reconsider. As it turned out, within the context of one chapter, this clearly was an overambitious objective. As a direct alternative, I have, in the respective footnotes, made reference to all the relevant literature, including the different trading strategies per agricultural commodity.

15.1 Introduction

If you are involved in the production of agricultural commodities, you can price your commodities using one or more combinations of these four alternatives (see Table 15.1):

- Cash sales
- Forward contracts

- Futures contracts
- Options on futures contracts

1. With cash sales you deliver your crop or livestock to the cash markets (such as the grain elevator or meat packer) and receive the price for the day. You get cash right away, and the transaction is easy to complete. But using this alternative, you have only one chance to sell. You take what you can get, but it involves quite some risk for producers.
2. A forward contract is negotiated now for delivery later. It is easy to understand. You enter a contract with the buyer who agrees to buy a specified quantity and quality of the commodity at a specified price at the time of delivery. The price is locked in, and you are protected if prices fall. However, you cannot take advantage of price increases, and you must deliver the specified amount, even if you have a crop failure. Both parties have some risk that the other will not honor the contract. A forward contract can therefore be defined as a private, cash-market agreement between a buyer and seller for the future delivery of a commodity at an agreed price. In contrast to futures contracts, forward contracts are not standardized and not transferable.
3. A futures contract is an agreement to buy or sell a commodity at a date in the future. You buy or sell through a brokerage firm that transacts the trade for you. Once you are set up with a firm, it is as easy as a phone call to make a trade. You must deposit a performance bond (a small percentage of the contract value) with the brokerage firm to guarantee any loss you may incur on the futures contract. If the value of the contract goes against your position, you will be asked to deposit more money. You also pay a broker a commission for every contract traded. A futures contract can therefore be defined as an obligation to deliver or to receive a specified quantity and grade of a commodity during a designated month at the designated price. Each futures contract is standardized by the exchange and specifies commodity, quality, quantity, delivery date, and settlement. Hedging is selling or buying a futures contract as a temporary substitute for selling or buying the commodity at a later date. For example, if you have a commodity to sell at a later date, you can sell a futures contract now. If prices fall, you sell your actual commodity at a lower cash price but realize a gain in the futures market by buying a futures contract at a lower price than you sold. If prices rise, your higher price in the cash market covers the loss when you buy a futures contract at a higher price than you sold. This may be considered a pure hedge, or a replacement hedge. It minimizes your risk and often earns you more than the forward-contract price. Hedging can therefore be defined as (1) taking a position in a futures market opposite to a position held in the cash market to minimize the risk of financial loss from an adverse price

change, and (2) a purchase or sale of futures as a temporary substitute for a cash transaction that will occur later.
4. Options on futures contracts are traded at futures exchanges too. An option is the right, but not the obligation, to buy or sell a futures contract at a specified price. You pay a premium when you buy an option, and you pay a commission to the broker.

For example, if you buy a put option and prices rise, you can let the option expire and sell in the cash markets at a higher price. If prices fall, you can protect yourself against the low cash price by

- offsetting the option (sell the same type of option), or
- exercising the option (exchange the option for the underlying futures contract).

An option can therefore be defined as the right, but not the obligation, to sell or buy the underlying (in this case, a futures contract) at a specified price on or before a certain expiration date. There are two types of options: call options and put options. Each offers an opportunity to take advantage of futures price moves without actually having a futures position.

As a producer of agri-commodities, you will eventually sell your commodity in the cash markets. You can sell directly in your local markets or negotiate a forward contract for sale at a later date. Even if you sell futures contracts or buy options to sell futures, you will close out your position and sell your commodity in the cash markets. Very few futures contracts are actually delivered. If

Table 15.1 Marketing possibilities for agricultural commodities

Marketing possibilities	Advantages	Disadvantages
Cash sales	• Easy to transact • Immediate payment • No set quantity	• Maximize risk • No price protection • Less flexible
Forward contract	• Easy to understand • Flexible quantity • Locked-in price • Minimize risk	• Must deliver in full • Opportunity loss if prices rise
Futures contract	• Easy to enter/exit • Minimize risk • Often better prices than forward contracts	• Opportunity loss if prices rise • Commission cost • Performance -bond calls • Set quantities
Options contract	• Price protection • Minimize risk • Benefit if prices rise • Easy to enter/exit	• Premium cost • Set quantities • Commission cost

you are selling grain or livestock on a cash basis, the terms are negotiated when you bring in the grain or livestock. The price is established then and there, and you make immediate delivery and receive payment. This type of sale occurs at elevators, terminals, packing houses, and auction markets. You can choose when to sell grains in the cash market. You can sell at harvest or store the grain until later when you expect prices to be better. Because of storage costs, there is risk involved in waiting for prices to rise. You can also make a cash sale with a deferred pricing agreement. You deliver the commodity and agree with the buyer to price it at a later time. For example, you may deliver corn in October and price it at any time between then and March. In this way, you transfer the physical risk of having the corn and the storage cost, and you may be able to get a higher price for the corn.

Alternatively, you can negotiate a forward contract with your local merchant for future delivery of your crop or livestock. You and the buyer agree on quantity, quality, delivery time, location, and price. This should be a written contract. Once you enter into this contract, you eliminate the risk of falling prices. However, if prices go higher at delivery time, you'll still receive the negotiated price. When you make delivery, it will be inspected before payment is made. There may be a premium or discount in price if quality or quantity varies. A basis contract is another method of forward contracting. In this case, you lock in a basis relating to a specified futures contract. When you deliver, the price you receive is the current price of the specified futures contract adjusted by the basis you agreed upon. For example, if a basis of $0.20 under was specified in the contract and the futures price is $3.04 on the delivery date, then the cash price you receive is $2.84 ($3.04 – $0.20 = $2.84). You need to know the local basis patterns before entering into this type of forward contract.

15.2 The basis and its calculation

The relationship of the local cash-market price and futures-market price is called basis. The value of basis is calculated by subtracting the price of the nearby futures contract from the local cash-market price. For example, if the cash price for corn is $2.80 and the futures price is $3.00, then the basis is $0.20 under ($2.80 – $3.00 = –$0.20). With a cash price of $2.95 and a futures price of $2.90, the basis is $0.05 over ($2.95 – $2.90 = $0.05).

Basis = cash price – futures price

The basis can therefore be defined as the difference between the spot or cash price and the futures price of the same or a related commodity. Basis is usually computed to the near future and may represent different time periods, product forms, qualities, and locations. The local cash-market price minus the price of the nearby futures contract equals the basis, that is, private, cash-market agreement between a buyer and seller for the future delivery of a commodity at an agreed price. In contrast to futures contracts, forward contracts are not standardized and not transferable. Storable commodity futures prices reflect the cost of delivering a commodity to a specific place. Cash prices reflect the cost of delivering (perhaps a different quality) to a different place. These costs include transportation, carrying charges such as storage costs for grain, and marketing costs such as weight shrinkage for livestock. Basis reflects supply and demand for a given commodity in a given location along with the cost of delivering (perhaps a different quality) to a different place. Basis varies from one location to another. Depending on the circumstances of the local market, the basis may be consistently positive (over) or negative (under). Each local market has its own pattern. Storable commodity basis also changes during the life of the futures contract. Basis tends to start wide, but the threat of delivery on the futures contract generally causes the basis to narrow. That is, the futures price moves closer to the delivery-point cash price during the delivery month.

15.3 The futures markets[3]

15.3.1 A little bit of history

It all started in Chicago. Chicago was a growing city in the 1830s and a center for the sale of grains grown nearby to be shipped east. By the early 1850s, when farmers had spread over the countryside, local merchants began to sell corn to the Chicago merchants on time contracts, or forward contracts, to minimize their risk. As soon as the forward contract became the usual way of doing business, speculators appeared. They did not intend to buy or sell the commodity. Instead, they traded contracts in hope of making a profit. Speculation itself became a business activity. Contracts could change hands many times before the actual delivery of the corn. During this time, contracts were negotiated and traded in public squares and on street curbs. The Board of Trade of the City of Chicago (CBOT) had been organized in 1848 with the intention to promote commerce. At first, there was little control over the trading of forward contracts. Sometimes people disappeared when the time came to settle contracts, and others could not pay. In 1865, the CBOT issued general rules setting forth the following:

- A requirement for a margin, or good faith, deposit
- Standardized contract terms for quantity and quality of the commodity and delivery procedures
- Payment terms

They called these standardized contracts futures contracts. All the ingredients for futures trading were now in place.

In 1874, merchants formed the Chicago Produce Exchange, which dealt primarily with butter, eggs, poultry, and other farm products. It was later named the Chicago Butter and Egg Board. In 1919, it became the Chicago Mercantile Exchange (CME). Many commodity exchanges have been organized since 1848. Some are still here today. Others have closed or merged with other exchanges. These days, only cash settlement occurs (no physical delivery is involved; only the change in price is settled at the contract maturity) and electronic trading. Futures trading is a global business and can be traded wherever one is in the world.

15.3.2 The exchanges and their role[4]

The essential functioning of these agricultural trading platforms includes the following:

1. The price discovery function: global supply and demand for agricultural commodities come together at the trading market, making possible the discovery of a global benchmark price.
2. Financial transparency and stability: clearing and processing of transactions via the world's largest clearing systems.
3. Liquidity: Global participants provide trading efficiencies and liquidity to the markets due to a variety of trading strategies.
4. Dual trading platforms with extended trading hours: an open-auction trading platform and a world-class electronic trading platform, often with 15 hours of trading per day
5. Different types of options: different types of options give participants an additional flexible choice to meet specific trading and risk-management objectives.

15.3.3 The agricultural futures market[5]

A futures contract is a standardized agreement to buy or sell a commodity at a date in the future. The futures contract specifies the following things:

- Commodity (live cattle, feeder cattle, lean hogs, corn, soybeans, wheat, milk, and so on)
- Quantity (the number of bushels of grain or pounds of livestock as well as the range of weight for individual animals)
- Quality (specific US grades)
- Delivery point (location at which to deliver the commodity or cash settlement)
- Delivery date (within the month that the contract terminates)

The only aspect of a futures contract that is not specified is the price at which the commodity is to be bought or sold. The price varies; it is determined on the floor, or electronically, as traders buy and sell the contracts. The prices they offer and bid reflect the supply and demand for the commodity as well as traders' expectations of whether the price will increase or decrease.

15.3.4 Offsetting futures[6]

Anyone can buy or sell futures contracts through the proper channels. For example, you can sell a CME Live Cattle futures contract even if you do not have any cattle to deliver. Although under the futures contract you are obligated to deliver, you can remove that obligation at any time before the delivery date by offsetting or buying the same type of futures contract. Similarly, you could buy a CME Live Cattle futures contract without the intention of taking delivery of the cattle. You remove the obligation to take delivery by offsetting or selling the same type of futures contract.

1. Two situations can occur:
    - You buy a futures contract and offset it by selling the same type of contract.
    - You sell a futures contract and offset it by buying the same type of contract.

Speculators have no intention of buying or selling actual commodities. They try to make money by buying futures contracts at a low price and selling back at a higher price or by selling futures contracts at a high price and buying back lower. They take on the risk that prices may change to their disadvantage.

As the delivery month of a contract approaches, the futures price tends to fall in line with the cash-market price of the commodity. Thus, most producers remove their obligation to deliver or take delivery on the futures contract just as speculators do. But producers will then sell or buy actual commodities in the cash markets.

15.3.5 Hedging with futures[7]

Hedging is buying or selling futures contracts as protection against the risk of loss due to changing prices in the cash markets. Hedging is a risk-management tool for the producer. If you have a crop of livestock to market, you want to protect yourself against falling prices in the cash markets. If you need to buy feed or feeder cattle, you want to protect yourself against rising prices in the cash markets. Either way, hedging provides you with that protection.

There are two basic types of hedges:[8]

- The short hedge, or selling hedge, used when you plan to sell a commodity. The short hedge protects the seller of a commodity against falling prices.

- The long hedge, or buying hedge, used when you plan to purchase a commodity. The long hedge protects the buyer of a commodity against rising prices.

If you are long futures, you bought a futures contract. If you made a long hedge, you bought a futures contract to protect against price increase. (You plan to buy the commodity.)

If you are short futures, you sold a futures contract. If you made a short hedge, you sold a futures contract to protect against price decrease. (You plan to sell a commodity.)

The short hedge protects a producer with a commodity to sell against falling prices. When you plan to sell a commodity, you can use a short hedge to lock in a price and protect against price decreases. The short hedger plans to sell a commodity and sells a futures contract to lock in a selling price and protect against falling prices. Thus:

1. Sell futures contract "then."
2. Buy back futures contract "then."
3. Sell commodity in cash market.

Short-hedge example: Suppose it is April. You are offered $67.50 by your packer for hogs to deliver in December, while the December CME Lean Hog futures price is $70.00. Your decision is to take the contract or hedge on your own. With December CME Lean Hog markets trading at $70.00 and expecting a $1.50 basis in December, you decide to hedge ($70.00 – $1.50 = $68.50).

Forward-contract offer	67.50
December futures	70.00
Expected basis	–1.50
Expected Hedged Return	**68.50**

Expected December basis	Futures price	Cash price	Futures gain/loss	Net price received
1.5	Buy back 67.00	Sell at 65.50	+3.00	= 68.50

Note: The profit in the futures market offsets the lower price in the cash market.

The results above show that the cash price has fallen to $65.50 and the futures price to $67.00. The basis is $1.50 under. You buy a December CME Lean Hog futures contract at $67.00. Because you sold it at $70.00, you receive a gain of $3.00 per pound ($70.00 – $67.00 = $3.00). Then you sell the hogs in the cash market at $65.50. The total price you received is the cash price of $65.50 plus the $3.00 futures gain, or $68.50. That is $1/cwt (hundredweight) more than the price you would have received if you had accepted the forward-contract price of April.

What if prices had risen? Suppose the December cash price is $71.00, the futures price is now $72.50, and the basis is $1.50 under. You buy a December CME Lean Hog futures contract at $72.50 at a loss of $2.50 ($70.00 – $72.50 = –$2.50). Then you sell in the cash market at $71.00. This time the total price you receive is the cash price of $71.00 plus –$2.50, the loss in the futures market, or $68.50, as estimated.

Expected December basis	Futures price	Cash price	Futures gain/ loss	Net price received
–1.5	Buy back 72.50	Sell at 71.00	+2.50	= 68.50

Note: The loss in the market is offset by the higher selling price in the cash market.

Both of these results assume that the basis in April and December is the same at $1.50 under. This is called a perfect hedge. We used the example only to show how the mathematics of the short hedge works. In real life, it is highly unlikely that the basis will remain the same as expected. Here are the results showing what happens to a $70.00 hedge when the basis strengthens (becomes more positive) and when prices fall or rise.

Results: Basis strengthens

Basis	Futures price	Cash price	Futures gain/loss	Net price received
–1.00	Buy back 67.00	Sell 66.00	+3.00	= 69.00
–0.50	Buy back 67.00	Sell 66.50	+3.00	= 69.50
–1.00	Buy back 72.50	Sell 64.00	–2.50	= 69.00
–0.50	Buy back 72.50	Sell 63.50	–2.50	= 69.50

Note: Whether the futures price rose or fell, your net price was higher with a stronger basis. (Compare to earlier results.) However, you cannot take full advantage of a price increase.

15.3.6 The long hedge

The long hedge protects against a rise in input costs such as those incurred by a packer procuring cattle and hogs, or a producer who needs to insure against higher feed costs. When you plan to buy a commodity, you can use a long hedge to lock in a price and protect against price increases. Thus:

1. Buy futures contract "then."
2. Sell back futures contract "then."
3. Buy commodity in cash market.

The long hedger plans to buy a commodity and buys a futures contract to lock in a purchase price and protect against rising prices.

Long-hedge example: Suppose it is December. A packer wants to protect procurement cost for hogs purchased in July. The July CME Lean Hog futures price is $71.00. The packer decides to buy a July CME Lean Hog futures contract at $71.00, expecting a –1.50 basis or $70.00 – 1.50 = $69.50 procurement protection price. In July, he will sell a July CME Lean Hog futures contract to offset his position and purchase the hogs in the cash market.

Futures	71.00
Expected basis	–1.50
Expected hedged return	69.50
Buy July futures	**71.00**

Results: Prices rise

Expected July basis	Futures price	Cash price	Futures gain/loss	Net price received
–1.5	Buy back 81.00	Buy at 79.50	–10.00	= 69.50

Note: The profit in the futures market offsets the higher price in the cash market.

The results above show that the cash price has risen to $79.50 and the futures price to $81.00. The basis is $1.50 under. The packer sells a July CME Lean Hog futures contract at $81.00. Because he bought it at $71.00, he receives a gain of $10.00 ($81.00 – $70.00 = $10.00). Then he buys the hogs in the cash market at $79.50. The total price he paid is the cash price of $79.50 minus the $10.00 futures gain, or $69.50. That is the same hedged procurement price estimated when the hedge was placed in December.

What if prices had fallen? Suppose the July cash price is $64.50, the futures price is now $66.00, and the basis is $1.50 under. The packer sells a July CME Lean Hog futures contract at $66.00 at a loss of $5.00 ($66.00 – $71.00 = –$5.00). Then he buys in the cash market at $64.50. This time the total price he pays is the cash price of $64.50 minus –$5.00, the loss in the futures market, or $69.50.

Expected December basis	Futures price	Cash price	Futures gain/ loss	Net price received
1.5	Sell back 66.00	Buy at 64.50	–5.00	= 69.50

Note: The loss in the futures market is offset by the lower purchase price in the cash market.

With a basis at $1.50 under in both of these examples, we are again talking about a perfect hedge. Actually, it is highly unlikely that the basis will be exactly the same as expected. When the packer quotes a forward-contract price to buy your lean hogs, he will offer a weaker basis than will most likely occur at contract delivery. Since there is a risk in the basis, the packer builds in protection by lowering the contract offer to you.

Results: Basis weakens

Basis	Futures price	Cash price	Futures gain/loss	Net price received
–2.00	Buy back 81.00	Sell 79.00	– +10.00	= 69.00
–2.50	Buy back 66.00	Sell 78.50	– +10.00	= 68.50
–2.00	Buy back 81.00	Sell 64.00	– –5.00	= 69.00
–2.50	Buy back 66.00	Sell 63.50	– –5.00	= 68.50

Note: Whether the futures price rose or fell, your net purchase price was lower with a weaker basis. (Compare to earlier results.) However, you cannot take full advantage of a price decrease.

15.3.7 Futures cash flow and performance bonds

You have to look at the cash required for futures trading. Before considering some practical hedging examples, we will take a look at the finances of hedging:

- The performance bond, or good faith, deposit
- Broker commission

The exchange clearinghouse requires that clearing members deposit performance bonds to guarantee performance on their customers' open futures contracts. Individuals trading in the market make the deposit with their brokerage houses. A performance bond is therefore the amount of money or collateral deposited by a client with his broker, or by a clearing firm with the exchange's clearing on open futures or options contracts before a customer can trade. The performance bond is not a part payment on a purchase. When you sell a futures contract, you do not receive payment. Instead, you deposit a performance bond (money) with your broker to guarantee payment of immediate losses you may suffer.

The value of your contract is calculated on a daily basis. If the futures price increases significantly and causes the value of your contract to increase beyond a certain point, you will get a performance-bond call and be asked to deposit more money to cover the loss in your account. A smaller maintenance performance-bond balance must be maintained to protect against the next day's possible losses.

Performance bond: A deposit to cover any loss you may incur on the futures contract.
Maintenance performance bond: A sum less than the initial performance bond that must be maintained in your account.
Performance-bond call: A demand for an additional deposit to bring your account up to the initial performance-bond level.

Your contract obligation is offset when you buy back (or sell back) a futures contract. The difference between the selling price and the buying price is your gain or loss. If the buying price is lower than the selling price, you earn a profit and receive the money. If it is higher, you suffer a loss, which is covered by the initial performance bond and any additional money you may have deposited with the broker. If you decide to hedge the sale or purchase of a commodity, be prepared for performance-bond calls. If your cash is tight, you may wish to have a lender finance the performance-bond deposit and potential performance-bond calls. If you close your position in the market with a gain, this deposit is yours – although you may want to leave it on deposit for your next hedge. A flat cost to producers who use the futures market is the commission charged by the broker for each contract traded for you. This cost is negotiable and depends on the level of service and the quantity of contracts traded.

15.4 Hedging strategies[9]

15.4.1 Short-hedge strategy

A wheat producer may sell wheat futures to hedge the sale if he or she thinks prices are heading down.

In September, you have planted winter wheat and you expect a crop of over 20,000 bushels. You would like to sell the crop soon after the June harvest. You are fairly certain that prices are heading down, so you want to lock in a price for July delivery. The performance bond deposit of $700.00 per contract and possible performance-bond calls will not cause you a cash-flow problem. You decide to sell four July wheat futures contracts (5,000 bushels each, or 20,000 bushels).

What price can you expect to get for your crop? The July futures price today is $3.90, and the local forward cash price for July is $3.63, or $0.27 under. Based on your experience, it is more likely to be about $0.16 under in July, so you set a target price of $3.74 ($3.90 – $0.16). You sell July wheat futures at $3.90.

Futures	3.90
Expected basis	–0.16
Bid price	**3.74**
Sell July futures	$3.90

In July, futures prices have fallen to \$3.65 and cash prices to \$3.50. The basis is \$0.15 under (\$0.01 better than you expected). You buy four July wheat futures contracts at \$3.65 to offset your position and experience a \$0.25 gain (\$3.90 – \$3.65). Then you sell the actual 20,000 bushels of wheat in the cash market at \$3.50. The total price you received per bushel is \$3.75 (\$3.50 + \$0.25).

Cash price received	3.50
+ Futures gain/loss	+0.25
Net price received	**3.75**

Looking at the overall picture, you have done \$5,000 (\$0.25 × 20,000 bushels) better than the cash price by hedging. You pay a broker a commission of \$50.00 for each contract, which totals \$200.00 for four contracts sold and bought, so your actual gain is \$4,800. A total of \$2,800 of your money has been tied up in the performance-bond account since last fall. You can choose to leave the deposit with the broker for your next transaction or have it returned to you.

15.4.2 Long-hedge strategy

A feedlot operator may buy feeder cattle futures to hedge against higher prices. You plan to buy 135 head of feeder cattle to place in the feedlot in March. Now in December, all indications are that prices will be rising, and you would like to lock in a low price for March. You decide to buy two March CME Feeder Cattle futures contracts (50,000 pounds each contract, or approximately 135 head total). You make arrangements with a lender for a performance-bond deposit of \$1,350 per contract and possible performance-bond calls.

How can you estimate a target purchase price? A local forward-contract bid may not be available to use as a guide in estimating basis. Cash prices and futures prices of livestock are largely independent of each other until the delivery period approaches. But based on the previous history of feeder cattle cash and futures relationship in March, you expect a basis of \$2.00 under. The futures price is at \$103.00/cwt, so you calculate a target price of \$101.00 (\$103.00 – \$2.00) in March. You buy March CME Feeder Cattle futures at \$103.00/cwt.

Buy March futures	103.00
Expected basis	–2.00
Target price	**101.00**

Expected basis	**Futures price**	**Cash price**	**Futures gain/loss**	**Net price received**
–1.5	Sell back 105.90	Buy at 104.00	–2.90	= 101.10

In March, the futures price has gone up to \$105.90, and the cash price is \$104.00. The basis is \$1.90 under (\$0.10 narrower than you expected). You sell back the two March CME Feeder Cattle futures contracts at \$105.90 and realize a gain of \$2.90/cwt (\$105.90 – \$103.00). Then you purchase the 135 head of feeder cattle in the cash market at \$104.00/cwt. The total price you paid per cwt is \$101.10 (\$104.00 – \$2.90).

Cash price paid	104.00
Futures gain/loss	–2.90
Net price received	**101.10**

By initiating a long hedge, you have reduced the cost of the feeder cattle by \$2,900 (\$2.90 × 500 cwt/contract × two contracts) from the cash price. After paying the commission of \$50.00 per contract, or \$100.00 total, you have saved \$2,800 on the purchase. Your performance-bond deposit of \$1,350 per contract has been tied up with the broker since December, but now you can choose to have it returned.

- The CME Feeder Cattle futures contract is 50,000 pounds, or 500 cwt.
- The CME Live Cattle futures contract is 40,000 pounds, or 400 cwt.
- The CME Lean Hogs futures contract is 40,000 pounds, or 400 cwt.

A hog producer may be said to be a pure hedger when placing a hedge to protect a target price that will cover the cost of production and allow for a profit. The producer is not concerned about the movement of prices. The concern is protecting the target price. The short hedge is maintained by the producer until he is ready to market his hogs. Then the producer buys futures contracts to offset a position and sells simultaneously in the cash market. The selective hedger is very common in real life. A milk producer who is a selective hedger would stay out of the futures market if he believes that prices will go up and hedge only if it is perceived that there will be a price decrease. Similarly, a hog producer who needs to purchase feed will stay out of the futures market if he thinks feed prices are falling and hedge only if a price increase is perceived. Both of them could be wrong. Some producers wait to place a hedge during the growing season. If they believe prices are going to rise and then fall, they want to sell futures contracts near the top of the market and buy back after prices fall again. Other variations are producers who hedge only part of their production, hedge it in stages, or combine cash-market sale, forward contracting, and hedging.

15.4.3 Types of orders

In either pit or electronic trading, there are several kinds of orders that can be placed with your broker. These are four of the most common:

- *Market order*: You instruct your broker to place your order as soon as possible at the best possible price. For example, you say, "Sell four July CME Lean Hog futures at the market." The trade is executed at the best price that can be obtained.
- *Price order*: You instruct your broker to place your order at a certain price or better. For example, you say, "Sell four July CME Lean Hog futures contracts at $70.00." The market has to trade at $70.00 or better (higher) before he or she can execute the trade. (Also called a limit order.)
- *Stop order*: You instruct your broker to place your order at a certain price level. A sell stop must be below the market; a buy stop must be above the market. For example, the futures price is at $70.00, and you say, "Sell four July CME Lean Hog futures contracts at a stop of $69.50." When the market falls to $69.50 or below, the stop order becomes a market order and is executed.
- *Stop close only*: This is a stop order that is executed only within the last minute of trading, during the close. You say, "Sell four July CME Lean Hog futures at a stop close only of $69.50," where this price is under the futures price at the time you place the order. You want the order to be filled only if the market is going to close at or below $69.50. The order will be placed during the closing period.

15.4.4 Supply-and-demand dynamics and how they affect prices

15.4.4.1 Prediction price action[10]

If you are going to be trading futures contracts, it is essential that you know how to make a reasonable estimate of what will happen to prices in the future. Of course, no one can know for certain what prices will be, but it pays to have an educated opinion as to whether prices will rise or fall.

Price expectation can be derived from two different approaches to analyzing the markets:

- Fundamental analysis uses supply-and-demand information to determine its anticipated impact on prices.
- Technical analysis interprets historical price movements to predict prices in the future.

While some speculators and hedgers may use only one of these approaches, others use a combination of fundamental and technical analysis to project prices.

Both supply and demand determine the market price for a commodity (see Table 15.2). External factors affect the supply and demand for any commodity.

Table 15.2 Demand-and-supply elements regarding agricultural commodities

Supply factors	Demand factors
• Weather	• Consumer tastes
• Yields	• Disposable income
• Carryover stocks	• Population size
• Exports and imports	• Cost of substitute good
• Cost of production	
• Market price	
• Government programs	

The price of a commodity is determined by how much of the commodity there is (supply), how much of it people want (demand), and what they are willing to pay for it. Supply can be defined as the quantity of a commodity that producers are willing to provide to the market at a given price. If grain prices are too low, producers may decide to keep the grain in storage, reducing the supply of the grain to the market. Weather is one of the primary factors influencing commodity supply. Clearly, conditions during the growing season and harvest will impact the short-run supply of a commodity. Weather can also affect supply indirectly, by affecting the cost of inputs into another commodity market: if feed prices are high, over the long run fewer livestock will be supplied to the market. But weather's effect on prices can be difficult to predict. For instance, an increase in feed-corn prices due to bad growing conditions may induce livestock producers to shift to wheat feed rather than reduce output altogether. Government efforts to promote domestic agriculture through subsidies, protective tariffs, support prices, preferential access to export markets, marketing boards, and so on have two principal effects on commodity prices. Because such policies tend to lead to overproduction of a protected commodity, prices will be lower overall. Finally, many commodities, like corn and soybeans, exhibit seasonal price patterns related to their yearly production cycle. Since prices after harvest must adjust in order to apportion the available supply over time until the next harvest, seasonal price patterns – the differences between current-year prices and long-run averages – are one of the primary fundamental price-forecasting tools.

Demand is defined as the quantity of a commodity that buyers are willing to purchase from the market at a given price.

Supply and demand together influence the market price of a commodity. When the supply of a commodity increases and demand decreases or stays the same, the price falls. When the supply decreases and demand increases or stays the same, the price of that commodity rises. Likewise, if the supply stays the same, changes in the demand will cause prices to rise or fall. Because the fundamentalist's objective is to predict which way prices are likely to move,

Table 15.3 Interaction of supply and demand on agricultural commodity prices

Supply +	**Demand >**	**Price**
Up	Same	Falls
Down	Same	Rises
Same	Up	Rises
Same	Down	Falls

those factors indicating changes in the supply and demand are going to be of interest. The fundamentalist studies how events change the value of the commodity – whether it becomes more valuable or less and whether prices can be expected to go up or down accordingly, as displayed in Table 15.3.

Carryover stocks are important in studying crop supply. The key to fundamental analysis is looking at the supply-and-demand number projections for the year and comparing them to last year and previous years. A corn, wheat, or soybean fundamentalist adds up last year's carryover and this year's production and then subtracts usage to find ending stocks. Ending stocks, which become the next year's carryover, are a cushion. Fundamentalists study this number carefully. Thus:

Carryover + Production – Usage = Ending stocks

Carryover = Projection usage If corn usage is projected at 7 billion bushels and carryover stocks are 7 billion, that means there is enough corn without producing any in that year. But producers would grow more, ending stocks would be very high, and prices would fall.	Prices would fall if carryover stocks would supply the needs for the year.
Carryover + Projection – Usage = 0 If carryover is 2 billion bushels, production is 7 billion, and usage is projected at 9 billion, then the ending stocks would be 0. That means there would not be a bushel of corn left in the country. Prices would rise dramatically.	Prices would rise if ending stocks were projected at 0.

These are two extreme examples of the relationship of the numbers in the formula above. Obviously, something between the two extremes will happen each year. Fundamentalists watch for the relationship of these numbers. One method of analysis is to look at this year's carryover and compare it to the carryover and related prices in other years. For example, if carryover is 2 billion bushels of corn and total usage is projected at 8 billion, than carryover is 25% of usage. You can go back over previous years where carryover was 25% of usage to study the prices in those years. Also look at years where carryover was in the 20% to 30% range to see how high and low prices went. From that information, you can make a good estimate for this year. Those data are made available by the USDA and other agricultural departments around the world, often at fixed dates throughout the year.

15.4.5 Crop specifics[11]

Fundamentalists study[12] specific supply-and-demand factors for each crop.

1. **Corn:**[13] Important fundamentals relating to corn supply are acreage and yields. Yields are dependent on moisture and temperature. Higher-than-normal rainfall causes higher yields. A combination of below-normal rainfall and above-normal temperatures can cause lower yields. The number of acres planted depends a lot on price. Low prices can encourage producers to plant other crops or participate in government production-control programs. A factor of corn demand is the number of livestock on feed. If the price of meat falls, fewer livestock will be produced, and when livestock numbers are lower, less corn is needed. If the price of corn rises too high, livestock producers will reduce their use of corn by reducing the number of animals on feed or the length of time livestock are on feed. Another demand factor is exports. A world bumper corn crop can reduce the demand for US exports. Also, the US dollar exchange rate affects exports – if the dollar is weak, foreigners can purchase more dollars for a given amount of their own currency and can buy more US corn.
2. **Wheat:**[14] As with corn, wheat supply is largely dependent on growing conditions.[15] Because there are basically two types of wheat – winter and spring – growing conditions should be watched during winter and summer. For example, a lack of adequate snow cover during the winter months can damage the winter wheat crop and reduce yields. On the demand side, exports play a large role. One-half to two-thirds of US wheat production is exported to other countries such as Japan and Russia each year.
3. **Soybeans:**[16] Most of the soybean crop is used for soybean meal and oil. Consequently, the demand for soybeans depends on the demand for meal, used for animal feed and in people food, and oil, used in edible oil products and in industry. A number to watch is called the crush margin, a relationship of the cost of soybeans to be crushed and the value of the resulting soybean oil and meal. If soybean prices are too high, the crush margin may decrease and cause a decline in processing. Then the demand for soybeans also decreases.
4. **Livestock:**[17] When studying livestock, fundamentalists focus on projected supply. Fundamental analysis for livestock is different from analysis of grains because there are no carryover stocks. Livestock cannot be stored as grains can. Basically, the livestock that are raised are marketed. Fundamental analysis for livestock is a matter of studying supply and demand to predict prices. Fundamental analysts will see if livestock numbers are projected

to be higher or lower than last year. They look at consumption patterns and decide if demand appears to be better or worse than before.[18] Based on supply-and-demand patterns over several years, many fundamentalists have developed their own formulas to calculate what effect a percentage increase or decrease in supply has on prices, given that other factors are equal. For example, a hog analyst may have determined that a 1% decrease in supply causes a 1% increase in prices. So, if supply is 98% of a year ago and the hog price was $50/cwt, then prices would be projected at $51/cwt. Livestock analysis is more difficult than crop analysis. On the supply side, cattle and hog producers are greatly influenced by the cost of feed and the current or expected livestock market price. If feed prices are too high or livestock prices fall, producers will take steps to raise fewer animals. Their first response to low prices is to bring more animals to slaughter, which actually increases current supply. This causes prices to fall even lower. Over a period of time, fewer offspring are produced and supplies decrease. Then prices rise again. The producers' response is to hold back breeding animals from slaughter. This causes supplies to decrease more and prices to rise higher. When the offspring are brought to slaughter, supplies increase. This is a repeating cycle with livestock supplies and prices. To estimate livestock demand, the fundamentalist will also study factors such as consumer tastes, the level of disposable income, population size, and the cost of competitive goods.

5. **Cattle:**[19] The average cattle cycle is 12 years. The production of cattle is a three-step process. Cattle are bred (nine-month gestation), and the calves are raised for six to ten months in a cow-calf operation. Next, they are fed roughage in a stocker operation for another six to ten months. Finally, they are sent to a feedlot to be brought to market weight, which takes another four to six months. Breeding cattle, of course, are held back by the cow-calf operation to produce more calves. Because the production cycle is so long, it takes a while for producers to respond to changing prices. The cattle cycle of increasing and decreasing supply takes from 9 to 16 years to complete. The average cattle cycle is 12 years. A fundamentalist is also aware of an annual seasonal pattern of cattle supply. The number of cattle available for slaughter is generally lower in the spring and rises in late summer and fall. Prices tend to rise and fall within the season to reflect this pattern, but in the opposite direction (see Table 15.4). The consumer demand for beef cannot be assumed to be constant, so the fundamentalist watches for changes in demand, the American public's perception of beef, and other sources of protein changes. Total consumption is a factor of both population and per capita use. While population has increased over the last ten years, per capita consumption of beef has fallen. At times, higher beef prices have caused a greater consumption of lower-price products such as pork, poultry, and fish.

Table 15.4 Cattle supply and pricing cycle

Increasing supply (more offspring)			
↓	Falling prices Increased slaughter (of females) Falling prices	Rising prices Decreased slaughter (of females) Rising prices	↑
Decreasing supply (fewer offspring)			

6. **(Lean) Hogs:**[20] The average hog cycle is four years. The typical hog farm is a farrow-to-finish operation, that is, a birth-to-final-marketing operation. Hogs are bred (four month gestation) and brought to market in five to seven months. Sows can be bred at about six months of age and generally once or twice before being sent to slaughter. Although the production cycle for hogs is not as long as for cattle, it still takes time for producers to react to market prices. The average hog supply cycle is about four years. Supplies will follow a pattern of increase and decrease over a period of every four years. A fundamentalist is familiar with an annual seasonal pattern of hog supply as well. Hog supply is greatest from August through December and lowest during June and July. Prices tend to rise and fall opposite to the supply trend. They are highest when the supply is low and lowest when the supply is high. The demand for pork is strongly affected by the price relative to income and the price of competitive meat products. If beef or poultry prices are low, the demand for pork will fall. If beef or poultry prices are high, pork demand will increase. Demand for certain pork products – such as ribs, bacon, and ham – display distinctly seasonal patterns, which influence the seasonal demand for the live animals.
7. **Dairy**[21]
8. **Cotton**[22]
9. **Coffee**[23]
10. **Indexes**[24]
11. **Lumber and forestry**[25]
12. **Sugar**[26]
13. **Cacao**[27]
14. **Frozen concentrated orange juice**[28]
15. **Canola (Canadian oil, i.e., a premium vegetable oil)**[29]

For most agricultural commodities, the USDA makes data available on a regular basis. Also the different exchanges make data sets available regularly and/or link external data sets through their websites.[30]

15.4.6 Analytical tools and technical charting[31]

Technical analysis uses historical price patterns to help predict futures prices. Technical analysts take a different approach to studying the markets. They believe that everything that can affect the market – including all the fundamental, psychological, and political factors – must ultimately be reflected in the price of the commodity. Instead of concerning themselves with the actual reasons why market prices are moving up or down, technical analysts study the market action itself. In this sense, they use a shortcut form of fundamental analysis. Prices move because of changes in supply and demand, which includes what traders hope or fear might happen. Day-by-day market action is the result of both the fundamental factors we have studied and the reaction of traders.

From the technical analyst's point of view, it is all reflected in the price. The analyst has studied historical price patterns, and he or she charts current market activity. Based on patterns in current activity, the analyst tries to predict where prices are going. As seen, technical analysis approaches price forecasting in a very different way than fundamental analysis. Technical analysis tools can be applied to many different markets, while a fundamental approach requires detailed knowledge about supply-and-demand factors and how those factors influence prices. Fundamental analysis assumes that it is new information about those supply-and-demand factors that will determine future prices. This kind of market behavior is described as efficient: current prices reflect all of the information available to the market (as well as expectations about the future based on that information). Only new, unexpected information will change prices, and past price movements and patterns will not be useful at all in predicting what prices will do. There is a continuing debate about whether prices are truly efficient, as the fundamental analyst believes, or whether price patterns and trading rules can consistently predict prices profitably.[32]

1. **Bar charts**: A bar chart shows a picture of what is happening in the market. A technical analyst uses these three important sources of market information in the process of predicting prices:
   - Price (the high, low, and close for the day)
   - Volume (the number of contracts traded that day)
   - Open interest (the number of outstanding contracts – that is, the number that have not been offset)

An analyst creates a picture of what is happening in the market by recording this information on a bar chart. It is important to realize that the analyst is recording history, or what has already happened. The bar chart does not show what is going to happen. The analyst is educated to read the visual patterns of price movement and is able to draw conclusions about the likely future direction of the market. A bar chart of price movement is made for a specific futures contract for a specific commodity. The vertical axis represents the price of the commodity; the horizontal axis represents time, but trading days only. The high, low, and closing prices are recorded each day. A vertical line is drawn to show the range of prices for the day. A sideways tic shows where the market closed.

2. **Point-and-figure chart**: Another chart is a point-and-figure chart. This kind of chart lets the analyst focus on the direction of price trends by ignoring time. Because of this, a point-and-figure chart can give a clearer signal to buy or sell than the bar chart, and many analysts use both. A point-and-figure chart is plotted on graph paper. The vertical axis represents price. The point-and-figure chart does not have time on the horizontal axis. The analyst uses, somewhat simplified, an X to indicate the day's price when prices are rising. When the prices reverse and fall, the analyst goes to the next column and records Os. When another reversal occurs, the analyst switches back to Xs and so on.
3. **Volume and open interest**:[33] The vertical lines at the bottom of the bar chart are a record of volume, or the number of contracts traded each day. If you study the prices in the previous chart from November through April, you will see that the volume rises and falls too. When buying or selling pressure is felt by traders, more contracts are traded. The volume, or trading pressure, on this chart increases when prices are changing trends. Look at the price lows in April and notice the increase in volume. The horizontal curve in the middle of the bar chart is a record of open interest, or the number of contracts outstanding at the end of each day. This number changes from day to day. Open interest is a measure of the commitment of money in the market. When open interest is rising, new money is flowing into the market, indicating that the current price trend will probably continue. When open interest is falling, money and commitment are leaving the market. The relationship of prices, volume, and open interest tells the analyst quite a bit about what can happen next. When prices are rising, rising volume and open interest indicate that the uptrend in prices may continue. When prices are rising, and volume and open interest are falling, the current uptrend in prices may be coming to an end. When prices are falling, rising volume and open interest tend to confirm the downtrend. When prices are falling and volume and open interest are falling, the analyst is alerted that prices may soon reach bottom and begin to rise again. Table 15.5 summarizes the relationship.

Table 15.5 Relationship between volume and open positions and impact on pricing

Prices	Volume	Open interest	Market direction
Rising	Rising	Rising	Bullish
Rising	Falling	Falling	Bearish
Falling	Rising	Rising	Bearish
Falling	Falling	Falling	Bullish

4. **Price patterns**: Technical analysts use certain price patterns to signal a change in the trend, or a price reversal. Some price patterns are named because of their shape. The head-and-shoulders formation is one of the most reliable patterns that indicate a major reversal in the market. The pattern is named because of its shape. It can occur at the top of an uptrending market or in inverted form at the bottom of a downtrend. The head-and-shoulders formation at the top of an uptrend shows a period during which the market changes from up to sideways and then from sideways to down. A line drawn under the two shoulders is called the neckline. The reversal of the trend is complete when prices pass the neckline on the right side.

   At the bottom of a downtrend, the head-and-shoulders formation shows a period when the market changes from down to sideways and then from sideways to up. The neckline is drawn above the two shoulders. The reversal of the trend is said to be complete when prices pass the neckline on the right side. Another common reversal pattern that occurs is the double top or double bottom. Double tops and bottoms signal a trend reversal just as the head-and-shoulders formation, but they usually take less time. At the top of the market, they are noted by two equal price peaks as prices hit the resistance level, dip, and hit the resistance level again before turning down.

   Just as with bar charts, there are formations in the point-and-figure chart that have meaning to the technical analyst. One type of formation is the double bottom. This formation is characterized by falling prices that try but fail to set new lows, after which prices begin moving up. The analyst considers it a buy signal when prices set a new high. With a double top, prices fail to set new highs and then begin moving down. In this case, the analyst would see the setting of a new low as a signal to sell. Another common pattern is a triangle. Triangles generally interrupt a trend and occur when supply and demand are temporarily balanced. Upward price movement is balanced by selling, downward price movement is countered by buying, and tops and bottoms become shallower. After this pause, prices tend to break out, generally continuing in the direction of the original trend.

5. **Price trends**: The technical analyst studies price movement to find price trends. When using the word "trend," the analyst simply means the general direction in which prices are moving. The analyst believes two things:
   - Prices do follow trends.
   - Once trends are established, they persist.

   The following price trends are observed:
   a. *Sideways trends*: In a sideways trend, prices tend to stay within a certain range. Prices can move in only three directions – up, down, and sideways. The sideways trend occurs during a relatively flat market. Although prices in a sideways trend fluctuate, they occur within a certain range. Prices tend not to go above a particular level or fall below a particular level. A resistance line occurs when prices are turned back (do not rise above a certain level) by selling pressure that stops a market advance. Prices do not fall below the support line where buying pressure stops a market decline. Once the sideways trend is established, prices tend to stay within these boundaries.
   b. *Uptrends (higher highs, higher lows)*: An uptrend is characterized by a series of higher highs and higher lows. The analyst watches carefully to see if the upward price movement will become a valid uptrend. A line is drawn under the rising bottoms in the fluctuating uptrend. It takes two rising bottoms to define an uptrend line and three to confirm it. A line above the rising tops in the uptrend helps the analyst to determine the range of fluctuation within the uptrend. The range, or channel, defines the support (line of lowest lows) and resistance (line of highest highs) for the uptrend. The analyst believes that generally prices will stay within this channel until the uptrend is reversed.
   c. *Downtrends (lower highs, lower lows)*: A downtrend is characterized by a series of lower highs and lower lows. The analyst watches the declining tops in the fluctuating downtrend. A line defining the downtrend connects the declining tops. It takes two declining tops to define a downtrend line and three to confirm it. A downtrend channel will define the range of prices within the downtrend.
   d. *Price reversals*: Some signals happen in a single day, while others take more time to develop. Technical analysts want to know when an uptrending market is going to start going down and when a downtrending market is going to start going up. They watch for several signals. The analyst watches the high, low, and close of each day for a key reversal signal that warns of a coming market turn. The bullish key reversal happens in a downtrending market. It is characterized by a high that is higher than the previous day's high, a low lower than the previous day's low, and a close above the previous day's close. The larger trading range for the day and the higher close may signal a coming uptrend in prices. If the

trading volume is unusually high, the signal is stronger. A bearish key reversal occurs in an uptrending market. It is characterized by a high that is higher than the previous day's high, a low lower than the previous day's low, and a close below the previous day's close. The bearish key reversal seen in an uptrending market can be a warning of a coming downtrend in prices.

6. **Gaps:** Gaps in prices from one day to the next can provide the analyst with information. The study of gaps – price areas at which the market did not trade from one day to the next – can help the analyst in three ways:

- To spot the beginning of a price move.
- To measure how far the move is going to go.
- To confirm the end of a move.

7. **Moving average charts**: Average prices provide a tool for identifying buy and sell signals. A moving average is the arithmetic average of prices over a period of time. For example, a three-day moving average of milk prices would be the average of the closing prices of the past three days. To find the average, you add the closing prices from the last three days and divide by three. The next day, you calculate a new average and so on. The moving averages can then be plotted on a graph. Analysts can use three-day, five-day, ten-day, or twenty-day moving averages – whatever suits them – to watch price moves. The analyst interested in short-term moves would use shorter-term moving averages.

Some of the typical rules the analyst will follow are these:

- Buy when the short-term (three-day) moving average moves above the long-term (ten-day moving average).
- Sell when the short-term (three-day) moving average moves below the long-term moving average.

The moving average can also be considered one of a number of technical indicators (basically a number calculated from a formula rather than a chart figure) whose quantitative value tells the analyst something about the character of the market. Some indicators, such relative strength or MACD (moving average convergence/divergence), can signal when a market is overbought or oversold, and are useful when markets are in a horizontal, or nontrending, trading range.

Overall, there is a wide variety of quantitative indicators used by technical analysts, and those listed above are only the most important and frequently used ones. Others calendar spreads[34] and/or seasonality[35] as a leading indicator. Managed futures are often used to lower overall portfolio volatility.[36]

Agricultural commodities can also be traded in block.[37] Block trading is a permissible, off-exchange, privately negotiated transaction either at or exceeding an exchange-determined minimum-threshold quantity of futures or options contracts that is executed apart and away from the open outcry or electronic markets.

15.4.7 Using options

Options on futures

Like futures, producers can use options on futures to lock in prices and benefit from price changes. Hedging with futures is a risk-management technique that provides protection against falling prices for those producers who have a commodity to sell and against rising prices for those who plan to buy a commodity. In both cases, producers can use futures to lock in a price. The hedger who plans to sell will be pleased with the locked-in price if prices fall, but on the other hand may wish that he or she had not hedged if those prices rise. The same is true about the hedger who plans to buy – happy with the locked-in price if prices rise, but not if prices fall. Buying options on futures is another market alternative. Options provide the producer with price protection just as futures do, but they also provide the opportunity to benefit from favorable price changes. And option buyers do not have to worry about performance bonds. An option on a futures contract is an agreement that grants the right, but not the obligation, to buy or sell an underlying futures contract at a specified price before a specified time. When you own an option, you have the right to buy or sell the underlying futures contract, but you do not have to. You can instead let the option expire. The ability to choose to use the option, or not, means that the option hedge will give producers the locked-in price they need, but, unlike a futures contract, it will also allow them to take advantage of price movements in their favor.

Constituting elements of an option

Each option specifies the following:

1. The type of option

There are two kinds of options. A put option grants you the right, but not the obligation, to sell a futures contract at the strike price specified. A call option grants you the right, but not the obligation, to buy a futures contract at the strike price specified.

2. The commodity and the contract month of the underlying futures contract

For instance, the February call option on CME Lean Hog futures gives the owner the right to a long position in the February CME Lean Hog futures contract.

3. The maturity of the option

The life of an option is determined by the futures contract month it covers. For example, a December option in a deliverable contract may expire one to two weeks before the first delivery of the December futures contract.

4. The price at which the futures will be bought or sold

An option on a futures contract states the price at which you may buy or sell the underlying futures contract. This price is called the strike price. The strike price of the option does not change. If the strike price is $70.00/cwt, that price stays the same throughout the life of the option. Since the option is a valuable asset in its own right, when you buy an option, you pay a price for it.

5. The premium of the option

The premium for each option is determined on the exchange floor by open outcry as the traders make bids and offers to buy and sell the options. Depending on the comparison of the strike price to the current futures price and other factors, a premium can be high or low. The seller of the option receives the premium.

Note that a put and a call are two entirely different agreements, not the opposite sides of the same agreement. If you buy a put, you can offset your position by selling a put at the same strike price. If you buy a call, you can offset your position by selling a call at the same strike price. Also, the seller of a put option must take the other side of the transaction and is obligated to buy the asset that the put owner is selling. Similarly, the seller of a call option must sell the asset to the call owner, who has the right to buy.

15.4.8 Options on futures versus options on actuals

As discussed above, exercising a commodity option involves taking a position in an underlying futures contract (see Table 15.6). On the other hand, most financial options, like options on shares or currencies, involve delivery of the underlying asset itself. For instance, the owner of a call option on company stock will receive delivery of 100 shares of the underlying stock at exercise. The advantage of an option on a futures contract is that it is generally cheaper to exercise, since it involves posting the performance bond for

Table 15.6 Comparison between options on futures and options on actuals

	Options on futures	Options on actuals
Delivery/ exercise	Buyer receives long (call) or short (put) futures position; seller takes short (call) or long (put) futures.	Buyer receives asset from seller.
Timing	Underlying futures position expires approximately when option matures.	Underlying asset does not expire.
Cash flows	Option owner and seller post performance bond; both positions marked to market.	Option owner pays (call) or receives (put) full value of strike price.
Supply of underlying	Supply of futures contracts virtually unlimited.	Limited supply of asset available for Delivery.

a futures contract rather than paying the full price of the underlying asset. Also, while the market supply of an asset can be limited, which may make delivery of the asset difficult upon exercise, there is a potentially unlimited supply of futures contracts available to fulfill the exercise of an option on a futures contract.

Buying put and call options

Once you buy a put or a call, you have three choices of action. A put option grants you the right to sell a futures contract at a specific price. For example, a December CME Milk Class III 11.25 put is an option that grants you the right to sell a December CME Milk Class III futures contract at $11.25 (the strike price). Several puts are offered for the same futures contract, differing only by the strike price. Another put available may be a December CME Milk Class III 11.00 put, which gives you the right to sell a December CME Milk Class III futures contract at $11.00. The 11.00 put would have a lower premium than the 11.25 put. You buy put options to lock in a minimum floor price for the sale of your commodity. Unlike hedging with futures, you can benefit if prices rise. This is how it works. As the buyer of an option, you have three choices:

- You can exercise the option (exchange it through your broker for the underlying futures contract).
- You can offset the option (sell a put with the same strike price and receive the current market value of the premium).
- You can let the option expire without any obligation (and lose the premium, of course).

Suppose you bought a December CME Lean Hog 74 put (strike price of $74/cwt). If the futures price falls below $74.00, you can offset your December CME Lean Hog 74 put and receive the premium. Or you can exercise the option and get a short futures position at the lock-in futures price of $74.00. If the December futures rise above $74.00, you can let the option expire and take advantage of the higher prices in the cash market. As we will see below, few producers ever exercise the option; most of them will either offset the option or let the option expire.

Call option

A call option grants you the right, but not the obligation, to buy a futures contract at a specific price. For example, a December CME Milk Class III 11.50 call is an option that grants you the right to buy a December CME Milk Class III futures contract at the strike price of $11.50. Call options allow you to lock in a maximum price for the purchase of a commodity. Again, you can exercise the option, offset it by selling a call with the same strike price, or let the option expire. If the futures price rises above your strike price, you can sell a call with the same strike price and receive the premium. If the futures price falls below, you can let the option expire and take advantage of the lower prices in the cash market. A comparison is provided in Table 15.7.

Table 15.7 Comparison between a put and call option

	Put option	Call option
What?	Right, but not the obligation, to sell a futures contract at the strike price	Right, but not the obligation, to buy a futures contract at the strike price
For whom?	Someone who PUTS the commodity on the market; sells	Someone who CALLS the commodity from the market; buys
Why use it?	Set minimum price; benefit from price increase	Set maximum price; benefit from price decrease
Then what?	Can offset, exercise, or let the option expire	Can offset, exercise, or let the option expire
a. Offset	Sell a put with the same strike price.	Sell a call with the same strike price.
b. Exercise	Exchange for short-futures contract at the strike price.	Exchange for long-futures contract at the strike price.
c. Expire	If the put is worthless, do nothing and sell in the cash market.	If the call is worthless, do nothing and buy in the cash market.

Table 15.8 From producer to the market

Buy put and pay premium	1. Offset: Sell put and get premium. 2. Exercise: Sell the underlying futures. 3. Expire: Do nothing and lose premium.	Buy back futures	Sell in cash market
Buy call and pay premium	1. Offset: Sell call and get premium. 2. Exercise: Buy the underlying futures. 3. Expire: Do nothing and lose premium.	Sell back futures	Buy in cash market

The three choices a producer has who buys a put or a call. In each case, the sale or purchase of a commodity is eventually carried out in the cash markets (see Table 15.8).

Options premium

Option premiums are determined during trading, reflecting supply and demand for the options. The premium for each option traded on the exchange is determined as the traders bid and offer to buy and sell options. The premium amount at which buyers are willing to pay and sellers are willing to sell depends on two factors:

Premium = intrinsic value + time value

Intrinsic value

When you buy an option, you pay a premium. The amount you pay depends in part on the value of the option if you exercised it now. For example, if the current futures price for CME Live Cattle is $80.00/cwt, you would pay a higher premium for a put option with a strike price of 84/cwt than one for 82/cwt. The 84 put would be worth $4.00 if exercised (you have a short futures position entered at $84.00 when prices have fallen to $80.00), but the 82 put would be worth only $2.00 if exercised. The relationship of the strike price to the current futures price is called the intrinsic value, or built-in value, of the option. You can find the intrinsic value easily by comparing the strike price to the futures price. The intrinsic value of a put is the strike price minus the futures price. There will be several options available for the same futures contract month, differing only in strike price. Some will be above the current futures price and some below. For example, February CME Live Cattle options may be available for 86, 88, 89, and so on. For example, if the strike price of a CME Milk Class III put option is 11.25 and the futures price is $11.20, then the intrinsic value of the put is $0.05. That is because you could exchange the put for a futures contract sold at $0.05 above the current price. If the futures price is equal to or greater than the strike price, the intrinsic value would be 0 – it cannot be negative.

PUT

Strike price	11.25	11.25	11.25
– Futures price	–11.20	–11.25	–11.40
Intrinsic value	0.05	0	0

The intrinsic value of a call option is the futures price minus the strike price. For example, if the strike price of a CME Milk Class III call option is 11.25 and the futures price is $11.60, then the intrinsic value of the call is $0.35. You could exchange the call for a futures contract bought at $0.35 below the current price. If the strike price is equal to or greater than the futures price, the intrinsic value is 0 – never negative.

CALL

Futures price	11.60	11.35	11.25
– Strike price	–11.25	–11.25	–11.25
Intrinsic value	0.35	0	0

Time value

The other part of the value of the option premium is the time value. This is the amount that buyers are willing to pay for the option in anticipation that, over time, a change in the futures price will cause the option to increase in value. It also represents the amount of risk the seller of the option is willing to take. The time value depends on two things: how long before the option expires, and how much futures prices are expected to move before the expiration date. The longer the option owner has to decide about exercising, the more valuable the option is. If there are wide swings in the futures price, the chance for the option buyer to benefit increases while the risk to the seller increases. Generally, time value is greatest at the beginning of the option period. That is when the buyer has the highest expectations and the seller is facing the greatest risk. The time value approaches zero as the option period draws to an end and always ends at zero.

An option premium depends on the option's strike price compared to the current futures price (its intrinsic value), the time until maturity, and the volatility of the futures price. Since the option premium includes both time value and intrinsic value, it is usually better to sell, or offset, the option in the market rather than exercise it before expiration. That is why few producers exercise an option: they usually prefer to offset (getting both the intrinsic value and the time value in the premium they receive) or let the option expire (when the time value becomes zero).

At-the-money

Traders have their own way of describing option value (see Table 15.9). When the strike price equals the futures price, the option is at-the-money and the intrinsic value of the premium is zero. However, time value is high at-the-money because the buyer has a high chance of prices moving to her benefit, and the seller has a high risk. As the option moves in-the-money, its intrinsic value increases. An out-of-the-money option has an intrinsic value of zero. In both of these cases, the time value decreases. The chance is less that an out-of-the-money option will end up in-the-money and that an in-the-money option will end up out-of-the-money.

You can see that the premium is based on the option's intrinsic value plus what the participants feel their opportunities and risks are. Knowing this can help you evaluate premiums at various strike prices and decide which to buy. Because time value always decays to zero when the option expires, you want to pay as little time value as possible.

	Call-options strike price	Cattle futures prices	Put-options strike prices	
Out-of-the-money	88	86	88	In-the-money
At-the-money	86		86	At-the-money
In-the-money	84		84	Out-of-the-money

- A put or a call that is the most in-the-money demands the highest premium.
- A put or a call that is the most out-of-the-money demands the lowest premium.

Table 15.9 Option value positions

	Call option	Put option
In-the-money	Strike price < futures price	Strike price > futures price
At-the-money	Strike price = futures price	Strike price = futures price
Out-of-the-money	Strike price > futures price	Strike price < futures price

Buying a put

When you have a commodity to sell, you can buy a put option to lock in a minimum selling price and take advantage of prices if they rise.

Buying a put example: Suppose it is March and you are raising feeder cattle. The November futures price is trading at $112.00. You want to protect yourself against falling prices, so you buy a November CME Feeder Cattle put with the right to sell a November CME Feeder Cattle futures contract at $112.00. You pay a premium of $3.00 for the option. With the option, you are effectively setting a floor protector price of $109.00 and expecting a –$2.00 basis in November. Your hedged net price is $107.00 ($109.00 – $2.00).

Action:	Buy November put	112.00
	Premium cost	3.00
	Basis Est.	–2.00
		107.00

Result: prices fall

Basis	Futures price	Options	Cash price	Options gain /loss	Net price received
–2.00	104.00	Sell 112, put 8.00	Sell 102	+ 8.00 – 3.00 = $5.00	= 107.00

When you are ready to sell your feeder cattle in November, the results are shown above. The futures price is now $104.00. You hold an option that grants you the right to sell a futures contract at $112.00, and this option now has a premium of $8.00. You sell a November CME Feeder Cattle put option at $112.00 and collect the $8.00 premium. Because you bought the put at $3.00, you receive a gain of $5.00 ($8.00 – $3.00). Then you sell the feeder cattle in the cash market at $102.00. The total price you receive is the cash price of $102.00 plus the $5.00 options gain, or $107.00. What happens if the November futures price goes up? Suppose the futures price goes up to $120.00 and the cash price is $118.00. Your November put has lost any value it had. You let the option expire and lose the $3.00/lb you paid out as a premium for the put. You sell your calves in the cash market at $118.00. The price you received for your calves is the cash price of $118.00 plus –$3.00, the loss on the option, for a total of $115.00.

Action	Buy Nov put	112.00
	Premium cost	–3.00
	Basis Est.	–2.00
		107.00

Result: prices fall

Basis	Futures price	Options	Cash price	Options gain/ loss	Net price received
–2.00	120.00	Let expire	Sell 118 +	–3.00	= 115.00

Very few producers will actually exercise the put option. If the value of the put increases, producers will offset the put by selling a put with the same strike price and receive a premium. They will let a worthless put expire. Here are more examples using a 112 Feeder Cattle put, a negative $2.00 basis, and varying futures prices. You can see how $107.00 is established as a minimum price and how the put option allows you to benefit from rising prices.

Basis	Futures	Options	Cash Price	Option gain/ Loss	Net cash received
–2	110	Sell put	Sell 108	+ 2.00 – 3.00	= 107.00
–2	106	Sell put	Sell 104	+ 6.00 – 3.00	= 107.00
–2	100	Sell put	Sell 98	+12.00 – 3.00	= 107.00
–2	118	Let expire	Sell 116	+ 0 – 3.00	= 113.00
–2	122	Let expire	Sell 120	+ 0 – 3.00	= 117.00
–2	126	Let expire	Sell 124	+ 0 – 3.00	= 121.00

Buying a call

Use a call to lock in a maximum purchase price and take advantage of prices if they fall. When you plan to buy a commodity, you can buy a call option to lock in a maximum purchase price and take advantage of prices if they rise.

Buying a call example: Suppose it is December and you are planning on buying corn for your hog operation in July. The July futures price is $2.40/bushel, and the local cash price for July corn is $2.19. You want to protect yourself against rising prices, so you buy a July corn 230 call – the right to buy a July corn futures contract at $2.30. You pay a premium of $0.26/bushel.

Action:	Buy July put	2.30
	Premium cost	+0.26
	Basis Est.	–0.21
		2.35

Result: prices rise

Basis	Futures price	Options	Cash price	Options gain/ loss	Net price received
–.21	2.75	Sell 230 call	Buy 2.54	+0.45 – 0.26 =$0.19	= 2.35

When you need the corn in July, the results are shown above. The futures price has risen to $2.75. Your 230 call that grants you the right to buy a futures contract at $2.30/bushel has increased in value. The option now has a premium of $0.45. You sell a July corn 230 call and collect the call premium of $0.45/bushel. Because you bought it at $0.26, you receive a gain of $0.19/bushel ($0.45 – $0.26). Then you buy the corn in the cash market at $2.54. The total price you paid for the corn is the cash price of $2.54 minus the $0.19 option gain, or a total of $2.35. What happens if prices fall? In July, the futures price has fallen to $2.10, and the cash price is at $1.89. Your July corn 230 call is out-of-the-money. You let the option expire and lose the $0.26/bushel premium you paid for the call. You buy corn in the cash market at $1.89. The price you paid for the corn is the cash price of $1.89 with a $0.26 loss of the option premium, for a total of $2.15.

Action:	Buy July call	2.30
	Premium cost	+0.26

Result: prices fall

Date	Futures price	Options	Cash price	Options gain/loss	Net price received
July	2.10	Buy expire	Buy 1.89	0.00 – 0.26	= 2.15

As with put options, very few producers would exercise a call option. Depending on the option's value, they would either offset the option or let it expire. Here are more examples using a 230 corn call option, a negative $0.21 basis, and varying futures prices. You can see how $2.35 is established as a maximum price and how the call option allows you to benefit from falling prices. Results: Options provide excellent opportunities for taking advantage of price decreases.

Basis	Futures	Options	Cash Price	Option gain/loss	Net cash received
–.21	2.40	Sell 2.30 call	Buy 2.19	0.10 + 0.26	= 2.35
–.21	2.80	Sell 2.30 call	Buy 2.59	0.50 + 0.26	= 2.35
–.21	3.00	Sell 2.30 call	Buy 2.79	0.70 + 0.26	= 2.35
–.21	2.15	Let expire	Buy 1.94	0 – 0.26	= 2.20
–.21	2.00	Let expire	Buy 1.79	0 – 0.26	= 2.05
–.21	1.90	Let expire	Buy 1.69	0 – 0.26	= 1.95

15.5 Option strategies

15.5.1 Options costs and hedging

There are costs associated with buying options.

The options strategies presented in this chapter include the practical financial aspects of using options such as

- Premium costs
- Broker commissions

As a buyer of a put option or a call option, a producer will pay a premium for an option up front; the seller of the option receives the premium. For example, a premium for a current live cattle put option may be $0.80/cwt, or a total of $320.00 per option ($0.80 × 400 cwt per option). A producer buying this live cattle put pays the premium of $320.00 at the time of purchase. The seller of the put is credited with $320.00.

Buyers of the live cattle put know their risk. They paid $320 for protecting a minimum sale price. If the option is worthless at the time they are ready to sell their live cattle, they will let the option expire and lose the premium. If the value of the option is above zero, they can offset their position by selling a put with the same strike price and perhaps earn a profit on the premium. Unlike with a futures contract, put buyers do not have to make a performance-bond deposit or face performance-bond calls while they hold the option.

Sellers or writers take the greater risk. They are subject to receiving an exercise notice initiated by an option buyer. Sellers can offset their positions by buying a put with the same strike price at any time before receiving an exercise notice. But it may not be to their benefit to offset at a particular time. Once an exercise notice is received, the seller of the put must take a long position in the underlying futures contract. Also, the seller has to deposit a performance bond and face performance-bond calls to guarantee any loss he or she may incur.

When an option buyer wants to exercise an option (and take the underlying futures position), an exercise notice is issued by the broker to the exchange clearing house, and the option is exchanged for the futures contract. An option seller will be assigned to take the opposite futures position. A cost to producers is the commission charged by the broker. This cost is negotiable and depends on the level of service, quantity traded, and sometimes level of premium. There are many approaches used in charging commissions on options.

Financial aspects of owning an option

- Pay the premium up front when you buy the option.
- Receive a premium when you offset the option.

- Lose the premium if you let the option expire.
- Pay a commission for each option traded.
- Make no performance bond deposit or calls.

15.5.2 Hedging with options compared to hedging with futures

Producers and purchasers can choose to manage risk using options or futures, so it is helpful to summarize the key differences between the two approaches. Fundamentally, a futures contract involves a promise by both participants to buy and sell. An options contract, on the other hand, gives the owner the right to buy or sell if he chooses to do so. This difference in contract terms leads to one important difference in the price fixed by the two contracts. A futures contract locks in one price (subject to basis risk). The option, on the other hand, establishes a floor price (through a put) or a ceiling price (through a call). Because the option owner can choose not to exercise, the option hedge allows the hedger to take advantage of price movements in his favor. For instance, a producer can fix a minimum floor price to sell his output by exercising his put option, yet take advantage of rising prices when they are available by choosing not to exercise. The other key difference relates to the payment for these two contracts. The option buyer must pay the total option price, or premium, up front at the beginning of the hedge, whereas the futures hedger does not pay a premium but only posts a performance bond. Note, however, that a futures position can be subject to a performance-bond call if prices move against the hedger, while the purchase price of the option is the maximum cash flow the hedger needs to make to hedge with an options contract. Finally, as will be seen, an options hedger can choose the level of protection he wants because he has available a series of options with different strike prices (see Table 15.10). For a futures hedger, there is only one futures price available (per futures delivery month) in the market at any one time.

15.5.3 Delivery price strategy

A hog producer uses put options to establish a minimum floor price. It is June and you expect to have 675 hogs ready for market in November. You are having

Table 15.10 Comparison of hedging with futures and hedging with options

	Hedging with futures	Hedging with options
Hedged price	One locked-in price	Floor or ceiling, with ability to improve if prices move in hedger's favor
Hedged price	One locked-in price	Choice of strike prices
Initial cash flow	Deposit performance bond	Pay (nonreturnable) premium
Hedge cash flow	Subject to performance bond calls	Payment of premium only

trouble reading the market and are not sure whether prices are headed up or down. If prices go down, you want some protection with a minimum floor price for December delivery of the hogs. If prices go up, you would like to be able to benefit. You decide to look at buying CME Lean Hog put options. What price can you expect to get for your hogs? Your first step is to set a target price. You calculate a target price as follows:

Target price for put option: Strike price – premium +/– expected basis

There are December CME Lean Hog put options available at various strike prices, so you can calculate a few target prices. You expect the basis in December to be $2.00 under. The current December futures price is $60.00.

Strike price	$60.00	$58.00	$56.00
– Premium cost	–$3.00	–$2.50	–$2.00
+ Expected basis	–$2.00	–$2.00	–$2.00
Target price	$55.00	$53.50	$52.00

You want to establish your minimum price at $55.00/cwt, so you buy the December CME Lean Hog 60 put.

You will need to buy three options to protect your hogs. The put with a strike price of 60 has a premium of $3.00/cwt, so the amount you have to pay up front is $1,200 per option ($3.00 × 400 cwt per option), or a total of $3,600 for three options. You have the $3,600 available to buy the puts and establish a minimum price of $55.00, so that is what you decide to do. You call your broker and ask to buy three December CME Lean Hog 60 puts at a premium of $3.00/cwt. You pay the premium total of $3,600 and the commission at this time.

Action:	Buy Dec puts	60.00
	Premium cost	–3.00
	Expected basis	–2.00
	Target price	$55.00

Result: prices fall

Basis	Futures price	Options	Cash price	Options gain/loss	Net price received
-2.00	55.00	Sell 60, put 5.00	Sell 53.00	+ 5.00 – 3.00, $2.00	= 55.00

In November, futures prices have fallen to $55.00, and the local cash price is $53.00. The basis is $2.00 under, just as you expected. The December CME

Lean Hog 60 puts now have a premium of $5.00. You sell three December CME 60 puts and receive the premium (the premium you paid originally was $3.00). You have earned $2.00/cwt on each of the options transactions ($5.00 – $3.00). You sell your hogs locally at $53.00/cwt. The total price you received is the cash price of $53.00 plus the $2.00 options gain, or a total of $55.00. Looking at the overall picture, you have done $2,400 ($2.00 gain × 400 cwt × three options) better with options. You paid $3,600 up front when you bought put options, and earned $2,400 back when you sold put options. You pay the broker a commission of $50.00 for each option traded, totaling $150.00 for three options bought and sold. Your actual gain is $2,350 ($2,400 – $150).

15.5.4 A real example

Hog prices plummeted at the end of 1998. In early December, the December CME Lean Hog futures contract was trading in the low $20.00 range after experiencing a contract high of $58.50 during the lifetime of the contract. Hog producers who had not used the futures market as protection against price risk would have sold their hogs at a great loss. What if prices rise? With futures at $61.00, the December CME Lean Hog 60 puts expire worthless, and you realize a loss on the premium of $3.00; you receive a total of $56.00 for the hogs.

Action:	Buy Dec puts	60.00
	Premium cost	–3.00
	Expected basis	–2.00
	Target price	$55.00

Result: prices rise

Basis	Futures price	Options	Cash price	Options gain/loss	Net price received
–2.00	61.00	Let expire	Sell 59.00	+ 0 – 3.00	= 56.00

15.5.5 A storage strategy

A producer uses put options to store corn and lock in a minimum selling price. It is November, and you have a corn crop of 35,000 bushels that could be sold. The local cash price now is $2.30/bushel, but you want to explore storing for July delivery. The local cash-forward bid for July corn is $2.50, which would give you a $0.20 storage gain but not cover your expected $0.28/bushel storage cost. You want to protect your downside risk and improve on the storage gain. And you would like to benefit if prices rise.

What can you expect to get for your corn? First, calculate target prices based on some of the available put options for July corn. The current July futures price is $2.70. You expect the basis to be $0.05.

Storage of grain works if the selling price received after storage minus the previous harvest selling price is greater than the cost of storing the crop.

Strike price	$2.90	$2.80	$2.70
– Premium cost	–$0.25	–$0.17	–$0.10
+ Expected basis	–$0.05	–$0.05	–$0.05
Target price	$2.60	$2.58	$2.55

You could also look at the storage gain for each strike price. You want to lower your storage costs, so you rule out buying 270 puts. However, in terms of storage gain, both the 290 puts and the 280 puts are satisfactory.

Strike price	$2.90	$2.80	$2.70
– Premium cost	–$0.25	–$0.17	–$0.10
+ Expected basis	–$0.05	–$0.05	–$0.05
Target price	$2.60	$2.58	$2.55
– Current cash price	–$2.30	–$2.30	–$2.30
Storage gain	$0.30	$0.28	$0.25

You will need to buy seven put options to protect your 35,000 bushels of corn. You would have to pay $8,750 total premium on seven July corn 290 puts ($0.25 × 35,000 bushels) or $5,950 on seven July corn 280 puts ($0.17 × 35,000 bushels). With your present cash flow, the $5,950 is better for you. The 280 puts will lock in a minimum price of $2.58/bushel. You call your broker and ask to buy seven July corn 280 puts at a premium of $0.17/bushel. You pay the premium total of $5,950 and commission at this time. Let's see what happens.

Action:	Buy July puts	2.80
	Premium cost	–0.17
	Expected basis	–0.05
	Target price	$2.58

Result: prices rise

Basis	Futures price	Options	Cash price	Options gain/loss	Net price received
–.05	3.10	Let expire	Sell 3.05	+ 0 – .17	= 2.88

In July, futures prices have risen considerably. The futures price stands at $3.10, and the local cash price is now $3.05. The basis is $0.05 under as you expected. The July corn 280 puts are out-of-the-money, so you let them expire and lose the $0.17/bushel premium you paid. You sell your corn in the cash market at $3.05/bushel. The total price you receive is the cash price plus the loss on the options, or a total of $2.88/bushel ($3.05 – $0.17).

If you had not placed this storage hedge, you would have sold your corn in November at $2.30/bushel. By buying put options, you have made a storage gain of $0.58/bushel ($2.88 – $2.30). Overall, you have earned $20,300 ($0.58 × 35,000 bushels) more by storing the corn and buying options. Subtracting your premium cost of $5,950, the commission cost of $350.00 ($50.00 per option × seven options), storage costs of $9,800 ($0.28/bushel storage cost × 35,000 bushels), the net gain is $4,200.

15.5.6 Calculating an options hedge ratio

In the prior example, the producer needed seven options to hedge his 35,000 bushel corn crop (7 contracts × 5,000 bu/contract = 35,000 bushels). The hedge position is based on the size, or units, of the position to be hedged. A different approach that is used, particularly in short-term hedges (ones that will be lifted before contract maturity), is to consider price changes rather than physical units. Ultimately, it is the dollar value of the purchase or sale that the hedger is trying to protect. If the hedger plans to close out his or her hedge before contract maturity and is looking for equal-option protection and a futures hedge, he or she may want to adjust the number of option contracts used as a hedge. Futures prices tend to move uniformly with the underlying cash position price: if the live hog cash price changes by $0.10/cwt, the futures price will tend to change by $0.10 as well. Options prices, on the other hand, do not usually change as much as the underlying cash price. If the hog cash price changes by $0.10, generally the options price will change by roughly half that, or $0.05/cwt. The technical term for this relative price change is called the option's delta. Here, the change in the options price compared to the change in the cash price was one-half, or 0.50, so the option's delta is 0.50. The hedge in this case would need two options contracts (2 contracts × $0.05/cwt change = $0.10) to hedge the loss in his cash position of ten cents. An option delta gives the short-term hedge ratio: the number of contracts is two (1/option delta, or 1/0.50). The delta of an option, and thus the size of the options position needed to hedge dollar value, can change over the option's life. For instance, an at-the-money option will have a delta close to 0.50, but as the option moves into the money, the delta rises toward a value of one. The reason behind this increase is that, as the option becomes more likely to be exercised, its price movements will mimic the underlying cash. Similarly, if the option moves very far out-of-the money, it is extremely unlikely that it will be exercised, and its delta moves toward zero.

15.5.7 A purchase strategy

A feedlot operator uses call options to lock in a maximum purchase price. As a purchaser of feeder cattle, you can buy call options to protect yourself against price increases while leaving yourself open to profit from price decreases. Suppose it is July and you are planning a purchase of 270 head of feeder cattle to place in the feedlot in December. Indications are that prices are going up, and you want to lock in a maximum purchase price. But prices have been fluctuating lately. You think there is a chance that prices may fall. In that case, you want to take advantage of falling prices. Because options provide both protection and opportunity, you take a look at buying January call options.

What can you expect to pay for the feeder cattle? Your first step is to set a target price for the feeder cattle purchase. You can calculate the target as follows:

Target price for a call: Strike price + premium +/– expected basis

There are January CME Feeder Cattle call options available at various strike prices, so you can calculate a few target prices. You expect the basis in January to be $2.00 under. The current January futures price is $111.50/cwt.

Strike price	$112.00	$114.00	$116.00
+ Premium cost	+$3.90	+$2.80	+$1.60
+ Expected basis	–$2.00	–$2.00	–$2.00
Target price	$113.90	$114.80	$115.60

You decide to establish your maximum purchase price at around $114.00/cwt, so you purchase the January CME Feeder Cattle 112 call option. For the feeder cattle you would pay a total of $7,800 ($3.90 × 500 cwt per option × four options) for the 112 calls. Your lender agrees to lend the $7,800 up-front premium payment, so you decide to purchase four January CME Feeder Cattle 112 calls. You call your broker and place an order to buy four January CME Feeder Cattle 112 calls at a premium of $3.90/cwt. You pay the premium and commission at this time. Let's see what happens.

Action:	Buy January calls	112.00
	Premium cost	+3.90
	Expected basis	–2.00
	Target price	$113.90

Basis	Futures price	Options	Cash price	Options gain/loss	Net price received
−2.00	90.00	Let expire	Buy 88.00	−0 − 3.90	= 91.90

In January, when you are ready to purchase feeder cattle, futures prices have fallen to $90.00. The basis is $2.00 under as you predicted, and the local cash price is $88.00. Your January CME Feeder Cattle 112 calls are out-of-the-money, so you let them expire and lose the $3.90/cwt premium you paid. You buy the feeder cattle in the cash market at $88.00/cwt. The total price you paid is the cash price of $88.00 plus the options loss of $3.90, or a total of $91.90 – a good improvement on your target price. Reviewing the overall transaction, if you had not hedged with call options, you could have bought feeder cattle at even less in the cash market. The total difference was the $7,800 premium loss plus $200.00 in commission, or $8,000. That is the price you paid for insurance for a price you could live with – your target of $113.90 or less. What if prices rise? The insurance that options provide covers you if prices rise. Suppose instead of falling, the futures price rose to $120.00 and the cash price to $118.00. Then hedging with call options would have held your purchase price at 113.90.

Action:	Buy January calls	112.00
	Premium cost	+3.90
	Expected basis	−2.00
	Target price	$113.90

Basis	Futures price	Options	Cash price	Options gain/loss	Net price received
−2.00	120.00	Sell 112 call, 8.00	Buy 118.00	− 8.00 + 3.90 = $4.10	= 113.90

15.5.8 Hedging a cattle sale

A cattle producer and a wheat farmer use options to hedge the sale of their commodity. It is November and you expect to have 140 steers ready for market in April. You suspect that prices are headed down, but it is hard to be sure. If prices go down, you want to protect a minimum sales price of $83.00 for the April delivery of cattle. If prices go up, you want to take advantage of it. You decide to buy April CME Live Cattle put options to hedge the sale. Here are the facts:

- April futures at $88.00 in November.
- Basis expected to be $2.00 under in April.
- An April CME Live Cattle 88 put at a premium cost of $3.00 will lock in a floor price of $83.00.

You call your broker and ask to buy four April CME Live Cattle 88 puts at the \$3.00/cwt premium. You pay the \$4,800 total premium (\$3.00 × 400 cwt × four puts) and the commission \$200 at this time.

Action:	Buy April puts	88.00
	Premium cost	–3.00
	Expected basis	–2.00
	Target price	\$83.00

Result: prices fall

Basis	Futures price	Options	Cash price	Options gain/loss	Net price received
–2.00	80.00	Sell 88 put, 8.00	Sell 78.00	+8.00 – 3.00	= 83.00

In April, futures prices have fallen to \$80.00, and the local cash price is \$78.00. The basis is \$2.00 under as you expected. The April CME Live Cattle 88 puts now have a premium of \$8.00. You sell four April CME Live Cattle 88 puts at the \$8.00/cwt premium. You have gained \$5.00/cwt on the transaction because you bought the puts for \$3.00. You sell the cattle locally at \$78.00/cwt. The total price you receive is \$83.00 (\$78.00 cash price + \$8.00 options gain – \$3.00 loss on premium) – the same as your target.

15.5.9 Hedging a wheat sale

It is March and you expect to have 25,000 bushels of wheat to sell in July. Prices have been volatile. There is a good chance they are going to fall. A local bid is offering a \$3.74/bushel selling price, but you want flexibility in having downside protection and capitalizing on higher prices if markets go up. If prices go down, you want to protect a minimum price of \$3.80 for the July delivery of wheat. If prices go up, you want to be able to take advantage of it. You make a decision to buy wheat put options to hedge the sale. Here are the facts:

- July futures at \$4.10 in March.
- Basis expected to be \$0.10 under in July.
- A wheat 410 put at a premium cost of \$0.20/bushel will lock in a floor price of \$3.80.

You call your broker and ask to buy five July wheat 410 puts at \$0.20/bushel premium. You pay \$5,000 total premium (\$0.20 × 5,000 bushels × five puts) and the commission at this time.

Action:	Buy March puts	4.10
	Premium cost	– 0.20
	Expected basis	–0.10
	Target price	\$3.80

Result: prices rise

Basis	Futures price	Options	Cash price	Options gain/ loss	Net price received
–10.00	4.40	Let expire	Sell 4.30	+ 0 – .20	= 4.10

To your surprise, in July futures prices have jumped to $4.40 and the local cash price is $4.30. The basis is $0.10 under, just as you expected. The July wheat 410 puts you bought are now worthless, and you let them expire. You have lost the $0.20/bushel premium you paid when you bought the puts. However, because the market has risen, the local cash market price is also very good. The total price you receive for the wheat is $4.10/bushel ($4.30 cash price – $0.20 options loss), which exceeds your target price. You might say that you would have been better off not to buy options in the first place. But you were buying insurance against the risk of falling prices. Because prices rose instead, the options allowed you to take advantage of it.

15.5.10 Hedging a hog sale

A meat packer hedges the purchase of hogs. A meat packer can use call options to hedge a planned purchase of hogs. Suppose it is November and the packer needs 1,200 hogs for slaughter in February. If prices go up, the packer wants insurance against the higher price. If prices go down, he or she wants to take advantage of it. The packer decides to buy CME Lean Hog call options to hedge the purchase. Here are the facts:

- February futures at $64 in November.
- Basis expected to be $2.00 under in February.
- A February CME Lean Hog 64 call at a premium cost of $3.00 will lock in a ceiling price of $65.00.

The packer calls a broker and asks to buy eight February CME Lean Hog 64 calls at $3.00/cwt premium. He or she pays the $9,600 total premium ($3.00 × 400 cwt × 8 calls) and the commission at this time.

Action:	Buy February calls	64.00
	Premium Cost	+3.00
	Expected Basis	–2.00
	Target Price	$65.00

Basis	Futures price	Options	Cash price	Options gain/ loss	Net price received
–2.00	70.00	Sell 64 call, 6.80	Buy 68.00	–6.00 + 3.00	= 65.00

In February, futures prices have gone up to $70.00, and the local cash price is $68.00. The basis is $2.00 under, as had been expected. The February CME Lean Hog 64 calls now have a premium of $6.00, and the eight February calls are sold at the $6.00/cwt premium. The packer's gain is $3.00/cwt on the transaction, because the calls were bought for $3.00. The hogs are purchased locally at $65.00/cwt. The total price paid is $65.00 ($68.00 cash price – $3.00 options gain). That is the same as the packer's target price.

Summary[38] – (data assumed in examples)

Basis	Cash price – Futures price Basis	Cash price – Basis Futures price	Futures price + Basis Cash price
What is the total value of the sale of the corn?	10,000 bushels × $2.40 dollars/bushel 24,000 dollars Answer: $24,000		
What will the storage costs be? And the storage selling price?	5 cents/bushel/month × 4 months 20 cents/bushel Answer: 20 cents/bushel	4 cents/bushel/month × 3 months 12 cents/bushel	2.50 current cash price + .12 storage cost 2.62 breakeven price Answer: Sell above $2.62/bushel
Setting a target price (for hedging)	Futures price + Expected basis Target price	2.25 futures – .10 expected basis 2.15 target purchasing price Answer: $2.15/bushels	
Hedging a sale or purchase? How many futures contracts do I need? And what size?	Expected to harvest 25,000 bushels One contract 5,000 bushels Answer: five contracts	Expect to buy 100,000 pounds of feeder cattle One future contract = 50.000 pound Answer: two contracts	
The total price you received or paid for a commodity?	Step 1 – gain or loss on futures transaction: Futures selling price – Futures buying price Futures gain or loss	Step 2 – The price you received for the sale of the commodity: a. Short Hedge Cash price received + Futures gain/loss Net price received	b. Long hedge Cash price paid – Futures gain/loss Net price paid

Continued

Summary – Continued

	1. Determine the gain or loss on the futures contracts. 65.00 futures selling price – 62.00 futures buying price 3.00 futures gain	2. Add the cash sales price and the futures gain. 61.00 cash price + 3.00 futures gain 64.00 net price received Answer: $64.00/cwt	
Calculating short-hedge results (result over cash price)	400 cwt/contract × 3 contracts 1,200 total cwt	1,200 cwt × 2.50 dollars/cwt 3,000 dollars total gain	3,000 total gain – 225 commission 2,775 net gain Answer: $2,775 better results
Setting a target price (for options)	Put target price Strike price – Premium + Basis Target floor price	Call target price Strike price + Premium + Basis Target ceiling price	
	1. Target floor price (selling at $45/cwt or better)	50.00 strike price – 2.75 basis (premium) 47.25	47.25 – 2.00 basis 45.25 target price Answer: Yes, $45.25 is acceptable.
	2. Target ceiling price for a call (buy at $67 max.). Will a call at a premium of $2.55 lock in an acceptable price?	64.00 strike price + 2.55 premium 66.55	66.55 + 1.00 basis 67.55 target price Answer: No, $67.55/cwt is not acceptable.
Calculating the total premium value	Premium for a corn call. You buy a March corn 200 call at a premium of 26 cents/bushel. What is the total amount you pay for the call?	5,000 bushels × .26 dollars/bushel 1,300 dollars Answer: $1,300 total premium	
	Premium for a Live Cattle put You buy an April Live Cattle 70 put at a premium of $4.50/cwt. What is the total amount you pay for the put?	4.50 dollars/bushel × 400 cwt 1,800 dollars Answer: $1,800 total premium	

Continued

Summary – Continued

	Premium for several puts You buy four July wheat 390 puts at a premium of 25 cents/bushel. What is the total amount you pay for the puts?	1. Multiply the number of put options by 5,000 bushels per contract. 5,000 bushels/option × 4 options 20,000 bushels	2. Multiply the total number of bushels by the 20,000 bushels. × .25 dollars/bushel 5,000 dollars Answer: $5,000 total premium
Calculating the total price you paid or received for a commodity	1. Calculate gain or loss on option Options selling price – Options buying price Options gain or loss	2. Add cash price you received Short-hedge cash price received + Options gain/loss Net price received	Long-hedge cash price paid – Options gain/loss Net price paid
	You bought soybean puts at a premium of 20 cents/bushel and sold them back for 18 cents/bushel. Then you sold the soybean crop in the cash market at $6.85/bushel. What was the total price you received for the soybeans?	.18 options selling price – .20 options buying price –.02 options loss	6.85 cash price – .02 options loss 6.83 total price received
	You bought feeder cattle calls at a premium of $2.75/cwt and offset the calls at $4.25/cwt. Then you bought the feeder cattle in the cash market at $58/cwt. What was the total price you paid for the feeder cattle?	4.25 options selling price – 2.75 options buying price 1.50 options gain	58.00 cash price – 1.50 options gain 56.50 total price paid Answer: $56.50/bushel

Continued

Summary – Continued

	You bought July wheat 360 puts at a premium of 30 cents/bushel. Now the 360 puts are worthless. You sell your wheat crop in the cash market at $3.65/bushel. What is the total price you received for the wheat? (out-of-the-money)	1. You let the option expire and lose the 30 cents/ bushel premium. –.30 options loss	2. Add the cash price and the options loss. 3.65 cash price – .30 options loss 3.35 total price received Answer: $3.35/ bushel If you let an option expire, the loss is the original premium you paid.
Calculate hedging results with options	You hedged your corn purchase by buying and then selling two corn calls. The total price you paid was 15 cents/bushel below the current cash price. The commission is $100 for the calls traded. What are your net savings compared to the cash price? (long hedge)	1. Option = 5,000 bushels 5,000 bushels/ options × 2 options 10,000 bushels 2. Multiply the total bushels by the saving under the cash price 10,000 bushels × .15 dollars/bushel 1,500 dollar savings	2. Subtract the commission from the total savings 1,500 total savings – 100 commission 1,400 net savings Answer: $1,400 better results

15.6 Application 1: types of traders on the agricultural commodity markets and their trading behavior

There are different types of traders active on these markets. Different in the sense that they have different objectives, different risk appetite, and a different focus on the end result (or what the end result should be). Here we will deal with the main categories. In the chapter on financial speculation we will refine the main categories (in particular those of the speculators)

In basic terms, there are two types of traders in the futures industry – hedgers and speculators. The primary distinction between them is that the hedger uses the futures and/or options market to reduce or eliminate price-level risk while the speculator uses futures and options to assume market risk for the potential profit opportunity. Another key distinction is that a hedger has an interest (either to purchase or sell) in the physical commodity, whereas the speculator does not.

15.6.1 Hedgers

Long hedgers (buyers of physical commodities) are at risk if prices rise between the current time and when they expect to purchase or price the physical commodity. To protect against this adverse price risk, the long hedger initiates a long-futures position, which is a temporary substitute for the eventual purchase of the physical commodity. If the market price increases over time, the higher purchase price of the physical commodity will be offset by a gain in their futures-market position. Conversely, if the market price decreases, the loss in their futures-market position will be offset by a lower purchase price of the physical commodity. Thus, the long hedger is able to establish a purchase price in advance of taking delivery of the physical commodity regardless of what direction the market moves. The long-hedge results will improve if the basis weakens (cash price declines relative to the futures price) between the time the hedge is initiated and the time it is offset when the physical commodity is delivered or priced. Since a basis is the relationship between two correlated prices, it can weaken whether the market price levels are going up or down. The following two examples[39] illustrate the basic mechanics of a long hedge, especially the relationship between the cash and futures markets. For simplicity, these examples keep the basis constant but note that the basis can fluctuate over time, and as such will impact the results of the hedge.

Example: long hedge with rising market conditions

Date	Cash market	Futures market	Basis
March 23	$2.70 bushel reference price*	Long (buy) July corn at $2.50/bu	+$0.20/bu
June 23	Buy corn at $3.00/bu	Sell (offset) July corn at $2.80/bu	+$0.20/bu
Result		.30/bu gain	No change

* Reference price is used to calculate the initial basis but is not an actual transaction.

The long hedger bought physical corn in July at $3.00/bu ($0.30/bu higher than the initial cash-market price). However, the hedger's futures position had a $0.30/bu gain to offset the higher cash-market price. The net purchasing price is $2.70 ($3.00 cash price – .30 futures gain). Note that the hedge result was the same as the initial cash-market price because the basis didn't change.

Example: long hedge with falling market conditions

Date	Cash market	Futures market	Basis
March 23	$2.70 bushel reference price*	Long (buy) July corn at $2.50/bu	+$0.20/bu
June 23	Buy corn at $2.50/bu	Sell (offset) July corn at $2.30/bu	+$0.20/bu
Result		.20/bu gain	No change

* Reference price is used to calculate the initial basis but is not an actual transaction.

The long hedger bought physical corn in July at $2.50/bu ($0.20/bu. lower than the initial cash-market price). However, the hedger's futures position had a $0.20/bu loss, which offset the lower cash-market price. The net purchasing price is $2.70 ($2.50 cash price + .20 futures loss). Note that the hedge result was the same as the initial cash-market price because the basis did not change.

Short hedgers (sellers of physical commodities) are at risk if prices fall between the current time and when they expect to deliver or price the physical commodity. To protect against this adverse price risk, the short hedger initiates a short-futures position, which is a temporary substitute for the eventual sale of the physical commodity. If the market price decreases over time, the lower selling price of the physical commodity will be offset by a gain in their futures market position. Conversely, if the market price increases, a loss in their futures market position will be offset by the higher selling price of the physical commodity. Thus, the short hedger is able to establish a selling price in advance of making delivery of the physical commodity regardless of what direction the market moves. The short-hedge results will improve if the basis strengthens (cash price increases relative to the futures price) between the time the hedge is initiated and when it is offset when the physical commodity is delivered or priced.

Since a basis is the relationship between two correlated prices, it can strengthen when the market price levels are going up or down. The following two examples illustrate the basic mechanics of a short hedge, especially the relationship between the cash and futures markets. In these examples the basis remains constant but note that basis can fluctuate over time, and as such will impact the results of the short hedge.

Example: short hedge with falling market conditions

Date	Cash market	Futures market	Basis
May 15	$5.60 bushel reference price*	Short (sell) November soybeans at $6.00/bu	–$0.40/bu
October 20	Sell soybeans at $5.20/bu	Buy (offset) November soybeans at $5.60/bu	–$0.40/bu
Result		.40/bu gain	No change

* Reference price is used to calculate the initial basis but is not an actual transaction.

The short hedger sold physical soybeans in November at $5.20/bu ($0.40/bu lower than the initial cash-market price). However, the hedger's futures position had a ($0.40/bu) gain to offset the lower cash-market price. The net selling price is $5.60 ($5.20 cash price + .40 futures gain). Note that the hedge result was the same as the initial market price because the basis did not change.

Example: short hedge with rising market conditions

Date	Cash market	Futures market	Basis
May 15	$5.60 bushel reference price*	Short (sell) November soybeans at $6.00/bu	–$0.40/bu
October 20	Sell soybeans at $6.00/bu	Buy (offset) November soybeans at $6.40/bu	–$0.40/ bu
Result		.40/bu gain	No change

* Reference price is used to calculate the initial basis but is not an actual transaction.

The short hedger sold physical soybeans in November at $6.00/bu ($0.40/bu higher than the initial cash-market price). However, the hedger's futures position had a ($0.40/bu) loss, which offset the higher cash-market price. The net selling price was $5.60 ($6.00 cash price – .40 futures loss). Note that the hedge result was the same as the initial cash-market price because the basis did not change.

Who are these agricultural hedgers?

As noted in the previous hedging examples, hedgers are classified as either long (buying) or short (selling) depending on whether they will be buying or selling the physical commodity at some later date. The following are a few types of individuals and firms that are potential agricultural hedgers. Soybean crushers use soybean, soybean oil, and soybean meal futures to hedge their gross processing margin – the difference between the cost of the soybeans and the eventual revenue from the sale of the soybean products: soybean meal and soybean oil. They purchase soybean futures to be protected against rising input costs and sell soybean oil and soybean meal futures to be protected against falling product prices. The soybean crusher could also use soybean crush options to protect their processing margin.

Food processors who face rising ingredient costs trade futures or options to manage their upside price exposure. A food company that produces salad dressings would buy futures contracts to establish a price level for ingredient purchases that will occur at a later date. A baker, who continuously buys flour, may buy wheat call options to establish a maximum (ceiling) price for their flour while retaining the opportunity to purchase at a better price if the wheat market declines. In addition to the traditional food manufacturer, restaurants, pet food manufacturers, and importers could also take advantage of the agricultural markets for managing their upside price risk.

Grain and oilseed producers plant their seeds in early spring and wait until the fall of the year to harvest their crops. In addition to the variables that could affect their production risk over this period of time, they may also face tremendous market risk in the form of potentially lower prices. To protect against the risk of falling market prices, a farmer could put on a short hedge by selling futures in the spring as a temporary substitute for the eventual sale of the physical commodity in the cash market during harvest, thus locking in (establishing) a sale price months in advance of the physical delivery. Farmers could also buy put options to lock in a minimum (floor) selling price while retaining the opportunity to sell their crops at a better price should the market rally. Livestock producers, like grain producers, could be both short (selling) hedgers for their livestock and long (buying) hedgers for their feed needs. Grain elevators and merchandisers are in the business of being middle men – buying and selling grain and oilseeds. Although their exposure is different than an outright buyer or seller of commodities, they still have risk exposure related to their margins – the difference between their purchase and sale prices. They can use futures and options in a variety of ways to protect their profit margins or to offer cash market contracts to their customers that are based off the futures and options contracts. Exporters could also use the markets in a similar fashion to the grain elevator or merchandiser.

15.6.2 Speculators

In addition to the hedger, the other general type of trader in the futures industry is the speculator. Speculators use a variety of trading techniques and strategies with the objective of making profits from expected movements in price levels or price relationships (spreads). Therefore, the speculators' role in the futures market is to assume market risk in return for potential profits. By taking positions in the futures and options market, the speculator is assuming market risk and providing valuable liquidity, which is a measure of a market's efficiency. Liquidity is the ease and efficiency with which a market participant can enter and exit a market position. If a speculator expects the market to rally at some later time, they would initially go long (buy) futures and then offset (sell back) their futures position at a later date. If the market did rally, as they expected, they would realize a gain. For example, if you go long on a wheat futures contract at $3.20/bu and later sell (offset) the same wheat futures contract at $3.50/bu, you would realize a $0.30/bu gain ($1,500 on a 5,000 bu contract).

If a speculator expects the market to decline, they would short (sell) a futures contract initially, and then if they offset (buy back) the contract at a later time when the market did decline, they would realize a market gain. For example, a speculator who sells oat futures at $2.00/bu and later offsets (buys back) the same oat contract at $1.85/bu will realize a $0.15/bu gain ($750 on a 5,000 bu contract). Note that if the market moves against the speculator's initial expectations, he or she will be exposed to a market loss.

So in summary, a speculator would achieve a gain in the futures market by buying low and selling high or by selling high and buying low. If they buy high and sell low or sell low and buy high, they will realize a market loss.

As said, an option is the right, but not the obligation, to sell or buy a futures contract at a certain price at any time on or before a specified date. The cost of an option is the premium, similar to an insurance premium, which is paid up front. Options are therefore very much like insurance policies. Insurance on a car or truck offers protection from the possible financial repercussions of an accident. People pay a premium for the insurance and hope that they'll never have to use it. Options on futures work the same way. A dairy producer, for example, who wants to guarantee he will sell his milk for a specific price in June can buy a CME Milk option in February at the price level he needs. He pays a premium for the option and, like insurance, hopes never to use it. But if in June the USDA Class III Milk price is below the price he planned on receiving for the milk, he can sell back his option and make up the difference. Just as if he had dented his truck fender and filed a claim with his insurance company, he would be reimbursed for the damages. Options provide protection from the fender benders of ever-changing economic conditions.

And, like insurance, people choose the level of coverage they want and thereby control their premium costs. Speculators buying options can only lose the premium paid, if wrong, but can profit substantially if right. Hedgers buying options also have limited loss, but unlimited gain potential. Thus, hedgers can use options to protect their cash position from adverse price moves, while retaining most of the gain in cash value from favorable price moves. CME, for example, offers options on all its Class III and Class IV Milk, Nonfat Dry Milk, and Butter futures contracts. Because CME dairy options are based on CME dairy futures, their contract and technical specifications are almost identical. Options on CME dairy futures are listed in the same trading months as the futures contracts.

15.7 Application 2: technicalities of options and futures[40]

Note: throughout this section, examples are used that are related to dairy products. This is for illustration purposes only and can therefore be replaced by other agricultural commodities or related commodities (lumber futures or fertilizer futures).

15.7.1 Options and types of options

An option is a choice. It is the right, but not the obligation, to buy or sell something, for example, a milk futures contract, at a specific price on or before a certain expiration date. An option allows its holder to take advantage of futures price moves without actually having a futures position and thereby dealing with the uncertainty of potential margin calls. There are two different types of options: puts and calls.

Put option

Put-option buyers have the right to sell (go short) a futures contract at a predetermined price on or before an expiration date. For example, a July $12.00 CME Milk put option gives the buyer the right to sell (be short) a July CME Milk futures contract at $12.00/cwt even if July CME Milk futures are trading at $11.00/cwt. Put options offer protection from falling prices.

Call option

Call-option buyers have the right to buy (go long) a futures contract at a predetermined price on or before an expiration date. It is important to note that puts and calls are separate option contracts. They are not the opposite side of the same transaction. For every put buyer there is a put seller; for every call buyer there is a call seller. The buyer pays a premium to the seller in each transaction.

15.7.2 Parties involved

Option buyer

Buyers – holders – of options can choose to exercise their rights and take futures positions. But most likely they will sell their options back into the market if the options have value. Producers who want to protect their milk selling prices or cheese manufacturers who want to protect their milk purchase costs would typically be option buyers. For every option buyer there is an option seller.

Option seller

An option seller is also called the writer or grantor. Option sellers are obligated to take the opposite futures position if the buyer exercises his right (long if the buyer exercises a put, short if he exercises a call). Most option sellers are speculators who are willing to take the risk of covering an adverse futures price move in return for the premium. For every option seller there is an option buyer.

Automatic exercise

If an option has not been offset by its expiration date and it has intrinsic value, the exchange's clearing house will automatically exercise it unless otherwise notified by the buyer's clearing firm.

Cash settlement

A way of settling certain futures contracts in which the buyer and the seller settle out in cash at expiration to a representative index of value for the

Table 15.11 Positions toward cash settlement

Put buyer	Put seller
• Pays premium	• Collects premium
• Has right to exercise against	• Has obligation if exercised
Call buyer	**Call seller**
• Pays premium	• Collects premium
• Has right to exercise against	• Has obligation if exercised

commodity. For example, CME Milk futures settle to the USDA Class III Milk price at 3.5% butterfat published for the month.

In summary, Table 15.11 represents the possible positions.

15.7.3 Underlying futures contract

The "underlying" is the contract that may be purchased or sold upon the exercise of the option. An option on a June Milk futures contract is the right to buy or sell one June Milk futures contract. Milk futures contracts are cash settled at expiration to the USDA Class III Milk price published for the month.

Strike price

The strike price, also known as the exercise price, is the price at which an option holder may buy or sell the underlying futures contract. Exercising the option results in a futures position at the designated strike price. Strike prices are set by the exchange at $.25 intervals for milk options. Strike prices are set around the existing futures prices. So if June futures were at 12.00/cwt, there may be strike prices listed at 12.50, 12.25, 12.00, 11.75, and so on. New strike prices are listed in addition to the initial strike prices, as milk futures move higher or lower.

15.7.4 Premium

The premium is the market-determined price paid for an option. The premium is ultimately determined by the willingness of buyers to purchase the option and sellers to sell it at each strike price. Factors that affect this willingness are strike-price level relative to futures-price level, time remaining until expiration, and market volatility.

15.7.5 Exercise

Options buyers exercise their options if they want to have the underlying futures position. Exercising a put gives the buyer a short (sell) futures position. Exercising a call gives the buyer a long (buy) futures position. Only the buyer has the right to exercise the option. The seller has the obligation to take the opposite, possibly adverse, futures position. There is an element of risk involved, so options sellers must post margin (performance-bond) funds.

15.7.6 Exercise style

There are two exercise styles – European and American. A European-style option can be exercised by the buyer only on the expiration date, but can be offset (sold back) on any day prior to expiration; an American-style option, on the other hand, can be exercised or offset (sold back) on any business day that the option is available for trading, up to and including the expiration date. CME dairy options are American style.

15.7.7 Expiration date

Expiration is the last day that buyers can buy back or exercise their options. It is important to know exactly when CME dairy options expire so strategies can be determined accordingly. For example, milk options expire on the same day as the futures, which is the business day immediately preceding the release of the USDA Class III Milk price.

Choice is the main feature of an option. Buying a dairy option provides the right, but not the obligation, to take a buy or sell (long or short) position in a dairy futures contract at a fixed price on or before an expiration date. For the right granted by the option contract the buyer pays a sum of money, called the premium, to the option seller. The option seller, or writer, keeps the premium whether the option is used or not. The seller, however, assumes some risk – he must fulfill the contract terms if the option is exercised by the buyer. Therefore he must post margin (performance-bond) funds. People who buy an option are buying a choice.

15.7.8 How are option premiums determined?

Who establishes the price or premium of an option? The buyers and sellers of options ultimately determine the price. Several factors affect futures option premiums:

1. The volatility of the underlying futures price. A (milk) futures market in which prices fluctuate significantly will command a higher premium for a milk option. Volatility is a measure of the degree of fluctuation in price on an annual basis. Volatile month-to-month USDA Class III Milk prices lead option buyers to think there is a greater chance of a major price change. Therefore, they are willing to pay more to protect against the likelihood of change (hedgers) or to capitalize on it (speculators). Sellers tend to see this situation as more risky and are only willing to accept that risk if they can receive a higher premium. For example

   Volatility comparison
   June CME Milk 13.00 put
   120 days to expiration
   Milk futures at $13.50

Volatility	17%	20%	24%
Approximate option premium	$0.30	$0.38	$0.5

2. The strike price compared to the futures price. The relationship between the strike price and the underlying futures price is a key influence on option premiums. Options can be in-, at-, or out-of-the-money. A put option is in-the-money when the price of the underlying futures contract is below the strike price. This makes sense because selling at a higher price has greater value than selling at a lower price. A put option is at-the-money when the underlying futures price is at or near the strike price. A put option is out-of-the money when the underlying futures price is higher than the strike price. For example, if May futures are at 11.25, an 11.50 put would be in-the-money, because the put gives the holder the right to sell futures at 11.50 and futures are now trading lower at 11.25. An 11.25 put would be at-the-money because it is at or the same as the price of the underlying futures contract. An 11.00 put would be out-of-the-money because it gives the holder the right to sell futures at 11.00, which is lower than 11.25, the price at which futures are currently trading.

   A call option is in-the-money when the price of the underlying futures contract is above the strike price. This makes sense because buying at a lower price has greater value than buying at a higher price. A call option is at-the-money when the underlying futures price is at or near the strike price. A call option is out-of-the money when the underlying futures price is lower than the strike price. For example, if May dairy futures are at 12.25, a 12.00 call would be in-the-money, because the call gives the holder the right to buy futures at 12.00 and futures are now trading higher at 12.25. A 12.25 call would be at-the-money because it is at or the same as the price of the underlying futures contract. A 12.50 call would be out-of-the-money because it gives the holder the right to buy futures at 12.50, which is higher than 12.25, the price at which futures are currently trading. Most hedgers buy options that are out-of-the-money because the premium is lower.

Put strike prices relative to futures CME Milk futures = $13.40	**Call strike prices relative to futures CME Milk futures = $13.40**
14.00 in-the-money	14.00 out-of-the-money
13.75 in-the-money	13.75 out-of-the-money
13.50 at-the-money	13.50 at-the-money
13.25 out-of-the-money	13.25 in-the-money
13.00 out-of-the-money	13.00 in-the-money

3. Time. The more time that remains to expiration, the higher the option's premium tends to be. This is because the longer the time remaining until expiration, the more chance that the underlying futures price will move to a point where the value of the option will increase enough to sell back and collect a profit. An option's time value erodes faster as the option moves toward the expiration date.

What effect does time have on options premiums?

E.g. 90 days to expiration, 17% volatility, Milk futures at $12.50
Strike price at 12.25

Approximate	Call	Put
Option premium	$.55	$.30

E.g. 30 days to expiration, 17% volatility
Milk futures at $12.50
Strike price at 12.25

Approximate	Call	Put
Option premium	$.38	$.14

Option specifications

Like futures contracts, CME dairy option contracts are standardized. There are five basic standard elements for each option contract:

1. The type or kind of option – that is, whether the option is a put or a call.
2. The underlying or corresponding futures contract.
3. The option month – the listed futures contract months on which options contracts will be based. Dairy options contracts expire the same time as the dairy futures contracts.
4. The strike price, which is set by the exchange.
5. The exercise style – American or European.

What is the premium?

An option's premium is its price – the price that the buyer of the option pays and the seller of the option receives for the rights given by the option. Generally the premium is made up of intrinsic value and time value.

What is intrinsic value?

Intrinsic value is the difference between the futures price and the strike price of an option. If a Milk option is out-of-the-money at expiration, its intrinsic value is 0; therefore since it has no value, it will expire worthless. If a Milk option is in-the-money at or near expiration, a holder can sell it back, exercise it into a futures position, or let it automatically exercise and cash settle. Its intrinsic value would be the difference between the strike price and the futures price, or the strike price and the USDA Class III Milk price, respectively. It is possible to offset an option position by selling it back prior to expiration.

What is time value?

Simply stated, time value is that part of an option's premium that exceeds the intrinsic value. Before expiration, most options will have time value, even if they're out-of-the-money. Time value reflects the probability that the option will move into-the-money or further into-the-money. The longer the time remaining until expiration, the greater the option's time value. Many options are sold back prior to expiration to capture both intrinsic value and time value. Once the option is exercised, the time value is lost.

What is time decay?

It is important to note that an option is a wasting asset; that is, its market value erodes as the option approaches expiration. This time decay normally accelerates the last 30 to 40 days to expiration. A similar analogy would be how a term insurance premium would erode in value as the policy approaches the renewal period.

Know the breakeven costs

It is important to know production costs to successfully hedge with options. Hedgers should be able to calculate all costs, both fixed and variable. Variable costs include feed and hay costs, hired labor, marketing charges, fuel and utility expenses, and so on. Fixed costs could include building depreciation, equipment, and taxes. If hedgers do not know their costs of production, they will not know if they are hedging at a profit or loss.

Expect convergence

To successfully use options contracts as a hedging tool, it is imperative that the underlying futures price and the cash price come together at expiration. Because (milk) futures are cash settled, it is pretty much guaranteed that on the last day of trading the futures price will converge to the, in this case, USDA Class III Milk price. This means that hedgers are less likely to experience the risk of variable cash-futures differential levels.

Calculate mailbox differential (basis)

(Dairy) options are based on monthly USDA prices. Final farm prices paid for dairy products often vary from the USDA prices on a dollars-per-hundredweight-of-product-sold basis. This is known as the cash/futures differential, mailbox differential, or basis. When using options to hedge (protect) a dairy price, it is important to recognize that variables in marketing orders will affect final farm pay prices. Hedgers need to understand that they will not receive the exact final farm pay price their options strike protection affords. For the best estimation of basis, it is necessary to keep accurate records of actual monthly final farm pay prices per hundredweight to see how they compare to the announced prices and thus the futures cash settlement price for that month. In most regions of the country, a final farm pay check exceeds the futures settlement price in that month, resulting in a positive basis differential. Consider the following example for estimating basis differential for milk to be marketed in August:

Farm historical pay price –	(August historical Class III price) Class III price	= Basis
August 2014 17.30	16.90	= +.60
August 2015 18.40	18.10	= +.30
August 2016 18.90	18.40	= +.50
3-year average farm pay price –	3-year average class III price	= 3-year average basis
54.60 divided by 3 = 18.2	53.4 divided by 3 = 17.8	= 0.47

15.7.9 Deciding which options to buy

Once hedgers have determined their cost of production and cash/futures differential, it is time to determine which option to buy. Remember there are two choices: puts or calls. Producers who have milk to sell at some future date and wish to establish a floor, or minimum price, use puts, thus leaving the opportunity open to capture additional profit from upward price movement. (Milk) producers who want to purchase feed in the future may also use calls to guarantee a ceiling on corn or soybean meal costs, for example, while leaving the potential to capture profit from downward price movement.

Hedgers must also determine the month they want to sell their milk or buy their feed and choose a put or call option corresponding to that month. Exchanges offer a (dairy) options contract for every calendar month, making them ideal hedging tools. It is possible to hedge more than one month's production by using put options for several consecutive months, for example, August, September, and October, all at the same time.

15.7.10 Choosing a strike price

There is no one right answer to the question of which strike price to choose. The choice depends on a hedger's ability to bear risk, which direction the hedger thinks the market is going, and how much the hedger is willing to pay for the option. For example, the higher the strike price (and resulting floor price) the more it is going to cost. If the market goes down by the time the milk is sold, the higher price has been worth the additional cost. However, if the market remains stable or goes up, the higher floor price would not have been needed and the higher premium would have been left on the table. It is up to each hedger, and possibly their lenders, to determine the amount of insurance or protection to take with options.

Example: choosing a strike price for milk options

	Market down	**Market up**	Milk price floor
Higher strike	Higher floor		strike price
Lower strike	Lower floor		– premium
Milk futures = $13.00			+/–cash/futures
120 days to expiration, 18% volatility			differential (optional)
			----------
			floor or minimum selling price
Strike price	**Premium ($)**	**Floor (not including cash/futures differential)**	Total premium cost Premium $.30 (hypothetical)
13.00	0.53	12.47	× milk option
12.75	0.41	12.34	contract 2,000 cwt
12.50	0.31	12.19	---------
			$600

Example: choosing a strike price for butter options

	Market down	**Market up**
Higher strike	Greater opportunity	Higher ceiling
Lower strike	Less opportunity	Lower ceiling
Butter Futures = $2.30		
120 days to expiration, 20% volatility		
Strike price	**Premium ($)**	**Ceiling**
2.50	.04	2.54
2.60	.02	2.62
2.70	.01	2.71

15.7.11 Choices after options have been purchased

Once hedgers have purchased options, they can offset or exit their positions in three ways:

1. Sell the option back if it has value. Typically, options holders will sell back (offset) their options prior to or at expiration and receive the current premium value. Prior to expiration, the premium value may include time value as well

as intrinsic value and could be higher or lower than the original purchase price, depending on how the underlying futures price has changed.

2. Exercise the option. Holders of options will do this if they want to take a short futures position (if they have bought a put) or a long futures position (if they have bought a call). They can also let the option be automatically exercised at expiration. Options holders never have to worry about forgetting to exercise. If their option has any intrinsic value, it is exercised automatically into a futures position, which is then cash settled to the USDA Class III Milk price.

Auto exercise at expiration
Strike price = $13.50
USDA Class III – 12.50
Difference 1.00 × 2,000 cwt.
$2,000

3. Let the option expire if it has no value at the time of expiration, without taking any action. In most cases, options are sold back prior to expiration if they have value, or left to expire if they have no value. Very few options are ever exercised early into a futures position; in the case of CME Milk options, they can be automatically exercised at expiration and cash settled to the USDA Class III Milk price.

In summary, four options:

- Purchased option has value – sell back.
- Purchased option has no value – let expire.
- A futures position desired – exercise.
- At expiration, cash-settled – auto exercise.

15.7.12 Establishing a minimum sale price for dairy: buying put options

Milk producers and processors may utilize dairy put options to create a floor price – minimum selling price – for dairy commodities. Choosing a particular strategy depends mainly on the level of protection desired. Consider these two short-hedging strategies using put options.

Example 1: buying a put option

Let's say it is January and a milk producer from the upper Midwest region of the United States decides to protect his average final farm pay price for August. CME Class III Milk futures are at $13.75/cwt. The producer buys an August put option with a strike price of 14.00 at a cost of $.62/cwt.

Historically, the upper Midwest final pay price for August has been $.88 higher than the August USDA Class III Milk price. The floor price would be the 14.00 strike, minus the premium of $.62, plus the cash/futures differential of $.88, or $14.26/cwt. Let's look at what happens if the market goes up, sideways, or down. The monthly August Class III Milk price comes out on September 4, the producer sold milk throughout the month of August, and actual basis was + $.88.

If August futures are ($):	Value of 14 put ($)	A. 14 put net gain/cost ($)	B. Cash/futures price ($)	C. Net realized price ($) (A + B = C)
16.00	.00	–.63	16.88	16.26
15.00	.00	–.62	15.88	15.26
13.75	.25	–.37	14.63	14.26
12.50	1.5	.88	13.38	14.26
11.50	2.5	1.88	12.38	14.26

As the table above demonstrates, when the futures price drops below the put strike price, the minimum selling price or insurance kicks in and protects the floor that was established when the 14.00 Milk put was purchased. Should the market go higher, the holder of the option will be able to realize the increase less the cost of the premium, while enjoying having had protection from a price drop.

Example 2: rolling up to a higher floor

Suppose a milk producer has purchased a put option similar to Example 1 and the market increases after the original purchase, but well before expiration. What are the alternatives? The producer still wants protection but also wants to establish a higher floor price. Here's how that can be done.

It is January and the producer purchases a June Milk 13.00 put option for $.30/cwt; at that time June futures are at $13.25/cwt. The upper Midwest estimated cash/futures differential for June is +$.97. The producer's estimated minimum selling price is $13.00, minus the premium of $.30, plus the estimated basis of $.97, or $13.67/cwt. Two months later, in March, June Milk futures are trading at $14.75/cwt. The producer buys a June 15.00 Milk put for $.70/cwt. He can either sell back, or offset, the original 13.00 put if it has value, or leave it in place for extra protection if the market should drop below the original strike price. In this example the producer will leave the 13.00 put in place. His new minimum price is the put strike price of 15.00, minus the premium of $.70, minus the $.30 premium of the 13.00 put, plus estimated basis of $.97, or $14.97/cwt. The producer has increased his floor or minimum selling price by $1.30. Let's take a look at what happens when the market goes up, sideways, or down from the

$14.75 roll-up futures position. The monthly Class III price comes out on July 3; the producer sold milk in June.

Without rolling up to higher floor

If June futures are ($)	Value of 13.00 put ($)	13.00 put net gain/cost (A) ($)	Cash/futures price (B) ($)	Net realized price (C = A + B) ($)
15.50	0	–.30	16.47	16.17
14.75	0	–.30	15.72	15.42
13.50	0	–.30	14.47	14.17
12.50	.50	+.20	13.47	13.67

Rolling up to higher floor

If June futures are ($)	13 put net gain/cost ($)	Value of 15.00 put (A) ($)	15.00 put net gain/ cost (B) ($)	Cash/futures price (C) ($)	Net realized price (D = A + B + C) ($)
15.50	–.30	0	–.70	+16.47	15.47
14.75	–.30	.25	–.45	+15.72	14.97
13.50	–.30	1.50	+.80	+14.47	14.97
12.50	+.20	2.50	+1.80	+13.47	15.47

Opportunities in which the market rallies strongly do not come along so often that a producer can afford to pass them up without at least considering a roll-up. Rolling up to a higher strike put is a good way to take advantage of a price increase while maintaining the downside price protection desired. As shown in the comparison above, the producer would only want to roll up and pay more premium if he thought there was a good chance that the market had peaked at $14.75/cwt and would fall back down to much lower levels. In the above situation, at $14.75 when the roll-up occurred, the 13.00 put had very little time value, so holding on to it would cost almost nothing. Also, if the market fell far enough, the producer would eventually make money on both the 15.00 put and the 13.00 put.

Other factors to consider before entering a trade

Option strategies can range from simple to complex. Individuals using options on (dairy) futures for the first time may want to simply buy a put to protect a minimum price on a portion of their production and then evaluate the results. Later, they may want to factor in their own mailbox differential and may then wish to experiment with rolling up to milk futures or higher-strike put options. In addition to understanding the risks and rewards of a particular option strategy before getting started, producers and processors of dairy products should keep several other factors in mind.

1. Cost of production

Unless producers and processors know their production costs, they will not know if they are pricing in a profit or loss. Successful pricing of milk requires a solid estimate of breakeven costs.

2. Futures and options contract specifications

It is necessary to be familiar with the major aspects of both the option contract and the underlying futures contract. Knowing the contract month, option type, and expiration date makes it possible to hedge purchases or sales in the appropriate contract.

3. Mailbox differential (basis or the difference between the cash and futures price)

The relationship of mailbox price to the USDA Class III Milk price is known as the mailbox differential. It is extremely important to estimate mailbox differential at sale time to more accurately predict whether the end result will be profit or loss. Of course, any estimated mailbox differential is just that – an estimate. The actual mailbox differential on a hedge is usually somewhat different from the estimate used when the put option was purchased.

4. Lenders and borrowers

Producers or processors who are borrowing money for their operations need lenders who understand what their marketing objectives are and who are willing to provide the necessary capital. They need knowledgeable brokers who understand the local dairy market, know what the mailbox differential is, and understand what their borrowers' goals are as hedgers.

15.8 (Soft) commodities as a hedge to your investment portfolio

In recent years it has been commented on or hinted at that there is an increase in the correlation between equity and commodity prices, and investment in commodity-related products has been blamed for this. First, such claims were investigated by looking at various measures of correlation. Next, they were assessed as to what the implications are of higher correlations between oil and equity prices for asset allocation. A time-varying Bayesian dynamic conditional correlation model was developed for volatilities and correlations and found that joint modeling commodity and equity prices produces more accurate point and density forecasts, which lead to substantial benefits in portfolio allocation. This, however, comes at the price of higher portfolio volatility. Therefore, the popular view that commodities are to be included in one's portfolio as a hedging device is not grounded when judging the performance since 2008.[41]

16

Commodities III: Speculation on (Agricultural) Commodity Derivatives[1] and Financialization of Commodity-Price Formation

16.1 Introduction

Billions of human beings rely on commodities to eat, heat, and commute. Brutal hikes in agricultural commodities in 2008 and 2011 – and to a certain degree on a continued basis ever since – caused malnutrition for hundreds of millions of people, and related "food riots." The rise of energy prices and agricultural-related commodity prices weighs on the daily lives of billions as well, as the part they take in monthly spending increases. For these reasons, the issue of excessive speculation in commodity derivatives has been covered by substantial reports and research.[2]

Everybody would agree with the fact that commodities futures markets exist solely for the benefit of bona fide physical hedgers, the producers and consumers of actual physical commodities. These markets do not exist for the purpose of speculation.

Agricultural markets represent one of the oldest forms of organized markets in the world, for obvious reasons. Metals and energy markets follow closely. Because of the crucial importance of these products, forward contracts (allowing for a deferred delivery) have existed for centuries as well. They allow producers and consumers of commodities to protect themselves against bad weather, price variations, and risk in general. One should look no further for an example of financial markets bringing value to society as a whole.

The hedging process goes as follows. Consumers or "transformers" (e.g., agro-industries or airline industries) are looking to secure their needs in agricultural commodities or oil. On the opposite side, producers want to protect their future revenues. A third category of participants, "traditional speculators." play the role of counterparties when consumers or producers do not find another commercial counterparty to hedge their risks. These speculators are looking for a remuneration of their risk by gaining from the underlying commodity's price fluctuation.

Central to protecting the price-formation mechanism of these markets is that speculators be restricted to a minority of participants: indeed, as long as this is the case, their projections will be based, although indirectly, on fundamental supply-and-demand factors, as these will determine the behavior of participants looking to hedge. When speculators gain a dominant position in a commodity-derivatives market, they base their projections on the potential behavior of other speculators, thereby disconnecting futures prices from fundamentals. Producers and consumers make commodities futures markets efficient, not speculators.

16.2 Impact of different players on the market

The orderly functioning of commodity derivatives markets, as described in the previous paragraph, is not just important to protect the price of instruments traded; it also has a direct impact on the price of the underlying (physical) commodity. Because commodity spot markets are so dispersed (due, among other factors, to the cost of transportation), they have for a long time relied on local supply and demand to determine prices. As consumption and production went global, the price on spot markets started to be based on futures prices. For most commodities today, the reference price is the futures price, adjusted to local supply-and-demand specificities.

This is a very important phenomenon to understand as it is different from what takes place on futures markets related to financial underlying assets. The price of a future contract related to a financial asset (equity, government bond) is derived from the price of the underlying asset and follows a relationship linked to the relative cost of carrying the future contract and the underlying financial asset. In the case of commodity futures, the relationship is, in most cases, inverted because buying the underlying physical commodity is much more difficult, cumbersome, and costly (transportation costs, storage costs, etc.) than buying a government bond or a basket of shares on the stock exchange. Contrary to financial futures on securities, commodity future prices find themselves in the position of driving the prices of the underlying assets.[3]

The fact that commodity futures markets drive the price of the underlying commodities, and in particular agricultural commodities, becomes highly problematic when, as shown by the data in Table 16.1, 69% of market open interest represents flows linked to speculation:[4] in such a situation, the fundamentals of the underlying markets become of second order in the price formation of commodity futures. Additionally, such speculative flows directly

Table 16.1 Market participants in the commodity markets

	Long open position, 1998 (%)	Long open position, 2008 (%)
Physical hedger	77	31
Index speculator	7	41
Traditional speculator	16	28

increase the price of the underlying physical commodities, as clearly demonstrated[5] by aggregate data.[6]

Thus speculative flows impact the price of commodities, physical and futures,[7] in two ways: by their proportion versus hedgers and by their very size (the more money is poured in, the higher the impact). Despite acknowledgement by investment firms themselves and conclusions from independent academic research, some industry members and trade associations deny that commodity-derivatives markets can affect the price of physical commodities.[8] They claim that the price of commodities is only driven by supply-and-demand fundamentals. While there is no denying that there is a long-term, gradual increase of commodities prices due to supply-and-demand fundamentals, the current speed of such increase and price volatility[9] is simply not linked to those fundamentals.

As regards volatility, empirical studies suggest that, in general, the introduction of derivatives trading in a commodity market leads to a reduction in the price volatility of the underlying product.[10] Derivatives markets allow investors to bring in their private information about the evolution of fundamentals; financial investors, although less informed, can improve market liquidity by acting as counterparts for hedgers. Both of these channels should contribute to the price-discovery process and reduce price volatility. However, an excessive level of financial trading could be suboptimal for market efficiency. As shown by anecdotal evidence and in survey studies,[11] a significant and growing proportion of financial traders may make their investment decisions irrespective of the physical market conditions – behavior that may add noise to the market and amplify price fluctuations. Holt and Irwin,[12] using private data provided by the Commodity Futures Trading Commission (CFTC), find a positive relationship between trading volumes of financial investors (large hedge funds and commodity-trading advisors) and price volatility in the commodity-derivatives markets. Nevertheless their analysis also confirms that speculative investors operate according to their private information on fundamentals, thus improving market efficiency,[13] while no support is found for the hypothesis of trend-following or noise-trading behavior.[14] A study by the IMF[15] identifies the following as the major determinants of persistent food-price volatility: the strength of real activity; the volatility of US inflation and exchange rate; and, with a much smaller effect, the total volumes of transactions in derivatives markets. Finally, Irwin and Sanders[16] find a negative relationship between index and swap-fund positions and market volatility.[17]

These divergent results are at least partially due to the lack of comprehensive and timely information on financial-investment flows and market fundamentals. Indeed, data on commodity investments are normally available only for organized US derivatives markets at weekly intervals (despite derivatives contracts being traded continually) and fail to differentiate positions according to their expiration date and type of investor. However, even the analyses conducted by the CFTC on the basis of their private information do not reach conclusive results (we can only say that they do not find any clear evidence in favor of an effect of financial investments on commodity futures prices).[18]

As a matter of fact, the economic relationship is hard to identify since a number of factors may simultaneously affect both investment decisions and prices, and furthermore, it is difficult to define a precise timing in the information flow (for instance, the statistics on market fundamentals are usually available less frequently than those relating to financial data). In the case of agricultural markets, controlling for new information and changes in fundamentals is further complicated by crops' seasonality and a wide range of potential demand shifters.

One of the main features of commodity futures markets over the last ten years has been the massive development of so-called structured products (often "wrapped" as exchange-traded funds (ETFs), exchange-traded notes (ETNs), or index swaps) linked to commodities, in particular to agricultural commodities. These products offer so-called investors an easy way to replicate passively the price evolution of a basket of underlying commodities. Notwithstanding the fact that to call money going into commodity-derivatives markets an "investment" is a misnomer (investment is about bringing money to productive use, and money going into commodity derivatives does not go to productive use) and should be called "betting" ("betting" being by definition a zero-sum game), commodity-index funds have two effects on commodity markets:

1. They have the mechanical effect of pushing prices up, as recognized by Wall Street firms themselves.[19] Commodity-index funds are sold by investment banks to all sorts of targets, from pension funds to mutual funds and to the retail market. Around $500 billion of commodity-index funds and structured products linked to commodities have been sold globally to these various pools of money, thereby converting them from their normal role as investors into commodity speculators, assisted by "zero-beta investment"[20] arguments and other marketing fallacies. At the risk of repeating ourselves, we want to reemphasize the point that no money coming from commodity-

index funds ever goes to commodity producers and that calling such funds "investment" funds is therefore a falsehood: the only proper name to describe commodity-index funds is "speculation" or "betting" funds. Assets allocated to commodity-index trading strategies have risen from $13 billion at the end of 2003 to $260 billion as of March 2008, and the prices of the 25 commodities that compose these indices have risen by an average of 183% in those five years.

2. They distort the price-discovery function of commodity futures markets, thereby making those markets significantly less useful for hedgers. This point is essential: the "financialization"[21] of commodity markets has the effect of making commodity futures markets less effective for their real economic purpose, which is the hedging of risk for natural (real) buyers and sellers of commodities. This phenomenon happens for the following reason: commodity-index speculators all behave according to one unique trading pattern, and this has a strong distorting impact on the price-discovery function of commodity futures markets as huge amounts of liquidity pour into passive long-only strategies. This in turn contributes to making commodity futures markets less and less economically useful for true hedgers.[22] While the traditional commodity speculator can bring liquidity[23] to the market, taking long and short positions based on price variations (thereby contributing to both increases and decreases in prices and being able to provide "the other side of the transaction" to hedgers), index funds always "consume" liquidity as they follow long-only strategies, buying systematically large quantities of commodity derivatives for long periods of time. Moreover, their replication strategy has the mechanical effect of pushing prices higher, thereby creating bubbles and feeding the self-fulfilling bullish prophecies. Another major impact of index funds, as demonstrated by the team of Professor Bar-Yam of the New England Complex System Institute, is the increase of volatility in physical markets. His research demonstrates that two factors play a special role in agricultural commodity-price increases: corn-to-ethanol conversion and speculation in commodity derivatives, which both distort supply-and-demand fundamentals. He goes on to demonstrate the link between hikes in food prices and social unrest in third-world countries. He claims: "We then analyzed the causes of food price increases. Many suggestions have been made: increased oil prices, adverse weather, rising meat consumption in China and dairy products in India, fluctuating currency exchange rates, conversion of corn to ethanol in the US, and investor speculation on commodities. We examined each of these factors and determined that only two were significant: corn-to-ethanol conversion and commodities speculation. To confirm our analysis, we built a non-equilibrium model of food prices that can represent the role of speculation. The model quantitatively agrees with the FAO price data. Recently we

> extended the model, enabling predictive validation against an additional 10 months' data. Ethanol production accounts for the underlying trend of price growth, while financial speculation is responsible for the price spikes in 2008 and 2011. Trend following by speculators causes prices to deviate from the supply-and-demand equilibrium. Supply and demand restoring forces eventually cause prices to reverse direction, but this only happens after a 6–12 month delay in agricultural futures markets. Our findings point to a need for urgent action to reduce hunger and prevent social disruption. Action is needed both to reduce ethanol conversion and the impact of speculation."[24]

It was highlighted before that there has been a boom of index funds, so-called "passive" investments in commodity derivatives in recent years. Evidence shows that these funds are being managed much more actively today as part of a recent increase in aggressive strategies, including structured products, hedge funds, and high-frequency trading funds. In recent research to highlight is Bicchetti and Maystre[25] from the United Nations who study the impact of high-frequency trading and trend-following strategies on commodity/equity markets correlations and increasing volatility in commodities markets. Reflecting on their findings and opinions, they write,

> Research and analyses increasingly support the view that the greater involvement of financial investors and their increased investments in commodities as financial assets have altered the functioning of commodity markets. Looking at intra-day and high frequency data allows us to get a better grasp of some recent technical developments that have affected the commodity markets. Considering the co-movement of the returns of the futures contracts of oil (WTI), corn, soybean, wheat, sugar, and live cattle with US stock market index (S&P 500 E-mini futures) at high frequencies, including 1-hour, 5-minute, 10-second and even 1-second, Bicchetti and Maystre (2012) find a synchronized structural break across commodity markets, which starts in September/October 2008 and continues until the latest observation of the dataset (end of 2011). Prior to 2008, high-frequency co-movements between commodity and equity markets did not usually differ from zero over a long-lasting period at such high frequencies. In the course of 2008, these correlations departed from zero and became strongly positive after the collapse of Lehman Brothers. The very existence of cross-market correlations at such high frequencies is consistent with the idea that recent financial innovations on commodity futures exchanges, in particular the high frequency trading activities and algorithm strategies, have an impact on these correlations. Trend-following strategies, for instance, typically try to benefit from upward and downward trends by herding.

> Although individually rational, the overall effect of trend following strategies may destabilize markets. This empirical evidence supports the idea that the financialization of commodity markets has an impact on the price determination process. Indeed, the recent price movements of commodities are hardly justified on the basis of changes of their own supply and demand. In fact, the strong correlations between different commodities and the S&P 500 at very high frequency are really unlikely to reflect economic fundamentals since these indicators do not vary at such speed. Moreover, given the large selection of commodities analyzed, one would expect to have different behaviors due to their seasonality, fundamentals and specific physical market dynamics. Yet, we do not observe these differences at any frequency. In addition, the fact that these correlations at high frequencies started during financial shocks provides additional support for financial-based factors behind this structural change. Therefore, the very existence of cross-market correlations at high frequencies favors the presence of automated trading strategies operated by robots on multiple assets.

They conclude, "This result is important for at least two reasons. First, it questions the diversification strategy and portfolio allocation in commodities pursued by financial investors. Second, it shows that, as commodity markets become financialised, they are more prone to external destabilising effects. In addition, their tendency to deviate from their fundamentals exposes them to sudden and sharp corrections."[26]

16.3 Response by the financial sector

No doubt, a response was expected from the other side. And since science used to validate arguments and science in matters of this nature is never fully conclusive, there is always room to maneuver. After all, correlation does not mean causality.[27]

The most recent and comprehensive response to the allegations formulated above came from the IFF (Institute for International Finance), the largest organization representing the aggregate interest of the global banking sector. In their 2011 report,[28] they answered most of the critiques reflected in the reporting backed by academic research findings.[29] In the report, they stress there is little convincing evidence linking financial investment with trends in commodity prices and volatility ("no causality"). While there have been periods of correlation (often attributed to "herding behavior"), in recent years, including among previously uncorrelated markets, researchers have not documented a clear causal link between financial investments and commodity prices. The report advances that there is however evidence for a strong link between commodity prices and fundamental supply-and-demand factors. Higher commodity prices and volatility should be seen in the context of a steady increase in demand (mainly from emerging markets) and periodic supply-chain constraints. The following key take-aways can be highlighted:

1. A review of the academic literature and studies by official-sector bodies suggests that despite periods of correlation, there is little evidence of a causal link between financial investment in commodities and trends in commodity prices and volatility.
2. These same studies broadly support the well-established view that commodity-price trends and volatility continue to be driven far more by market fundamentals. In recent years, rising demand from emerging markets has contributed to the trajectory of commodity prices and volatility, exacerbated by periodic or structural supply constraints (e.g., impacts of inclement weather, demand for biofuels, infrastructure bottlenecks).
3. Financial investment (sometimes referred to as "speculation") does not take place in the absence of fundamentals; rather, it allows new information in spot and futures commodity markets, such as changes in fundamental supply-and-demand-related factors, to be processed. Constraints on financial investment could dampen price signals to suppliers – and hence the supply response.
4. It is important to bear in mind that financial investment is an integral part of commodities trading.
5. Measures to enhance the transparency of data provision to regulators on prices, trading activity, and factors affecting the supply and demand of individual commodities are broadly welcomed. However, the imposition of additional regulations such as position limits on trading activity could impair market liquidity and efficiency. Commercial-market participants who need to hedge their exposures (e.g., heavy users of commodities such as airlines or food manufacturers) will tend to hold net short positions; on the other side of the transaction, financial investors such as commodity-index traders generally take long-only positions as a hedge against inflation and for portfolio diversification. Both sides thus benefit – financial investment provides essential market liquidity and counterbalancing positions. The ability to hedge against inflation and invest in commodities as part of a diversified portfolio reflects the social utility of financial investment in commodities and is of particular value for long-term investment vehicles, such as retirement plans.
6. By far the most effective way to tackle the problem of excessive commodity-price volatility and upward pressure on prices is to directly address underlying supply/demand imbalances. Measures should include steps to alleviate supply constraints, for example, by removing restrictions on the supply or export of key commodities, or by investment in productive capacity.

The bottom line is that financial investing (so not the hedging part) in (agricultural) commodities is good, as liquidity is needed: it contributes to the price-discovery function; and it has no structural impact on commodity prices, although temporary volatility is possible due to herding behavior in the market. It remains however to be seen if the additional liquidity provided is required and leads to an enhanced price-discovery function.

16.4 The debate rages on

In a last leg of the debate for now, the UN responded to the arguments, not directly but within the framework of future policy design.[30] Growing demand from large developing economies and frequent supply shocks, such as adverse weather and export bans, are generally accepted as more tangible factors that explain volatility rather than the hundreds of billions of dollars of bets placed on expectations of temporarily rising prices. Despite a growing body of evidence on the destabilizing influences emanating from financial markets, the "real economy" explanations still dominate the debate. It is not commonly recognized that demand from financial investors in the commodity markets has become overwhelming during the last decade. Of course, supply-and-demand shocks can still move commodity prices time and again. But with the volumes of exchange-traded derivatives on commodity markets now being 20 to 30 times larger than physical production, the influence of financial markets has systematically transformed these real markets into financial markets.

Let's break the arguments down a bit. Price volatility has long been a major feature of commodity markets, given the tightness in many global commodity markets and the inelasticity of demand. While commodity-specific shocks have played a key role in the past, especially on the supply side and in the oil market, this factor lacks persuasive power today. When political shocks occur, the biggest oil producers undertake remarkable efforts to stabilize prices and compensate for falling supply by stepping up production in other areas. Rapidly but steadily growing demand for a range of commodities, especially in emerging economies, does not explain the huge swings recorded in many of these markets from quarter to quarter.

The new element is the presence of the financial investor, and that is happening at an accelerating pace. This resulted in a significant increase of commodity assets under such management, from less than $10 billion around the end of the last century to a record high of $450 billion in April 2011 (based on the report they are responding to[31]). Consequently, the volumes of exchange-traded derivatives on commodity markets are now 20 to 30 times greater than physical production.[32] These investors treat commodities as an asset class, which means that they are betting on a certain price trend during the period they are invested in commodity assets. They do not trade systematically on the basis of fundamental supply-and-demand relationships in single markets, even if shocks in those markets may influence their behavior temporarily. In general, however, their decisions to buy and sell are rather uniform (herding) and are driven by the same kind of information that is available for other financial markets. As they hold by far the largest positions in the commodity markets, it is undeniable that they exert considerable influence on the price movements of those markets. I guess that is, in my humble opinion, the weakest point in the argument. I can perfectly imagine that high volumes dictate certain price trends, as they do in other financial markets as well. However, those price trends are always temporary and always reverse when cyclical aspects or externalities come into the equation. I cannot, however, explain the longer-term persistent trends. It could be that financialization accentuates existing underlying dynamics and volatility. Maybe that is an argument in itself for regulators to master the market more than currently is the case. The question that remains is whether there a long-term impact on prices and volatility in agricultural commodity markets even if financialization and financial investors imply significant amounts moving in and out of the markets on a regular basis. Proving such a causal relationship is almost impossible, and therefore the argument has been built around correlation. There is however a UNCTAD discussion paper revealing the co-movements between returns on several commodity markets and those on the US stock market over the 1997–2011 period using tick data (i.e., market data that show the price and volume of every print), although such co-movements are what lawyers call circumstantial evidence. However, they can explain the demand/supply relation when dealing with short-term (ST) volatility.[33] In essence the paper documents striking similarities in the evolution of rolling correlations between the returns on several commodity futures and ones on the US stock market, computed at high frequencies. It also highlights a structural change that took place recently in these markets. Prior to 2008, high-frequency co-movements between commodity and equity markets did not usually differ from zero over a long time period at such high frequencies. In the course of 2008, these correlations departed from zero and became strongly positive after the collapse of Lehman Brothers. The authors indicate,

> The persistence of this trend until December 2011 – except for crude oil in early 2011, which coincides with the uprising in Libya – remains difficult to explain. Further research is needed to get a complete understanding of the mechanisms at work behind this structural change. Yet, given the high frequencies, we think that HFT strategies, in particular the trend-following ones, are playing a key role. We believe a conjunction of factors made that change possible. First, financial technical innovation spurred HFT through the gradual introduction of full electronic trading on exchange platforms

> since 2005. Second, investors moved away from passive strategies and opted for active ones when the rising trends on equity and commodity markets stopped, in particular since the fall of 2008. Third, lasting uncertainties and positive feedback effects reinforced this trend.
>
> In our view, this finding adds to the growing empirical evidence supporting the idea that the financialization of commodity markets has an impact on the price determination process. Indeed, the recent price movements of commodities are hardly justified on the basis of changes of their own supply and demand.
>
> In fact, the strong correlations between different commodities and the S&P 500 at very high frequency are really unlikely to reflect economic fundamentals since these indicators do not vary at such speed. Moreover, given the large selection of commodities we analyze, we would expect to have different behaviors due to their seasonality, fundamentals and specific physical market dynamics. Yet, we do not observe these differences at any frequency. In addition, the fact that these correlations at high frequencies started during the financial shocks provides additional support for financial-based factors behind this structural change. Therefore, the very existence of cross-market correlations at high frequencies favors the presence of automated trading strategies operated by computers on multiple assets. Our analysis suggests that commodity markets are more and more prone to events in global financial markets and likely to deviate from their fundamentals.
>
> This result is important for at least two reasons. First, it questions the diversification strategy and portfolio allocation in commodities pursued by financial investors. Second, it shows that, as commodity markets become financialized, they are more prone to external destabilizing effects. In addition, their tendency to deviate from their fundamentals exposed them to sudden and sharp corrections.[34]

The research of Bicchetti and Maystre already mentioned, as well as UNCTAD proprietary research,[35] is therefore built around correlation. Significant positive co-movements of the returns of futures contracts for oil and a broad range of other soft commodities with futures contracts on the US stock market at high frequencies appear at up to one-second intervals. Clearly these very short-term commodity-price movements cannot be justified by changes in supply and demand in specific markets. Fundamentals for the US stock market and commodities markets differ greatly, and different fundamentals cannot induce similar price movements simultaneously, continuously, and consistently for the past few years across all the markets investigated. Indeed, given the large selection of commodities considered, different behaviors would have been expected due to the seasonality, industrial usage, and specific physical-commodity market dynamics. However these differences have not been observed. While fundamentals cannot explain these price co-movements, the stock market and commodities do share one common, critical feature: the dominant position of financial investors.

Algieri (2012)[36] investigated the dynamics of primary commodity prices and the role of speculation over time. In particular, the relationship between speculation and price volatility on the one side, and the linkage between excessive speculation and price volatility on the other side, was examined, with the scope to establish whether volatility drives speculation or speculation drives price volatility, or whether there are no linkages between the two variables. In order to identify the presence of any lead-lag relationships, two batteries of Granger causality tests were carried out for the period 1995–2012. Algieri concludes, in contrast to several academic studies (supra) that reject any causal relationship between the two variables, that excessive speculation drives price volatility, and that often bilateral relationships exist between price volatility and speculation. In addition, the lead-lag relationships are found not for the entire sample period from 1995 to 2012, but only when small subperiods are taken into account. It turns out, in fact, that excessive speculation has driven price volatility for maize, rice, soybeans, and wheat[37] in particular time frames, but the relationships are not always overlapping for all the considered commodities.

Although it might not be possible to evidence the long-term impact of financial investors on commodity prices, the sources of short-term volatility can be traced back to financial investors. That creates another problem, which goes back to the essential functioning of the capital markets, that is, as an effective price-discovery function. Due to the increased participation of financial players in those markets, the nature of information that drives commodity-price formation has changed. Contrary to the assumptions of the efficient-market hypothesis, the majority of market participants do not base their trading decisions purely and independently on the fundamentals of supply and demand; they also consider aspects related to other markets or to portfolio diversification. This introduces spurious price signals into the market.

Moreover, in a situation of widespread herding in financial markets, the assumption of an atomistic market, in which participants trade individually and independently of each other on the basis of their own interpretation of fundamentals, no longer holds. The price-discovery market mechanism is seriously distorted.[38] Prices can move far from levels justified by the fundamentals for extended periods. Because of these distortions, commodity prices in financialized markets do not provide correct signals about the relative scarcity of commodities. This impairs the allocation of resources and has negative effects on the real economy.

For the sake of completeness, it can be mentioned that there are two types of volatility in focus during the analysis of agricultural commodity prices:

- Historical (realized) volatility, indicating a volatility of an asset in the past, is based on observed (realized) movements of price over a historical period. It represents past price movements and reflects the resolution of supply-and-demand factors.
- Implicit volatility,[39] is the market's view on how volatile an asset will be in the future. It represents the market's expectation of how much the price of a commodity is likely to move and tends to be more responsive to current market conditions.

Intermittently in this discussion a wider set of literature has been released contributing to the debate.[40]

I may add that a similar dynamic has taken place in real-estate markets in recent decades. It used to be so that real-estate investments were illiquid and an inflation hedge relative to other asset classes from which it distinctly behaved. As we entered the 1980s and 1990s, more and more real estate became liquefied as real-estate companies went public, and more importantly real estate as an asset class became available through REITs (real-estate investment trust') and other listed vehicles. Over time, this has been leading to a real-estate market that behaves and moves similar to other public markets.[41]

16.5 Getting technical: Correlation versus causality

The existence of a simultaneous correlation between futures returns and changes in investment positions held by financial investors does not prove any causal relationship (the typical risk included in "post hoc ergo propter hoc"). No clear conclusion can be drawn without an underlying structural model. Nonetheless, detecting whether positions tend to lead price movements or instead follow them surely helps to identify the underlying economic mechanism. For instance, trend-following strategies (i.e., investors buy when the price rises and sell when it declines) signal some degree of market inefficiency, as future investments are decided on the basis of past information; on the other hand, futures returns (which are obtained through market-clearing prices, for which long positions are the same as short positions) could follow investments made by a certain group of investors because they may have private information, and other investors tend to follow their actions. Based on preliminary analysis, there is no evidence that demonstrates that position changes by financial traders may have triggered large price swings, as futures returns were found to be uncorrelated with their lagged values (if a self-reinforcing mechanism going from prices to positions and then back to prices was in place, returns would have been found to be autocorrelated).

There is a need for an empirical framework able to accommodate the necessary conjectures and test them. Lacking theoretical restrictions, the relationship is explored between prices and investments by means of a vector autoregressive (VAR) model containing futures returns and changes in positions held by money managers and swap dealers. Long and short positions are reported separately. The optimal number of lags has been determined according to selection criteria (the Akaike information criterion and Schwarz criterion), on the basis of which we have included only one-period (one-week) lagged explanatory variables for all commodity markets. Consequently, the model is:

$$y_t = \gamma 0 + \Gamma y_{t-1} + e_t$$

With the vector of endogenous variables:

Yt = (rt, MMlong t, M short t, Swaplong t, Swapshort t),

where Mmlong t (Mmshort t) represent changes in the shares of money managers' long (short) positions (relative to the open interest) at time *t* and swaplong t (swapshort t) are similarly defined for swap dealers' long (short) positions; *rt* is for futures returns; *et* is the reduced-form error term.

Following that, there is a need to perform a battery of Granger causality tests[42] in the attempt to identify the temporal direction of the relationship which breaks down in two parts, (1) the direction of causality going from investor positions to futures prices, and (2) the effects in the opposite direction. Granger causality exists when one fails to reject the null hypothesis.

At first glance, the most recent results show quite sparse evidence of causation going from investor positions to futures prices, with no systematic pattern emerging across the various markets. Changes in long positions held by money managers tend to reduce future returns in the market of feeder cattle and live cattle, but they tend to raise them in the corn and wheat markets. A rise of swap dealers' share on total long positions leads the increase of futures returns in three markets (cocoa, soybeans, and soybean oil). Overall results confirm previous findings[43] of no systematic influence of financial investments on agricultural commodity futures prices.

With respect to the possibility of reverse causality, Granger tests identify many more markets where future returns tend to anticipate changes of financial investors' positions, especially those of money managers. In six markets (cocoa, coffee, feeder cattle, soybean oil, sugar, and wheat-Kansas), the group of money managers increased its long exposure along with future return improvements. Swap dealers instead appear less influenced by returns evolution; just in the market of live cattle, following a future returns improvement, they increase their net exposure. These results seem in line with the conclusions reached by the IMF,[44] according to which speculative investments tend to follow price variations in a number of agricultural and industrial commodity markets.

Results from Granger tests may also suggest the existence of reciprocal causality: the existence of a self-reinforcing mechanism that goes from returns to investments and then back from the latter to the former, with this feedback effect triggering possible price spirals. Based on the evidence, this seems to occur very rarely: only between positions of money managers and futures returns in the cocoa market. The empirical analysis is limited by the impossibility of going beyond statistical causality without theoretically based structural restrictions. Nonetheless, overall the evidence seems to support the idea that financial investors, by following price trends, could implement possibly noisy investment strategies, with their behavior revealing the existence of some degree of market inefficiency.

For a given level of uncertainty about fundamentals, we expect smaller price variability in more efficient markets, as all the available information will be promptly incorporated in equilibrium prices. This is why a vast stream of literature has considered price volatility as the main benchmark for evaluating the efficiency performance of trading in the derivatives market.

If investments by financial investors contribute to a prompt inclusion of new information in the market, we should expect a stabilizing effect on futures price dynamics. However, trend-following behavior, of which some evidence is found consistently across research, might induce price overshooting (or undershooting) following an unexpected shock before prices adjust toward the new fundamental equilibrium. In a recent theoretical contribution, Basak and Pavlova[45] show that index investors, by taking higher exposure to risky assets, increase the volatility of the assets included in the indices.

Borin and Nino (2012) first show that for all of the 11 agricultural commodity-derivatives markets investigated (over the period from June 2006 to September 2011), futures price returns appear to be simultaneously correlated with money-manager investments; for swap dealers' positions, however, evidence of such a correlation is much weaker. Then, as in previous studies, they examine this relationship more closely by means of Granger causality tests in a VAR model. They find that higher future returns lead to additional money-manager investments in around half of the 11 markets while evidence of the opposite effect, going from higher positions to futures returns, is rather scant. This result seems to indicate that "traditional speculators" react to price changes rather than cause them; it could also imply some trend-following behavior by money managers, which in turn may uncover a certain degree of market inefficiency. Finally, concerning the possible relationship between investment positions held by financial traders and futures price volatility, they employed a GARCH model where the lagged values of investment positions enter both the mean and the variance equations. This analysis confirms that money managers and swap dealers may play different roles: while the positions held by money managers tend to reduce price volatility in a large number of markets, the evidence on swap dealers is more mixed, and their investments seem to amplify price volatility in some markets. Although the analysis suffers from severe problems of data limitations and the above evidence is not overwhelming, nonetheless this result gives some support to the idea that swap dealers, whose growing weight in the regulated exchanges tends to reflect the large exposures of "commodity-index investors" in the OTC markets, may have a destabilizing impact on futures prices, at least in the short run. On the contrary, the activity of more traditional speculators seems to favor price stability, probably enhancing market liquidity.

Even assuming that an excess of financial investments in commodity markets might be harmful, one still needs to devise policy responses that could be effective in curbing its undesirable effects[46] while preserving market efficiency. Effective regulation may be developed only by improving our knowledge of market mechanisms, which is now limited by the lack of appropriate statistics. As agreed by policy makers in international fora, this can only be achieved by increasing transparency and making available more detailed information both on regulated and OTC financial markets, as well as on physical fundamentals.[47] Transparency and media coverage itself have been looked at as a driver of commodity prices.[48] More specifically, the effects of media coverage of commodity-price increases and decreases on the price of the commodity were examined and how media coverage in other commodities affects prices. Evidence was provided of the relationship between media coverage and its intensity to the price level of agricultural commodities and oil futures.

Further conclusions were that price movements are correlated with the media coverage of up movements, or increases in prices. The direction of the correlation is robust and positive for media coverage of increases in prices, and negative for decreases in prices. These results point to increases in prices being exacerbated by media attention by 8%. In addition, interesting countervailing effects of this reinforcing price pressure were identified due to media activity in the previous days. Finally, even though volatility is higher for the set of days where there is media coverage, this hides important dynamics between media coverage and volatility. The volatility of market-adjusted returns is negatively correlated with media coverage, both up and down. Market days with intense media coverage of commodity prices tend to have lower volatility. Overall, the study concluded, there seems to be no direct relationship ("no robust evidence") observable between speculation of agricultural futures and a potentially destabilizing effect on spot prices.[49]

A very recent study looked at bubble formation in the food-commodity markets over a 42-year period,[50] which is much longer than studies performed up until then. They used the daily prices from individual futures contracts to test whether speculative bubbles exist in agricultural futures markets and to identify whether patterns of bubble behavior exist over time. The test results demonstrate that all 12 agricultural markets experienced multiple periods of price explosiveness. However, bubble episodes only represent a very small portion of the price behavior for the 42-year period. In addition, most of the bubbles are short-lived, lasting fewer than 20 days. They also found that explosive periods are more common with longer durations in the first half of the analyzed sample period, indicating that the most recent bubble episodes (2007–2008) may not have been as severe as the mid-1970s episode. Though receiving far less public attention, results from this study suggest that negative bubbles contribute significantly to price behavior, accounting for more than one-third of total bubble episodes. In general, the size of the bubble (return from bubble extremum to end of bubble) increases as the bubble signal (return from start to bubble extremum) gets larger for both positive and negative bubbles. It appears that during the last half of the sample period, with a couple of exceptions, agricultural futures markets have adjusted faster to stochastic shocks than in the first half of the sample period, resulting in fewer bubbles of shorter length and less overreaction.

16.6 Is there something bigger going on?

The answer to this question is maybe yes. Everyone who has been following the financial markets for more than 15 years has been experiencing that in recent years, stock market crashes are coming in at a faster pace. After 1987, we had to wait until 2002, then to be followed by the 2008 meltdown. In the run-up to the 2002 crisis, large amounts of liquidity were absorbed by the tech space. After 2002, central banks and in particular the Federal Reserve has been pushing massive liquidity in the market to stimulate economic activity. That liquidity had to find a home, and it did. In particular emerging markets and real estate were the favored domains. It was also these domains that were hit hard when the 2008 Great Recession arrived. Once again, the crisis was managed with even larger amounts of liquidity through all sorts of facilities (QE 1, 2, and 3, QE infinity, Operation Twist, OMT, LTRO, ESF, etc.). Once again that liquidity had to find a home, and once again it did so in the commodity and in particular the agricultural commodity space. And not without impact, as Figure 16.1 displays. In particular, observe the price impact and increasing volatility in 1972, the year in which the Bretton Woods system was abolished as well as the period starting in 2006 when Bernanke took office for his first period as chairman of the Federal Reserve after President George W. Bush appointed him earlier that year. That was followed by a second four-year tenure endorsed by President Barack Obama starting in early 2010.

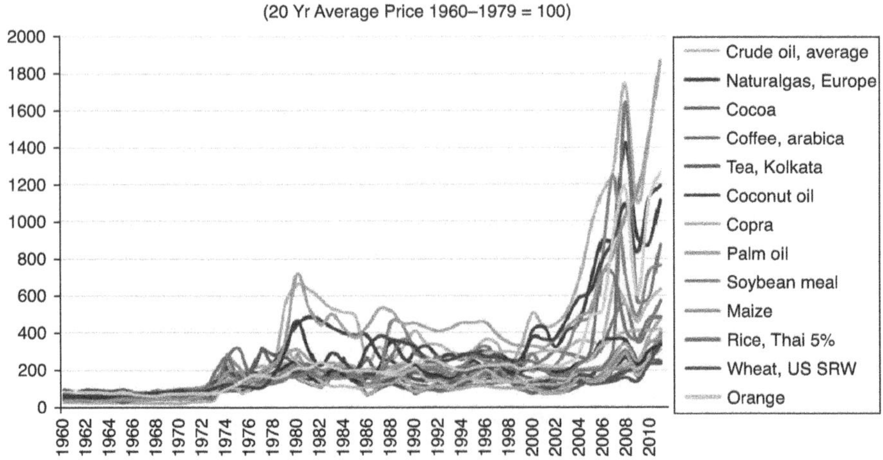

Figure 16.1 Commodity-price evolution 1960–2011
Source: World bank Commodity Price data (Pink Sheet).

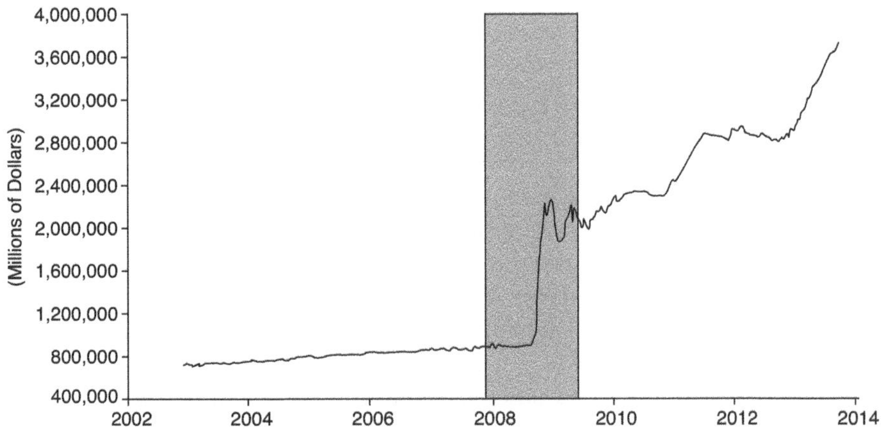

Figure 16.2 Balance-sheet evolution of the FED until September 2013
Note: Shaded areas indicate US recessions.
Source: Board of Governors of the Federal Reserve System.

It therefore felt somewhat disconnected when Ben Bernanke before Congress (March 2013) claimed that he had the best inflation record of all recent FED chairs. But even further, it helps to explain the upward pressure on commodity price. Behind the financial investor, there is an even larger force operating that has an (unintended) impact on (agricultural) commodity markets. For

Table 16.2 ECB and other major CBs' balance-sheet evolution

Central banks' balance sheets as a % of IMF nominal GDP (year-end)					
	2005	2007	2009	2011	2013 (YTD)
US	6, 5	6, 2	15, 3	18, 1	22, 6
UK	6, 6	7, 3	17, 9	15, 3	24, 5
Japan	23, 2	21, 2	25, 3	30, 2	36, 2
Eurozone	13, 3	17, 4	21, 1	29, 2	32, 4

Source: Thomson Reuters data stream and proprietary research.

reference, the evolving balance sheet of the Federal Reserve, ECB, and some relevant countries has been included in Figure 16.2

16.7 Regulatory responses and initiatives

Initially within the G20 and later per country or on a regional (EU) basis, policy and regulatory responses have been formulated as follows.

16.7.1 The EU

After the financial crisis, the EU responded in its traditional white-heated way, prepared to further push the markets into its regulatory hell. In its 2010 road map for reform, the European Commission said the key causes of the financial crisis included the following:

1. an unstable and inadequately supervised financial system;
2. opaque financial operations and products;
3. irresponsibility in some financial institutions, which pursued short-term profits, neglected risk management, and paid unjustifiable bonuses; and
4. a financial system that overlooked the fact that it was supposed to serve the real economy and society as a whole, contributed to the creation of bubbles, and often disregarded consumer interests.[51]

More specifically for the agricultural space, the European Commission published the working paper "High Prices on Agricultural Commodity Markets: Situation and Prospects. A Review of Causes of High Prices and Outlook for World Agricultural Markets,"[52] on translated in more general terms in the proposal for a regulation of the European Parliament and of the Council on OTC derivatives, central counterparties and trade repositories.[53] The proposal, after modification, was approved in 2012[54] and became Regulation (EU) No. 648/2012 of the European Parliament and the Council on July 4, 2012, and came into effect in August 2012, known as EMIR. In December 2012, the

European Commission further adopted nine regulatory and technical standards to complement the obligations defined under the regulation on OTC derivatives, central counterparties, and trade.[55] They were developed by the European Supervisory Authorities and have been endorsed by the European Commission without modification. The Commission Delegated Regulations (EU) Nos. 148/2013 to 153/2013 of December 19, 2012, supplementing EMIR were published in the *Official Journal* on February 23, 2013, and entered into force on March 15, 2013.

It needs to be stressed that there were no specific rules implemented to tackle any of the issues discussed before in the commodities space. There were a number of reasons for this: (1) inadequate performance by the regulators worldwide; (2) the problem needs to be tackled in coordination between regulators, which we know is a problem because commodity markets are global; and (3) most agricultural commodities are traded outside the EU and therefore are outside the regulatory reach. The aforementioned regulation, as was indicated, only covers the OTC market. There are very few agricultural commodities traded outside the normal exchange-based platforms. That is because the agricultural futures and forwards are standardized contracts and the exchange also functions as a clearinghouse that monitors counterparty risk. Consequently, the OTC regulation adds little to nothing to the issues discussed before as they emerged in the agricultural commodities space.

The most common suggested measures to tackle the aforementioned issues follow:

1. Implement a European consolidated regulatory position-reporting system, including positions resulting from OTC and regulated trading of commodities and commodity derivatives.
2. Define hedging positions – objectively reducing risk directly relating to commercial activities – as opposed to speculative positions.
3. Define and implement ex ante individual limits on speculative positions on commodity-derivatives markets (resulting from a transaction executed OTC or on a regulated venue) as a percentage of the total market – for example, 2.5%, as a means to have at least 40 market participants, limiting the risk of market abuse.
4. Define and implement ex ante market limits on speculative positions on commodity-derivatives markets (resulting from a transaction executed OTC or on a regulated venue), as a percentage of the total market – for example, 30%, as a means to protect the hedging function and the quality of the price-formation mechanism of these markets from the detrimental effect of excessive speculation.
5. Limits (individual and market) must apply for all months of the contract (not just the spot month) and to both cash and physically settled contracts.

6. Position-management arrangements have failed to prevent market abuse and do not have the purpose of limiting speculation. They are thus an inadequate alternative to position limits. However, used alongside limits, it may provide regulators with an additional tool with which to oversee the markets.
7. Prohibit all financial products offering commodity-index replication.

The EMIR directives objectives that focus on financial derivatives in general are as follows:

- The new rules objectives are to increase transparency in the OTC derivatives market and to make it safer by reducing counterparty credit risk and operational risk.
- To increase transparency. The new rules requires that (1) detailed information on OTC derivative contracts entered into by EU financial and nonfinancial firms be reported to trade repositories and made accessible to supervisory authorities, and that (2) trade repositories* publish aggregate positions by class of derivatives accessible to all market participants. In the course of the negotiations, the scope of the proposal has been widened to cover the reporting of both listed (i.e., non-OTC) and OTC derivatives.
- To reduce counterparty credit risk. The new rules introduce (1) stringent rules on prudential (e.g., how much capital they need to hold), organizational (e.g., the role of risk committees), and conduct-of-business standards (e.g., disclosure of prices) for CCPs**; (2) mandatory CCP clearing for contracts that have been standardized (i.e., they have met predefined eligibility criteria); and (3) risk-mitigation standards for contracts not cleared by a CCP (e.g., exchange of collateral).
- To reduce operational risk. Operational risk is the risk of loss resulting from inadequate or failed internal processes, people and systems, or external events. The proposal requires the use of electronic means for the timely confirmation of the terms of OTC derivatives contracts. This allows counterparties to net the confirmed transaction against other transactions and ensure accurate bookkeeping.

*A trade repository is a central data center where details of derivatives transactions are reported. Trade repositories are commercial firms. There are global trade repositories for credit, interest rate, and equity OTC derivatives, and soon for commodities and foreign exchange.

**A CCP (central counterparty) is an entity that interposes itself between the two counterparties to a transaction, becoming the buyer to every seller and the seller to every buyer. A CCP's main purpose is to manage the risk that could arise if one counterparty is not able to make the required payments when they are due – that is, defaults on the deal.

CCPs are commercial firms. There are currently about a dozen CCPs, all but one located in Europe or the United States, clearing interest rates, credit, equity, and commodities OTC derivatives. There is currently no CCP clearing foreign exchange OTC derivatives.

The main obligations under EMIR are the following:

- Central clearing for certain classes of OTC derivatives
- Application of risk-mitigation techniques for non-centrally cleared OTC derivatives
- Reporting to trade repositories
- Application of organizational, conduct-of-business, and prudential requirements for CCPs
- Application of requirements for trade repositories, including the duty to make certain data available to the public and relevant authorities

The following *entities* are covered by the different provisions in EMIR:

Financial counterparties and nonfinancial counterparties above the clearing threshold	>	Clearing obligation Risk-mitigation techniques Reporting obligation
Nonfinancial counterparties below the clearing threshold	>	Reporting obligation Certain risk-mitigation techniques (timely confirmation, portfolio reconciliation and compression, dispute resolution)
CCPs	>	CCP requirements
Trade repositories	>	TR requirements

The following *instruments* are covered under the different provisions of EMIR:

OTC derivatives	>	Clearing obligation and risk-mitigation techniques
All derivatives	>	Reporting obligation
All financial instruments	>	CCP requirements

The obligation to clear OTC derivatives contracts through a CCP and report derivatives to trade repositories will apply to financial firms (banks – both universal banks and investment banks, insurance companies, funds, etc.) and to nonfinancial firms (energy companies, airlines, manufacturers, etc.) that have large positions in OTC derivatives. New contracts, changes to existing contracts, and termination of contracts need to be reported no later than the working day following the relevant event (execution/modification/termination).

The proposal provides for some limited exemptions from clearing and reporting requirements for nonfinancial firms.

- Contracts by nonfinancial firms

Contracts by nonfinancial firms below a "clearing threshold" will not have to be cleared through a CCP. "Commercial and treasury hedging activities," that is, when these firms use OTC derivatives to hedge risks related to their activities, will be subtracted from the firm's overall position, which means that they will not count toward the threshold set for the clearing obligation. These activities do not need to be cleared. For example, commercial hedging could be when airlines using OTC derivatives to secure the price at which they buy fuel, or when exporters use OTC derivatives to shield themselves from fluctuations of exchange rates.

These thresholds are not set out in the proposal. ESMA, the European Securities and Markets Authority, together with ESRB, the European Systemic Risk Board, and other relevant authorities have drafted technical standards on what these thresholds should be.

The value of the clearing thresholds were defined as follows:[56]

- EUR 1 billion:* Credit derivative contracts
- EUR 1 billion:* Equity derivative contracts
- EUR 3 billion:* Interest-rate derivative contracts
- EUR 3 billion:* Foreign-exchange derivative contracts
- EUR 3 billion:* Commodity derivative contracts and others

* in gross notional value

When setting these thresholds, ESMA will have to take into account the systemic relevance of the sum of net position and exposures by counterparty per class of derivatives (i.e., looking at how much overall risk they pose to the system).

Not all OTC derivative contracts count toward the clearing threshold. Those OTC derivative contracts entered into in order to reduce risks relating to the commercial or treasury financing activity of the nonfinancial entity, or of nonfinancials of the group it belongs to, are excluded from the calculation of the clearing threshold. Criteria to determine those contracts are specified in the technical standards mentioned above. All other OTC derivative contracts entered into by the nonfinancial or other nonfinancial entities of the group are taken into account for the calculation of the clearing threshold. When the amount for one class of OTC derivative contracts is surpassed, you exceed the clearing threshold.

Example of application of the clearing threshold (in gross notional value):

Nonfinancial company A

Total credit-derivative contracts 2 billion

- Hedging 1.8 billion
- Nonhedging 0.2 billion

Company A does not exceed the clearing threshold. Its nonhedging credit-derivative contracts (2 billion) is below the value of the clearing threshold for that class of OTC derivatives (1 billion).

Nonfinancial company B

Total equity derivative contracts 1.5 billion

- Hedging 1.2 billion
- Nonhedging 0.3 billion
- Total interest-rate derivative contracts 3.5 billion
- Hedging 0
- Nonhedging 3.5 billion

Company B exceeds the clearing threshold. Its nonhedging activity in interest-rate derivative contracts is above the value of the clearing threshold for that class of OTC derivatives (3 billion).

- Financial institutions involved in the management of public debt

Members of the European System of Central Banks (ESCB), public bodies charged with or intervening in the management of the public debt, and the Bank for International Settlements will not be subject to the clearing, reporting, or bilateral risk-mitigation obligations in order to avoid limiting their powers to intervene to stabilize the market if and when required.

- Pension funds

Besides, a temporary exemption from central clearing for pension funds (until August 5, 2015) has been introduced in the course of the negotiations. This is aimed at ensuring that pension funds do not incur disproportionate costs that could ultimately impact EU pensioners. Once the industry has developed the appropriate technical solutions for the provision of noncash collateral as variation margins by pension funds, they will be subject to central clearing. In the interim, pension funds will have to exchange collateral for their OTC derivatives.

- Intragroup transactions

In addition, an exemption from the clearing obligation for transactions entered within a group of financial firms, nonfinancial firms, or a mix of financial and nonfinancial firms has been introduced ("intragroup exemption"). This exemption is necessary because requiring clearing of intragroup transactions could substantially increase the capital and liquidity required by firms that centralize risk management in certain entities as well as increase operational complexity. However, to ensure that the exemption does not increase systemic risk, EMIR will require that intragroup exempted transactions be subject to bilateral collateralization unless two conditions are met: there is no current or foreseen practical or legal impediment to the prompt transfer of own funds and repayment of liabilities between the counterparties, and the risk-management procedures of the counterparties are adequately sound, robust, and consistent with the level of complexity of the derivative transactions.

Which OTC derivatives contracts are earmarked for mandatory clearing

To have as many OTC contracts as possible cleared through a CCP, the regulation introduces two approaches to determine which contracts must be cleared:

- A "bottom-up" approach – Where a competent authority has authorized a CCP to clear a class of derivatives, it will inform ESMA, who will assess whether a clearing obligation should apply to that class of derivatives in the EU, and develop draft regulatory technical standards that will have to be adopted by the Commission.
- A "top-down" approach – ESMA, on its own initiative and in consultation with the European Systemic Risk Board, will identify contracts that should be subject to the clearing obligation but for which no CCP has yet received authorization.

The top-down approach will ensure that if no CCP clears a product that should be subject to the clearing obligation, there are tools available to regulators to get the product cleared through a CCP. It will also ensure that new products will not fall through the net.

ESMA will use the following criteria when determining eligibility for the clearing obligation: the degree of standardization of the contract and operational processes; liquidity and the volume of contracts; and availability of fair, reliable, and generally accepted pricing information.

Counterparty credit risk, as indicated, is one of the main detriments of the OTC market and has created a very concentrated market, as only large institutions that know each other well have regular dealings that engage in those transactions with each other.

Before the regulation was put in place, participants in the OTC derivatives market do not collect sufficient collateral (i.e., guarantees; usually they are in the form of cash or securities) to mitigate counterparty credit risk, which refers to the risk of loss arising from one party not making the required payments when they are due.

The regulation requires that OTC derivatives that are standardized will have to be cleared through CCPs. Since an OTC derivative contract cleared by a CCP usually involves the posting of higher amounts of collateral than an equivalent contract that is not cleared by a CCP, this will increase the amount of collateral held in the system.

As collateral will be held in a few places, there is an argument that risk will be concentrated there. To avoid CCPs becoming a source of risk to the financial system in themselves, the proposal introduces stringent conduct-of-business, organizational, and prudential requirements so that CCPs manage risk properly and are therefore safe to use. The final text defines stringent requirements on CCPs' liquidity risk management, which will be specified in technical standards. CCPs' access to liquidity could result from access to central-bank or to credit-worthy and reliable commercial-bank liquidity, or a combination of both.

If a contract is not deemed eligible (e.g., prices are not available or the contract is not liquid), or if one of the parties to an eligible contract is not subject to the clearing obligation, then that contract will in all likelihood not be cleared by a CCP. For such contracts, the proposal will require the institutions subject to the clearing obligation to apply robust bilateral risk-management techniques, including marked-to-market on a daily basis of outstanding contracts and holding of additional capital.

No derivatives were excluded. The regulation covers all segments of the OTC derivatives market (interest rate, credit, equity, foreign exchange, and commodities). At this stage, while the segments differ in their characteristics, the Commission considers that there is no strong evidence to exclude any of them from the scope of the proposal.

There is, on the contrary, a strong incentive to adopt a comprehensive policy on OTC derivatives, as the failures uncovered by the financial crisis are present in all segments of the OTC derivatives market. Furthermore, there is a risk that excluding any segment from the outset would create a loophole that could be exploited by market participants. This is because any derivative contract can be partitioned and reconstructed into different but economically equivalent contracts. For example, if a specific contract is eligible for clearing but can be reconstructed into two other types of derivatives that are not covered by the regulation and do not have to be cleared through a CCP, market participants would be able to avoid clearing requirements by modifying the original contract.

Beyond this regulation, other measures relevant to OTC derivatives have been taken, notably the revision of the Capital Requirements Directive (e.g., differentiation of capital charges between CCP-cleared and non-CCP-cleared contracts); the Market in Financial Instruments Directive (e.g., ensuring trading of standardized contracts on organized trading venues, enhancing trade and price transparency across venues and OTC markets as appropriate); and the Market Abuse Directive (extending the scope of the directive to OTC derivatives).

16.7.2 Agricultural commodities and the Dodd-Frank Act

Under the Dodd-Frank Act enacted in the United States, chapter VII is the one that interests us from an agricultural point of view. chapter VII was designed to deal with derivatives in general terms. It requires the following:

- Mandatory clearing
- Mandatory trade execution
- Registration of swap dealers/major participants

The Dodd-Frank Act is aimed at improving transparency in derivatives markets and limiting the activity of pure financial traders. In particular, the new legislation extends the limits on the number of positions held by a single financial trader in each market and, by the second quarter of 2012, has imposed central clearing for current OTC transactions.

For uncleared derivatives/swaps, the following requirements were put in place:

- Mandatory margin requirements
- Segregation of margin

Their philosophy was to create a smaller, more homogeneous swap market. When implementing the Dodd-Frank Act,[57] the CFTC – Commodity Futures Trading Commission acknowledged that – the main instrument covered by the commission's proposals are agricultural commodity "swaps".[58] Basically, swaps empower farmers, ranchers, agribusinesses, and other food suppliers to hedge against certain risks common to their trade, such as bad weather or a crash in the price of a food item they sell. The commission's new rules affect agricultural commodity swaps that aren't sold on exchanges and don't run through third-party intermediaries – so-called over-the-counter swaps. There is a major concern that the CFTC has cast its net too wide on this and will affect real economy transactions, which was not embedded in the initial understanding.[59]

16.8 Financialization of commodity-price formation[60]

16.8.1 Introduction

The term "financialization of commodity trading" indicates the increasing role of financial motives, financial markets, and financial actors in the operation of commodity markets. Or more specifically, financialization is the process of alignment of commodities returns with pure financial assets ("pooling effect"), thus increasing co-movements among asset classes that have been historically seen as following opposite causal pattern. This process began well before the 2008 financial crisis, but it has speeded up over the years. It occurred while international trade in the world increased[61] and technological advancements made access to commodities markets easier than ever before.

As mentioned in previous chapters, there is a clear link between the easy access to credit and the accommodating monetary policy we have been experiencing in recent years and the fact that real prices of commodities during that period have been pushed up by an increased interest in piling up inventory (carry trade).[62] It is the combined effect of all three elements that allowed for the globalization of commodity markets and raises three policy concerns: (1) Does the pooling effect increase the pro-cyclicality and cause channeling shocks? (2) Do financial positions direct commercial positions in the commodity markets? and (3) Should we be concerned about the size of the financial-investor pool versus the physical investor pool?

Indeed, investors have been engaging in commodities trading for the purpose of portfolio diversification ever since it became evident that commodity futures contracts exhibited the same average returns as investments in equities, while over the business cycle their returns were negatively correlated with those on equities and bonds. The empirical evidence for this finding emerged from an analysis of data stretching over a long period from 1959 to 2004.[63] That analysis also shows that the returns on commodities were less volatile than those on equities or bonds, because the pairwise correlations between returns on futures contracts for various commodities (e.g., oil and copper, or oil and maize) were relatively low. Commodity futures contracts were also found to have good hedging properties against inflation (i.e., their return was positively correlated with inflation).[64]

Such portfolio diversification considerations gained further impetus in the early 2000s with the increasing recognition both in academic circles[65] and among potential investors[66] that commodities were entering a new supercycle. It was believed that rapidly growing demand associated with urbanization and industrialization, as well as changes in dietary habits toward more protein-rich diets in major emerging economies, particularly China and India, had triggered a new, prolonged increase in real commodity prices.[67] It has, anno 2013, been extensively argued that the commodity supercycle is over and prices will reverse to the mean. Arguments provided include foremost the slowing growth in China, which is now trying to reinvent itself as a consumption rather than an export hub.

16.8.2 Financial investors use a (wider) range of instruments[68]

Investment in commodity indexes has probably attracted the most attention over the past few years. Index investment tracks returns on weighted commodity baskets (e.g., the Standard & Poor's Goldman Sachs Commodity Index [S&P GSCI] and the Dow Jones-Union Bank of Switzerland Commodity Index [DJ-UBSCI]).[69] These indexes are composites of futures contracts on a broad range of commodities (including energy products, agricultural products, and metals) traded on commodity exchanges. Investing in a predetermined basket of commodities, as is done in index investment, rests on the assumption that commodities have a unique risk premium that is not replicable by combining other asset classes, and that they form a fairly homogeneous class that can be represented by a few positions (Scherer and He, 2008).[70] These characteristics are likely to be accentuated in periods of commodity supercycles. During these periods, commodity-specific market intelligence, as generally gathered by investors that focus on specific commodities, may be considered unnecessary. As a result, the fees associated with investing in commodity indexes are fairly low.

Financial investors gain exposure in commodity indexes by entering into a bilateral financial agreement, usually a swap, with a bank or another large financial institution. The investor purchases parts in a commodity index from the bank, and the bank in turn hedges its exposure resulting from the swap agreement through futures contracts on a commodity exchange. Financial investment in commodity indexes involves only "long" positions (i.e., pledges to buy commodities) and relates to forward positions (i.e., no physical ownership of commodities is involved at any time).[71] This process – known as "rolling" – gives rise to a roll yield that is positive in a "backwardated" market and negative in a "contango" market.[72]

This specific characteristic of index trading implies that roll yields are of particular importance to position taking by index traders. Financial investors that follow a more active trading strategy, such as money managers (see below), are unlikely to rely on long-term-oriented index investment; rather, they tend to operate on the basis of more short-term investment horizons and take positions on both sides of the market through futures and options contracts. This enables them to earn positive returns in both rising and declining markets.

Since about 2009, a third basic instrument has gained considerable importance, namely so-called exchange-traded products (ETPs). Most ETPs, which comprise exchange-traded notes (ETNs) and exchange-traded funds

(ETFs), replicate the return on a single commodity, while a few track commodity groups. The shares of ETPs are traded on equity markets.

Financial investors are also increasingly using structured products. These products can take different forms but typically combine an underlying asset with a derivative (such as an option). The addition of a derivative is often aimed at protecting the capital invested in the underlying asset and thereby reducing risk while maintaining the possibility of benefiting from the current price trend. This option gives the right (but not the obligation) to buy (or sell) an asset at a specified price within a given time frame. Given the generally nonstandardized character of structured products, they are typically traded OTC between an investment bank and a financial investor. Structured products on commodity indexes first appeared on the market in 2006. They compete with the traditional, broad-based commodity indexes but make the rolling process more flexible, thereby reducing roll losses.

It is difficult to assess the size of the financialization of commodity trading due to the lack of comprehensive data. But it is reflected, for example, by the strong increase, starting around 2004, in the number of futures and options contracts outstanding on commodity exchanges and in the amount of outstanding OTC commodity derivatives. The number of contracts outstanding on commodity exchanges has continued to increase since the collapse of commodity prices in mid-2008 and is now about 50% higher (with a certain level of variation in the period 2010–2013 but consistently higher compared to 2008) than in the first half of 2008, when commodity prices peaked. In contrast, the notional amount of outstanding OTC derivatives has dropped to about one-third, which corresponds to roughly half of its level in 2005–2006, but also to about five times its level in 1999.[73] For agricultural commodities specifically, the amounts outstanding were $384 billion (June 2010), $461 billion (December 2010), $402 billion (June 2011), $391 billion (December 2011), and $319 billion (June 2011).

16.8.3 Market participants

Several categories of market participants are active in commodity markets. These categories are usually distinguished on the basis of the reports on traders' positions that are published in anonymous and summary form by the CFTC – in its weekly Commitment of Traders (COT) reports. The main purpose of these reports is to improve transparency about activity in futures markets. The CFTC used to distinguish only between two categories of market participants: those that hedge an existing exposure, which it categorized as "commercial," and those that do not hedge, which it categorized as "noncommercial."[74] However, it became widely perceived that, as a consequence of the growing diversity of market participants in futures exchanges and the greater complexity of their activities, the traditional COT data may fail to fully reflect such activity

(CFTC, 2006). This is because those hedging, and therefore defined as commercial market participants, have normally been considered as entities that use transactions in futures contracts to reduce risk in the conduct of a commercial enterprise. However, many market participants who report positions as hedges, and who therefore fall under the "commercial" category, are in fact commodity swap dealers, who hedge to offset financial positions.

If their underlying positions were held directly as commodity futures contracts (rather than being intermediated through OTC swap agreements), they would be categorized as "noncommercial." Responding to these concerns, in 2007 the CFTC introduced a new and better categorization in its Supplementary Commodity Index Traders (CIT) reports with data on positions of index traders for 12 agricultural commodities.[75] The index-trader positions include those taken by both pension funds, previously classified as noncommercial traders, and swap dealers that had been classified as commercial traders. According to the CFTC (2009), CITs generally replicate a commodity index but may belong to either the commercial or noncommercial category.

In September 2009, the CFTC went even further and started to publish Disaggregated Commitment of Traders (DCOT) reports. These reports have been providing weekly data beginning in June 2006 for the 12 agricultural commodities covered by the CIT reports plus a range of energy commodities and metals, such as crude oil, natural gas, copper, and gold. The DCOT reports distinguish five trader categories.

Trader categories in the CFTC's Disaggregated Commitment of Traders reports

1. Producers, merchants, processors, users (PMPU)	Entities that predominantly engage in the physical commodity markets and use the futures markets to manage or hedge risks associated with those activities.
2. Swap dealers	Entities that deal primarily in swaps for a commodity and use the futures markets to manage or hedge the risks associated with those swap transactions. The bulk of these traders' clients are index investors who invest in commodity indexes such as the S&P GSCI and the DJ-UBSCI.
3. Money managers	Entities that manage and conduct organized futures trading on behalf of their clients. This category includes registered commodity trading advisors (CTAs), registered commodity pool advisors (CPOs), and unregistered funds identified by the CFTC. Hedge funds and large ETFs are part of this category.
4. Other reporting traders	Every other reportable trader that is not included in one of the other three categories.
5. Nonreporting traders	Smaller traders who are not obliged to report their positions.

The DCOT reports consider the first two trader categories (i.e., PMPU and swap dealers) as "commercial" traders, and the other two categories as "non-commercial" traders. By contrast, the index-trader category of the CIT reports does not coincide with the swap-dealer category in the DCOT reports. This is because the swap-dealer category of the DCOT reports includes swap dealers who do not have commodity-index-related positions and therefore are not included in the index-trader category of the CIT reports. Also, the index-trader category of the CIT reports includes pension and other investment funds that place index investments directly into the futures markets rather than going through a swap dealer; these traders are classified as managed money or other reportables in the DCOT reports.[76]

Money managers generally have a short-term perspective and adopt an active investment strategy. This strategy goes beyond the consideration of commodities as a fairly homogenous asset class with a unique risk premium, which is characteristic of broad-based passive index investment; it also takes into account factors such as different short-term supply-and-demand dynamics, as between industrial metals and energy. Perhaps more importantly, active trading strategies try to take advantage of profitable investment opportunities arising (1) in declining markets (by taking "short" in addition to "long" positions); (2) from taking longer-dated futures positions than those usually included in readily available indexes; (3) from trading commodities that are barely, if at all, included in the popular commodity indexes (e.g., soybean oil is not included in the S&P GSCI, while cocoa is not included in the DJ-UBSCI); and (4) from employing a "relative value" approach, such as by exploiting differences in quality, regional dynamics, intracommodity dynamics (e.g., soybeans versus soybean oil), and cross-commodity dynamics (e.g., trading oil and feedstock used for biofuel production against other food commodities).

The money-manager category includes a range of investors, such as hedge funds and institutional investors, which follow different trading strategies based on macroeconomic fundamentals, detailed commodity research, algorithmic trading or trend following, and general financial portfolio-diversification considerations. Thus they are able to adjust their exposure in commodity markets according to changes in asset prices with a view to stabilizing the structure of their portfolio.

As a conclusion to the topic, a summary[77] of the characteristics of the different product groups and trading strategies is presented in Table 16.3.

16.8.4 Financialization and pricing dynamics

The financialization of commodity trading has made the functioning of commodity exchanges controversial. Their traditional functions have been to facilitate price discovery and allow the transfer of price risk from producers and consumers to other agents that are prepared to assume the price risk.

Table 16.3 Different product groups and trading strategies

	Spot contract	Forward contract	Futures contract
Nature of transaction	Bilateral (OTC)	Bilateral (OTC)	Multilateral (exchange)
Terms of transaction (delivery date, contract size)	Customized	Customized	Standardized
Price	FOB	FOB/CIF	FOB (or in warehouse)
Settlement	Physical	Cash/physical (to maturity/ shipment)	Cash (daily) Offset/physical (To maturity, "physical" if requested)
Typical holding period	To delivery	To delivery	Before delivery
Delivery	Spot	Customized	Selected months
Storage cost	No	Yes	Yes
Transaction cost	Medium	Medium/high	Low
Leverage	No	No	Yes
Counterparty risk	Limited (spot)	High	Limited (daily mark-to-market)
Currency risk	No	Limited (choice)	Yes
Price risk	No	Yes	Yes
Interest rate risk	No	No	Yes
Regulation and supervision	Limited	Limited	High

These functions are impaired to the extent that trading by financial investors increases price volatility and drives prices away from levels that would be determined by physical commodity supply-and-demand relationships.

As a result, commodity-price developments no longer merely reflect changes in fundamentals: they also become subject to influences from financial markets. Consequently, market participants with a commercial interest in physical commodities (i.e., producers, merchants, and consumers) face greater uncertainty about the reliability of signals emanating from commodity exchanges. Thus, managing the risk of market positions and making storage, investment, and trading decisions becomes more complex. This may discourage long-term hedging by commercial users. Moreover, with greater price volatility, hedging becomes more expensive, and perhaps unaffordable for developing-country users, as well as riskier.[78]

Price formation[79]

The way prices are formed in markets for physical commodities and futures contracts is the result of complex interactions between idiosyncratic factors, such as product characteristics (quality, storability, substitutability, etc.);

supply-and-demand factors (capital intensity, industry concentration, production facilities, average personal-income level, technological developments, etc.); and exogenous factors such as access to finance, public subsidies and interventions, and the weather. Price formation relies on the efficient functioning of market organization for physical commodities and linked futures contracts. Market microstructure developments, such as market liberalization, the development of futures-market infrastructure, and the expansion of international trade, have significantly altered the organization of commodity markets over the last decade. In general, supply factors (such as capital intensity) are more important drivers of price formation for energy commodities and industrial metals, while agricultural and soft commodities markets are more influenced by demand factors (such as income growth) and exogenous factors that can cause supply shocks (such as weather events or government policies).

Market dynamics

Volatile spot-price levels across several commodities and a growing correlation between returns of financial and nonfinancial assets have raised concerns over the role of factors that are unrelated to market fundamentals in price formation. Exogenous factors, such as greater interaction with the financial system and supply constraints in the freight markets, have become increasingly important over the last decade. Recent empirical analysis[80] confirms that demand-and-supply fundamentals remain solid drivers of futures price formation across all the commodities markets covered by the report. By channeling information about supply-and-demand fundamentals to the physical and futures markets, together with ensuring smooth management and aggregate transparency of inventories, the functioning of commodities price-formation mechanisms can be improved.

The growth of emerging economies (in particular, of Chinese industrial consumption) lies behind the structural shift in prices, which – through the astonishing growth of international markets – has contributed to greater interconnection between physical commodities markets and so to higher responsiveness to pro-cyclical global demand factors. Despite the fact that growth in demand has slowed down across commodities markets, demand levels are still reaching new historical peaks thanks also to product and market characteristics. For instance, technological changes have promoted the widespread use of some commodities for alternative applications, such as corn for fuels or soybean oil for pharmaceutical products. New fundamental factors may therefore affect the use of a commodity and its price formation, which may ultimately increase the correlation with other factors that are not directly linked to the underlying physical commodity (the weight of crude-oil prices in the price formation of corn, for example).

In fact, some commodities may be very responsive to crude-oil prices. First, responsiveness is the result of the (exogenous) link to transport fuels or costs of fertilizers for agricultural commodities, for instance. Second, responsiveness to crude-oil prices may be linked to direct government interventions to promote biofuels. This is the case for corn, for instance. However, the evidence points to only a weak (but strengthening) link between corn and crude oil, which rules out for the moment any transmission of the instability of energy policies to the market for corn.

In aggregate, demand has been constantly growing across all commodities markets for more than a decade. This has led to a general fall in stock-to-use ratios, in particular for agricultural and soft commodities. Without significant investments in new technologies, questions remain over the ability of current supply to satisfy growing demand in the long term.

In line with the historical trend, commodities are a volatile asset class, and price volatility is on average within a stable range in the long term. However, the growing interconnection between financial and nonfinancial assets, and between regional physical markets, has amplified the reaction to market shocks, such as the recent financial crisis and the global economic downturn, and thus created volatility peaks in the short term. As a consequence, short-term volatility remains above precrisis levels, in particular for agricultural commodities.

In summary, the key drivers of commodity-price formation are[81] displayed in Table 16.4.

The weighting of the drivers behind each (agricultural) commodity is different.[82]

Low costs of financing and lower opportunity costs (returns on alternative asset classes) have favored storage of commodities (carry trades), especially those with good "store-of-value" properties, such as metals. These circumstances have increased the opportunities for financial participants to enter these markets and the opportunities for commodity-trading houses to use financial leverage to expand their physical interests. As a result, returns from commodities are increasingly pooled with returns from pure financial assets (a "pooling effect"). The process increases co-movements among asset classes that have historically been seen to follow opposite causal patterns. This situation is the result of the combined effects of multiple circumstances, including the growth of international trade and cross-border interaction among physical markets, reinforced by easier access to international finance and credit partly due to widespread expansionary monetary policies, a favorable regulatory framework with the deregulation in the United States, and technological changes favoring electronic trading and promoting accessibility to futures markets from any remote location around the globe. In fact, empirical evidence suggests that a strong positive correlation between

Table 16.4 Drivers of commodity-price formation

Product characteristics	Supply factors
• Quality • Storability • Renewability • Recyclability • Sustainability • (Final) usability	• Production convertibility and capital intensity • Horizontal and vertical integration • Storability and transportability • Industry concentration • Geographical concentration (emerging markets) • Technological developments • Supply peaks and future trends
Demand factors	**Exogenous factors[a]**
• Income growth and urbanization • Technological developments and alternative uses • Long-term habits and demographics • Economic cycle	• Financialization process and monetary policies • Subsidy programs • General government interventions (e.g., export bans) • The economic cycle and other macroeconomic events • Technological developments • Unpredictable events (e.g., weather)
Market organization	
• Microstructural developments (e.g., competitive setting) • Functioning of international recognized benchmark futures or physical prices • International trade • Expansion of commodity future markets and non-commercial investors • Futures-market infrastructure	

Key product characteristics	Exogenous factors	Examples
Seasonality	Weather and currency	Drought
Transportability	Freight market/mobility restrictions	Freight capacity
Alternative uses/ substitutability	Other commodity markets	Biofuel policies
Storability	Market/warehouse location	Pipeline disruption
Production yields	External incentives for long-term investments or technological shock	Price subsidies

Note: [a] See also for the presentation accompanying the report indicated ibid. Valiante http://www.eurocapitalmarkets.org/system/files/Presentation_Commodities.pdf.

commodities prices and financial indices emerged in the early 2000s, when all of the factors mentioned above came together with renewed strength. Since then, the correlation has remained strongly positive across all commodities markets.

Overall, the financialization process has increased pro-cyclicality, that is, responsiveness to the economic cycle and vulnerability of commodities markets to short-term shocks also coming from the financial system. However, the latter has been instrumental to the growth of international commodities markets. Unless governments want to push back on international trade, financialization is a natural outcome of the new environment we live in. Despite the growing interconnection, fundamentals remain key drivers of futures price formation. Well before the financial crisis erupted in 2008, commercial participants (e.g., commodity producers and merchants) were responding to strong demand pressures by quickly expanding their physical business activities on a global level, and so laying the path for the growth of futures markets and the entry of noncommercial participants (e.g., investment funds) who were attracted by high returns.

Technological developments in trading (e.g., algorithmic trading), financial innovations (e.g., commodities indexes), and easy access to international finance, prompted by accommodating monetary policies, fueled this expansion. The value of international trade in commodities futures has soared, together with the size of commercial participants and their interests in futures markets, which have ultimately favored the arrival of purely financial participants. The empirical analysis confirms that the expansion of commercial futures positions has been leading price formation in futures markets through the steady increase in futures positions and OTC financial activities. Noncommercial futures positions have, in the meantime, become by far the biggest component of futures markets, though evidence still points to commercial participants leading price formation in futures markets.

Commodity trading houses with interests across different commodities markets and significant financial exposure have boosted their physical holdings in international markets and may become "too physical to fail." The use of financial leverage to increase physical holdings, through easy access to international finance and helped by accommodating monetary policies, may have systemic implications. Aggregate data on physical holdings, coupled with a minimum set of information confidentially disclosed to regulators, for example, may reduce risks of moral hazard for national governments that have to cope with the sheer size of these entities in case of trouble.

Technological developments have changed the infrastructure of commodities markets and prompted innovation and sophistication in risk management. While these changes provided tools for (some) trading practices by noncommercial participants, bundled in very high intraday volumes, which can theoretically damage price formation in the short term through herding behaviors, evidence (see infra) suggests that to date the role of noncommercial participants in commodities markets has been generally benign. The growth of index investments has so far not caused distortions in price formation. An indiscriminate ban on legitimate trading practices may result in liquidity losses at the expense of the efficiency of price formation, although this report does not perform an ex ante quantification. The actions of supervisors should target damaging trading practices, such as cornering attempts, rather than specific categories of traders. Proper surveillance mechanisms and supervision of exchange policies are essential, in particular when it comes to dealing with complex algorithmic or pure high-frequency trading. More time and data (e.g., aggregate data on volumes by category of trader) are needed, however, to improve the analysis of trading practices in the short term and the long-term effects of financial participants on price formation.

Market organization

Futures markets are an essential infrastructure to support risk management in physical markets and, therefore, their price formation. Futures markets have supported the development of international trade and the consolidation of commercial participants fueled by the opening up of international trade. Transparent and stable futures markets promote healthy interaction between the physical and financial spheres of commodities markets, which today are inextricably linked. As a result of greater interconnectedness, market infrastructure also allows faster circulation of information by increasing accessibility, and thus the resilience of price-formation mechanisms. However, as market infrastructure adapts to a more global and interconnected environment after demutualization, exposure to global risks requires a sophisticated surveillance mechanism and more coordination between supervisory authorities at the international level.

As the industry pushes for consolidation at the regional and global levels, a minimum set of requirements to ensure accessibility and interaction with competitors while preserving rights on key intellectual properties may be beneficial for innovation around new products and services to attract liquidity and, ultimately, serve the interests of commodity users. The implications of financial reforms on the market power of market-infrastructure operators should be carefully assessed.

Warehousing and delivery systems linked to futures exchanges are an important element of efficient price formation, which helps the convergence of futures to spot (physical) prices. Both loading-out capacity and the location of warehouses depend on the nature of the commodity. For example, industrial-metal warehouses are typically needed close to net-consumption areas, while for agricultural commodities a location close to net-production areas is often preferable, as the product requires immediate storage and delivery. Expanding points of delivery and/or increasing delivery capacity should depend on the characteristics of the underlying physical markets in order to limit supply bottlenecks (i.e., delivery queues) and improve the functioning of international benchmarks.

Internal management of positions by the exchange, linked to the actual delivery capacity of the infrastructure, may also be helpful to avoid artificial shortages if significant positions suddenly take delivery, as occurred in 2010 when the Armajaro fund took delivery of roughly 5% of global yearly production of cocoa in just a few days, creating a supply shortage among the exchange's sponsored warehouses. This would require periodic assessment of the rules set by the infrastructure, whether they still fit structural developments in the underlying physical market.

Issues with the delivery system or liquidity problems with the underlying physical markets of the futures contracts that are recognized international benchmark prices can affect the functioning of commodities markets' organization and ultimately the convergence between futures (forward) and spot prices. Moreover, a well-functioning delivery system provides an efficient tool to support supply adjustments when disequilibrium between physical demand and supply emerges. For instance, problems with the physical delivery of LME aluminum forwards are increasing the reliance on more opaque regional premiums assessments (on average more than 15% of the nominal LME price in 2012), which are partially compensating for the fall in price of the official benchmark following a period of oversupply. Excess or shortage of supply in the physical market of the futures contract can also increase reliance on regional premiums. The West Texas Intermediate and the Brent futures contracts for crude oil have been suffering from (regional) supply excess and shortage, respectively, in their underlying physical markets. Tackling the underlying supply balance and delivery issues is crucial for price formation. There is therefore a risk that by adding financial layers (e.g., the use of derivatives) and price assessments as a substitute for prices formed with arm's-length transactions or replacing transparent exchange-based price-formation mechanisms with a pricing system reliant on assessed regional premiums, the actual conditions of underlying physical markets may no longer be reflected.

More broadly, a recognized international benchmark should (1) have enough supply in the underlying reference physical market (supply security), (2) provide market access and an efficient price-discovery system (demand security), and (3) promote competition in the upstream and downstream physical market, and where possible develop secondary markets for underlying forward contracts. For markets such as crude oil, initiatives would need to be undertaken at the global level by the relevant forum to achieve these objectives.

Conflicts of interests in commodities markets can have harmful effects, with strong implications for physical flows and market competition. Therefore rules for sponsored warehouses, for example, should be set by the exchange only once the interests of its shareholders (often represented in the board of the exchange) in the external market infrastructure (e.g., ownership of sponsored warehouses) are properly disclosed and ultimately managed. Conflicts may arise, in particular, when financial and nonfinancial activities are combined in the same entity. Conflicts of interests between the ownership of market infrastructures and/or of physical/futures/other financial holdings of market participants therefore need to be appropriately identified, disclosed, and ultimately managed by the parties involved under the coordinated international supervision of competent authorities.

Finally, claims that the size of futures markets is many times larger than physical markets and thus may distort price formation based on underlying physical transactions cannot be proven, but also cannot be ruled out. Further data and analysis is required to substantiate such claims. When looking at liquidity curves in futures markets, the size of open interest is only a fraction of the corresponding physical markets size, with high peaks only for cocoa and coffee (respectively at around 80% and 210%). However, when looking at yearly volumes of contracts compared to yearly production, futures markets are many times larger than the corresponding physical production (up to nine times larger for the main corn futures contract). But comparisons between volumes of transactions that are only carried out to exploit information about physical trades in the trading of different futures maturities (e.g., calendar spread) with actual physical production (which is not a measure of the intensity of physical trade) may ultimately overestimate the weight of futures over physical markets. Physical production is an inaccurate and conservative proxy of underlying physical market transactions.

The CEPS-ECMI Task Force Report[83] estimates the total notional value of outstanding (open-interest) over-the-counter and exchange-traded financial transactions in commodities (e.g., futures and options) at around $5.58 trillion in 2012. Over-the-counter transactions make up roughly 38% of the total outstanding value (open interest).

16.8.5 Herding behavior and arbitrage[84]

The availability and processing of information plays a key role in the determination of asset prices. This role has traditionally been examined on the basis of the efficient-market hypothesis (EMH), whereby prices perfectly and instantaneously respond to all available information relevant to a freely operating market. Market participants continuously update their expectations from inflowing public and private information. This means that prices will move either when new information becomes publicly available (in the case of commodities, for example, following announcements of harvest forecasts or changes in oil production), or when private information leads to transactions that affect prices.

Crucial assumptions of the EMH are that market participants evaluate assets on the basis of fundamentals, act fully rationally, base their actions on publicly available or their own private information, and do so independently of each other. However, some circumstances can cause individuals to deviate from this

Table 16.5 Taxonomy of the different types of herding behavior

Rational herding	Spurious herding	
	Intentional herding	Conformity based Compensation based Reputation based Information based
Irrational herding	Noise trading	Pseudo-signals Beliefs/sentiments/ positive feedback strategies/algorithms

assumed behavioral pattern and to engage in herd behavior. Herd behavior frequently occurs when decisions need to be taken in situations of uncertainty.[85] It may be defined as the tendency of individuals to mimic the actions of a larger group, rather than acting independently and on the basis of their own information.

Herd behavior can take various forms and may be rooted in irrational behavior, but it may also be fully rational. Table 16.5[86] provides a taxonomy of different types of herd behavior. Early models of herd behavior were based on assumed deviations from perfect rationality, or so-called "noise trading."

16.8.6 The price effects of financialization in the (agricultural) commodity markets

The impact of financial traders on commodity prices is difficult to quantify. Part of this difficulty is due to the fact that the financialization of commodity trading became a major factor roughly at the same time as demand for physical commodities from emerging economies started to increase rapidly. These roughly simultaneous developments make it difficult to disentangle their relative price impacts. Accordingly, most empirical assessments of the impact of financialization on commodity prices have emphasized either fundamental supply-and-demand factors or variables that reflect the financialization of commodity trading. Given that commodity prices have been influenced by both factors, both of these groups of studies have found a significant impact on commodity prices of the variables they selected. Hence, those that attribute most of the development of commodity prices over the past few years to fundamental factors (e.g., Sanders and Irwin, 2010), as well as those that point to an additional impact from increased financial investment (e.g., Gilbert, 2010),[87] have been able to provide empirical support for their point of view. A prominent recent empirical study has included both fundamental and financial variables (Tang and Xiong, 2010).[88] The results of this analysis refute the contention that growing demand from emerging economies was the only driver of the commodity price hike in 2006–2008. They show that variables reflecting financialization remain significant even after controlling for fundamental factors.

This finding suggests that the process of financialization has caused commodity prices to be determined no longer simply by supply and demand, but also by a wide range of financial factors and financial investors. The resulting change in commodity-price dynamics is likely to persist and seriously affect commodity producers' hedging strategies, as well as many countries' food and energy policies.

A further analysis by Gilbert (2010)[89] conducts Granger causality tests that relate returns on futures contracts to changes in the positions of index investors for the period from January 2006 to March 2009. For the seven commodities in the sample, the study finds that changes in index-trader positions Granger-cause price changes for aluminum, copper, crude oil, and maize, while no such impact is detected for soybeans and wheat. To the extent that changes in index-trader positions are perceived by other traders as conveying information – similar to the informational cascades discussed above – these price effects should persist. In an additional step, Gilbert (2010)[90] employs a regression analysis to examine whether the uncovered price impacts following position changes of index traders are persistent effects. The results suggest such persistent effects to be present for copper, crude oil, and wheat.

In a final step, Gilbert estimates the price impact of index-based investments by comparing the actual price developments with those that would have prevailed had there been no index investments. This hypothetical price development is estimated based on the econometric exercises just mentioned.

Overall, evidence indicates that active investment strategies are increasingly gaining importance at the expense of the more passive, broad-based index-investment strategies. It also indicates a close correlation between commodity prices and the positions of financial investors that pursue an active trading strategy.[91]

17
The Global Water Challenge

Water – as we know it – is a commodity, but that is about to change. Consumption patterns and levels, population growth, the need for energy, and changing climate patterns will all be arguments in that mix.[1] This will ultimately mean an increase in demand for water – in the face of a limited supply of this key resource. Usage conflicts are inevitable and will become more acute on account of wasteful use and pollution. Investments in water infrastructure as a whole will have to increase over the next few decades to be able to satisfy the growing demand. Naturally, the focus of investment in industrialized countries differs from that in developing countries and emerging markets. Further, the question is how the relationship between public and private levels will emerge in terms of the investments needed to support this evolution and create sufficient supply. Also a mix of technological solutions will be needed to enhance supply levels and support reuse. Conflicting use will be almost inevitable, and the geopolitical dimensions are the great "unknown unknowns" in this story.[2]

17.1 Some initial data about global water markets[3]

The water market differs in many respects from other markets. One characteristic on the supply side is that the amount of water in the world always remains the same, since water exists in a cycle of evaporation and precipitation. Water covers over 70% of the Earth's surface. But of the nearly 1.4 billion km^3 of water on Earth, only about 2.5% is freshwater. In turn, close to 70% of this water is captured in the form of ice, especially at the polar caps and in glaciers and thus is not immediately available for human use. Groundwater makes up about 30% of the world's freshwater resources, so in many countries it is the most important source of supply. Only 1% or so of the freshwater on Earth is found in lakes, rivers, wetlands, and different types of soil (including permafrost). Nevertheless, surface water from lakes and rivers (and thus also from precipitation) plays the dominant role in humankind's water supply. On a global average, over 70% of the water used by humans is drawn from surface sources. In several countries, especially in the oil-rich Arab states, a significant proportion of the water supply is based moreover on seawater desalination. As indicated in the chapter on agriculture, 70% of the water available for human consumption is used for agricultural purposes.

However, in basic terms there is sufficient water to satisfy all human needs.[4] According to Germany's Federal Institute for Geosciences and Natural Resources (BGR), annual worldwide withdrawals of groundwater equal less than 10% of groundwater recharge.[5] Nonetheless, many countries and regions suffer from a scarcity of water or even shortages. This is mainly attributable to several demand components (overextraction, waste, and pollution). A major supply-side reason for a shortage of water is, of course, climatic conditions. The reason is that the volume of surface water is closely correlated with precipitation, which in turn varies by season and by region;[6] this holds true not only globally, but also for individual countries. As a good, water is not substitutable. This is a major difference compared to most other economic goods. Direct consumption of water (drinking or preparation of food) by humankind is negligible in terms of total water consumption. But the absence of safe drinking water (potable water) and sanitation facilities is an existential problem affecting many people.

17.2 Users of water: agriculture on top

The biggest user of freshwater by far is irrigation farming (as indicated, roughly 70% of global water consumption). In many developing countries and emerging markets, this value at times noticeably exceeds 90%. The irrigated agricultural area on Earth has more or less doubled over the past 50 years. The second most important users are the industrial and energy sectors, with a combined share of around 20%. Finally, private households account for about 10% of use. Much of this household water is used for body care, flushing toilets, and doing the laundry, especially in industrial countries.

The key drivers of global water demand[7] are closely correlated with population growth. This means that most of the prospective increase in demand for water is likely to be accounted for by developing countries and emerging markets, as 90% of the population growth forecast up to 2050 will be found in these countries.

More people will require not only more drinking water but also more food.[8] For this reason, irrigation farming will continue to gain significance in efforts to boost agricultural yields; moreover, the quantity of pollutants finding their way into groundwater will probably increase in many countries owing to inefficient use of pesticides and fertilizers. A further reason for the increase in water consumption is the growing level of urbanization, which is usually linked with a higher standard of living and higher water consumption. The problem of water scarcity or shortages (and also water stress) is caused, or exacerbated, in many places by wasteful use or pollution. The losses due to seepage and leaky pipes both in public supply systems and irrigation farming are in some cases far in excess of 50%, even in industrial countries. In many emerging nations, the majority of water used flows untreated into lakes, rivers, and seas. The increasing need for energy is a further driver behind increasing water demand. The cooling of power-generating plants is one factor that plays a part, but naturally also hydropower per se. A further factor worth mentioning is the increasing importance of biofuels, using not only arable land but also consuming water supply at increasing rates.

Also businesses rely on water. The spectrum of potentially affected sectors is huge, because no sector can get by without any water at all. Besides agriculture, the food industry (including the beverage industry) is one of the sectors that requires large quantities of high-quality water. Further examples of intensive-use sectors are energy (e.g., cooling), mining (e.g., mines), chemicals, shale gas exploration, pharmaceuticals, paper and cellulose, textiles and clothing, the tourism sector, and semiconductors.

All in all, water consumption is expected to rise faster than it did in the past half century. If there is an increase in the water stress of many countries, it follows that the risk of political conflicts between individual states over shared water resources will also grow (claiming water rights, border tensions, and even outright violent conflicts). It's no surprise that the World Economic Forum (WEF) has listed water scarcity as one of the major global risks for many years now.

Governments have a sizeable influence on the sector in every country of the world, which is understandable considering the existential importance of water as an economic good. The government functions not only as a regulator but also as an operator. Water is supplied and wastewater disposed of largely by the public sector, or at least it regulates these activities. Besides, government is responsible for the bulk of investment activity in the sector (estimated at about 70%). One very major difference between the water market and other goods markets is the pricing aspect. In the water sector, the price is usually set by the government, and it does not reflect supply and demand. In most countries, drinking-water and wastewater prices (or charges) do not cover costs – particularly taking into consideration the scarcity rent of water as a limited resource. There are various reasons. For one, the social component is particularly important. Also, low water prices are also used very deliberately

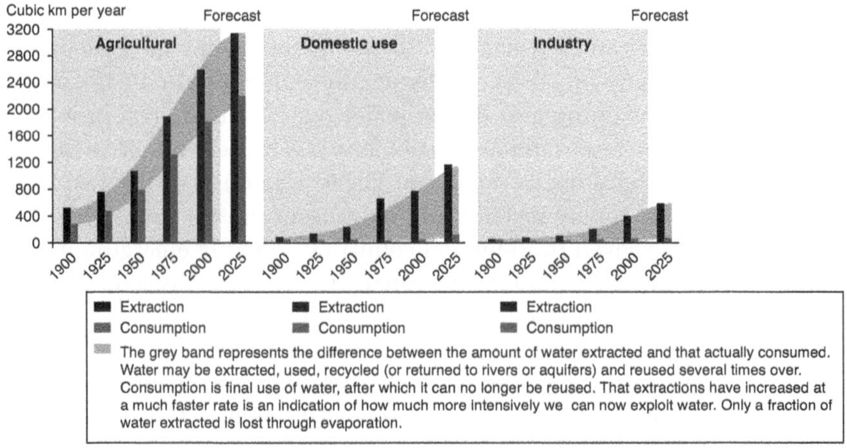

Figure 17.1 Evolution of global water use

Note: Domestic water consumption in developed countries (500–800 liters per person per day) is about six times greater than in developing countries (60–150 liters per person per day).

Source: Igor A. Shiklomanov (1999), State Hydrological Institute (SHI, St. Petersburg) and United Nations Educational, Scientific and Cultural Organisation (UNESCO, Paris).

as an instrument to foster certain sectors. This applies especially to agriculture (subsidies for irrigation, no payments for use of groundwater, free disposal of wastewater, etc.).[9]

A summary of the evolution of water consumption per category can be visualized as shown in Graph 17.1.[10]

17.3 The water market: not a slam dunk in investment terms

Given the large role of the government in the water sector and in particular in the pricing mechanism, investing in the water sector is not very straightforward. Prices being charged do not reflect the scarcity of water as a resource and cause waste. The absence of ownership rights for water resources carries a similar effect. Moreover, the low prices are the main reason that the actual investments in the water sector fall short of the real requirements. The underinvestment impacts the reliability of the systems. This type of pricing has a daunting effect also on private investors – at least if the income needed for the amortization of such investments cannot be raised via other instruments.[11] Another problem for investors is the fact that the water market is not only large and global but that consequently data are sensitive to uncertainty.

Only very few countries keep statistics on the volume of water market revenues. Revenues in the sector are likely to grow more rapidly than the pure demand for water (+/–3% p.a.) because water prices are set to climb over the next few years.

The following areas require attention:

1. Agriculture. Growing needs for irrigation due to increased demand for food, changing dietary habits, and so forth will lead to conflicting use. In some developing countries today, already more than 80% of the water supply is used for agriculture, even though only a fraction of the farmland is irrigated.[12] Water availability is only one of the problem areas.[13] Suggested measures are (1) raising water prices for farmers or at least phasing out subsidies in order to curb the amount of water wasted (alternatively: usage rights or quotas for withdrawal); (2) deploying more efficient irrigation technologies (simple catchment and storage systems, drip irrigation, sealed greenhouses) to reduce water consumption and avoid irrigation's negative side effects (soil salination) via government grants financed with higher water prices;[14] (3) converting agricultural production in arid countries to products requiring little water or suitable for planting in the location;[15] and (4[16]) encouraging trade[17] in agricultural products[18] (using microfinance).
2. Urbanization and industrialization. The numbers behind the trend were discussed elsewhere in this book, but in short, by 2030 about 60% of the total population will live in cities. In 2030 about 80% of the urban population will live in cities in developing countries and emerging markets. In parallel, the degree of industrialization, particularly in developing countries and emerging markets, is also on the increase. These two trends will mean a steady increase in the demand for water in urban areas over the next few years. In cities around the world, infrastructural shortcomings have reached gigantic proportions. Public funding for infrastructure often cannot cover the actual investment requirements. The infrastructure is adapted more slowly than population growth in cities. Leaks and water loss are the consequence.[19] But also in developed economies, where infrastructure was built 100 years ago or so, re-upping is needed, but governments are cash-strapped, so that is a similar problem in the making. Water charges should therefore increasingly enable full cost recovery. Even in most OECD countries, the prices charged for water are too low to completely cover operating and maintenance costs.[20] A key aspect of pricing reform, moreover, has to be the conversion to cost-oriented basic prices and operating prices (i.e., fixed and variable shares in the overall price).[21] The

use of public-private partnerships (PPPs) should be used to enhance efficiency or expand infrastructure.

3. Treatment of wastewater. A missing or inadequate wastewater-disposal system is the cause of a multitude of illnesses that develop via contact with pathogenic organisms.[22] Eutrophication (overfertilization) is caused by both inefficient fertilization in agriculture and the influx of nutrient-containing urban and industrial effluents; the most significant nutrients are phosphorus and nitrogen.[23] Ecological wastewater-management practices[24] are being introduced and are based on the recycling of materials. Household wastewater management[25] can include a number of ecological sanitation methods. Heymann[26] suggests a number of them: (1) Separation toilets and gravity system: with this system, yellow water and brown water are channeled into separate containers. The yellow water is used for agricultural purposes. The brown water composts in the container. The ripe compost can be used for agriculture, too. Gray water can be treated separately in, for example, a constructed wetland or a membrane filtration system. The purified gray water can subsequently be released into the household water system or the environment. (2) Vacuum toilet with biogas plant: in this model, yellow water and brown water are channeled from a vacuum toilet to a biogas plant. There, the wastewater is treated anaerobically. The biogas extracted is used to generate electricity or heat. The digested sludge from the biogas plant can be used in agriculture. The gray water is treated separately as in the previous method. (3) Adaptation of the existing wastewater system to incorporate separation toilets: the use of separation toilets in the conventional sewage system means yellow and brown water can be captured separately. One can opt for centralized or decentralized collection. In decentralized collection, yellow water is stored on a decentralized basis and drained off. In centralized collection, by contrast, yellow water with a low pollution content is released into the sewage system and captured and stored separately in the treatment plant. Treatment of industrial wastewater is largely guaranteed already in industrial countries, but in many developing countries and emerging markets this poses a huge challenge, because in these regions much of the wastewater is discharged untreated into surface water. Regulation sets or should set limits on the level of pollution in effluents that industrial companies are allowed to release into surface water or the public sewage system. If the limit values are exceeded, heavy sanctions should be imposed.
4. Cheap specific solutions for the poorest nations and poorest inhabitants. The complexity is in the diverse climatic and topographical conditions in those countries. This can include storage of water in countries or regions with little precipitation. Collection basins or sand dams are all options. The local groundwater deposits can also be tapped by drilling wells. Other

options are increasing agricultural yield and recycling water (using ceramic filters[27]) as discussed. Low-tech systems in poor rural areas can include, for example, dry-toilet systems that store the collected matter in septic tanks and subsequently recycle it for agricultural use. The advantage here is that nutrient- and substrate-rich human excrement is reused for agricultural purposes, allowing the substitution of biological fertilizers for conventional types. Dry toilets involve relatively low investment and installation costs and require little maintenance.[28]

All of these initiatives come at a cost, a cost that often cannot be absorbed by the poorest rural inhabitants in emerging economies. Often, traditional development aid (tax revenues and microcredits are alternatives) plays a significant role. The World Health Organization (WHO) calculated that for every euro invested in developing countries in water-supply and sanitation facilities, it is possible to generate a return in the form of mitigated economic costs of 4 to 12 euros. If these numbers hold up, it is a wonder that many more measures haven't been taken to improve the supply situation. Probably a key reason for this is that a potential investor does not benefit from the decrease in macroeconomic costs.[29]

17.4 Climate change and water

Local and global water systems will be impacted by climate change.[30] Impacts[31] can and will include the following:

- Long-lasting droughts may lead to a decline in the water table, or surface water (e.g., rivers) may dry up.
- After long drought phases, the soil is often unable to absorb any rainfall. In such cases, heavy rainfalls flow off rapidly aboveground, accelerate erosion of the soil, and may trigger large-scale flooding.
- The rise in sea level will result in coastal groundwater being displaced or mixed with salt water. This will narrow the scope of agricultural production and worsen soil salination.
- The gradual melting of glaciers induces, among other things, pronounced changes in seasonal and long-term river-flow quantities (increases initially, decreases over the longer term). This has an influence on all areas whose water supply hinges on the existence of glaciers or on rivers fed by glaciers or precipitation in mountainous regions.

It is estimated that, in general, the poorest countries are the ones hit hardest by climate change. This is because the huge importance of agriculture there makes their economy more dependent on climatic conditions than is the case in the industrial countries. In addition, most of the developing countries and emerging markets are found outside the temperate climate zones, which is why they are more vulnerable to weather extremes anyway. Finally, the possibilities for poor countries to adapt to climate change are limited owing to an absence of financial and technological resources. Heymann concludes, "The tragic part for these worst-hit countries is that because they themselves emit minor amounts of greenhouse gases they can only be blamed to a small degree for climate change itself."[32] Although that in itself is true, eyebrows can be raised by the position of some of these emerging nations within the context of the UN climate change annual conferences (most recently in Copenhagen, Durban, and Cancun) where the follow-up for the Kyoto Protocol was discussed (among many climate-related issues). Their willingness to prioritize economic growth (equaling more emissions or at least no commitment to reductions of GHGs) over long-term sustainability is a decision that, as said, will fly them in the face in financial, ecological, and humanitarian terms over time. In the OECD world, the cost of agricultural production has become more costly, and yields will decline from time to time.

Mainstream climate-smart measures are needed. One can differentiate between two adaptation strategies. First of all, the availability of water can be increased by, for example, transferring water from areas with adequate resources or else by purifying seawater or wastewater. Projects that include transferring or even transporting water by means of vessels have so far proven unrealistic, as the price and ecological cost of doing so is prohibitive and cannot be offset by the benefits. Pipelines, in contrast, have demonstrated better economics. And secondly, efforts can focus on reducing the amount wasted. That includes the following measures:

- Water transfers: importing water via pipelines or canals from areas with adequate water resources to dry regions is a suitable means of boosting the availability of supply in regions with shortages.
- Dam systems can be used in locations with extreme periods of drought and rainfall to protect against flooding. Another advantage of dams is that electricity can be generated via hydropower. But there are also negatives: upstream, the rates of flow decline, and this causes slower transports of sediment and nutrients. This may culminate in the dammed section silting up. Downstream from the dam, by contrast, erosion increases.
- Seawater desalination:[33] a major hurdle at this point is the energy intensity of the process and some membrane-technology hurdles (see infra this chapter).
- Wastewater recycling: a benefit is that one can apply it everywhere. The most important aspect is wastewater as a resource for irrigation purposes. The use of wastewater for industrial processes (e.g., in refrigeration or process water

cycles) is already widespread. Particularly expensive treatment procedures can turn urban wastewater into drinking water.
- Use of rainwater.
- Minimization of water consumption using efficient appliances and technology (e.g., sanitary engineering).[34]

All in all, these necessary steps toward guaranteeing sufficient water supply going forward require billions of euros of investments.[35] Nevertheless, investing in the water markets is not straightforward, not the least because of the dominant position of the government. They, in turn, are struggling with limited public financial capacity and policies that support low water prices to please electorates. PPPs might help to unlock some of the potential. However, there are very few countries that have a proven track record of timely and efficient PPP projects, and the procedures are often untidy. Turning the water market private is a very sensitive debate, and most countries are not ready for it (and it still comes with many risks and complexities for the companies managing the concessions). Beyond that, the private sector can suggest business models that come at the lowest cost to taxpayers but with higher efficiency than the current system. And ultimately there is the question, what is the WTP (willingness to pay) of residents? Awareness that only higher prices only guarantee solutions that embrace continued and secured water supply is necessary.

Water infrastructure and water technology seem the two easiest entry points for investments. There are many technologies high in demand but also high in evolution. Most of them are smaller private-market companies and therefore not always easy to identify.[36] The demand for efficient irrigation technologies, seawater desalination and sewage treatment facilities, technical equipment (e.g., pumps, compressors, and fittings), filter systems or disinfection processes (e.g., using ozone or ultraviolet light), and efficient sanitation facilities will probably rise sharply.[37] Often the bottleneck is to make scientists, companies, and policy makers work together efficiently in bringing these solutions to the market.[38] Which countries embody the most opportunities can be built around, for example, population (growth); environmental sensitivity; available water resources per capita; connectivity of people to the public sewage, wastewater, and drinking-water system; and geographical location, as well as the regulatory environment.[39]

17.5 Water investments

Water may very well turn out to be the critical commodity story of the 21st century. When you look at other industries such as timber, technology, mining, and consumer goods, the critical element to sustainability for these businesses is the availability of water in some usable form. To address the challenges, the global water industry will have no choice but to undergo a fundamental transformation from its present usage and pricing practices. Water has no economic substitute, at any price – the only substance or commodity in the world of which this is true. We cannot create nor destroy water (it is therefore better to talk of water "use" rather than water "consumption").

The investible "water industry" is really a collection of fundamentally different businesses which come from various economic sectors but all have something to do with the delivery/treatment of clean water. A more apt descriptor is "hydro-commerce," which includes utility companies plus their diverse supply chain of technology providers, engineering firms, and heavy-equipment manufacturers. The companies in this hydro-commerce universe are united not characterized by business type but by a far more important investible characteristic: consistently growing sales and earnings almost irrespective of external economic conditions. The opportunity to generate noncyclical, low-correlated returns on public hydro-commerce equities remains the most compelling fundamental for many investors in this area, given its relatively consistent returns.

Hydro-commerce may be more explicitly characterized as those companies that provide products and services enabling the flow of usable water from (1) initial raw supply sources, through (2) collection and treatment, to (3) distribution among the various types of end users, and finally through (4) wastewater treatment and disposal. These companies can be divided into two key sectors – (1) the water utilities themselves, which actually provide drinking-water and wastewater services to end users, and (2) the vast array of supporting industrial companies that provide the technologies, services, and "razor-blade" products that the utilities need in order to continue operating each and every day.[40]

17.5.1 Water rights and infrastructure

Additional and improved water-supply, sanitation, and irrigation infrastructure will be needed for countries to achieve their development goals, from human health and food security to energy security and climate resilience. While data on current water-infrastructure stocks and sector financing is sparse, estimates of needed investment globally are about 500 billion euros per year. To pay for this infrastructure, developing countries in particular will face serious obstacles, including low levels of government support and an inability to attract private finance. The recent financial crisis has made financing the gap in global water infrastructure even more difficult. All water-sector stakeholders can contribute to reforms that help close this gap. Service providers can improve performance and maintain accountability to customers. Governments and donors can stretch their dollars and collaborate to reduce risks for potential private-sector investors. The water-reform strategy in the Philippines provides an example of how each stakeholder can make its contribution to the reform process, helping to bridge the space between what the public needs and what private investors can provide. All stakeholders should focus on helping service providers achieve financial sustainability by improving cost recovery and public expenditure. Cost recovery that enables private investment is the optimal strategy for service providers to pay for current and future water infrastructure. However, it is politically difficult to implement even in developed countries like the United States. Recovering the full cost of delivering water services is complex. Reconciling the economic and financial objectives of cost recovery is hard enough, and this is exacerbated by considerations of environmental sustainability and social affordability. The mechanisms and instruments designed to manage these different objectives determine the role that tariffs and user charges can play, which will differ from water service to water service. Service providers can partially close the gap by lowering costs and making efficiency gains. Such efficiencies, even in the absence of full cost recovery, will improve the ability of utilities to adapt to future risk and will make them less dependent on external funding. Efficiencies will also translate into improved service, which can start the positive cycle of increasing revenues (for farms) or productivity (for people), prompting a higher ability and willingness to pay for services. Tariffs that reflect the cost of inflation can assist in maintaining the trend toward lower per capita consumption, while those that account for environmental externalities can go one step further by addressing water scarcity and supporting a green economy. Each and every case faces different challenges requiring a different solution.[41]

Another investment thesis is in water rights, which is fundamentally about buying at lower values and selling at higher prices that the market can bear. Water rights are legally authorized rights to use water from a water source and generally emerge from property ownership and the right to use adjacent bodies of water. The rights can also be created by contract when one person transfers rights to another. Apart from buying and selling water rights, success highly depends on stewardship of the water resource. Stewardship entails respecting the resource and its value and providing solutions in a way where everybody wins. There are opportunities for water-rights investors to work with municipalities in helping to aggregate water rights or provide annual allocations. The private sector can often bring nimbleness and value-add to negotiations with families and individuals who own the rights and help allocate the resource to higher-value needs.

17.5.2 Filtration

Following are some examples of filtration systems that are becoming more and more popular:

- Backflow preventers, which stop contaminated water from flowing back down pipes and polluting the potable source water

- Membrane treatment to filter out contaminants
- Pumps and filters to treat commercial and domestic water
- Ultraviolet disinfection to treat drinking water and cleanse wastewater

This part of the solution spectrum has been focused on for a long time, and therefore significant investment results cannot be expected.

17.5.3 Technology and innovation

Low water tariffs make, as indicated, adoption of technologies less attractive. Despite the constraints, there are increasing opportunities for technologies that enable greater efficiency in existing water-infrastructure systems and solutions that convert wastewater to a resource:

- *Increasing water efficiency.* Measuring and reducing non-revenue water, the gap between the amount of water injected into the distribution system and the amount of water billed to consumers, is important in addressing our water inefficiencies and an increasing opportunity for the sector. High levels of non-revenue water reflect a mixture of physical water loss through infrastructure leaks or metering and data issues. There is a direct relationship between high water pressure and leakage in municipal water systems. Therefore, appropriately managing water pressure results in fewer leaks and decreased consumption. Technology such as Netbase Water Management software, which provides municipalities with the data monitoring and analytics needed to manage water flow, system pressure, and water loss – ultimately promote more informed decision making.[42]
- *Water reuse and energy-efficient water.* Given increasing supply constraints, companies are turning to technologies that can enable wastewater to become a resource. These recycle and re-use solutions can meet water demand needs for non-potable water intensive sectors such as power generation. Technology solutions are also addressing the energy intensity associated with water delivery and treatment. One example is membrane bioreactor technologies (MBR), which filter water, recover by-products, and treat wastewater for recycling. MBR technology[43] is 30% more energy efficient than conventional water treatment technologies. Another example is a technology where the complete elimination of liquid discharge from an industrial manufacturing process can reduce both the waste and energy typically required for wastewater treatment. This process recovers distilled water for reuse while simultaneously creating beneficial salt byproducts, which can be used for de-icing roads or softening gray water.
- *Decentralization and distribution systems.* Decentralized wastewater systems can also reduce the energy intensity of large concentrated water systems, which require significant energy to deliver water to points of consumption.

> Urban systems rely on pressurized water tanks that pump water into a tower and allow gravity to maintain a constant system pressure as water is forced through the pipes. Decentralized water networks are adopted in developing countries where legacy infrastructure is absent. Regulations in many developing countries also facilitate the distributed-water model. For example in China, new buildings with more than 100 inhabitants are required to have self-contained wastewater networks. In the United States and other parts of the world, this is challenging given existing infrastructure, but there are opportunities for technology to provide more modular on-site systems, particularly to address industrial needs. For example, wastewater plants are addressing some of the water needs in the Pacific Southwest as much of the reusable water in large quantities can be redirected toward commercial purposes such as cooling water for power plants.[44]

Emerging best practices and new models are tested and developed every single day – often in a scenario where private parties and the (local) government are working together.[45]

17.6 The role of water in energy

There is an inextricable relationship between water and energy – each is exceedingly dependent on the other. Approximately 50% of water withdrawal is from the power-generation sector, resulting in extreme susceptibility to constraints in water availability.[46] The vast majority of water used in the energy sector is for cooling at thermal power plants. Carbon pricing can play a role in incentivizing less carbon-intensive power sources, such as renewables, which tend to be less water demanding.

With the rapid expansion of shale/hydraulic exploration in the United States, it is proving critical to develop technologies that address water and wastewater management in hydraulic fracturing ("fracking"). Though with greater use of natural gas for power generation there have been benefits of both lower greenhouse gas emissions and lower water use relative to coal-fired power generation, upstream fracking uses significant amounts of water and produces wastewater with fracking chemicals and high salinity. According to the Shale Gas Information Platform (SHIP), drilling and fracturing a typical horizontal shale gas well requires roughly 10 to 30 million liters of water, equal to eight times the volume of an Olympic-size swimming pool. Each well can also result in 20% of the water flowing back within the first 60 days. Technologies are helping to address the water challenge by enabling return water, also known as brine, to be recycled and reused. The scale-up of exploration and production efforts has resulted in multiple companies often leasing and drilling wells in the same area. A compelling case can be made for optimizing cooperation among oil and gas operators to pool water needs and facilitate local wastewater treatment and reclamation opportunities.

17.7 Desalination[47]

Already in 2004 a panel of experts concluded that desalination might have a role in achieving food security, although its major use will continue to be for drinking water.

- Applying water-desalination technology to agriculture is generally rather cost ineffective.
- To date, the application of water desalination in agriculture is limited to a small number of areas, only for certain high-value crops and with government subsidies in capital costs.
- Insofar as the technology for desalination will be expensive, limiting its application in agriculture, the focus should be on the safe reuse of treated wastewater to meet water demand prior to desalination.[48]

The best desalination technologies are distillation[49] (multistage flash, MSF) and membrane technologies (reverse osmosis, RO, and electrodialysis reversal, EDR). RO and EDR are applied to desalinate brackish water, with salt concentrations of less than 10 g/liter, while RO and distillation are applied for seawater, with a salt concentration of more than 30 g/liter. Distillation plants treat large volumes of water (55,000 m^3/d), almost exclusively seawater, and they are often built together with power plants (dual purpose). Current trends show that distillation costs are falling because of economies of scale (large plants), and RO costs are decreasing more rapidly because of new technology developments, competition, and economies of scale. Water desalination has positive impacts on the environment, such as increasing water availability and recycling poor-quality water. However, it also has several negative impacts, mainly brine disposal of residues from desalination;[50] chemical additives used for antifouling, anticorrosivity, etc.; visual impact on the landscape; noise; and emission of greenhouse gases. The complexity of brine disposal from inland areas is greater than that for coastal areas. Distillation impacts are considered to be the worst. Desalinated water is more expensive than conventional water resources, and it is not (yet) affordable for most crops.[51] However, desalinated water might be affordable for high-value crops, especially where subsidies on capital costs are provided. Desalinated water is of high quality and can have less negative impact on soils and crops in comparison with direct use of brackish water.

For cost considerations, brackish-water desalination is more suitable for agricultural production than is seawater desalination. Moreover, desalination facilities near the point of use are preferred in order to minimize transfer costs. In terms of operation and maintenance (O&M), small to medium plants are more problematic. Wastewater and water desalination constitute potential sources of water for agriculture and other uses. Technologies for tertiary wastewater treatment and desalination have very much in common. However, the cost of treatment varies depending on the type of treatment and the intended final use of product water. Treated wastewater reuse in agriculture is less expensive than is desalinated water. With its associated benefits, treated wastewater reuse also has problems in terms of public acceptance and potential health and environmental risks.

The story of desalination is a story that is still being written. Not surprisingly, those countries that are freshwater strapped are those that see most technological and financial attention for the field – Israel[52] was mentioned but also Saudi Arabia,[53] and in fact the whole Middle East & North Africa (MENA) region enjoys increasing attention, and this both for consumption as well as agricultural purposes.

17.7.1 Desalination technology

Desalination is a separation process that produces two streams: freshwater and saline solution (brine). Saline water is classified as either brackish water or seawater depending on the water source, brackish water being less saline than seawater. Two main commercial desalination technologies have gained acceptance throughout the world, namely those based on thermal and on membrane processes. Thermal processes, except freezing, mimic the natural process of producing rain. Saline water is heated, producing water vapor, which in turn condenses to form freshwater, thus producing freshwater by distillation. These processes include multistage flash, multiple-effect distillation (MED), and vapor-compression (VC) distillation. In all these processes, condensing steam is used to supply the latent heat needed to vaporize the water. Owing to their high-energy requirements, thermal processes are normally used for seawater desalination and in dual-power and water-production plants. In addition, thermal processes are capable of producing high-purity water, do not require sophisticated pretreatment, and are not sensitive to water contamination by oil or other organic matter.[54] One can further distinguish thermal and membrane desalination. Technological trends include integrated membrane solutions, increased energy efficiency, and increased recovery ratio for seawater RO. New developments have and will also witness lower use of materials, fewer chemicals, and smaller footprints.

17.7.2 Water desalination and agriculture

Although there is a broad experience in the use of desalinated water for agricultural purposes, desalination technologies are only regularly used in some countries because of the costs involved in this process. A cost analysis is essential in order to determine whether water desalination may be feasible at the beginning of the 21st century to produce a water resource that could be used to complement or substitute natural water resources in areas with water shortages. Therefore, it is necessary to analyze the factors influencing the water desalination costs of the different desalination technologies. To obtain an average cost of desalinated water, it is necessary to consider three aspects:

- the technology to be used for desalination;
- the quality of the feed water to be desalinated; and
- the required quality of the product water.

The cost structure of desalination differs somewhat per country, but a fair average reflection is as follows:

- Power: 37–43%
- Payback: 33–43%
- Membranes: 3.5–4.5%
- Chemical cleaning: 0.2–0.3%
- Membrane replacement: 2–5%
- Labor: 4–11%
- Chemicals: 4–7.5%

Distillation costs are higher than RO costs, even accepting variations among the different distillation technologies. For this reason,[55] distillation technologies are not (yet) feasible (from an economic point of view – in reality desalination is often used for agriculture in many parts of the world, often publicly funded) for producing water to be used in agriculture, although exceptions exist in places where cost structures and economies of scale work together in harmony. If the costs of desalinated water for agriculture have to be as low as possible, the trends in the last years to reduce costs should be considered:

- Improved membrane performance and efficiency
- Enlarged capacity of desalination plants in order to achieve an economy of scale and reduce investment costs
- Falling electricity tariffs
- Savings in labor costs, which is the second of the main operating costs
- Enlarged capacity of the production lines and racks
- Increased efficiency of mechanical equipment
- Falling equipment costs

The major issues for discussion are whether this type of desalination plant is appropriate to produce water for use in agriculture, as well as the factors that can increase the costs of the desalinated water and decrease the profitability of the crops to be grown, such as the following:

- the available area to be cultivated,
- the distance from that area to the desalination plant, and
- the existing infrastructure for water distribution.[56]

Desalination of brackish water has long been used as a way of improving the quality of water applied in agriculture. In this case, the energy costs are also the main operating costs. However, as the differences in water salinity are higher than in seawater, these costs can vary considerably, and consequently the total production costs can also vary. A second aspect of desalination of brackish waters is that different designs can be applied according to the quality of the feed water and the required salt content of the desalinated water. Therefore, costs can have large variations.

In addition to the favorable effect of producing water of good quality and preventing the soil-salinization hazard of irrigating with brackish water, water desalination technologies have certain environmentally unfavorable impacts:

- Carbon dioxide (CO_2) production related to energy consumption, although it is lower than other domestic and industrial processes. It is also lower in RO and EDR than in distillation technologies.
- Contamination by chemicals used in pretreatments that are rejected by the membranes, by products used for membrane and tube cleaning, and by coagulants and aids in filters.
- Impact of buildings on the landscape.
- Influence of the brine discharge on marine flora and fauna and on endangered species. Temperature (more relevant in distillation technologies), pH, and salt concentration are the major parameters to consider.

To reduce CO_2 emissions, reducing energy consumption is the main issue. With RO, energy savings of more than 80% have been achieved compared with distillation. Buildings need to be integrated into the landscape. Brine discharge in coastal areas is diluted by mixing with freshwater. Depth of discharge and other design aspects of the disposal pipeline (one or several exits, areas of agitation, and angle of the rejected flow) enable unfavorable environmental impacts to be reduced. For inland disposals, injection into deep wells of contaminated aquifers or discharge into large-flow rivers could be the solution.

Desalination of water for agriculture is technically feasible, and the appropriate technology is available. Therefore, only economic and environmental considerations can limit its application, an aspects being furiously worked on.

Water quality and agriculture interactions are many and complex. The development of large irrigation schemes has been an important contributor to global food security, particularly in arid areas, but it has also been associated with land- and water-salinity problems. Both expansion and intensification of agriculture have led to an increasing use of fertilizers and pesticides that, when not well managed, has degraded the water quality of rivers, lakes, and marine water bodies. Intensification of livestock farming systems is a case in point: concentrating inputs increases the potential transmission of pollution from both animal waste and fodder production, and if not managed or regulated well, eutrophication of freshwater bodies can easily result. At the same time, such is the level of water scarcity and pollution that millions of farmers worldwide are driven to irrigate with marginal-quality water such as wastewater from urban areas or saline agricultural drainage water. Minimizing both the production and food safety risks, and at the same time maximizing benefits when using such water, is an enormous challenge. Additionally, concerns about the use of naturally occurring arsenic-laden groundwater in agriculture are growing, and therefore this emerging issue will need special attention. Salinity is the most important criterion for evaluating the quality of irrigation water because of the potential crop-yield reductions that can result from the use of saline water, which inhibits water uptake by plants. Agricultural practice tends to induce accumulation of salt in land and water. Salts accumulated in soils can be mobilized by irrigation practice through the modification of water circulation across land. In addition, the pumping of groundwater can induce saline intrusion in coastal aquifers or the migration of low-quality water from underlying aquifers.[57]

17.7.3 Current status and the way forward for desalination

As discussed, there are a number of methods for removing salt from seawater or brackish groundwater to provide water for consumption and agricultural purposes. The two most common processes, thermal distillation and reverse osmosis, are described below somewhat more extensively than before. Their descriptions are followed by descriptions of some of the more innovative and alternative desalination technologies.[58] The earliest commercial plants used thermal techniques. Improvements in membrane technology have reduced costs, and membrane technology is less energy intense than thermal desalination (although it is more energy intense than most other water-supply options). Reverse osmosis and other membrane systems account, for example, for nearly 96% of the total US desalination capacity and 100% of the municipal desalination capacity.

17.7.4 Reverse Osmosis

Reverse osmosis forces salty water through a semipermeable membrane that traps salt on one side and lets purified water through. Reverse-osmosis plants have fewer problems with corrosion and usually have lower energy requirements than thermal processes.

Examples of how research advances in the traditional desalination technologies of reverse osmosis have the potential for improving the competitiveness and use of desalination are nanocomposite and nanotube membranes and chlorine-resistant membranes. Nanocomposite membranes appear to have the potential to reduce energy use within the reverse-osmosis process by 20%, and nanotube membranes may yield a 30% to 50% energy savings.[59] Membranes are susceptible to fouling by biological growth (i.e., biofouling), which reduces the performance of the membranes and increases energy use. The most widely used biocide is chlorine because it is inexpensive and highly effective. The most common membranes used in reverse osmosis, however, do not hold up well to exposure to oxidizing agents like chlorine. Advancements in chlorine-resistant membranes would increase the resiliency of membranes and expand their applications and operational flexibility.[60]

17.7.5 Distillation

In distillation, saline water is heated, separating out dissolved minerals, and the purified vapor is condensed. There are three prominent ways to perform distillation: multistage flash, multiple-effect distillation, and solar distillation. In general, distillation plants require less maintenance and pretreatment before the desalination process than reverse osmosis facilities. While solar distillation is an ancient means for separating freshwater from salt using solar energy,[61] research into improving the technology is increasing. In large part the interest stems from the potential application for the technology to supply freshwater to small remote settlements where saline supplies are the only source and power is scarce or expensive.

17.7.6 Innovative and Alternative Desalination Processes

A. Capacitive Deionization

Capacitive deionization desalinates saline waters by absorbing salts out of the water using electrically charged porous electrodes. The technology uses the fact that salts are ionic compounds with opposite charges to separate the salts from the water. The limiting factor for this technology is often the salt absorption capacity of the electrodes. The technology shows promise for energy-efficient desalination using electrodes of optimized pore size.

B. Electrodialysis

Electrodialysis and capacitive deionization technologies depend on the ability of electrically charged ions in saline water to migrate to positive or negative poles in an electrolytic cell. Two different types of ion-selective membranes are used – one that allows passage of positive ions and one that allows negative ions to pass between the electrodes of the cell. When an electric current is applied to drive the ions, freshwater is left between the membranes. The amount of electricity required for electrodialysis, and therefore its cost, increase with increasing salinity of feed water. Thus, electrodialysis is less economically competitive for desalting seawater compared to less saline, brackish water.

C. Forward osmosis

Forward osmosis is an increasingly used but relatively new membrane-based separation process that uses an osmotic pressure difference between a concentrated "draw" solution and the saline source water; the osmotic pressure drives the water to be treated across a semipermeable membrane into the draw solution. The level of salt removal can be competitive with reverse osmosis. A primary challenge is in the selection of a draw solute; the solute needs to either be desirable in the water supply or be easily and economically removed. Research is being conducted on whether a combination of ammonia and carbon dioxide gases can be used as the draw solution. The attractiveness of forward osmosis is that its energy costs can be significantly less than for reverse osmosis when combined with industrial or power-production processes.[62] A disadvantage of this technology is that it yields a lower quantity of freshwater per unit of water treated and a larger quantity of brine that requires disposal.[63]

D. Freezing processes

Freezing processes involve three basic steps: (1) partial freezing of the feed water in which ice crystals of fresh water form an ice-brine slurry, (2) separating the ice crystals from the brine, and (3) melting the ice. Freezing has some inherent advantages over distillation in that less energy is required and there is a minimum of corrosion and scale formation problems because of the low temperatures involved. Freezing processes have the potential to concentrate waste streams to higher concentration than other processes, and the energy requirements are comparable to reverse osmosis. While the feasibility of freeze desalination has been demonstrated, further research and development remains before the technology will be widely available.

E. Ion exchange

In ion exchange, resins substitute hydrogen and hydroxide ions for salt ions. For example, cation exchange resins are commonly used in home water softeners to remove calcium and magnesium from "hard" water. A number of municipalities use ion exchange for water softening, and industries requiring extremely pure water commonly use ion exchange resins as a final treatment following reverse osmosis or electrodialysis. The primary cost associated with ion exchange is in regenerating or replacing the resins. The higher the concentration of dissolved salts in the water, the more often the resins need to be renewed. In general, ion exchange is rarely used for salt removal on a large scale.

Given its widespread use, ion exchange accounts for more than 2.4% of total US municipal and industrial freshwater use.[64] Two-thirds of the US desalination capacity is used for municipal water supply; industry uses about 18% of the total capacity.[65] Open questions however include how to focus research, at what level to support desalination research and projects, and how to provide a regulatory context that protects the environment and public health without disadvantaging the technology. Coastal communities look to seawater or estuarine water, while interior communities look to brackish aquifers when considering desalination. The most common desalination technology in the United States is reverse osmosis, which uses permeable membranes to separate the freshwater from the saline water supply. Membrane technologies are also effective for other water-treatment applications. Many communities and industries use membranes to remove contaminants from drinking water, treat contaminated water for disposal, and reuse industrial wastewater (e.g., saline waters co-produced from oil and gas development). For some applications, there are few competitive technological substitutes. Wider adoption of desalination is constrained by financial, environmental, and regulatory issues. Although desalination costs dropped steadily in recent decades, significant further decline may not happen with existing technologies. Electricity expenses represent from one-third to one-half (see infra) of the operating cost of desalination. Its energy intensity also raises concerns about associated greenhouse gas emissions and its usefulness as a climate-change-adaptation measure. Substantial uncertainty also remains about the technology's environmental impacts, in particular management of the saline waste concentrate and the effect of surface-water intake facilities on aquatic organisms. Desalination facilities require a significant number of local, state, and federal approvals and permits. Emerging technologies (e.g., forward osmosis, nanocomposite, and chlorine-resistant membranes) show promise for reducing desalination costs. Research to support development of emerging technologies and to reduce desalination's environmental and social impacts is particularly relevant to the debate on the future level and nature of federal desalination assistance. The federal government generally has been involved primarily in desalination research and development (including for military applications), some demonstration projects, and select full-scale facilities. For the most part, local governments, sometimes with state-level involvement, are responsible for planning, testing, building, and operating desalination facilities. Some states, universities, and private entities also undertake and support desalination research. While interest in desalination persists among some, especially with drought concerns high, efforts to maintain or expand federal activities and investment are challenged by the domestic fiscal climate and differing views on federal roles and priorities.

Desalination treatment costs have dropped steadily in recent decades, making it more competitive with other water-supply augmentation and treatment options. Electricity expenses vary from one-third to one-half of the cost of operating desalination facilities.[66] A rise in electricity prices could reverse the declining trend in desalination costs; similarly, drops in electricity costs (e.g., due to falling costs associated with natural gas–fueled electric generation) would improve desalination's competitiveness. Costs and cost uncertainties remain among the most significant challenges to implementing large-scale desalination facilities, especially seawater desalination plants.[67] Desalination's energy intensity also raises concerns about the greenhouse gas emitted and its usefulness as part of a climate-change-adaptation strategy.[68] Substantial uncertainty also remains about the environmental impacts of large-scale desalination facilities.[69]

17.7.7 Energy concerns and alternative-energy use

The cost of desalination for consumption water remains a barrier to adoption. Like nearly all new freshwater sources, desalinated water comes at substantially higher costs than existing municipal water sources. The competitiveness of reverse-osmosis seawater desalination is highly dependent on the price of electricity. Reverse osmosis pushes water through a membrane to separate the freshwater from the salts; this requires considerable energy input. Currently the typical energy intensity for seawater desalination with energy recovery devices is three to seven kilowatt-hours of electricity per cubic meter of water (kWh/m^3). The typical energy intensity of brackish desalination is less than seawater desalination, at 0.5 to 3.0 kWh/m^3, because the energy required for desalination is a function of the salinity of the source water.[70] Uncertainty in whether electricity prices will rise or fall creates significant uncertainty in the cost of desalinated water. If electricity becomes more expensive, less electricity-intensive water supply options (which may include conservation, water purchases, and changes in water pricing) become comparatively more attractive. Recent drops in natural-gas prices and little to no growth in electricity demand has increased the cost competitiveness of existing desalination technologies in recent years.

Cost-effectively reducing desalination's energy requirements could help reduce overall costs. In recent decades, one of the ways that desalination cost reductions were achieved was through reduced energy requirements of reverse-osmosis processes. Now the energy used in the reverse-osmosis portion of new desalination facilities is close to the theoretical minimum energy required for separation of the salts from the water.[71] Therefore, although there still is some room for energy-efficiency improvements in using desalination as a water supply, dramatic improvements are not likely to be achieved through enhancements to standard reverse-osmosis membranes. Instead energy-efficiency improvements are more likely to come from other components of desalination facilities, such as pretreatment[72] of the water before it enters the reverse-osmosis process, enhanced facility and system designs, or the use and development of a new generation of technologies (supra).

For example, energy-efficiency advances in the nonmembrane portions of water systems and the use of energy-recovery technologies are reducing energy use per unit of freshwater produced at desalination facilities. Pumps are responsible for more than 40% of total energy costs at a desalination facility.[73] Energy efficiency advances in a type of pump that is useful for smaller applications (called a positive displacement pump) have made desalination more cost effective for some applications and locations and less sensitive to electricity price increases.[74]

Desalination's electricity consumption has greenhouse gas and other emissions associated with it if the electricity is generated using fossil fuels. The use of desalination as a climate-change-adaptation strategy is questioned because of its potential fossil-fuel intensity relative to other adaptation and water-supply options.[75] Electricity price uncertainty and emissions considerations have driven many desalination proponents to investigate renewable energy supplies and collocation with power plants.[76] The extent to which desalination technologies can be coupled with intermittent renewable or geothermal electric generation,[77] use off-peak electricity, and operate in areas of limited electric generation or transmission capacity but with renewable energy resources is increasingly receiving attention. Desalinating more water when wind energy is available (which requires facilities that can operate with a variable water inflow) and storing the treated water for when water is demanded can almost be viewed as a means of electricity storage and reduction of peak demand.[78] Efforts to jointly manage water and energy supply and demand and to integrate renewable energy with desalination may bolster support for desalination.

The present deployment of renewable-based desalination – less than 1% of desalination capacity based on conventional fossil fuels[79] – does not reflect the advantages of this technology option. Renewable desalination (see Table 17.1) is mostly based on the RO process (62%), followed by thermal processes such as MSF and MED. The dominant energy source is solar photovoltaics (PV), which are used in some 43% of the existing applications, followed by solar thermal and wind energy.[80]

Table 17.1 Possible combinations between desalination technologies and renewable energy[a]

	Membrane technologies				
Thermal technologies	**Multistage flash, MSF**	**Multi-effect distillation, MED**	**Vapor compression, VC[b]**	**Reverse osmosis, RO**	**Electrodialysis, ED**
Renewable technologies	x	x	x	x	x
Solar thermal			x	x	x
Solar PV			x	x	x
Wind	x	x	x	x	x
Geothermal	x	x	x	x	x

Note: [a] Adapted from A. A. Al-Karaghouli et al. (2011), "Renewable energy opportunities in water desalination, desalination, trends and technologies", M. Schorr (ed.), ISBN: 978-953-307-311-8, InTech (retrieved: http://cdn.intechopen.com/pdfs/13758/ InTech-Renewable_Energy_opportunities_in_water_desalination.pdf). For an extensive write-up of the individual renewable technologies see IEA-ETSAP and IRENA Technology Brief I12, p. 9–12.

[b] Vapor compression (VC) distillation process, where the heat for water evaporation comes from compression rather than from direct heating. This process is generally used in combination with other processes (MED) to improve overall efficiency

As energy cost is a large chunk of the total cost, the feasibility of these technologies is a matter of the energy cost involved. Different parameters will determine the outcome.[81]

As a final note, water supply is new in most parts of the world, which means the health and environmental regulations, guidelines, and policies regarding its use are still being developed. Existing federal, state, and local laws and policies often do not address unique issues raised by desalination. This creates uncertainty for those considering adopting desalination and membrane technologies. Environmental and human health concerns are often raised in the context of obtaining the permits required to site, construct, and operate the facility and dispose of the waste concentrate. While the quality of desalinated water is typically very high, some health concerns remain regarding its use as a drinking-water supply. The source water used in desalination may introduce biological and chemical contaminants to drinking-water supplies that are hazardous to human health, or desalination may remove minerals essential for human health. For example, boron, which is an uncommon concern for traditional water sources, is a significant constituent of seawater and can also be present in brackish groundwater extracted from aquifers comprised of marine deposits. Boron levels after basic reverse osmosis of seawater commonly exceed current WHO health guidelines. While the effect of boron on humans remains under investigation, boron is known to cause reproductive and developmental toxicity in animals and irritation of the digestive tract, and it accumulates in plants, which may be a concern for agricultural applications.[82] Boron can be removed through treatment optimization, but that treatment could increase the cost of desalted seawater.

Similarly, demineralization (particularly the removal of the essential minerals calcium and magnesium) by desalination processes can also raise health concerns.[83] This has prompted researchers to promote the remineralization of desalinated water prior to the water entering the distribution system in communities that are highly dependent on desalinated water.[84] Another health-related concern is the extent to which microorganisms unique to seawater and algal toxins may pass through reverse osmosis membranes and enter the water supply, and how facilities may need to be operated differently when these organisms and algal toxins are present. Algal toxins are a consideration for desalination facilities in locations affected or potentially affected by harmful ocean algal blooms that can produce a range of substances ranging from toxic to neurotoxic (e.g., domoic acid). How to effectively manage desalination facilities in order to avoid public health threats from algal blooms is an emerging area of interest and research.[85] For inland brackish desalination, significant constraints on adoption of the technologies are the uncertainties and the cost of waste-concentrate disposal. For coastal desalination projects, the concentrate management options are often greater because of surface-water disposal opportunities. The disposal option selected is largely determined by which alternatives are appropriate for the specific characteristics of the concentrate and the cost effectiveness of the alternatives, which is largely shaped by the proximity of the disposal option and the infrastructure, land, and treatment investments required. Disposal options typically include land application, evaporative ponds, surface-water disposal, or deep-well injection. Land application can include spraying concentrate on salt-tolerant plants or infiltration; land application typically is used for small volumes of brackish-water concentrate. Evaporation ponds use solar radiation to precipitate salt crystals, which are then harvested and typically disposed; in some cases the salts or other constituents may be beneficially reused. Surface-water disposal of waste concentrate is permitted on a project-specific basis based on predicted acute and chronic effects on the environment.[86] Inland surface-water disposal is particularly challenging because of the limited capacity of inland water bodies to be able to tolerate the concentrate's salinity. In some cases a limited amount of concentrate can be sent to a large-volume wastewater treatment facility. Most regulatory bodies generally classify waste concentrate as an industrial waste, thus requiring that the concentrate be disposed of in deep wells appropriate for industrial waste. Desalination proponents argue that desalination's concentrate is sufficiently different from most industrial waste that it should be reclassified to increase the surface and injection-well disposal opportunities.[87] Most

Table 17.2 Summary of major desalination technologies

Thermal technologies	Membrane technologies
Multistage flash, MSF	Reverse osmosis, RO
Multi-effect distillation, MED	Electrodialysis, ED
Vapor compression, VC	

countries have made efforts to promote the beneficial use of waste concentrate (e.g., use as liquids in enhanced oil and gas recovery) and facilitate its disposal, including land-application techniques.

It should be noted that there has been a fallback in the desalination market due to the following:

- The completion of major desalination-plant-building programs in Algeria, Spain, and Australia.
- The weakness of the real-estate market in the UAE, southern Spain, and the western United States.
- Project delays and postponements related to a variety of issues such as elections, financing difficulties, heavy rains, and administrative issues.
- Increased energy prices starting 2011 will probably continue to restrain market recovery. Despite some level of easing in the period 2012–2014, they still are in many countries hovering around all-time highs[88]

18
Future Agricultural Dynamics

During the last few decades we have been experiencing production surpluses in the developed world and stagnating growth in the developing part of the world – most of it being the result of (intended) policies. It is the policy reforms combined with the economic growth in emerging economies that the world is now and has been experiencing which led to changing patterns in the demand/supply relationships for most agricultural commodities. It has turned the agricultural industry into a market-driven environment ready to accept investments. This has also led to an increase in agricultural trade relations and volumes – a phenomenon the developing nations of the world have been benefiting from most. Despite the technology improvements in the industry, it is likely to experience a decline of productivity in the short to medium term. Although projections conclude that supply should keep track of increasing global demand, the margin is minimal, and therefore it is to be expected that agricultural prices will stay relatively high.

A minimal margin is not comforting in an environment where the world is still trying to figure out what the impact will be of changing climate patterns on output, availability of arable land, changing patterns in consumer behavior, etc. Within this context, the world will consider food-waste or food-loss reducing measures to expand that margin of safety. It will even be forced to rethink the way agricultural products are distributed from producer to end customer and change some of the unholy practices that have slipped into the buying practices of large distribution operators (products should look nice, not too small or large, need to fit in their box sizes, etc. – the rest will be destroyed and priced into the goods that are available for purchase in the stores).

Judged on a global scale, we can't get around what will happen in China, which, just as has been the case with other commodities (oil, copper, aluminum, etc.), will import agricultural products as it sees fit to supply a rising demand across the nation. With changing dietary habits, rising income levels, and urbanization, China's demand levels will most certainly outpace its anticipated future supply. Even if that shortage can be sourced on the international markets (Russia seems like a great first-line supplier), it inevitably will lead to additional food-price inflation both in China as well as on the international markets.

In what follows I will discuss the thematic issues or components in the mix that will determine the outlook for agricultural products and their pricing on the international markets. I will follow the technical criteria as set out in the most recent OECD-FAO Agricultural Outlook 2013,[1] which can also be referred to for further numerical data support.

18.1 Price evolution of agricultural commodities

At the closing date of the manuscript of this book, most agricultural commodities were trading at high levels judged from a historical perspective.[2] Although some of them are priced somewhat below their 2009–2010 and 2012 peak prices, others are trailing that peak very well, and others are pushing the boundaries even higher compared to past pricing levels. In review, the agricultural markets have been holding up very well in recent years despite the slowdown we have experienced in economic terms. It is expected that in the years to come, prices will come back down somewhat due to improved output numbers and normalization of production after a somewhat extraordinary 2012–2013 season. Having said that, prices will hold tight at relatively high levels, and peak prices will be tested every now and then when we move through the economic cycles and agricultural-production cycles.

Livestock will be a notable exception to that trend as global livestock inventory allow for only limited supply response to significantly rising demand from emerging economies, keeping meat prices (very) high for the years to come. However, for certain commodities, such as sugar, cotton, cereals, and oilseeds, prices are expected to stay on par in real terms relative to the last decade.

Judged over longer periods of time, both crop and meat prices are expected to rise due to a lower productivity growth (relative to demand) and stronger demand coming from emerging markets and alternative energy sources like biofuels. The pace at which the global economic levels of activity will pick up will magnify or mitigate those dynamics. Compared to the last decade (2003–2012), the next decade (2013–2022) will experience a reduction in the growth (relative to the last decade) of agricultural output globally. To be precise: –3 % for the OECD region, –50% for the BRICS region, –33% in frontier markets, –5% in the remaining part of the world, and –26% on a global basis. Global agricultural production will slow to 1.5% from 2.1% in the last decade. Dynamics that feed into this are increasing cost levels, resource constraints, and the already mentioned environmental pressures which impact the supply dynamics across all agricultural commodities.

Although agricultural markets have proven to be very resilient in general to economic downturns, a significant macroeconomic vulnerability remains in the form of changing oil prices. Rising oil prices make every assumption in terms of expected agricultural output uncertain – hereby ignoring the additional impact of a possible strengthening/depreciating US dollar and the impact of global levels of competitiveness in the industry. Particularly in developing economies where 30% to 50% of budgets go to food purchases, rising food inflation will be a persistent problem.

As is the case in other parts of the global economy, there is somewhat of a bifurcation manifesting between developing and developed nations. In agricultural terms, the output of emerging economies is increasing due to changing policies and the application of newer technologies that enhance productivity. This albeit with varying levels across the world, where Sub-Saharan Africa is a notable laggard on the productivity spectrum relative to other parts of the world.

Overall, these assumptions are prone to externalities, as assumptions or projections always are. Externalities can be as simple as productivity shortfalls, price volatility, and trade interruptions as well as geopolitical dynamics and significantly changing or extraordinary weather patterns and will all have an impact on pricing, most likely in an upward fashion. Statistical-variability analysis demonstrates that wheat, grains, rice, and oilseeds are the commodities most vulnerable to changing crop-yield variability, whereas meat, dairy, and biofuels are most vulnerable to macroeconomic dynamics such as growth and exchange rates. Overall, energy prices and world-trade patterns of agricultural commodities tend to magnify these dynamics through input costing.

Net exports (exports minus imports) will rise globally but very asymmetrically, with only the Americas (both North and Latin America) being and staying the largest net exporters. Asia, the Middle East, and North Africa will increase their trade deficits in net terms.

18.2 Commodity-specific demand/supply dynamics

- Wheat/grains: production growth expected of 16% to 22%, but lower yield growth, which will not be compensated by expansion. Demand driven by human consumption (68% of total demand).
- Rice: slowing of yield growth to 1%.
- Cereal: production expected to grow, but vulnerable due to low stock-to-use ratios. Major producers in emerging economies (more than 50% of global output).
- Oilseed (vegetable and palm oil): expected increase in output due to yield gains (26%). Will compete arable land away from production of other

agricultural commodities. Production of palm oil is also projected to expand to reach a 34% share of total vegetable oil production. Faster growth (above the rise in growth due to enhanced human consumption) of vegetable oil use for biodiesel is expected in developed countries, particularly in the European Union to 2022 as a result of the Renewable Energy Directive (RED).

- Sugar: production expected to increase (globally about 2% per annum) – Brazil (50% of global exports) and India as leading producers. Developing economies are the largest producers and are expected to experience the largest growth in consumption compared to the developed world (2.5% growth versus 0.5%). Moderate-yield growth expected. Additional production that is projected to come from sugarcane rather than sugar beets.
- Cotton: cotton will lose market share to man-made fibers (also due to its explosion in spot prices in recent years). India is the largest producer with an expected rise of 25% in output while China's output drops by an expected 17%. World cotton production is expected to grow by 1.6% per annum, marginally more slowly than consumption at 1.7% per annum.
- Ethanol: production is expected to rise, with nearly 70% in the next decade, and the same is expect for biodiesel. Problematic as it might be, biofuel production will account in ten years' time for about 28% of sugarcane, 15% of vegetable oil, and 12% of grain consumption globally.
- Meat: developing economies will deliver a growth of about 80% in global meat production. Growth in per capita meat consumption globally will slow as consumption in developing countries will reach the levels of developed nations. Poultry will account for most of the expected growth in consumption. World production growth will slow from 2.3% during the last decade to 1.6%. Larger meat production is expected to originate mostly from productivity growth, notably from improved genetics, animal health, and feeding practices. Despite the projected faster growth in meat consumption in developing countries, per capita consumption of meats will remain much higher. Per capita consumption growth is projected to be spread over all the meat types in developing countries, with the largest increase expected for poultry consumption, followed by pig meat and beef.
- Milk: production is expected to decline (expected growth rates to decline from 2.3% to 1.8% the last decade) due to high feed costs for feed-based operations, while pasture-based systems face water shortages and land competition. Incremental output the next decade will come three-fourths from developing nations (50% of that incremental output will come from India and China only). Consumption of dairy products will grow faster in developing economies, with the United States, Oceania, and Europe as the main exporters. Consumption of dairy products in developing countries is

expected to increase on average at about 2.2% p.a. and to basically match the growth in supplies. The expansion in demand reflects robust income growth and growing affluence, increasing populations, further Westernization of diets, and greater access to refrigeration facilities.

- Capture fisheries: Expected growth in output only 5% (aquaculture's expected growth 35%) during the next decade. Major issues: high feeding costs and limited availability of production locations. Aquaculture will surpass capture as the main source for human consumption in the next two to three years. Total harvesting in 2013: 180 million tonnes. Consumption of fish products is expected to increase strongly over the coming decade. Per capita fish consumption will continue to be higher in the more advanced economies. Aquaculture's share of human consumption of fish is projected to exceed that of capture fisheries in 2015 and to reach 53% of total human consumption by 2022. Consumption of fishmeal and fish oil is expected to be constrained by production, which will continue to be dependent mostly on the highly regulated capture fisheries.

Note: Capture fisheries are also known as wild fisheries. Capture fishery refers to all kinds of harvesting of naturally occurring living resources in both marine and freshwater environments. The aquatic life they support is not controlled in any meaningful way and needs to be "captured." Aquaculture, also known as aquafarming, is the farming of aquatic organisms. Aquaculture involves cultivating freshwater and saltwater populations under controlled conditions and can be contrasted with commercial fishing, which is the harvesting of wild fish. According to the FAO, aquaculture "is understood to mean the farming of aquatic organisms including fish, mollusks, crustaceans and aquatic plants. Farming implies some form of intervention in the rearing process to enhance production, such as regular stocking, feeding, protection from predators, etc. Farming also implies individual or corporate ownership of the stock being cultivated."[3]

Most of the production gains are due to improved productivity. A breakdown of expected productivity changes per product compared to the last decade yields the following results (in percentage, area/yield):

- Cotton: 2/2
- Rice: 2/12
- Wheat: 5/11
- Sugar: 13/4
- Grains: 8/15
- Oilseeds: 10/14
- Milk: 16/7

18.3 China: the 800-pound gorilla in the room

Despite its enormous landmass, China holds very little agricultural land and has access to limited water resources. A country in transition, with massive urbanization and the like, the 4.5 times increase in output it experienced during the last three decades is about to slow in coming years due to resource and rural labor constraints. The Communist Party understands that food security, mitigating food inflation, and self-sufficiency as a country in this respect are key for the country to stay socially stable to some degree. Although China's agricultural imports and exports increased from 27.9 to 155.7 billion US dollars over the last decade, its import dependence increased from 6.2% to 12.9%, leading to a net agricultural trade deficit of 31 billion US dollars at the end of 2012.

Going forward it is expected that China's increasing consumption will outpace its production output by about 0.3% per annum, which is broadly speaking in line with the actual situation of the last 10 years. In its latest five-year plans, the government has set specific targets with respect to areas and production of the most important commodities (wheat, rice, soybeans, and grain) and focused on preventing land from exiting the agricultural space. While most imports of agricultural commodities are expected to rise, it is expected to drop more than 20% for cotton. This is mainly caused by the fact that a lot of textile production has move outside the country into India and other Southeast Asian low-labor-cost countries.

Meat and dairy sectors are expanding in terms of consumption and consequently require imports. For some of the meat categories, China will become the largest consumer worldwide over the next ten years. The same story holds true for milk production, which is expected to slow, although a 38% rise in consumption is expected over the same period. Imports of milk are expected to rise 20%. It will maintain its global role in the fisheries sphere and will see its output rise, but at a much lower rate than in the last ten years. Nevertheless, it is expected that it can maintain its leadership position with about 68% of global output as the largest fisheries exporter globally.

Many of the assumptions about China's agricultural development are built around wider economic factors, such as overall economic growth and its sustainability, increasing resource constraints on production, and potential vulnerability with respect to increased production due to climate change and severe weather events in particular.

18.4 Food-price inflation

As indicated earlier, prices are hovering just below or at their peak levels of 2009–2011. Although it is expected that prices will come down somewhat for most agricultural commodities except for livestock, the message is clear that we will have to deal with consistently higher food prices going forward. In the short term prices will come down due to higher supply levels and stock replenishments due to higher prices (higher prices triggered more investments), but in the longer term increased energy prices, industrial use, growing demand, and changing diets combined with more frequent price volatility, often policy induced, will keep prices consistently high. Fundamentally, agricultural markets have changed in the last decade, where global production tries to adjust to demand signals it received from the market, often responding in cycles. Nevertheless, increased production has not been able to drive down prices to pre-2007 levels. In fact, medium- to long-term analysis shows enhanced volatility and wider uncertainty in forecasts, which could put significant upward price pressure on some categories. The FAO-OECD outlook 2013[4] indicates, "These price projections are predicated on the key assumption of normal production conditions and the absence of unforeseen market shocks such as droughts and animal disease outbreaks. When this assumption is relaxed, as in the discussion of Outlook uncertainties in the last section of the chapter, agricultural commodity price prospects clearly become much more variable. Moreover, with stocks-to-use ratios anticipated at or near historical lows for many commodities in later years of the outlook period, there is more risk of price rises than falls with further periods of price surges and bouts of enhanced volatility in the coming decade." Price volatility will increase each time there is a production shortage, restriction on commodity flows, extreme-weather event, or outbreak of diseases. A look at recent years shows us that food inflation has been very variable, and at different levels when comparing the OECD region with developing economies. That differential is often caused by the different methods for calculating the index as well as by the different costs of production, on top of exogenous supply shocks. In 2008–2009 the food inflation in the OECD was about 6% compared to about 15% in the developing part of the world. Another observation is that food-price inflation varies significantly among countries even if these countries are part of a homogenous bloc as in the EU.

In general, there is no direct relationship (or causality) between food-price inflation measured through the CPI (consumer-price index) and those of international agricultural commodity prices. The input-output dynamics of the food industry of a country can have a significant impact disturbing that relationship. Due to the fact that commodities undergo a significant transformation through transportation, storage, processing, and marketing/distribution before reaching the end consumer, the price link between the two components is far from straightforward. The key driver that determines the relationship between food prices and international commodity prices is the structure of the food basket and the agricultural commodity share in its composition.

18.5 Macro and policy dynamics

The global agricultural markets will also continue to be dominated by broader macroeconomy dynamics in a globalized world. As most emerging economies have been bouncing back better after the economic crisis in economic terms, the agricultural dynamics in consumption terms are gearing toward that part of the world. As time progresses, the dichotomy in economic terms between the OECD region and the developing economies will be supportive for increasing world demand and trade in agricultural products. Also, exchange rates are a dynamic to be blended into the analysis. Exchange rates do deeply influence the relative competitiveness of exporters and the affordability of purchases for importers. That is not only true for the trade of agricultural products but also for input prices like energy and oil which also, due to global growth, are expected to stay at historically high pricing levels. High energy and oil prices in particular will have an effect on both demand and supply of agricultural products on the one hand through higher agricultural supply costs and on the other hand through increased demand for agricultural feedstock being used for biofuel production.

This is beyond the increased demand due to expected population growth. Despite stronger economic growth leading to (or contributing to) slowing growth in population due to lower birth rates, globally the pace of the growth of global population is expected to decline by 1% (only). Emerging economies account for the largest part of the population growth, with Africa topping that rank, with an expected growth of 2.3 % per annum during the next decade. The most recent FAO-OECD Outlook reports, "The changes underway in the developing countries with an increasing share of world economic activity and their greater responsiveness of consumption to income growth along with more rapid population gains and large middle classes, are expected to support increased demand and trade for food, feed, fibers and fish over the projection period. When combined with a weaker US dollar, the high economic growth in developing countries should underpin continued high agricultural prices in the coming decade. Low levels of general inflation should sustain food demand in the OECD area while high inflation will dampen demand growth in some emerging countries."[5]

This assumes a status quo of agricultural policies and trade policies and their underlying dynamics. The most important assumptions are

- Differentiated growth rates in OECD region versus developing economies for the short to medium terms, having their impact on spendable income, unemployment rates, etc.

- Slowing population growth in the next decade (to 1% per annum), which will manifest itself in all regions of the world.
- Inflation differentials are measured vis-à-vis the United States (because of the pricing of agricultural commodities in US dollars) with constant exchange rates in real terms. A depreciation in some developing market currencies versus the US dollar can significantly enhance their competitive position. Within the OECD region, no major inflation differentials are expected in the next decade, which is something that can be questioned given the recent monetary interventions in the OECD region.
- Increasing oil and energy prices at a rate (2.6% per annum) above expected overall inflation. Temporary price actions outside the indicated bandwidth are to be expected due to energy, economic, and geopolitical drivers.
- No major changes in the direction of trade policies on a global level.

18.6 Expected market development

In general, and as already pointed out, prices of most agricultural and fisheries are expected to stay stubbornly high for the mid- to long run. Short term it is more a function of the specific commodity, although most agricultural commodities are in 2013 hovering around their peak levels and will test those peaks on a regular basis based on their demand/supply dynamics and influenced by possible exogenous shocks. The market drivers behind those pricing levels have been discussed on a variety of occasions throughout the book. The nature of the pricing dynamics will change over time, however, from demand/supply dynamics, input costing levels, resource constraints, and inventory levels determining the pricing levels in the following years to pricing movements in later years determined by improved profitability and increased production.

The expected pricing trend per agricultural commodity over the next ten years is reflected in Table 18.1 (from their mid-2013 pricing levels).

In relative terms against each other, the pricing dynamics can be described as follows: "In comparing the different product groups, prices of vegetable oils are projected to rise relative to protein meals. Oilseed and sugar prices rise more than cereal prices and cotton prices. Meat and dairy product prices rise relative to the costs of feeds of protein meals and cereals. Fish prices rise relative to meat prices and biofuel prices rise relative to the costs of the main agricultural feedstocks of cereals, vegetable oils and sugar crops... In making the comparison with the average level of prices in the base period (2010–2012), real price increases over the projection period are only projected for beef (3%), pig meat (2%) and fish products (1%). For all other commodities, prices are projected to be lower in real terms. Nonetheless, growth in demand for food and feed products, along with continuing biofuel demand, is projected to hold

Table 18.1 Expected pricing trend per agricultural commodity during the next decade (2013–2023)

Commodity	ST outlook (2–3 years	LT outlook (> 3 years)	Commodity	ST outlook (2–3 years)	LT outlook (> 3 years)
Wheat	Downward (due to pick-up in production)	Gradually upward	Milk/diary	Upward	Gradually upward
Maize	Downward (due to pick-up in production)	Gradually upward	Cotton	Flat	Downward
Oilseeds	Downward (due to pick-up in production)	Gradually upward	Livestock	Red meat: Flat (high pricing at starting point)	Flat/slightly upward (high pricing at starting point)
Rice	Flat (with periods of volatility)	Flat (with periods of volatility)		Pig meat: upward	Pig meat: Flat/ slightly upward
Sugar	Continue their decline	Upward due to biofuel (ethanol) use		Poultry: flat	upward
Fish/fish meal/fish oil	Strongly upward	Strongly upward	Biofuel Ethanol Biodiesel	Upward Upward Slightly upward	Upward Upward Slightly upward

the prices for cereals, oilseeds, sugar, cotton, fish, and livestock products on a higher plane in the coming decade, and above levels achieved in the pre-2007 decade, in both nominal and real terms."[6]

All agricultural commodities are expected to move higher in nominal terms versus the last decade, but only beef, pig meat, and fish prices are expected to rise in real terms.

18.7 Food demand evolution in frontier and emerging economies

A fundamental driver of agricultural markets over the mid- to long term is the rising food demand in frontier and emerging economies. Even in the last five years during weak economic conditions, agricultural products are holding up remarkably well. In those economies, rising levels of economic growth combined with reduced levels of unemployment and consequently higher average income levels, the emergence of a middle class, and urbanization are the fundamental drivers of higher demand for agricultural commodities. Also changes in dietary habits away from basic staples and grains to higher-protein foods, including meats and dairy products, contribute to that phenomenon.

Higher levels of expected consumption of meat and dairy products will also lead to enhanced indirect demand for grains and growth in protein meals use. Nevertheless, consumption per capita is still a lot higher in the OECD region. That difference will narrow over the next decade, although considerable variations will continue to exist among consumption levels in developing nations. The largest increase in demand is expected in Central and Eastern Europe, Central Asia, and Latin America, while less so in Africa due to the wider disparity and distribution of income growth.

In developing nations where food expenditures represent a large part of household budgets, changes in income or prices of commodities can have a significant impact on buying/spending habits. That phenomenon is unknown in the developed world where food demand is rather inelastic and not driven by changes in income or prices of agricultural commodities. In developed nations, changing demand is a consequence of population growth or reduction (demographics) and lifestyle changes.

The expected differential in the next years compared to the last three years (2010–2012) between increasing crop consumption in the developing world (EM) versus the OECD region breaks down per commodity as indicated in Table 18.2, demonstrating the relative saturation in the OECD region.

Table 18.2 Expected differential between production and consumption for the different parts of the world per agricultural commodity

Commodity	EM (%)	OECD (%)	Commodity	EM (%)	OECD (%)
Wheat	20	7	Beef	22	3
Grains	33	10	Pig meat	21	3
Rice	18	5	Poultry	32	11
Oilseeds	30	16	Sheep	18	–3
Protein meals	36	14	Butter	49	9
Veg. oil	32	14	Cheese	25	12
Sugar	33	3	Fish	20	6
Cotton	18	16			

18.8 Global agricultural output growth barely keeping up with increased demand

The recent increase in commodity prices should fuel sufficient profitability and generate interest for additional investment in the sector and in particular in agricultural production and technological advancements in order for the total global output to continue to grow in the next decade – to be precise, an expected average growth of 1.5%, compared to 2.1% during the last decade. That is slower than in the past but still faster than the expected population growth so that growth in output per capita is expected to be 0.5%. Following are some arguments leading to an expected decline in production:

- Average reduced availability of agricultural land for production due to intense competition for other uses (some emerging countries like Russia and Brazil will see available land volumes increase). In developing nations there will be competition for land between specific agricultural commodities based on financial dynamics but no total land-volume expansion. The increase in expected output will therefore come from higher yields. The change in expected output per commodity is shown in Table 18.3.
- Global average crop yields have been declining over the last two decades due to a reduced level of investments in crop R&D. Despite the fact that in recent years there has been renewed attention for crop R&D, dissemination of improved varieties and seed-quality improvement techniques, that renewed attention will take time to manifest. Combined with an increased cost pattern for production and growing resource constraints, this will lead to expected slower yield growth. The expected changes in yield and area per commodity where already provided earlier in this chapter.

Table 18.3 Change in expected output per agricultural commodity and region of the world

Commodity	EM (%)	OECD (%)	Commodity	EM (%)	OECD (%)
Wheat	24	5	Beef	23	2
Grains	28	16	Pig meat	22	5
Rice	14	4	Poultry	32	13
Oilseeds	28	21	Sheep	17	3
Protein meals	26	15	Butter	50	13
Veg. oil	22	16	Cheese	21	12
Sugar	27	9	Fish	18	4
Cotton	8	–7			

Table 18.4 Expected growth differential (output) on a regional basis

Region	Growth last decade (%)	Expected growth next decade (%)
Aggregate world	2.1	1.5
Developed world/OECD	0.9	0.7
Frontier economies	4.1	2.7
BRICS	3.1	1.6
Other developing nations	2.7	2.0

- Given that the prices of fertilizers, farm chemicals, and farm machinery are closely related to oil prices, a gradual upward pressure is to be expected with respect to their future price levels. Water will become more difficult to access for agricultural purposes and more costly to supply. These factors all lead to higher agricultural supply costs. This in turn will reduce the return on invested capital, which can be expected to further slow the growth in production, especially with respect to input-intensive agriculture. Less input-intense farming practices will benefit, such as pasture-based dairy and meat operations. Exchange-rate depreciations will help some countries to offset some of the increasing cost patterns. Achieving high levels of productivity will be key in all aspects of the food chain.
- The reduction of output growth will manifest itself in both the developed world (in particular the large emerging economies) as well as in the OECD region. The OECD region has historically been faster in adjusting to price changes due to highly capital- and input-intensive farming practices, which is likely to continue, although absolute growth increases over the longer term will come predominantly from developing economies. A breakdown per region of this slower annual growth output is reflected in Table 18.4.

It will be essential that developing economies close the yield gap with their developed-world counterparts (combined with area expansion). Even then it will not be a walk in the park for world food markets: "High yield variability, due in part to more variable weather conditions and low input farming practices in some developing countries, is expected to be one of the factors behind continued market and price variability in the next decade."[7]

18.9 Risk analysis

We all understand the limited relevance of (long-term) forecasts – limited, that is, unless conditional on a specific set of assumptions about policies and possible policy changes, the response of market participants, and the impact of exogenous market drivers. Statistics and stochastic analysis in particular do not fully help in understanding and measuring the implications of all possible uncertainties and risks that might emerge in the future and how they respond to each other (reenforcing or neutralizing). A more meaningful approach is to select and isolate a specific number of uncertainties and analyze them based on their fundamental dynamics, including the quantification of the likely range of variation around the baseline values of the identified uncertainty. This is also known as the partial-stochastic model, or the Aglink-Cosimo model.[8] The objective is to assess and understand uncertainty surrounding major assumptions regarding macroeconomics (e.g., GDP, CPI, and exchange-rate evolutions against the US dollar) and crop yields (for a variety of crops and different countries in the world) and to what degree and via what mechanisms those uncertainties impact the development of the agricultural market.

The questions we would like to see answered when evaluating these outcomes as investors trying to understand a possible investment opportunity will undoubtedly sound as follows: Which of the projections are more vulnerable to the analyzed uncertainty in these drivers? Are some commodities more vulnerable than others? Are the projections for some countries more subject to uncertainty than others, and if so, why? Is more uncertainty transmitted to market outcomes from the macroeconomic drivers or from the drivers of crop yields?

The extent of the variability in the macroeconomic drivers obtained by repeated sampling of errors from the assumed distributions is measured by the average annual coefficient of variation (ACV). The most important results are the ones between the tenth and ninetieth percentiles, leaving out the extreme outcomes, which are then averaged across the relevant observed period. That exercise is relatively simple for macroeconomic variables such as GDP and CPI but a lot less straightforward for currency fluctuations and oil-price dynamics. Errors and possible outcome bandwidth are therefore (considerably) larger. This is observed in the study described in the 2013 OECD-FAO report mentioned earlier.[9] This study also shows that the ACV is considerably higher when judged for emerging economies versus the EU/US. Although I do not doubt the research findings, I wonder if the enhanced macroeconomic instability of emerging economies will manifest itself in the years to come judged against the backdrop of our experiences in recent years, the relative health of state finance, and improving political stability in many emerging economies.

The crop-yield analysis shows similar dynamics, but with variation judged across different agricultural commodities (Table 18.5). Average yield uncertainty at the global level is much smaller than at the country and regional levels. But which of the two dynamics analyzed has the strongest impact on what commodities?

Table 18.5 Crop-yield analysis and uncertainty assessment

Commodity	Uncertainty analysis	Commodity	Uncertainty analysis
Wheat	Mostly yields	Beef	Macroeconomic uncertainty comparable or greater than those of yield uncertainty
Grain	Mostly yields	Pig	Macroeconomic uncertainty comparable or greater than those of yield uncertainty
Oilseeds	Mostly yields	Poultry	Macroeconomic uncertainty comparable or greater than those of yield uncertainty
Rice	Mostly yields	Pork	More crop yield
Cheese	Much more sensitive to uncertainty from macroeconomic conditions (and demand fluctuations) than from yields	Protein meals and vegetable oils	More sensitive to uncertainty from crop yield than macroeconomic Conditions. Additionally, for vegetable oils, macro-economic uncertainty dominates due to the use of this commodity as a bio-diesel feedstock.
Biofuels	More responsive to macroeconomic uncertainty (GDP and crude-oil price) transmitted from the demand side of the market		

In all instances the normal distribution is used to reflect the spread of uncertainties. It therefore assumes that a particular driver's uncertainty is symmetrically distributed around its most likely value. But what about the emergence of asymmetric risk? It requires further analysis[10] but needs to be based around predetermined scenarios. For example, the mentioned OECD-FAO paper analyzed two reports: (1) lower-than-normal yields for three major crops in 2022 and (2) a lower-than-expected world crude-oil price in three consecutive years, 2020–2022 and the impact it will subsequently have on currency pairs).

Scenario 1: World market prices for grains and oilseeds are on average considerably higher in this scenario (wheat +12%, coarse grains +16%, oilseeds +14%) despite the fact that global production in that scenario is only 2% below the baseline level, displaying the tightness in the grain markets. Average world market prices of other agricultural commodities are also considerably higher than in the baseline, between 5% and 9% for the main meats and 4% for dairy products. World market prices are on average 6% above baseline levels for ethanol and 3% higher for biodiesel, and that of vegetable oils is 4% higher.

Scenario 2: World market prices for some biofuel feedstocks are lower than in the baseline (coarse grains –3%, vegetable oils –2%, wheat –4% [three-year averages of subset averages]). World market prices for biofuels are also lower in this subset of runs than in the baseline (ethanol –7%, biodiesel –5%).

18.10 Food security as a function of productivity growth and reduction of waste

The global population-growth rate is expected to slow to only 1% per year during the next decade. Nevertheless, that adds close to 800 million people to global population numbers, which will put additional strain on food production and supply systems. Biofuel production will account for additional demand whereas the global scope for area expansion remains limited and geographically concentrated in a few regions. Focus therefore needs to be on increased productivity, as the targets are steep. It is estimated that by 2050, agricultural production needs to increase by 60% to meet the rising demand for food. Closing the productivity and yield gap between developed and developing nations will be a significant part of the required solution and acts as a risk mitigant: "The increasing role of developing countries in agricultural production, consumption and trade should enhance food security as it spreads global production risk while generating higher incomes and domestic production in many net food importing countries...At the same time there is a growing need to improve the sustainable use of available land, water, marine ecosystems, fish stocks, forests and biodiversity. It is estimated that some 25% of all agricultural land is highly degraded, with growing water scarcity a fact for many countries. Many fish stocks are over-exploited, or in risk of being over-exploited. There is also a growing consensus that climate change and extreme-weather events will increase."[11]

Notes

1 Introduction

1. For an in-depth review of the different streams, the contributors, and the sectors targeted, see figure 4 in the 2012 FAO report "The state of food and agriculture 2012" (retrieved: http://www.fao.org/publications/sofa/en).
2. See for more details, P. Campbell (2011), "The future prospects of global arable land", Global Food and Water Crises Research Program (retrieved: http://www.futuredirections.org.au/publications/food-and-water-crises/53-the-future-prospects-for-global-arable-land.html).
3. See also European Commission (2011) "Questions and answers on food waste minimisation and food packaging optimisation", MEMO/11/598, September 14, 2011 (retrieved: http://europa.eu/rapid/pressReleasesAction.do?reference=MEMO/11/598).
4. Source: "Eat. Think. Save", initiative of the United Nations Environment Program. See also "An overview of global food losses and waste", Future Directions International (retrieved: http://www.futuredirections.org.au/files/FDI%20Strategic%20Analysis%20Paper%20-%2025%20November%202011%20-%20An%20Overview%20of%20Global%20Food%20Losses%20and%20Waste.pdf), including regional analysis. Conclusions of the overall research program were as follows:
   - Almost one-third of food produced for human consumption – approximately 1.3 billion tons per year, which could feed the total global population – is either lost or wasted.
   - Overall food loss primarily occurs in the production-to-retail phase of the food chain. Consumers in industrialized nations waste significantly more food than their counterparts in developing countries.
   - Postharvest losses occur mainly due to corruption and meager infrastructure in the developing world.
   - Preharvest losses are another major factor in the developing world, as a result of insufficient investment in biosecurity practices.
   - A voluntary platform for retailers to report food waste statistics and a less over-conservative system of setting use-by dates would aid in combating waste.
   - An increasing world population, and its progressively scarce resources, makes reduction in food losses and waste a key component in any strategy for a sustainable future global food supply.
5. See T. Davy (2013), "Food wastage: The irony of global gluttony", Future Directions International, p. 5 (retrieved: http://www.futuredirections.org.au/files/FDI_Food_Waste_SAP_20_June_2013.pdf). Also see FAO (2012), "Towards the future we want: End hunger and make the transition to sustainable agricultural and food systems" (retrieved: http://www.fao.org/docrep/015/an894e/an894e00.pdf); J. Gustavsson, C. Cederberg, U. Sonesson, R. van Otterdijk, and A. Meybeck, FAO (2011), "Global food losses and food waste" (retrieved: http://www.fao.org/docrep/014/mb060e/mb060e00.pdf); and FAO (2013), "Food wastage footprint: Impacts on natural resources", summary report, Rome (retrieved: http://www.fao.org/docrep/018/i3347e/i3347e.pdf). The global volume of food wastage is estimated to be 1.6 Gtonnes of

"primary product equivalents," while the total wastage for the edible part of food is 1.3 Gtonnes. This amount can be weighed against total agricultural production (for food and nonfood uses), which is about 6 Gtonnes. Without accounting for GHG emissions from land use change, the carbon footprint of food produced and not eaten is estimated to 3.3 Gtonnes of CO_2 equivalent: as such, food wastage ranks as the third top emitter after the United States and China. Globally, the blue-water footprint (i.e., the consumption of surface- and ground-water resources) of food wastage is about 250 km^3, which is equivalent to the annual water discharge of the Volga River, or three times the volume of Lake Geneva. Finally, produced but uneaten food vainly occupies almost 1.4 billion hectares of land; this represents close to 30% of the world's agricultural land area. While it is difficult to estimate impacts on biodiversity at a global level, food wastage unduly compounds the negative externalities that monocropping and agriculture expansion into wild areas create on biodiversity loss, including mammals, birds, fish, and amphibians. The loss of land, water, and biodiversity, as well as the negative impacts of climate change, represent huge costs to society that are yet to be quantified. The direct economic cost of food wastage of agricultural products (excluding fish and seafood), based on producer prices only, is about USD 750 billion, equivalent to the GDP of Switzerland.

6. See, for a full overview, K. Anderson, G. Rausser, and J. Swinnen (2013), "Political economy of public policies insights from distortions to agricultural and food markets", Policy Research Working Paper 6433 (retrieved: http://elibrary.worldbank.org/docserver/download/6433.pdf?expires=1380805779&id=id&accname=guest&checksum=EB01EF80A641E87714F363DB5BCF1EEB).
7. See further, D. W. te Velde, S. Page, N. Cantore, A. Matthews, M. King, O. Boysen, and N. Keijzer (2012), "The EU's Common Agricultural Policy and development" (retrieved: http://www.odi.org.uk/sites/odi.org.uk/files/odi-assets/publications-opinion-files/7906.pdf). O. Jones (2012), "The CAP towards 2020: Legal proposals" (retrieved: http://thinkandtalkwithtods.files.wordpress.com/2012/07/cap-2020.pdf). DG Agriculture and Rural Development, "The Common Agricultural Policy explained" (retrieved: http://ec.europa.eu/agriculture/cap-overview/2012_En.pdf).
8. See I. Solovyeva and E.-A. Nuppenau (2013), "Improving measures for targeting agri-environmental payments: The case of high nature value farming" (retrieved: http://ageconsearch.umn.edu/bitstream/125648/2/Solovyeva_Nuppenau.pdf).
9. See http://ec.europa.eu/agriculture/cap-post-2013/
10. The three pillars of CAP are as follows:
    1. CAP – first pillar
        - Direct aids: Council Regulation (EC) Nr. 73/2009 on single payments schemes
        - Market interventions: Council Regulation (EC) Nr. 72/2009 on single common market organization
    2. CAP – second pillar
        - Rural development policy: Council Regulation (EC) Nr. 1698/2005 on support for rural development European fund for rural development (EARFD)
    3. Horizontal regulations
        - CAP financing: Council Regulation (EC) Nr. 1290/2005 on financing of CAP.
11. See further A. Monteleone and F. Pierangeli (2013), "The reform of the CAP post-2013: Allocation criteria in the second pillar" (retrieved: http://ageconsearch.umn.edu/bitstream/154653/2/Monteleone.pdf).

12. See further, F. Mantino (2013), "What is going to change in EU rural development policies after 2013? Main implications in different national contexts", *Bio-based and Applied Economics* 2(2): 191–206 (retrieved: http://ageconsearch.umn.edu/bitstream/156474/2/12989-25204-1-PB.pdf).
13. See extensively, D. Laborde and W. Martin (2012), "Agricultural trade: What matters in the Doha Round?", *Annual Review of Resource Economics* 4: 265–283.
14. See extensively on the topic, World Trade Organization (2003), "3.15 Agriculture – Dispute settlement, course on dispute settlement in international trade, investment and intellectual property," includes discussions about the special position of agriculture in WTO/UN and GATT agreements, the content on the agreement on agriculture, treatment of agricultural subsidies and other provisions, and other WTO agreements with relevance to agriculture.
15. G. Guastella, D. Moro, P. Sckokai, and M. Veneziani (2013), "Investment behaviour of EU arable crop farms in selected EU countries and the impact of policy reforms" (retrieved: http://ageconsearch.umn.edu/bitstream/152083/2/FM%20WP42%20Investmetn%20Behaviour%20of%20EU%20Farms%20final-1.pdf).
16. Studies have developed a new set of more satisfactory indicators for that purpose, drawing on the recent literature on trade restrictiveness indexes. It then exploited a global agricultural distortions database recently compiled by the World Bank to generate the first set of estimates of those two indicators for each of 28 key agricultural commodities from 1960 to 2004, based on a sample of 75 countries that together account for more than three-quarters of the world's production of those agricultural commodities. See further, P. J. Lloyd, J. L. Croser, and K. Anderson (2009), "How do agricultural policy restrictions to global trade and welfare differ across commodities?", Policy Research Working Paper 4864 (retrieved: http://elibrary.worldbank.org/docserver/download/4864.pdf?expires=1379154463&id=id&accname=guest&checksum=760A9338D52CEF327A833541CF35AB2A).
17. See S. Romano, M. Cozzi, P. Giglio, and G. Catullo (2013), "Post-2013 EU common agricultural policy: predictive models of land use change", *Bio-based and Applied Economics* 2(2): 151–172 (retrieved: http://ageconsearch.umn.edu/bitstream/156472/2/11142-25202-1-PB.pdf).
18. See further:
    - "FAO organic agriculture" (www.fao.org/organicag/en).
    - "FAO organic agriculture and environmental stability of the food supply" (ftp://ftp.fao.org/docrep/fao/meeting/012/ah950e.pdf).
19. See further, "FAO bio-energy (www.fao.org/bioenergy).
20. P. Evans (2009), "Local food no green panacea", CBC (retrieved: http://www.cbc.ca/news/story/2009/07/22/consumer-local-food.html).
21. J. Roach (2009), "High-rise farm: The future of food?" (retrieved: http://news.nationalgeographic.com/news/2009/06/090630-farm-towers-locally-grown.html); G. Monbiot (2010), "Towering lunacy" (retrieved: http://www.monbiot.com/2010/08/16/towering-lunacy).
22. See, for example, N. Heydon (2013), "Fishy project pays off" (retrieved: http://www.theland.com.au/news/agriculture/general/news/fishy-project-pays-off/2670920.aspx?src=rss).
23. Further definitions can or will include the following:
    - Kriging: A sophisticated interpolation methodology which, in addition to estimating data values at unsampled points, also gives a measure of the precision of the estimate.

- Remote sensing: An all-encompassing term for a range of technologies which allow indirect measurement of something, often from afar. Examples include satellite or airborne imagery, or electromagnetic (EM38) soil survey.
- Response curve: A curve or equation to a curve describing the marginal increase in crop yield obtained as a result of marginal increases in the application of fertilizer, spray, or soil amendment.
- Soil testing: The process of sampling and (chemically) analyzing soils with a view to making informed fertilizer management decisions.
- Strips: The basic building blocks of many simple experimental designs. These may be one-way or two-way.
- Treatments: The net result of combining variables and levels. An experiment in which five levels of a single variable that is being applied has five treatments.
- Yield response: The marginal increase in yield obtained from application of (usually) fertilizer. See also "response curve."

24. K. O. Fuglie, J. M. MacDonald, and E. Ball (2007), "Productivity growth in U.S. agriculture", EB-9, US Department of Agriculture, Economic Research Service.
25. Source: P. Andrade- Sanchez and J. T. Heun (2010), "Things to know about applying precision agriculture technologies in Arizona", working document, pp. 2–3.
26. L. D. Whipker and J. T. Akridge (2009), "Precision agricultural services dealership survey results", Working Paper 09–16, *CropLife* magazine and Center for Food and Agricultural Business, Department of Agricultural Economics, Purdue University.
27. See further, A. K. Singh, "Precision agriculture", Water Technology Center, New Delhi.
28. V. I. Adamchuk, J. W. Hummel, M. T. Morgan, and S. K. Upadhyaya (2004), "On-the-go soil sensors for precision agriculture", *Computers and Electronics in Agriculture* 44: 71–91.
29. J. Berntsen, A. Thomsen, K. Schelde, O. M. Hansen, L. Knudsen, N. Broge, H. Hougaard, and R. Hørfarter (2006), "Algorithms for sensor-based redistribution of nitrogen fertilizer in winter wheat", *Precision Agriculture* 7: 65–83.
30. R. Ferguson, A. Dobermann, and J. Schepers (2007), "Precision agriculture: Site-specific nitrogen management for irrigated corn", University of Nebraska Lincoln Extension, Bulletin EC/163, pp. 1–7.
31. R. Taylor and J. Fulton (2010), "Sensor-based variable-rate application for cotton", Oklahoma Cooperative Extension Service, Oklahoma State University, Stillwater, Oklahoma.
32. S. A. O'Shaughnessy and S. R. Evett (2010), "Developing wireless sensor networks for monitoring crop canopy temperature using a moving sprinkler system as a platform", *Applied Engineering in Agriculture* 26(2): 331–341.
33. See for further updates on technology and news, www.precisionag.com, www.insidegnss.com, and gpsinformation.org/dale/nmea.htm.
34. See further, T. Griffin (2011), "Precision agriculture adoption and profitability: Fact vs. myth beyond pretty maps", University of Arkansas, Department of Agricultural Economics and Agribusiness (retrieved: http://www.infoag.org/abstract_papers/papers/abstract_103.pdf).
35. See further, S. M. Swinton and J. Lowenberg-DeBoer (2001), "Global adoption of precision agricultural technologies: Who, when and why?" in G. Grenier and S. Blackmore, eds, *Third European Conference on Precision Agriculture* (Montpellier, France: Agro Montpellier [ENSAM]), pp. 557–562.
36. S. M. Swinton and J. Lowenberg-DeBoer (1998), "Evaluating the profitability of site specific farming", *Journal of Production Agriculture* 11: 439–446.

37. See M. Robertson, P. Carberry, and L. Brennan (2007), "The economic benefits of precision agriculture: Case studies from Australian grain farms", Grains Research and Development Corporation (retrieved: http://www.grdc.com.au/uploads/documents/Economics%20of%20Precision%20agriculture%20Report%20to%20GRDC%20final.pdf).
38. See further, "A brief history of agriculture", John Hopkins School of Public Health (retrieved: http://ocw.jhsph.edu/courses/nutritionalhealthfoodproductionandenvironment/PDFs/FoodEnv-sec2b_McKenzieOCW.pdf).
39. See further, D. Keeney and L. Kemp (2003), "A new agricultural policy for the US" (retrieved: http://www.mnproject.org/pdf/A%20New%20Agriculture%20Policy%20for%20the%20U.S.%20by%20Dennis%20Keeney%20%20Lo.pdf).
40. See for updates, Stockholm International Water Institute annual reports.
41. Leopold Center study by E. M. Tegtmeier and M. Duffy (2004), "External costs of agricultural production in the United States", *International Journal of Agricultural Sustainability* 2(1): 1–20. D. Pimentel (2005), "Environmental and economic costs of the application of pesticides primarily in the United States", *Environment, Development and Sustainability* 7: 229–252.
42. See D. A. Bessler (2013), "On agricultural econometrics", *Journal of Agricultural and Applied Economics* 45 (3): 341–348 (retrieved: http://ageconsearch.umn.edu/bitstream/155449/2/jaae453life1b.pdf).
43. See further, K. Deininger, D. Nizalov, and S. K. Singh (2013), "Are mega-farms the future of global agriculture? Exploring the farm size-productivity relationship for large commercial farms in Ukraine", Policy Research Working Paper 6544 (retrieved: http://elibrary.worldbank.org/deliver/6544.pdf?itemId=/content/workingpaper/10.1596/1813–9450–6544&mimeType=pdf).
44. See further, L. Yarui, J. D. Woodard, and D. J. Leatham (2013), "Causality among foreign direct investment and economic growth: A directed acyclic graph approach", *Journal of Agricultural and Applied Economics* 45(4): 617–637 (retrieved: http://ageconsearch.umn.edu/bitstream/157308/2/jaae597.pdf).
45. See extensively, C. Brouwer, K. Prins, M. Kay, and M. Heibloem (FAO Land and Water Development Division) (2005), "Irrigation water management: Irrigation methods training manual number 5" (retrieved: ftp://ftp.fao.org/agl/aglw/fwm/Manual5.pdf).
46. See further in detail, A. Phocaides (2000), "Technical handbook on pressurized irrigation techniques". It includes a comprehensive review on the topic, including techniques, equipment, scheduling, etc. Also included is an interesting irrigation glossary in chapter 16, pp. 175–194 (retrieved: ftp://ftp.fao.org/agl/aglw/docs/pressirrig.pdf).
47. See for an extensive review, ibid., Brouwer et al., chapter 7, pp. 57–64.
48. See extensively, G. D. Schaible and M. P. Aillery (2012), "Water conservation in irrigated agriculture: Trends and challenges in the face of emerging demands", USDA, Economic Information Bulletin Number 99 (retrieved: http://www.ers.usda.gov/media/884158/eib99.pdf).
49. See further, for more information on the topic, www.fao.org/ag/ca.

2 The Investible Agricultural Space

1. Deutsche Bank Research (2009), "Investing in agriculture: Far-reaching challenge, significant opportunity. An asset management perspective" (retrieved: https://www.db.com/us/docs/Ag_whitepaper_062409.pdf).
2. List derived from ibid., note 1, p. 8.

3. "Human land use practices, especially those tied to agriculture, have transformed the biosphere. The 40% of terrestrial photosynthesis appropriated for human use, the 50% of the global nitrogen flux from synthetic fertilizers and fossil fuels combustion and the freshwater withdrawals that exceed 50% of the accessible supply lead to the same conclusion: Never has a single species had such a planetary presence. Although we know much about the magnitude of the human transformation to the biosphere, we know little about its exact location and geographic manifestation." C. Monfreda et al. (2008), "Farming the planet: 2. Geographic distribution of crop areas, yields, physiological types, and net primary production in the year 2000", *Global Biogeochemical Cycles* 22(1).
4. Derived from, World Population Growth in the 21st Century (2012), "How many are coming to dinner? I think we'll need more folding chairs", Global AgInvesting, Research and Insights, HighQuest Partners LLC, p. 2 (retrieved: http://www.globalaginvesting.com/downloads/files/World-Population-Growth-in-the-21st-Century-277F.pdf).
5. Ibid., note 4, p. 3.
6. Population data set, United Nations, Department of Economics and Social Affairs, Population Division (2013), "World population prospects".
7. MIT Emissions Prediction and Policy Analysis (EPPA) model.
8. See further, W. J. Kiernan (2012), "The closing of the consumption gap – Rising global incomes, increasing consumption, changing diets and agriculture in the 21st century", HighQuest Partners LLC (retrieved: http://www.globalaginvesting.com/downloads/files/Agriculture-Fundamental-II-The-Closing-of-the-Consumption-Gap.pdf).
9. See ibid., note 1, p. 19.
10. See ibid., note 1, p. 24.
11. See further, D. A. Lobell, M. B. Burke, C. Tebaldi, M. D. Mastrandrea, W. P. Falcon, and R. L. Naylor (2008), "Prioritizing climate change adaptation needs for food security in 2030", *Science* 319: 607–610. W. Schlenker, W. M. Hanemann, and A. C. Fisher (2005), "Will U.S. agriculture really benefit from global warming? Accounting for irrigation in the hedonic approach", *American Economic Review* 95(1): 395–406. W. Schlenker and M. J. Roberts **forthcoming**"Why climate change impacts on agriculture could be economically substantial", North Carolina State University, Columbia University, and NBER. Losses have been estimated, and indications are that a number of developing economies would be most vulnerable to a change in climate patterns and rising temperatures.
12. See ibid., note 1, pp. 24–25.
13. See more extensively, J. Lehman (2007), "Bio-energy in the black", *Frontiers in Ecology and Environment* 5(7): 381–387. J. Pretty (2008), "Agricultural sustainability: Concepts, principles and evidence", *Philosophical Transactions of the Royal Society* 363: 447–465. According to Pretty, reduction initiatives can be taken along three major initiatives: (1) increase carbon sinks in soil organic matter and aboveground biomass, (2) reduce direct and indirect energy use to avoid greenhouse gas emissions (CO_2, CH_4, and N_2O), and (3) increase biomass-based renewable energy production to avoid carbon emissions.
14. See further, WRI (2013), "Agricultural offsets markets", www.wri.org.
15. See further, FAO (2008), "Climate change, biofuels and land", Rome. World Bank (2006), "Indicators of sustainable development", Climate Change Series. Both contain a lists of adaptation measures.

16. See, for example, G. Branca, T. Tennigkeit, W. Mann, and L. Lipper (2012), "Identifying opportunities for climate-smart agriculture investments in Africa", Rome, FAO, and EPIC.
17. D. Mitchell (2008), "A note on rising food prices", World Bank. R. Trostle (2008), "Global agricultural supply and demand: Factors contributing to the recent increase in food commodity prices", USDA Economic Research Service.
18. See ibid., note 1, p. 34.
19. See ibid., note 16.
20. K. G. Cassman et al. (2003), "Meeting cereal demand while protecting natural resources and improving environmental quality", *Annual Review of Environment and Resources* 28: 315–358.
21. See ibid., note 4.
22. R. Licker et al. (2010), "From the ground up: The role of climate versus management on global crop yield patterns", *Global Ecology and Biogeograph* 1. M. Johnston et al. (2009), "Resetting global expectations from agricultural biofuels", *Environmental Research Letters* 4(1).
23. As 100% maximum production capacity is unrealistic, just as that is the case in a normal manufacturing plant due to outages, maintenance and repairs, etc.
24. See World Resources Institute 2013, www.wri.org.
25. See J. A. Foley et al. (2005), "Global consequences of land use", *Science* 309(5734): 570–574. Foley et al. (2005) present a conceptual framework for assessing these trade-offs by comparing natural ecosystems, intensive croplands, and a hybrid system of both natural and cropped lands. The ability for each landscape to provide vital ecosystem services varies greatly, and with future increases in agricultural production, the full portfolio of ecosystem services will need to be considered with each decision made.
26. See, for a regional breakdown of water withdrawals, S. Siebert et al. (2005), "Development and validation of the global map of irrigation areas", *Hydrology and Earth System Sciences* 9(5): 535–547.
27. FAOSTAT (2013), Food and Agriculture Organization of the United Nations.
28. Ibid., note 21. N. Ramankutty et al. (2008), "Farming the planet: 1. Geographic distribution of global agricultural lands in the year 2000", *Global Biogeochemical Cycles* 22(1).
29. J. Galloway et al. (2002), "Optimizing nitrogen management in food and energy production, and environment change – Response", *Ambio* 31(6): 497–498.
30. Ibid., note 1, pp. 57–58.
31. J. A. Foley et al. (2005), "Global consequences of land use", *Science* 309(5734): 570–574.
32. By the IPCC.
33. Ibid., note 1, p. 60.
34. Ibid.
35. Ibid., note 1, p. 61.
36. Ibid., note 1, p. 62.
37. Ibid., note 36.
38. See, for an example, ibid., note 1, pp. 63–64.
39. Ibid., note 1, p. 64.
40. V. Khosla (2008), "Where will biofuels and biomass feedstocks come from?" (retrieved: http://www.khoslaventures.com/presentations/WhereWillBiomassComeFrom.pdf).
41. Ibid., note 1, p. 65.
42. Ibid., note 1, p. 66.

43. Ibid., note 21 Cassman et al.
44. Ibid., note 1, p. 67.
45. D. Tilman et al. (2006), "Carbon-negative biofuels from low-input high-diversity grassland biomass", *Science* 314(5805): 1598–1600.
46. T. Searchinger et al. (2008), "Use of US croplands for biofuels increases greenhouse gases through emissions from land-use change", *Science* 319(5867): 1238–1240. H. K. Gibbs, M. Johnston, J. A. Foley, T. Holloway, C. Monfreda, N. Ramankutty, and D. Zaks (2008), "Carbon payback times for crop-based biofuel expansion in the tropics: The effects of changing yield and technology", *Environmental Research Letters* 3: 034001. J. Fargione et al. (2008), "Land clearing and the biofuel carbon debt", *Science* 319(5867): 1235–1238.
47. Ibid., note 1, p. 69.
48. Ibid., note 58.
49. Ibid., note 1, p. 70.
50. Ibid., note 1, p. 71.
51. Union of Concerned Scientists, D. Gurian-Sherman (2009), *Failure to yield: Evaluating the Performance of Genetically Engineered Crops* (Cambridge, MA: UCS Publications).
52. Listing the benefits and challenges.

Genetically modified foods

(Potential) benefits	Controversies
Crops • Enhanced taste and quality • Reduced maturation time • Increased nutrients, yields, and stress tolerance • Improved resistance to disease, pests, and herbicides • New products and growing techniques • Animals • Increased resistance, productivity, hardiness, and feed efficiency • Better yields of meat, eggs, and milk • Improved animal health and diagnostic methods • Environment • "Friendly" bioherbicides and bioinsecticides • Conservation of soil, water, and energy • Bioprocessing for forestry products • Better natural waste management • More efficient processing • Society • Increased food security for growing populations	**Safety** • Potential human health impacts, including allergies, transfer of antibiotic resistance markers, unknown effects • Potential environmental impacts, including unintended transfer of transgenes through cross-pollination, unknown effects on other organisms (e.g., soil microbes), and loss of flora and fauna biodiversity • Increased use of herbicides due to weed resistance and also reverting to use of more toxic herbicides • Access and Intellectual Property • Domination of world food production by a few companies • Increasing dependence on industrialized nations by developing countries • Biopiracy, or foreign exploitation of natural resources • Violation or elimination of traditional seed-sharing rights • Ethics • Violation of natural organisms' intrinsic values • Tampering with nature by mixing genes among species • Objections to consuming animal genes in plants and vice versa • Stress for animals • Labeling • Not mandatory in some countries • Mixing GM crops with non-GM products confounds labeling attempts • Society • New advances may be skewed to interests of rich countries. • New advances serve needs of private entities only, not societies broadly.

Source: The Genome Project/US DOE.

53. Ibid., note 1, p. 75.
54. Policy initiatives that develop infrastructure, such as transport, ports, telecommunications, and energy and irrigation facilities, as well as management skills, labor supply, and skills to use modern technology such as GPS, will all help raise productivity. Governmental coordination of agricultural research and public and private partnerships to develop new technologies such as improved seed and crop varieties and water resources management, while often found in separate policy instruments, has major implications for agricultural productivity. Source: Ibid., note 1, p. 76. A list of research areas were drafted and quantified. See J. Von Braun, S. Fan, R. Meinzen-Dick, M. W. Rosegrant, and A. N. Pratt 2008, "What to expect from scaling up CGIAR investments and Best Bet Program", International Food Policy Research Institute.
55. State of food and agriculture report 2013, FAO, Rome, p. x.

3 Climate Change and Agriculture

1. See further for a general review, F. Tubiello, J. Schmidhuber, M. Howden, P. G. Neofotis, S. P. E. Fernandes, and D. Thapa (2008), "Climate change response strategies for agriculture: Challenges and opportunities for the 21st century", IBRD/World Bank (retrieved: http://www.fao.org/waicent/faoinfo/economic/esd/dp42Combined_web.pdf).
2. See S. Hachigonta, G. C. Nelson, T. S. Thomas, and L. M. Sibanda (2013), "Southern African agriculture and climate change", IFPRI (retrieved: http://www.ifpri.org/sites/default/files/publications/rr179.pdf). The eight chapters of this detailed 370-page report look at the likely effects of climate change on eight countries in southern Africa: Botswana, Lesotho, Malawi, Mozambique, South Africa, Swaziland, Zambia, and Zimbabwe, looking at possible impacts including:
   1. Direct – Likely range of effects on crop and livestock productivity based on different climate-change scenarios.
   2. Indirect – Resultant effect on income generated from agricultural production at farm and national levels and the implications for availability and prices of food, both in domestic and international markets.
3. See further:
   - FAO Climate Change (www.fao.org/climatechange)
   - Intergovernmental Panel on Climate Change (IPCC) (www. ipcc.ch)
   - NASA (http://data.giss.nasa.gov/gistemp)
4. With a 95% probability.
5. See further, IPCC (2013), "Climate change 2013: The physical science basis", September 27, 2013, 5th assessment report (retrieved: http://www.ipcc.ch/report/ar5/wg1/#.Uka8W9J4rJc). This report was front-runned by the (updated) Greenpeace report (2013), "Dealing in doubt: The climate denial machine vs. the climate science" (retrieved: http://www.greenpeace.org/usa/Global/usa/report/Dealing%20in%20Doubt%202013%20-%20Greenpeace%20report%20on%20Climate%20Change%20Denial%20Machine.pdf). This report from Greenpeace provides an interesting and reasonably concise overview of the attacks on climate science and climate scientists ahead of the publication of the next Intergovernmental Panel on Climate Change (IPCC) report. Meanwhile, the scientific consensus – and evidence – continues to grow that climate change is both real and anthropogenic, and no amount of denial will change the fact that this has very real implications for farmers and investors in the years to come. The report documents

1. The history of organized attacks on climate science, scientists, and scientific institutions like the UN IPCC.
2. The key moments in the campaign of climate denial started by the fossil-fuel industry, and traces them to their sources, uncovering the key players driving and funding the denial movement.
3. The range of tricks used by the denial movement, from "pal review" instead of peer review to "self-publishing" and personal attacks on scientists.

6. See R. M. Rejesus, M. Mutuc-Hensley, P. D. Mitchell, K. H. Coble, and T. O. Knight (2013), "U.S. agricultural producer perceptions of climate change", *Journal of Agricultural and Applied Economics* 45(4): 701–718 (retrieved: http://ageconsearch.umn.edu/bitstream/157312/2/jaae580.pdf). This study examines US crop producers' perceptions of climate change, its effects on crop agriculture, and likely ways farmers would adapt to weather extremes. Based on a survey of crop producers in four states, we find that a significant proportion of farmers do not perceive that climate change has been scientifically proven and do not believe that it will adversely affect average crop yields and yield variability. Farmers are likely to diversify crops, buy crop insurance, modify lease arrangements, and exit farming in response to extreme weather caused by climate change.
7. See further:
   - IPCC (www.ipcc.ch)
   - FAOSTAT (http://faostat.fao.org)
   - Monitoring and Assessment of GHG Emissions and Mitigation Potentials in Agriculture, MICCA (www.fao.org/climatechange/micca/ghg/en)
8. With cutoff date December 31, 2012.
9. FAO statistical yearbook 2013, Rome (retrieved: http://www.fao.org/docrep/018/i3107e/i3107e.pdf); see part 4 and global data sets.
10. European Commission (2006), "Environmental impact of products: Analysis of life-cycle environmental impacts", Technical Report EUR 22284, Brussels (retrieved: http://ec.europa.eu/environment/ipp/pdf/eipro_report.pdf).
11. K. Venkat (2012), "The climate change and economic impacts of food waste in the United States", Clean Metrics Corp. (retrieved: http://www.cleanmetrics.com/pages/ClimateChangeImpactofUSFoodWaste.pdf).
12. P. Smith, D. Martino, et al. (2007), "Climate change 2007: Mitigation of climate change", Contribution of Working Group III to the Intergovernmental Panel on Climate Change (IPCC), 4th assessment report (Cambridge: Cambridge University Press); and A. Wreford, D. Moran, and N. Adger (2010), "Climate change and Agriculture", OECD (retrieved: http://www.agri.ankara.edu.tr/fcrops/1289__IKLIM_DEGISIKLIGI_VE_TARIM.pdf).
13. Ibid., note 8; Smith et al.
14. J. Wright (2010), "Feeding nine billion in a low emissions economy", review for the Overseas Development Institute and Oxfam. UK Government Office for Science (2011), "Foresight: The future of food and farming", London. US-EPA (2006), "Global mitigation of non-CO_2 greenhouse gas emissions, 1990–2020" (Washington, DC: United States Environment Protection Agency).
15. See further in detail, C. Schaffnit-Chatterjee (2011), "Mitigating climate change through agriculture: An untapped potential", Deutsche Bank Research (retrieved: http://www.dbresearch.com/PROD/DBR_INTERNET_EN-PROD/PROD0000000000278568/Mitigating+climate+change+through+agriculture%3A+An+untapped+potential.pdf), in particular subsection 1.3, pp. 5–6.
16. See comprehensively, FAO (2013), "Climate-smart agriculture sourcebook" (retrieved: http://www.fao.org/docrep/018/i3325e/i3325e.pdf).

17. See specifically on this point, ibid., note 5, pp. 375–413, including a lengthy and interesting literature list at pp. 408–412.
18. N. Brooks, S. Anderson, J. Ayers, I. Burton, and I. Tellam (2011), "Tracking adaptation and measuring development", International Institute for Environment and Development Climate Change Working Paper No. 1, London, IIED (retrieved: http://pubs.iied.org/pdfs/10031IIED.pdf).
19. Which equates the following ministerial declaration. Meeting G20 Agricultural Ministers, Paris, June 2011: "We recall the triple challenge for agriculture: meeting food security objectives while adapting to climate change and reducing its contribution to greenhouse gas emissions."
20. See ibid., note 8, pp. 386–387, and FAO (2009), "Food security and agricultural mitigation in developing countries: Options for capturing synergies", Rome (retrieved: http://www.fao.org/docrep/012/i1318e/i1318e00.pdf).
21. The most relevant literature in this respect is P. Smith, D. Martino, et al. (2007), "Climate change 2007: Mitigation of climate change", Contribution of Working Group III to the Intergovernmental Panel on Climate Change (IPCC), 4th assessment report (Cambridge: Cambridge University Press). J. Bellarby, B. Foereid, A. Hastings, and P. Smith (2008), "Cool farming: Climate impacts of agriculture and mitigation potential", report for Greenpeace International, Amsterdam. S. J. Scherr and S. Sthapit (2009), "Mitigating climate change through food and land use", Ecoagriculture Partners and Worldwatch Institute, Worldwatch Report 179. Von Thünen-Institut (Johann Heinrich) (2009), "Erfassung, Bewertung und Minderung von Treibhausgasemissionen des deutschen Agrar- und Ernährungssektors", Arbeitsberichte aus der vtI_Agrarökonomie, June 2009. J. Wright (2010), "Feeding nine billion in a low emissions economy", review for the Overseas Development Institute and Oxfam.
22. Based on and see further, ibid., note 15, pp. 7–9.
23. See further, J. D. Glover, C. M. Cox, and J. P. Reganold (2007), "Future farming: A return to roots?", *Scientific American*, August 2007. J. D. Glover and J. P. Reganold (2010), "Perennial grains: Food security for the future", *Science and Technology*, Winter 2010. Scherr and Sthapit (2009).
24. Ibid., note 19; Scherr and Sthapit (2009).
25. J. Lehmann, J. Graunt, and M. Rondon (2006), "Bio-char sequestration in terrestrial ecosystems: A review", *Mitigation and Adaptation Strategies for Global Change*, March 2006, pp. 403–427. Ibid., note 11; Scherr and Sthapit (2009).
26. R. D. Schuiling and O. Tickell (2010), "Olivine against climate change and ocean acidification" (retrieved: http://www.innovationconcepts.eu/res).
27. See also, FAO (2013), "Tackling climate change through livestock", Rome (retrieved: http://www.fao.org/docrep/018/i3437e/i3437e.pdf). The global livestock sector is a significant emitter of the greenhouse gases that contribute to anthropogenic climate change. This comprehensive study released as part of the FAO's Livestock Environmental Assessment and Performance Partnership assesses the environmental performance of the sector and considers its potential to deliver a significant share of the necessary mitigation effort. An informative read for any livestock investor looking to achieve sustainability improvements, the report addresses such questions as
    1. What parts of livestock supply chains are the most significant contributors to climate change?
    2. Where do the greatest opportunities to reduce emissions exist on a species, production system, and regional basis?
    3. What sets of technologies exit to reduce emissions and what is their viability and mitigation potential?

4. What sector-specific guidelines and metrics would be most appropriate for assessing and monitoring the environmental impact of the livestock sector and measuring the effectiveness of new policies on sustainable food production, economic growth, and poverty alleviation?

28. Ibid., notes 15 and 18; Scherr and Sthapit (2009). IFAD (2009), "Livestock and climate change", November 2009.
29. Ibid., note 17; Smith et al. (2007).
30. UN-REDD website.
31. Blaser and Robledo (2007) consider six direct factors, not taking into account infrastructure development, urban development, and mining – which is also a direct driver but generally limited in area.
32. H. K. Gibbs et al. (2010), "Tropical forests were the primary sources of new agricultural land in the 1980s and 1990s", PNAS, September 2010. A secondary forest is one that has regrown after a major disturbance such as fire, insect infestation, timber harvest, or windthrow.
33. Climate Change Agriculture and Food Security CCAFS (2011), "The hidden climate costs of chocolate", April 2011 (retrieved: http://www.ccafs.cgiar.org/blog/hidden-climate-costs-chocolate).
34. Ibid., notes 15 and 19; Scherr and Sthapit (2009).
35. REDD stands for Reducing Emissions from Deforestation and Forest Degradation. REDD+ includes the increasing of carbon stocks through sustainable management of forests and planting. Its principle is to give developing and emerging countries incentives to conserve their forests through financial transfers.

    The REDD mechanism is an effort to give a financial value to the carbon stored in forests. As part of the offset scheme of the carbon markets, it is designed to produce carbon credits via emissions-saving projects. These credits can be traded within the carbon markets and used by industrialized governments and corporations to meet their targets, in addition to cutting their emissions. The UN forecasts that financial flows for GHG reductions from REDD+ could reach up to USD 30 billion a year, thus rewarding a meaningful reduction in GHGs while supporting pro-poor development, biodiversity conservation, and ecosystem services. Source: UN-REDD website.
36. U. Hoffmann (2011), "Assuring food security in developing countries under the challenges of climate change: Key trade and development issues of a profound development in agriculture", Discussion Paper No. 201, UNCTAD, February 2011.
37. Ibid., note 19; Scherr and Sthapit (2009).
38. Most of these methodologies will or have been discussed more extensively in different chapters and with a different focal point, hence the reduced attention they are given in this context.
39. See further, A. Kasterine (2010), "Organic agriculture and carbon markets", in FiBL & IFOAM 2010: The World of Organic Agriculture, pp. 87–90, Statistics and Emerging Trends 2010, Bonn and Frick (retrieved: http://www.intracen.org/uploadedFiles/intracenorg/Content/Exporters/Sectors/Fair_trade_and_Environmental_Exports/Climate_change/kasterine-2010-carbon-markets.pdf) (full report retrieved: http://www.organic-world.net/fileadmin/documents/yearbook/2010/world-of-organic-agriculture-2010.pdf). FAO (ed. M. Müller-Lindenlauf) (2009), "Organic agriculture and carbon sequestration" (retrieved: ftp://ftp.fao.org/docrep/fao/012/ak998e/ak998e00.pdf).
40. Source: FAO (U. Niggli et al.) (2009), "Low GHG agriculture: Mitigation and adaptation potential of sustainable farming systems", Food and Agriculture Organization of the United Nations, Rome. Organic agriculture can be described as a holistic production management system that avoids use of agrochemicals and genetically

modified organisms; minimizes pollution of air, soil, and water; and optimizes the health and productivity of plants, animals, and people. (Retrieved: ftp://ftp.fao.org/docrep/fao/010/ai781e/ai781e00.pdf).

41. See further, J. Wright (2010), "Feeding nine billion in a low emissions economy", review for the Overseas Development Institute and Oxfam. T. De Ponti, H. C. A. Rijk, M. K. Van Ittersum (2011), "The crop yield gap between organic and conventional agriculture", *Agricultural Systems* 108: 1–9. U. Hoffmann (2011) "Assuring food security in developing countries under the challenges of climate change: Key trade and development issues of a profound development in agriculture", Discussion. Paper No. 201, UNCTAD, February 2011. Rat für Nachhaltige Entwicklung (2011), "Gold-Standard Ökolandbau", Rat für Nachhaltige Entwicklung (RNE), July 2011.
42. O. De Schutter (2010), "Agro-ecology and the right to food", United Nations, December 2010. The core principles of agro-ecology: recycling nutrients and energy on the farm, rather than introducing external inputs; integrating crops and livestocks; diversifying species and genetic resources over time and space and focusing on interactions and productivity across the agricultural system rather than on individual species.
43. The principles of conservation agriculture: (1) continuous minimum mechanical disturbance, (2) permanent organic soil cover, and (3) diversification of crop species, grown in sequences and/or associations.
44. FAO (2010), "Climate-smart agriculture", FAO, Rome (retrieved: http://www.fao.org/docrep/013/i1881e/i1881e00.pdf).
45. Integrated food-energy systems examples: (1) biogas from livestock residues, (2) animal feed from by-products of corn ethanol, and (3) bagasse for energy as a by-product of sugarcane production for food purposes.
46. Some estimations put that number even higher.
47. Urban agriculture: vegetables, fruits, mushrooms, herbs, meat, eggs, milk, and even fish are being produced in community gardens, private backyards, schools, hospitals, rooftops, window boxes, and vacant public lands. This home production can provide a significant share of a family's food requirements. Source: FAO (2009), "Organic agriculture and carbon sequestration", note 36. See also, UK Government Office for Science (2011), "Foresight: The future of food and farming", London; and T. Garnett, H. Charless, and D. Godfray (2012), "Sustainable intensification in agriculture: Navigating a course through competing food system priorities" (retrieved: http://www.futureoffood.ox.ac.uk/sites/futureoffood.ox.ac.uk/files/SI%20report%20-%20final.pdf).
48. Amongst other places FAO (2011), "Global food losses and food waste", FAO, Rome. Other (and more recent) research on food waste is indicated in the section of food waste. Some put the food waste rate at 1/3 to 1/2. (retrieved: http://www.fao.org/docrep/014/mb060e/mb060e00.pdf).
49. First-generation biofuels are not sustainable. They compete with food crops, and agro-bioenergy in general does not match the criteria of sufficiency. Arable land resources are limited, and further expansion into forest, grassland, and woodland areas will result in significant carbon emissions, which offset the primary justification for using biofuels. Source: European Commission SCAR (2011), "Sustainable food consumption and production in a resource-constrained world", Standing Committee for Agricultural Research (SCAR), February 2011 (retrieved: http://www.scp-knowledge.eu/sites/default/files/knowledge/attachments/Umwelt-Wissenschaft-EU-Landwirtschaft2050.pdf).
50. See further for extensive and comprehensive coverage, FAO (2012), "Sustainable diets and biodiversity: Directions and solutions for policy, research and action", eds

B. Burlingame and S. Dernini (retrieved: http://www.fao.org/docrep/016/i3004e/i3004e.pdf).

51. Evidence does suggest that food should be produced closer to consumers if it can be produced efficiently in terms of productivity and GHG emissions. This is usually the case with seasonal food. However, the GHG impact of production from most favorable locations (in terms of requirements of land and other resources) may be lower than that of less efficient local production if it is shipped efficiently, even over long distances. Benefits in income and food security incurred to low-income farming communities from international exports should also not be neglected. Source: Hoffmann (2011).
52. See for some first-line experiences, J. Parks (2013), "The effects of food labeling and dietary guidance on nutrition in the United States" (retrieved: http://ageconsearch.umn.edu/bitstream/149667/2/AAEA%20Nut-Label-DRAFT-IV-5–23–13.pdf). In general she finds that using food labels may reduce CVD risk and facilitate the management of body weight and diet-related health conditions such as hypertension. The results provide evidence that food labels and dietary guidance have helped consumers make healthier choices and that they have the potential to help those who do not yet use these tools.
53. CIRAD-INRA (2011), "Agrimonde: Scenarios and challenges for feeding the world in 2050", Editions Quae, Versailles, 2011. Ibid. EU SCAR (2011). OECD (2008), "Promoting sustainable consumption: Good practices in OECD countries", Paris, 2008.
54. Sources: www.waterfootprint.org, UK DEFRA (2006), National Geographic, USDA National Nutrient Database, and Oxfam.
55. See for further details, C. Schaffnit-Chatterjee (2011), "Mitigating climate change through agriculture: An untapped potential", Deutsche Bank Research, chapter 4, pp. 16–20.
56. See also ibid. C. Schaffnit-Chatterjee (2011), chapter 5, pp. 20–22.
57. Scherr and Sthapit (2009). FAO (2009), "Enabling agriculture to contribute to climate change mitigation", FAO, Rome (retrieved: http://unfccc.int/resource/docs/2008/smsn/igo/036.pdf).
58. This is one of the most complex and controversial parts of the discussion around GHG emissions. See for a branching out of the topic to other industries, N. Winchester, D. McConnachie, C. Wollersheim, and I. A. Waitz (2013), "Economic and emissions impacts of renewable fuel goals for aviation in the US" (retrieved: http://ageconsearch.umn.edu/bitstream/155003/2/winchester-etal_AAEA_2013.pdf); and M. Khanna and A. Oliver (2013), "The welfare costs of GHG reduction with renewable energy policies in the US" (retrieved: http://ageconsearch.umn.edu/bitstream/154999/2/khanna%2Boliver_AAEA_2013.pdf).
59. Scherr and Sthapit (2009).
60. Hoffmann (2011); C. Schaffnit-Chatterjee (2010), "Risk management in agriculture: Towards market solutions in the EU", Deutsche Bank Research, September 2010. C. Schaffnit-Chatterjee (2011), "Where are food prices heading?", Deutsche Bank Research, March 2011. Weltsichten (2011), "Die Agrarpolitik endet nicht an Europas Grenzen", Ein Dossier vom Evangelischen Entwicklungsdienst und – Brot für die Welt, July 2011.
61. K. Deutsch (2011), "Doha or Dada: The world trade regime at an historic crossroads", Deutsche Bank Research, June 2011.
62. Hoffmann (2011).
63. Hoffmann (2011).

64. See further, C. Schaffnit-Chatterjee (2010).
65. See Ibid., note 5, p. x.
66. See C. Schaffnit-Chatterjee (2011), "Mitigating climate change through agriculture: An untapped potential", Deutsche Bank Research, chapter 6, pp. 23–26.
67. Cap and trade vs. carbon tax.

    Principle: A cap-and-trade scheme, also called emissions trading scheme (ETS), sets a limit (cap) on emission levels and allows the price of the emissions to vary. A carbon tax puts a price on emissions but allows the emission levels to change.

    Pros and cons: A carbon tax is more flexible; it can be increased if emission levels are still too high, whereas permits are allocated for the duration of a cap-and-trade scheme. However a basic tax does not differentiate between various levels of ability to pay and tends to disproportionately affect low-income groups.

    Use: Carbon taxes have been introduced in a number of countries for specific sectors (e.g., Finland, France, the Netherlands, Norway, Sweden, and some Canadian provinces). ETS have been implemented by a number of governments and municipalities and the EU. Large companies are allocated a permit to re-lease a set amount of GHGs and can trade it. Source: A. Kasterine and D. Vanzetti (2010), "The effectiveness, efficiency and equity of market-based mechanisms to reduce greenhouse gases in the agri-food sector", Trade and Environment Review 2010, United Nations Conference on Trade and Development (UNCTAD), Geneva.
68. Carbon footprint labeling: The assessment will consider
    - Farming practices (fertilizer usage, soil coverage, etc.)
    - Energy on the farm (source and usage)
    - Transport
    - Energy for transformation and storage
    - Packaging (material and amount)

    Source: Von Thünen-Institut (Johann Heinrich) (2009), "Erfassung, Bewertung und Minderung von Treibhausgasemissionen des deutschen Agrar- und Ernährungssektors", Arbeitsberichte aus der vtI_Agrarökonomie, June 2009.
69. For a review of the mechanism and efficiency carbon taxes, environment payment services, and emission trading systems (ETS) and cap-and-trade systems, see C. Schaffnit-Chatterjee (2011), Ibid., chapter 6, pp. 23–24. I do, however, disagree with some of the statements made about the efficiency of ETS and cap-and-trade systems.
70. Hoffmann (2011) and Ibid. FAO (2009) and FAO (2010). D. F. Larson, A. Dinar, and J. A. Frisbie (2011), "Agriculture and the clean development mechanism", Policy Research Working Paper 5621, World Bank, April 2011 (retrieved: http://elibrary.worldbank.org/docserver/download/5621.pdf?expires=1379259034&id=id&accname=guest&checksum=D9E56C59F27EF20C590253E124D22CE6). A. Kasterine and D. Vanzetti (2010), "The effectiveness, efficiency and equity of market-based and voluntary measures to mitigate GHG from the agri-food sector", Trade and Environment Review, UNCTAD, Geneva (retrieved: http://www.intracen.org/uploadedFiles/intracenorg/Content/Exporters/Sectors/Fair_trade_and_Environmental_Exports/Climate_change/TER_UNCTAD_KasterineVanzetti.pdf).
71. Schaffnit-Chatterjee (2011), p. 25 points out that "a range of financing mechanisms are needed, from market-based efforts to public sector funds, with the following issues to be addressed: scaling up of funding and delivery mechanisms, reducing transaction costs and improving the contribution to sustainable development by reaching producers including smallholders. Public finance has often acted as a

catalyst for action, or a way to fund activities or areas which are neglected by the private sector. Private-sector engagement is essential, especially to scale up government-financed development projects and to sustain these projects once government funding is reduced or withdrawn," with further reference to C. Hebebrand (2011), "Leveraging private sector investment in developing country agrifood systems", Policy Paper Series, Chicago Council on Global Affairs (retrieved: http://www.thechicagocouncil.org/UserFiles/File/GlobalAgDevelopment/Report/CCGA%20GADI%20Private%20Sector%20Policy%20Paper%20FINAL%20VERSION.pdf). For a suggested road map, see Schaffnit-Chatterjee (2011), p. 26. The road map includes amongst others the usual suspects: (1) emission reduction, (2) internalization of GHG costs, (3) remove perverse policy incentives, (4) reduce waste, (5) change consumer behavior, (6) exchange information and transfer skills.

72. Principles for responsible agro-investment (RAI): (1) land and resource rights; (2) food security; (3) transparency, good governance, and enabling environment; (4) consultation and participation; (5) economic viability and responsible agro-enterprise investing; (6) social sustainability; and (7) environmental sustainability.

    Principles for responsible investment in farmland (PRI): (1) promoting environmental sustainability, (2) respecting labor and human rights, (3) respecting existing land and resource rights, (4) upholding high business and ethical standards, and (5) reporting on activities and progress towards implementing and promoting the principles. Source: RAI, Knowledge exchange platform, UNPRI (2011). For the most recent (2013) annual report, see http://www.unpri.org/viewer/?file=wp-content/uploads/AnnualReport20131.pdf.

73. FAO developed a series of reports on climate change and agricultural initiatives and reporting/monitoring. See Mitigation of Climate Change in Agricultural Series. The most recent feedback on the database results can be found in the report (2013), "Agriculture, forestry and other land use mitigation project database: Second assessment of the current status of land-based sectors in the carbon markets (retrieved: http://www.fao.org/docrep/017/i3176e/i3176e.pdf). The updated AFOLU ("Forestry and Other Land Use") MP ("Mitigation Project") database confirms trends already identified in the previous paper, but it also notes certain changes. The regulatory markets continue to dominate carbon markets. Voluntary carbon projects make only a small contribution to the total. Clean Development Mechanism (CDM) projects still constitute the highest number of AFOLU projects, the majority of which are manure-treatment projects and, to a lesser extent, afforestation/reforestation projects.

74. See extensively, "Global survey of agricultural mitigation practices" (2010), Mitigation of Climate Change in Agricultural Series 1 (retrieved: http://www.fao.org/docrep/012/al388e/al388e00.pdf); "National integrated mitigation planning in agriculture: A review paper" (2013), Mitigation of Climate Change in Agricultural Series 7 (retrieved: http://www.fao.org/docrep/017/i3237e/i3237e.pdf); and "National planning for GHG mitigation in agriculture: A guidance document" (2013), Mitigation of Climate Change in Agricultural Series 8 (retrieved: http://www.fao.org/docrep/018/i3324e/i3324e.pdf).

75. See extensively: FAO (2013) animal production and health paper Nr. 177, Mitigation of greenhouse gas emissions in livestock production. A review of technical options for non-CO_2 emissions. (Retrieved: http://www.fao.org/docrep/018/i3288e/i3288e.pdf).

76. IPCC, FAO, EPA, or others.

77. See further, A. N. Hristov et al. (2013), "Nutritional and management strategies to mitigate animal greenhouse gas emissions" (retrieved: http://dairy.ifas.ufl.edu/rns/2013/9_hristov.pdf).
78. Source and see further, E. Heymann (2011), "Talking point: Climate change: Forest conservation must be given higher priority", Deutsche Bank Research, and further FAO, Forests and Climate Change (2012), "Working with countries to mitigate and adapt to climate change through sustainable forest management", Rome; as well as Forests and Climate Change: Adaptation and Mitigation (2009), "European tropical research network" (retrieved: www.etfrn.org/file.php/18).
79. See further, FAO (2012), "Forests and climate change working paper 10 – Forest Management and Climate Change: A literature review", Rome (retrieved: http://www.fao.org/docrep/015/md012e/md012e00.pdf).
80. See FAO (2012), "State of the world's forests", Rome (retrieved: http://www.fao.org/docrep/016/i3010e/i3010e.pdf).
81. Adapted from source: "Investing in climate change 2012. Investment markets and strategic asset allocation. Broadening and diversifying the approach", p. 8, Deutsche bank climate change advisors (retrieved: https://www.dbadvisors.com/content/_media/Inv_in_CC_2012.pdf).
82. See further, FAO (2011), "Linking climate change financing and sustainability implications for agriculture" (retrieved: http://www.fao.org/docrep/015/i2480e/i2480e00.pdf).
83. See, for example, Land Commodities AG (2012), "Climate risk primer: Australian broadacre cropping. A primer on the facts and current state of the science relating to the spatial variability of climate risk and the implications for Australian broadacre cropping investors. Including implications of El Niño and La Niña" (retrieved: http://www.landcommodities.com/wp-content/Assets/Downloads/Climate%20Risk%20Primer.pdf). They conclude that for the majority of extreme-weather years occurring in Australia over the last century, owning farms in different parts of Western Australia, as opposed to different parts of Australia as whole, would have been a more effective diversification strategy from the perspective of mitigating climate-related risk (p. 11).

4 Agricultural Risk Management and Insurance

1. "Risk and uncertainty are ubiquitous and varied within agriculture and agricultural supply chains. This stems from a range of factors including the vagaries of weather, the unpredictable nature of biological processes, the pronounced seasonality of production and market cycles, the geographical separation of production and end uses, and the unique and uncertain political economy of food and agriculture sectors, both domestic and international." Source: See S. Jaffee, P. Siegel, and C. Andrews (2010), "Rapid agricultural supply chain risk assessment: A conceptual framework", World Bank, Washington, DC (retrieved: http://siteresources.worldbank.org/INTCOMRISMAN/Resources/RapidAgriculturalSupplyChainRiskAssessmentConceptualFramework.pdf).
2. Categorization based on (2011), "Weather-index insurance for agriculture: Guidance for development practitioners", Agricultural and Rural Development Discussion Paper 50 (retrieved: https://www.agriskmanagementforum.org/sites/agriskmanagementforum.org/files/WEATHER%20INDEX%20INSURANCE%20FOR%20AGRICULTURE_WB_WII_Paper_Nov_2011.pdf).

3. For a synthesis of risk-management literature, see B. Barnett and K. Coble (2008), "Poverty traps and index-based risk transfer products", *World Development* 36(10): 1766–1785 (retrieved: http://www.agriskmanagementforum.org/sites/agriskmanagementforum.org/files/Documents/Poverty%20Traps%20Index-Based%20Risk%20Transfer.pdf).
4. Ibid., note 2, p. 2.
5. Example derived from ibid., note 2, p. 2.
6. Ibid., note 5, p. 2.
7. See for a similar overview, S. Jaffee, P. Siegel, and C. Andrews (2010), "Rapid agricultural supply chain risk assessment: A conceptual framework", World Bank, Washington. DC.
8. See extensively, World Bank (2005), "Managing agricultural production risk: Innovations in developing countries", Report 32727-GLB (retrieved: http://siteresources.worldbank.org/INTARD/Resources/Managing_Ag_Risk_FINAL.pdf).
9. C. Schaffnit-Chatterjee (2010), "Risk management in agriculture", p. 4, Deutsche Bank Research (retrieved: http://www.dbresearch.com/PROD/DBR_INTERNET_EN-PROD/PROD0000000000262553.PDF).
10. See, for example, J. White (2009), "Managing risk: The big 5 types of agricultural risk" (retrieved: http://hortmgt.dyson.cornell.edu/pdf/risk_mgt/newsletters/The_Big_Five.pdf). Alternative version can be found in OECD (2009), "Managing risks in agriculture: a holistic approach", chapter 2, pp. 15 (retrieved: http://www.ti.bund.de/fileadmin/dam_uploads/Institute/MA/ma_de/ma-allgemein/OECD-Seminar/Additional_Information/Managing%20Risk%20in%20Agr%20Holistic%20approach.pdf). See also, J. Harwood, R. Heifner, K. Coble, J. Perry, and A. Somwaru (1999), "Managing risk in farming: Concepts research and analysis", Agricultural Economic Report No. 774, Economic Research Service, USDA. R. Holzman and S. Jorgensen (2001), "Social risk management: A new conceptual framework for social protection, and beyond", *International Tax and Public Finance* 8: 529–556. R. B. M. Huirne, M. Meuwissen, J. B. Hardaker, and J. R. Anderson (2000), "Risk and risk management in agriculture: An overview and empirical results", *International Journal of Risk Assessment and Management* 1: 125–136. B. Hardaker, (2000), "Some issues in dealing with risk in agriculture", Working Paper 2000–03, University of New England. G. Moschini and D. A. Hennessy (2001), "Uncertainty, risk aversion, and risk management for agricultural producers", chapter 2 in B. L. Gardner and G. C. Rausser, eds (2001), *Handbook of Agricultural Economics*, Elsevier, 1st ed., 1(1). W. N. Musser and G. F. Patrick (2001), "How much does risk really matter to farmers?", chapter 24 in Just and Pope Assessment of the Role of Risk in U.S. Agriculture (2002) Kluwer academic publishers, Massachusetts, Norwell. J. B. Hardaker, R. Huirne, J. R. Anderson, and G. Lien (2004), "Coping with risk in agriculture", CABI Publishing.
11. See also, FAO Policy Learning Program (2008), "Module 3: Investment and resource mobilization, session 5: Risk mitigation in agricultural investment" (retrieved: http://www.fao.org/docs/up/easypol/531/3-5_investment-risk_mitigation_157en.pdf). C. Miller (2007), "Risk mitigation and management for agricultural investment", workshop background paper, "Module: Investment and resource mobilization". See also Calvin Miller (2008), "FAO, Risk mitigation and management for agricultural investment and resource mobilization" (retrieved: http://www.fao.org/docs/up/easypol/526/35_risk_management_background_155en.pdf).

    Further reading could include:

    - World Bank, *Agricultural Investment Sourcebook*, "Module 10: Managing agricultural risk, vulnerability, and disaster" (retrieved: http://www.worldbank.org/agsourcebook).

- Myong Goo Kang and Nayana Mahajan (2007), "An introduction to market-based instruments for agricultural price risk management", FAO Agricultural Management, Marketing and Finance Working Document 12.
- R. A. J. Roberts (2005), "Insurance of crops in developing countries", FAO Agricultural Services Bulletin 159.
- C. Eaton and A. W. Shepherd (2001), "Contract farming: Partnerships for growth", FAO *Agricultural Services Bulletin* 145.
- World Bank, *Agricultural Investment Sourcebook*, "Module 7: Investments in rural finance for agriculture" (retrieved: http://www.worldbank.org/agsourcebook).
- FAO (2004), "Twelve key challenges in rural finance", paper by Calvin Miller.
- USDA Risk Management Agency (1997), "Introduction to Risk Management.

12. **Mitigation** is the lessening or limitation of the adverse impacts of hazards and related disasters. Risk-mitigation options are numerous and varied (for example, crop and livestock diversification, income diversification, soil drainage, mulching, use of resistant seeds, avoidance of risky practices, and crop calendars).

    Transfer refers to the transfer of the potential financial consequences of particular risks from one party to another. While insurance is the best-known form of risk transfer, in developing countries the use of informal risk transfer within families and communities is extremely important.

    Coping refers to improving the resilience to withstand and manage events, through ex ante preparation and making use of informal and formal mechanisms in order to sustain production and livelihoods following an event. Although we have noted that coping is an ex-post activity, it is possible to plan and to prepare for coping activities on an ex ante basis. This is often fiscally beneficial, as the ability to quickly respond to events often reduces losses. (Source of definition: ibid., note 2, p. 4).
13. J. Cordier and S. Gouin (2000), "Gestion territoriale du risque de contractualisation", paper presented at the Colloque SFER/AIEA2 "L'agroalimentaire entre local et global", Paris, June 15 and 16.
14. See extensively on this matter, "Farmers use of marketing and production contracts", USDA Economic Research Service (retrieved: http://www.ers.usda.gov/media/523263/aer747a_1_.pdf). The most important arguments are (1) income stability, (2) improved efficiency, (3) market security, (4) access to capital, (5) controlling input supply, (6) improving response to consumer demand, and (7) expanding and diversifying operations.
15. See further, M. Vassalos, C. A. Dillon, and T. Coolong (2013), "Optimal land allocation and production timing for fresh vegetable growers under price and production uncertainty", *Journal of Agricultural and Applied Economics* 45(4): 683–699 (retrieved: http://ageconsearch.umn.edu/bitstream/157391/2/jaae595.pdf). Production timing is an essential element in fresh vegetable growers' efforts to maximize profitability and reduce income risks. The present study uses biophysical simulation modeling coupled with a dual-crop (tomatoes, sweet corn) whole-farm economic formulation to analyze the effects of growers' risk-aversion levels and price consideration (seasonal or annual price consideration) in expected net returns and production practices. The findings indicate that consideration of seasonal price trends results in higher expected net returns and greater opportunities to mitigate risk. Furthermore, risk-aversion levels substantially influence production timing when seasonal price trends are considered.
16. See in detail ibid., note 8, p. 33.
17. See extensively on this topic, J. MacDonald, J. Perry, M. Ahearn, D. Banker, W. Chambers, C. Dimitri, N. Key, K. Nelson, and L. Southard (2004), "Contract, markets and prices: Organizing the production and use of agricultural commodities",

USDA Economic Research Service, Agricultural Economic Report 837 (retrieved: http://www.ers.usda.gov/media/284610/aer837_1_.pdf).

18. OECD (2008), "An assessment of risk exposure in agriculture: A literature review", Working Party on Agricultural Policies and Markets, December 2008, Paris (retrieved: http://www.oecd.org/agriculture/agricultural-policies/42749727.pdf).
19. K. Balcombe (2009), The nature and determinants of volatility in agricultural prices", Technical Report, FAO (retrieved: http://mpra.ub.uni-muenchen.de/24819/1).
20. S. K. Roache (2010), "What explains the rise in food price volatility?", International Monetary Fund Working Paper, May 2010 (retrieved: http://chede.org/chede/wp-content/uploads/2010/06/IMF-Reasons-for-Food-Price-volatility.pdf).
21. C. Schaffnit-Chatterjee (2012), "Food prices: Research briefing", DB Research (retrieved: http://www.dbresearch.de/PROD/DBR_INTERNET_EN-PROD/PROD0000000000289023/Food+prices%3A+Food+price+index+up+in+Q1+but+wheat+and+rice+under+downward+pressure.PDF).
22. See, for a full list, the chapter on commodity markets. Most of them were all modeled according to the CBOT setup and dynamics.
23. S. Kimura, J. Antón, and C. LeThi (2010), "Farm level analysis of risk and risk management strategies and policies: Cross country analysis", OECD Food, Agriculture and Fisheries Papers, No. 26, OECD Publishing (retrieved: http://www.oecd-ilibrary.org/docserver/download/5kmd6b5rl5kd.pdf?expires=1371826609&id=id&accname=guest&checksum=699283CD38AFA360C0161267E08B7F93).
24. OECD (2009), "Risk management in agriculture: A holistic conceptual framework", Working Party on Agricultural Policies and Markets, March 2009, Paris (retrieved: http://www.oecd.org/agriculture/agricultural-policies/42750215.pdf).
25. See for updated data, http://www.futuresindustry.org/volume-.asp.
26. It was Sarris in 1997 who already listed the key sector factors for an options and futures markets to thrive. A. Sarris (1997), "Price variability in cereal markets and risk management strategies for market participants", report presented to the OECD Group on Cereals, Animal Feeds, and Sugar of the Working Party on Agricultural Policies and Markets, Paris. He indicated the following conditions to be absolutely key:
    - Substantial commodity price variability
    - Large number of traders and speculators to ensure liquidity
    - Products with standardized grades and quality
    - Limited government intervention in pricing and trade
    - A regulatory body to ensure and safeguard the integrity of the markets
    - Good transportation and communication systems
    - A well-functioning financial system overall
    - An effective legal environment
    - Political and macroeconomic stability
27. See, for example (extensive literature to be found in relevant chapter), R. Kappel, R. Pfeiffer, and J. Werner (2010), "What became of the food price crisis in 2008?", *Aussenwirtschaft* 65(I): 21–47, Zurich (retrieved: http://www.nadel.ethz.ch/publikationen/food_price_crisis.pdf). S. K. Roache (2010), "What explains the rise in food price volatility?", International Monetary Fund Working Paper, May 2010 (retrieved: http://chede.org/chede/wp-content/uploads/2010/06/IMF-Reasons-for-Food-Price-volatility.pdf). M. Robles, M. Torero, and J. von Braun (2009), "When speculation matters", International Food Policy Research Institute, Issue Brief 57, February 2009, Washington, DC (retrieved: http://www.ifpri.org/sites/default/files/publications/ib57.pdf).

28. A good example of this risk layering and risk retention in a different industry is traditionally to be found in auto insurance, where the policyholder has a deductible and a no-claims bonus. The deductible and bonus serve two functions. First, policyholders remain liable for minor damages to their vehicles, which are the most frequent sources of loss. Second, the policyholder is provided an incentive to drive in a responsible manner, avoiding causing losses to others. The net effect of this is that insurers are seeking only to accept transfer of the second-tier, larger (but much more infrequent) losses on which they are able to charge reasonable premiums and for which they are able to make sufficient financial provisioning (source: ibid., note 2, p. 5).
29. See, for example, A. Tarasov (2013), "Impact of interest rates on the decision to insure in agricultural production", *Studies in Agricultural Economics* 115: 1–7 (retrieved: http://dx.doi.org/10.7896/j.1226). The paper seeks to define the relationship between interest rates and decisions to insure among agricultural producers using the financial methodology. The choices are ultimately reduced to two options: to insure or to limit and absorb risk. Each choice produces a complex cash flow that is compared to the alternative and discounted by several factors. The difference between the options produces a quantitative measure of the financial incentive to insure. Some discounting factors of the cash flows follow the key interest rate to an extent for the latter to influence the decision to insure along with demand for insurance. The proposed method is tested on data from the emerging economy of Ukraine and the United States for the period 2002–2011. All participants in agricultural insurance markets can use the proposed methods to maximize efficiency. The research shows that, ceteris paribus, agricultural insurance requires bigger government subsidies to be viable under higher interest rates.
30. In order for a risk to be insurable, two basic requirements have to be met: managing the adverse effects of "asymmetric information" and overcoming the implications of "systemic risks". J. R. Skees and B. J. Barnett (1999), "Conceptual and practical applications for sharing catastrophic/systemic risk", *Review of Agricultural Economics*, are adding a few more item to that list:
    - Symmetric information: The insurer and the insured have (nearly) the same information as regards the probability distribution of the risk (the probability of a bad outcome). This is normally not the case, the main problems being moral hazard and adverse selection. Therefore, insurance solutions are only viable, and can be offered at reasonable cost, if these problems can be adequately dealt with (see box).
    - Independent risks: Risks should be (nearly) independent across insured individuals. If risks are systemic (dependent), special measures have to be taken in order to make insurance solutions viable (see box above).
    - Large number of exposure units: The law of large numbers allows an accurate prediction of average future losses and the calculation of the premium.
    - Calculable chance of loss: In order to fix the premium rates, the insurance company must be able to estimate both average frequency and average severity of loss. For low-probability risks with potentially catastrophic outcomes, it is difficult to fix a rate.
    - Actual losses occurring must be determinable and measurable.
    - In the perception of the potential buyer of a policy, potential losses must be significant; otherwise he will bear the risk himself. At the same time, premiums must be economically affordable.

31. In late 2013, Australia seems to be moving in the direction of introducing a multi-peril crop insurance system: http://www.queenslandcountrylife.com.au/news/agriculture/cropping/general-news/spark-of-hope-for-mpci/2670396.aspx.
32. See further, R. Dismukes, K. H. Coble, C. Miller, and E. O'Donoghue (2013), "The effects of area-based revenue protection on producers' choices of farm-level revenue insurance", presentation at the Agricultural and Applied Economics Association 2013 Annual Meeting, Washington, DC, August 4–6, 2013 (retrieved: http://ageconsearch.umn.edu/bitstream/149545/2/AAEA2575DismukesCoble%20(1).pdf). Producers' increased reliance on crop insurance has led to concerns about losses producers could incur that are not covered by crop insurance. In the current farm bill debate, several proposals that would be based on area (county) revenue and are intended to cover a portion of producers' crop insurance deductibles, referred to as "shallow-loss" programs, have been advanced. The analysis demonstrates, using an empirically based simulation model and a certainty equivalent criterion, how shallow-loss coverages might affect optimal coverage levels of farm-level revenue insurance for a moderately risk-averse producer. The analysis suggests that area-based revenue insurance designs have some potential for causing producers to reduce coverage levels for farm-level revenue insurance, though the marginal differences in the certainty equivalents are often relatively small on a percentage basis.
33. Weather Futures and Options Block Trading: block trades of weather products allow eligible parties to manage exposure by region to weather-related risk or access unique speculation opportunities related to temperatures, snowfall, frost and hurricanes. Executed apart from the public auction market, block trades help market participants secure a fair and reasonable single price for large transactions.
34. Definitions derived from ibid., note 2, pp. 9, 11. It also contains a very good summary of the benefits, challenges, and features of all types of insurance products discussed. Ibid., note 2, p. 10.
35. See further: J. B. Tack and D. Ubilava (2013), "El Niño southern oscillation impacts on crop insurance", presentation at the Agricultural & Applied Economics Association's 2013 AAEA and CAES Joint Annual Meeting, Washington, DC, August 4–6, 2013 (retrieved: http://ageconsearch.umn.edu/bitstream/151429/2/AAEA_Poster_Combined.pdf).
36. See further, World Bank (2005), "Managing agricultural production risks: Innovations in developing countries", World Bank Agriculture and Rural Development Department, Report 32727, World Bank, Washington, DC. O. Mahul and C. J. Stutley (2010), "Government support to agricultural insurance: Challenges and options for developing countries", World Bank, Washington, DC (retrieved: https://openknowledge.worldbank.org/bitstream/handle/10986/2432/538810PUB0Gove101Official0Use0Only1.pdf?sequence=1).
37. For further comments and details with respect to the weighting and qualification of risks within the index, see ibid., note 2, p. 17.
38. The index is based on Pacific Ocean surface temperatures measured by the National Oceanic and Atmospheric Administration. Increases in the sea surface temperatures are a good indication of an El Niño weather pattern that brings torrential rain and catastrophic flooding to parts of northern Peru. See further on this topic, J. R. Skees and B. Collier (2010), "New approaches for index insurance: ENSO insurance in Peru", in R. Kloeppinger-Todd and M. Sharma, eds, *2020 Vision for Food, Agriculture, and the Environment, Focus 18 Innovations in Rural and Agriculture Finance* (Washington, DC: International Food Policy Research Institute).

39. See for more analyses on the differences between the two product types, ibid., note 2, p. 19. In essence it comes down to this: Derivative contracts derive their value by looking at an underlying index. They are not necessarily associated with any physical loss and simply base their payouts on the performance of the index. Currently, the majority of weather trading is in the derivative market. There are two different types of derivative products, exchange-traded contracts (ETCs) and over-the-counter contracts (OTCs). Insurance is the transfer of risk by a client (the insured) to a third party (normally an insurer). In exchange for a "consideration" (premium paid by the insured to the insurer), the insurer agrees to pay valid claims that occur during the period of the policy, in accordance with the terms and conditions of the insurance policy. Weather risk will be structured as insurance, while derivatives more often serve large-scale buyers of weather insurance, such as large corporations or major national macrolevel transactions. There is a major international market for the reinsurance of traditional agricultural insurances.
40. See for a recent evolution, R. Cai, J. D. Mullen, J. C. Bergstrom, W. D. Shurley, and M. E. Wetzstein (2013), "Using a climate index to measure crop yield response", *Journal of Agricultural and Applied Economics* 45(4): 719–737 (retrieved: http://ageconsearch.umn.edu/bitstream/157314/2/jaae582.pdf). Using principal component analysis, a climate index is developed to estimate the linkage between climate and crop yields. The indices based on three climate projections are then applied to forecast future crop-yield responses. We identify spatial heterogeneity of crop-yield responses to future climate change across a number of US northern and southern states. The results indicate that future hotter/drier weather conditions will likely have significant negative impacts on southern states, whereas only mild impacts are expected in most northern states.
41. See further, M. Brunette and S. Couture (2013), "Risk management behavior of a forest owner to address growth risk", *Agricultural and Resource Economics Review* 42(2): 349–364 (retrieved: http://ageconsearch.umn.edu/bitstream/155526/2/ARER%2042x2%202013%20Couture.pdf). They analyze risk-management behavior (financial savings versus physical savings) of a private forest owner who values amenities in relation to uncertainty about timber growth.
42. See for extensive coverage: World Bank discussion paper series "Weather-index insurance for agriculture: Guidance for development practitioners" (2011), Agriculture and Rural Development Discussion Paper 5, in particular chapters 3–6. See also "Weather-index-based insurance in agricultural development: A technical guide" (2011) (retrieved: http://www.ifad.org/ruralfinance/pub/WII_tech_guide.pdf).
43. See further, "Guide to agricultural risk management" (2013) (retrieved: http://www.ag-risk.org/NCISPUBS/LAIPPUB/2012/Guide_2013.pdf).
44. See J. Dolginow, R. Massey, B. Myers, and N. Kitchen (2013), "Adjusting crop insurance APH calculation to accommodate biomass production", selected paper prepared for presentation at the Agricultural & Applied Economics Association's 2013 Crop Insurance and the Farm Bill Symposium, Louisville, KY, October 8–9, 2013 (retrieved: http://ageconsearch.umn.edu/bitstream/156945/2/APH_Biomass_Massey.pdf).
45. See, for example, E. Belasco, S. Galinato, T. Marsh, C. Miles, and R. Wallace (2013), "High tunnels are my crop insurance: An assessment of risk management tools for small-scale specialty crop producers" (retrieved: http://ageconsearch.umn.edu/bitstream/155319/2/ARER%2042x2%202013%20Belasco.pdf).

46. See extensively on this, J. R. Skees (2012), "Agricultural risk management or income enhancement?", *Regulation* 22(1) (retrieved: http://www.cato.org/sites/cato.org/files/serials/files/regulation/1999/4/agrisk.pdf). This article analyzes how the current method for calculating actual production history (APH) may deter producers from planting perennial grasses and penalizes those producers who convert some of their row cropland to perennial grasses. An alternative APH calculation method suggested here would continue to provide a safety net to producers, reduce indemnity payments by insurance companies, and reduce an impediment to planting perennial grasses.
47. An extensive series of articles was issued by Bloomberg in September 2013 (http://go.bloomberg.com/special-reports/report/doomed-crops-record-profits) at a point in time when the farm bill 2013 (the House legislation is H.R. 2642; Senate is S.954) was under review and crop-insurance claims were up due to planting delays (http://www.bloomberg.com/news/2013-09-17/u-s-crop-insurance-claims-jump-amid-planting-delays-usda-says.html). See further, J. L. Novak (2013), "2013 farm bill: Does passage depend on long-dead philosophers?", *Journal of Agricultural and Applied Economics* 45(3): 375–379 (retrieved: http://ageconsearch.umn.edu/bitstream/155459/2/jaae453life5b.pdf).
48. By O. A. Ramirez, C. E. Carpio, and R. M. Rejesus (2011), "Can crop insurance premiums be reliably estimated?", *Agricultural and Resource Economics Review* 40(1): 81–94.
49. G. Colson, O. Ramirez, and S. Fu (2013), "Crop insurance savings accounts: A viable alternative to crop insurance?", p. 4 (retrieved: http://ageconsearch.umn.edu/bitstream/124739/2/CISA%20Paper%2006-04-2012.pdf). The goal of their research was thus to explore the potential viability of a different insurance design that could be an effective risk-management tool for many farmers and operate without major government subsidization. Specifically, they formally considered one of the more controversial approaches – a system based upon farmer-owned savings accounts. See further, R. Dismukes and R. Durst (2006), "Whole-farm approaches to a safety net", *Economic Information Bulletin* 15, US Department of Agriculture, Washington, DC. D. K. Enahoro and B. A. Gloy (2006), "Examining the potential benefits of federally subsidized farm savings accounts for dairy farmers", paper presented at AAEA annual meeting, Long Beach, CA, July 23–26, 2006. B. Gloy and M. L. Cheng (2006), "Farm savings accounts for specialty crop Growers", *Department of Applied Economics and Management Research Bulletin* RB 2006–03, Cornell University. J. R. Skees, J. R. Black, and B. J. Barnett (1997), "Designing and rating an area yield crop insurance contract", *American Journal of Agricultural Economics* 79(2): 430–438. O. A. Ramirez, S. K. Misra, and J. E. Field (2003), "Crop yield distributions revisited", *American Journal of Agricultural Economics* 85(1): 108–120.
50. See for their full reporting: OECD (2011), "Managing risk in agriculture: Policy assessment and design", Paris.
51. J. R. Skees and B. J. Barnett (1999), ibid., note 9.
52. See further, "Risk management tools for EU agriculture – with a special focus on insurance" (2001), Working Document (retrieved: http://ec.europa.eu/agriculture/publi/insurance/text_En.pdf).
53. M. P. M. Meuwissen, B. M. Huirne, and J. B. Hardaker (1999), "Income insurance in European agriculture", *European Economy* 2, Luxembourg.
54. Ibid., note 47.
55. C. Schaffnit-Chatterjee (2009), "The global food equation", Deutsche Bank Research, *Current Issues*, September (retrieved: http://www.dbresearch.com/PROD/DBR_INTERNET_EN-PROD/PROD0000000000247631.PDF).

56. OECD-FAO (2009), "Agricultural outlook" (retrieved: http://www.oecd.org/site/oecd-faoagriculturaloutlook/45599621.pdf).
57. J. Von Braun and M. Torero (2009), "Implementing the physical and virtual food reserves to protect the poor and prevent market failure", Policy Brief 10, International Food Policy Research Institute, February 2009, Washington, DC (retrieved: http://www.ifpri.org/sites/default/files/publications/bp010.pdf).
58. See further, J. Drakeford and A. Benfield (2013), "Agricultural insurance: A global class of business globally" (retrieved: http://www.usda.gov/oce/forum/presentations/Drakeford.pdf).
59. See for an overview of the different crop insurance systems in the world, http://ec.europa.eu/agriculture/analysis/external/insurance/existing_En.pdf.
60. M. Meuwissen, M. Asseldonk, and R. Huirne (2008), "Income stabilisation in European agriculture", Wageningen Academic Publishers.
61. M. B. Diaz-Caneja, C. G. Conte, F. J. Gallego Pinilla, J. Stroblmair, R. Catanero, and C. Dittman (2009), "Risk management and agricultural insurance schemes in Europe", Joint Research Centre, European Communities, Luxembourg (retrieved: http://ec.europa.eu/dgs/jrc/downloads/jrc_reference_report_2009_09_agrI_ins.pdf).
62. Ibid., note 40, pp. 27–28.
63. J. B. Hardaker, R. B. M. Huirne, and J. R. Anderson (1997), "Coping with risk in agriculture", book review, *Journal of Agricultural and Applied Economics* 29(2): 437–438 (retrieved: http://ageconsearch.umn.edu/bitstream/15067/1/29020437.pdf).
64. Ibid., note 47.
65. J. Harwood, R. Heifner, K. Coble, J. Perry, and A. Somwaru (1999), "Managing risk in farming: Concepts, research, and analysis", Agricultural Economic Report 774, Market and Trade Economics Division and Resource Economics Division, Economic Research Service, US Department of Agriculture (retrieved: http://www.agriskmanagementforum.org/sites/agriskmanagementforum.org/files/Documents/Managing%20Risk%20in%20Farming.pdf).
66. Thus, one measure of price risk for winter wheat at planting time is the standard deviation (or coefficient of variation) of price changes from September to July in the July wheat futures price. That is, the difference between the September 1 quote and the next July 1 quote on the July futures contract can be obtained for several years, and the standard deviation (or coefficient of variation) calculated on that annual series of price-difference observations.
67. For a further in-depth analysis, see ibid., note 47, pp. 14–59.
68. Leveraging refers to the producer's use of debt to finance the operation. Increasing the degree of leverage increases the likelihood that in a year of low farm returns the producer will be unable to meet his or her financial obligations and heightens the potential for bankruptcy. Thus, in general, highly leveraged producers operate in an environment of greater financial risk than do producers who choose a less highly leveraged farm structure. Increasing the farm's leverage (that is, borrowing) increases the capital available for production, allowing expansion of the business, but it also entails incurring a repayment obligation and creates the risk of loan default because of the risks inherent in the farming operation. Because of these many factors, a farmer's use of debt to finance the operation interacts with both the production and marketing risks faced by the producer. See P. J. Barry and C. B. Baker, "Financial responses to risk in agriculture", in Peter J. Barry, ed., *Risk Management in Agriculture* (Ames: Iowa State University Press, 1984), pp. 183–199. See further, ibid., note 50, pp. 38–43.
69. Some of the methods that farmers use to manage liquidity, and hence their financial risk, include the following: (1) selling assets, (2) managing the pacing of

investments and withdrawals, and (3) holding liquid credit reserves. See further, ibid., note 50, pp. 44–46.

70. Producers can also manage their farming risks by either leasing inputs (including land) or hiring workers during harvest or other peak months. Leasing refers to a capital transfer agreement that provides the lessee (the actual operator) with control over assets owned by someone else (the lessor) for a given period, using a mutually agreed-upon rental arrangement. Farmers can lease land, machinery, equipment, or livestock. There are different lease models (financial/capital lease versus operational lease, dry, wet, and moist lease variants).

    Leasing has similarities with leveraging, in that both are methods used to expand control over resources. In addition, both commit the farmer to regular payments. Leasing appears, however, to have some advantages.

    One advantage is that control can be gained over long-life inputs (such as land and machinery), without making long-term payment commitments. In addition, leasing provides producers with flexibility in allocating their asset portfolios – a producer can be in either the farming business or the land-ownership business, without being in both. Leasing improves the renter's flexibility to respond to changing market conditions. In addition, leasing reduces the long-term fixed payments on borrowed capital that may strain liquidity in years of reduced output, and it can reduce both financial and production risk for the lessee. See further, ibid., note 50, pp. 46–47.

71. As discussed before, a futures contract is an agreement priced and entered on an exchange to trade at a specified future time a commodity or other asset with specified attributes (or in the case of cash settlement, an equivalent amount of money). The US exchanges that trade agricultural futures contracts are the Chicago Board of Trade; the Chicago Mercantile Exchange; the Kansas City Board of Trade; the Minneapolis Grain Exchange; the New York Coffee, Sugar, and Cocoa Exchange; and the New York Cotton Exchange. Trading is conducted either through "open outcry" on the floor of the exchange or electronically. Contracts for major field crops (including corn, wheat, soybeans [with variations in meal, oil, and crush], and cotton); four types of livestock and animal products (live cattle, feeder cattle, live hogs, and pork bellies); and sugar and frozen concentrated orange juice have been traded for years. In more recent decades, futures contracts for coffee, (rough) rice, oats, ethanol, cocoa, boneless beef, and dairy products have been introduced.
72. Basis is sometimes calculated as cash price minus futures price and sometimes as futures price minus cash price, as mutually understood by the parties involved. Cash prices may be quoted relative to the futures, such as "10 cents over" or "20 cents under" the futures price.
73. Source: ibid., note 50, p. 36 – but adjusted for 2013 price and cost levels.
74. Source: R. G. Heifner, B. H. Wright, and G. E. Plato (1993), "Using cash, futures, and options contracts in the farm business", AIB-665, US Department of Agriculture, Economic Research Service, April 1993 (retrieved: http://naldc.nal.usda.gov/download/CAT93973963/PDF).
75. Example derived from USDA adapted by Deutsche Bank Research in C. Schaffnit-Chatterjee (2010), "Risk management in agriculture", p. 20.
76. Information about anticipated price variability within a given year can be obtained from commodity options quotes. The value of a commodity option depends on the volatility of the underlying future price, the future-price level, the strike price, the interest rate, and the time to maturity (Black). Holding the last four variables

constant, a higher volatility implies a higher price for both puts and calls. Volatility cannot be observed until after the fact. However, if an options price is observed along with the last four variables, an "implied volatility" can be calculated. Such implied volatilities embody the current judgments of traders – who have money on the line – as to the actual volatility likely to be realized.

77. Source: ibid., note 50, pp. 37–38, but adjusted for 2013 price and cost levels.

5 Biofuels and Agriculture

1. See further, *OECD-FAO Agricultural Outlook 2011–2020* (2011), chapter 3, "Biofuels" (retrieved: http://www.oecd.org/site/oecd-faoagriculturaloutlook/48202074.pdf).
2. See further, European Climate Foundation (2010), "Biomass for heat and power: Opportunity and economics" (retrieved: http://www.europeanclimate.org/documents/Biomass_report_-_Final.pdf).
3. See FAO (2008), "The state of food and agriculture 2008", and specifically "Biofuels and agriculture: A technical overview (part 2)" (retrieved: ftp://ftp.fao.org/docrep/fao/011/i0100e/i0100e.pdf), pp. 10–22.
4. See extensively on the topic, C. M. Rendleman and H. Shapouri (2007), "New technologies in ethanol production", USDA, Agricultural Economic Report 842 (retrieved: http://www.usda.gov/oce/reports/energy/aer842_Ethanol.pdf), and further, K. H. Mathews Jr. and M. J. McConnell (2009), "Ethanol co-product use in U.S. cattle feeding: Lessons learned and considerations", FDS-09D-01 (retrieved: http://www.ers.usda.gov/media/147398/fds09d01.pdf), and H. Shapouri, J. A. Duffield, and M. Wang (2003), "The energy balance of corn ethanol revisited", *American Society of Agricultural Engineers* 46(4): 959–968 (retrieved: http://naldc.nal.usda.gov/download/48999/PDF).
5. Ibid., FAO (2008), p. 11.
6. See R. Miao (2013), "Impact of ethanol plants on local land use change", *Agricultural and Resource Economics Review* 42(2): 291–309 (retrieved: http://ageconsearch.umn.edu/bitstream/155314/2/ARER%2042x2%202013%20MiaoRevised.pdf).
7. For further technical details, ibid., FAO (2008), p. 13.
8. For a write-up of the production process, see ibid., FAO (2008), p. 13.
9. See for an illustrative overview of the production process, ibid., FAO (2008), p. 14.
10. See R. Wisner, D. Anderson, R. Plain, D. Hofstrand, and D. O'Brien (2012), "Future patterns of U.S. grains, biofuels, and livestock and poultry feeding", (retrieved: http://ageconsearch.umn.edu/bitstream/156196/2/Future%20Feeding%20Patterns%20fullrpt.pdf). This report looks at what the future may hold for these crops and the livestock sector in the face of rising demand from food, feed, and fuel sources, shedding light on such questions as How is demand from the energy sector likely to affect demand going forward? What would this mean for the cost and availability of these grains? and How might this determine the future patterns and direction of the feed, livestock, poultry, and dairy industries?
11. See GBEP (Global Bioenergy Partnership) 2007, "A review of the current state of bioenergy development in G8+5 countries", Rome, GBEP Secretariat, FAO (retrieved: http://www.globalbioenergy.org/fileadmin/user_upload/gbep/docs/BIOENERGY_INFO/0805_GBEP_Report.pdf).
12. See infra in the next chapter on the sustainability question. For the environmental impact, see ibid., FAO (2008), chapter 5, pp. 55–77.

13. In an internal World Bank document leaked to the *Guardian* newspaper in 2008, its author, Don Mitchell, lead economist for the Development Prospects Group of the World Bank, concluded that "most of the remaining 70–75% increase in food commodities prices [between January 2002 and June 2008] was due to biofuels." D. Mitchell (July 2008), "A note on rising food prices", World Bank Development Prospects Group (retrieved: http://wwwwds.worldbank.org/servlet/WDSContentServer/WDSP/IB/2008/07/28/000020439_20080728103002/Rendered/PDF/WP4682.pdf).
14. D. Rajagopal and D. Zilberman (2007), "Review of environmental, economic and policy aspects of biofuels", World Bank Policy Research Working Paper 4341, World Bank, Washington, DC. D. Rajagopal, S. E. Sexton, D. Roland-Host, and D. Zilberman (2007), "Challenge of biofuel: Filling the tank without emptying the stomach?", *Environmental Research Letters* 2 (30 November), for global data; R. Naylor, A. J. Liska, M. B. Burke, W. P. Falcon, J. C. Gaskell, S. D. Rozelle, and K. G. Cassman (2007), "The ripple effect: Biofuels, food security, and the environment", *Environment* 49(9): 31–43, for national data and further the Worldwatch Institute.
15. Ibid., FAO (2008), p. 17, figure 7, for a summary of the studies referred to.
16. See further, USDA and US Department of Energy (2005), "Biomass as feedstock for a bio-energy and bio-products industry: The technical feasibility of a billion-ton annual supply" (retrieved: http://www.osti.gov/bridge).
17. Ibid., FAO (2008), p. 18. Chapter 5 of the report assesses the impact of biofuels on GHGs and concludes that the impact is not so positive as initially assumed. See also the next chapter for more updated data sets in this regard. For a good introduction on the life cycle of GHG balances, see chapter 5, p. 6. However, the data sets presented are somewhat outdated (2008) and reference can be made to the more updated FAO data sets in the next chapter under the heading "Biofuels and the Sustainability Conundrum". Nevertheless, the 2008 report remains recommended reading for its good and comprehensive introduction and reflection of how historical dynamics have led to the current situation.
18. See further, ibid., FAO (2008), pp. 18–21.

6 Biofuels and the Sustainability Conundrum

1. A. Elbehri, A. Segerstedt, and P. Liu (2013), "Biofuels and the sustainability challenge: A global assessment of sustainability issues, trends and policies for biofuels and related feedstocks", FAO, Rome (retrieved: http://www.fao.org/docrep/017/i3126e/i3126e.pdf). Highly recommended is also its literature list, which is both comprehensive and up to date (pp. 149–169). See also World Energy Council (2010), "Biofuels: Policies, standards and technologies" (retrieved: http://www.worldenergy.org/documents/biofuelsformatedmaster.pdf).
2. See ibid., note 1.
3. See also J. Ferrell and V. Sarisky-Reed (2010), "National algal technology roadmap", US Department of Energy (retrieved: http://www1.eere.energy.gov/bioenergy/pdfs/algal_biofuels_roadmap.pdf).
4. See ibid., note 1.
5. See for that in detail, M. Solecki, A. Scodel, and B. Epstein (2013), "Advanced biofuel market report 2013" (retrieved: http://www.fuelinggrowth.org/wp-content/uploads/2013/09/E2-Biofuel-Market-Report-2013.Final_.pdf).
6. See ibid., note 1.
7. See ibid., note 1, pp. xiii–xvi.

8. J. W. A. Langeveld, J. Dixon, H. van Keulen, and P. M. F. Quist-Wessel (2013), "Analysing the effect of biofuel expansion on land use in major producing countries", Biomass Research Report 1301 (retrieved: http://www.biomassresearch.eu/Biomass%20Research%201301_Analysing%20the%20effect%20of%20biofuel%20expansion%20on%20land%20use.pdf). This study based on land-use data from the Food and Agriculture Organisation of the United Nations casts some doubt on commonly accepted views about the indirect land-use change impacts of biofuels. The study analyzes data on land-use change over the last decade in the countries representing 97% of global ethanol production.

   The data suggests that current models assessing the impact on land use of crops grown for biofuel production do not accurately take account of the real-world evidence of changes in agricultural production and land use, most notably increases in yields and cropping intensity.
9. See also the following:
   - K. Aschworth, O. Wild, and C. N. Hewitt (2013), "Impacts of biofuel cultivation on mortality and crop yields", Nature Climate Change.
   - C. Bowyer (2011), "Anticipated indirect land use change associated with expanded use of biofuels and bioliquids in the EU: An analysis of the national renewable energy action plans", update March 2011, Institute for European Environmental Policy (IEEP), London.
   - C. Brannigan, G. Gibson, N. Hill, M. Dittrich, A. Schroten, H. van Essen et al. (2012), "Development of a better understanding of the scale of co-benefits associated with transport sector GHG reduction policies", Brussels: European Commission (retrieved: http://www.eutransportghg2050.eu/cms/assets/Uploads/Reports/EU-Transport-GHG-2050-II-Task-1-FINAL-12Jul12.pdf).
   - W. Britz and T. W. Hertel (2011), "Impacts of EU biofuels directives on global markets and EU environmental quality: An integrated PE, global CGE analysis", *Agriculture, Ecosystems & Environment* 142(1–2): 102–109.
   - S. Broch, K. Hoekman, and S. Unnasch (2013), "A review of variability in indirect land use change assessment and modeling in biofuel policy", *Environmental Science and Policy* 29: 147–157.
   - J. Bruinsma (2011), "The resources outlook: By how much do land, water and crop yields need to increase by 2050?", in P. Conforti, ed., *Looking Ahead in World Food and Agriculture: Perspectives to 2050* (Rome: Food and Agriculture Organization of the United Nations).
   - A. Carpentier and E. Letort (2012), "Accounting for heterogeneity in multicrop micro-econometric models: Implications for variable input modeling", *American Journal of Agricultural Economics* 94: 209–224.
   - G. Davies (2012), "Removing biofuel support policies: An assessment of projected impacts on global agricultural markets using the AGLINK-COSIMO model", analytical piece of work by Defra Economics, URN: 12D/084 (retrieved: http://www.decc.gov.uk/assets/decc/11/meeting-energy-demand/bio-energy/5134-removing-biofuel-support-policies-an-assessment-o.pdf).
   - S. De Cara, A. Goussebaile, R. Grateau, F. Levert, J. Quemener, and B. Vermont (2012), "Revue critique des études évaluant l'effet des changements d'affectation des sols sur les bilans environnementaux des biocarburatns ", rapport ADEME (retrieved: http://www2.ademe.fr/servlet/getBin?name=7AC5DFA02A2CE66DFDE000D7FA33AA56_tomcatlocal1333626720098.pdf).
   - J. Dumortier, D. Hayes, Miguel M. Carriquiry, F. Dong, X. Du, A. Elobeid, J. F. Fabiosa, and S. Tokgoz (2011), "Sensitivity of carbon emission estimates from indirect land-use change", *Applied Economic Perspectives and Policy* 33(3): 428–448.

- Ecofys (2012), "Assessing grandfathering options under an EU ILUC policy" (by D. Peters, A. van den Bos, and J. van de Staaij) study commissioned by Transport & Environment (retrieved: http://www.transportenvironment.org/sites/default/files/media/ecofys_2012_grandfathering_iluc.pdf).
- Ecofys, Agra CEAS, Chalmers University, IIASA, and Winrock (2011), "Biofuels baseline 2008" (by C. Hamelinck, M. Koper, G. Berndes, O. Englund, R. Diaz-Chavez, E. Kunen, D. Walden et al.) study for the European Commission DG Energy under Tender No. TREN/D1/458/2009 (retrieved:http://ec.europa.eu/energy/renewables/studies/doc/biofuels/2011_biofuels_baseline_2008.pdf).
- R. Edwards, D. Mulligan, and L. Marelli (2010), "Indirect land use change from increased biofuels demand: Comparison of models and results for marginal biofuels production from different feedstock", European Commission Joint Research Centre, Ispra (retrieved: http://ec.europa.eu/energy/renewables/studies/doc/land_use_change/study_4_iluc_modelling_comparison.pdf).
- R. Edwards, D. Mulligan, and L. Marelli (2010), "Indirect land use change from increased biofuels demand: Comparison of models and results for marginal biofuels production from different feedstocks", Scientific and Technical Reports of the Joint Research Center, European Commission (retrieved: http://ec.europa.eu/energy/renewables/consultations/doc/public_consultation_iluc/study_4_iluc_modelling_comparison.pdf).
- E. Elobeid, M. A. Carriquiry, and J. F. Fabiosa (2012), "Land-use change and green house gas emissions in the FAPRI-CARD model system: Addressing bias and uncertainty", *Climate Change Economics* 3(3): 1–26.
- G. Fischer (2011), "How can climate change and the development of bioenergy alter the long-term outlook for food and agriculture?", in P. Conforti, ed., *Looking Ahead in World Food and Agriculture: Perspectives to 2050* (Rome: Food and Agriculture Organization of the United Nations).
- U. R. Fritsche, H. van Steen, J.-E. Petersen, L. Marelli, K. Andersson, G. Berndes, and N. Urbancic (2012), "Proceedings of the workshop on biofuels and indirect land use change, Brussels, 25 January 2012", report for the European Parliament's Directorate General for Internal Policies, Policy Department A: Economic and Scientific Policy, IP/A/ENVI/WS/2011–18 (retrieved: http://www.europarl.europa.eu/document/activities/cont/201202/20120229ATT39558/20120229ATT39558EN.pdf).
- A. Golub and T. W. Hertel (2012), "Modeling land-use change impacts of biofuels in the GTAP-BIO framework", *Climate Change Economics* 3(3): 1–30.
- D. Laborde and H. Valin (2012), "Modeling land-use changes in a global CGE: Assessing the EU biofuel mandates with the MIRAGE-BioF model", *Climate Change Economics* 3(3): 1–39.

The amount of research that has been published and/or produced in the last three years on this topic is of such a volume and magnitude that it is impossible to list them all here. Only a relevant selection was provided above or elsewhere in the notes. It is therefore highly recommended to consult the literature list of the reports mentioned in the notes, especially for those looking for an encyclopedic overview on the topic and the current state of affairs. Please also consider that undoubtedly the debate about this topic will rage on after the closing date of the manuscript and that further future consultations will be required to stay fully up to date. It requires no reconfirmation that the topic is headbangingly complicated and prone to many sensitivities and borderlining some of the traditional scientific capabilities.

10. L. W. M. Beurskens, M. Hekkenberg, and P. Vethman (2011), "Renewable energy projections as published in the national renewable energy action plans of the European member states – covering all 27 EU member states with updates for 20 member states", ECN-E-10–069, Energy Research Centre of the Netherlands, Petten.

11. See further, R. Govinda, J. Timilsina, C. Beghin, D. van der Mensbrugghe, and S. Mevel (2010), "The impacts of biofuel targets on land-use change and food supply", A Global CGE Assessment, Policy Research Working Paper 5513, World Bank (retrieved: http://elibrary.worldbank.org/docserver/download/5513.pdf?expires=1379781832&id=id&accname=guest&checksum=32E2E77D25D04F595F5A1ED60E76FE23). The study finds that an expansion of global biofuel production to meet currently articulated or even higher national targets in various countries for biofuel use would reduce gross domestic product at the global level; however, the gross domestic product impacts are mixed across countries or regions. The expansion of biofuels would cause significant land reallocation, with notable decreases in forest- and pasturelands in a few countries. The results also suggest that the expansion of biofuels would cause a reduction in food supply. Although the magnitude of the impact on food supply at the global level is not as large as perceived earlier, it would be significant in developing countries such as India and those in Sub-Saharan Africa.
12. M. Blanco Fonseca, A. Burrell, H. Gay, M. Henseler, A. Kavallari, R. M'barek, I. Pérez Domínguez, and A. Tonini (2010), "Impacts of the EU biofuel target on agricultural markets and land use: A comparative modeling assessment", JRC Scientific and Technical Reports EUR 24449 EN, European Commission Joint Research Centre, Institute for Prospective Technological Studies, Seville.
13. See for a full review of all the studies done and a breakdown per agricultural commodity, B. Kretschmer, C. Bowyer, and A. Buckwell (2012), "EU biofuel use and agricultural commodity prices: A review of the evidence base", report prepared for action aid, section 5.3, pp. 48–49, and the relevant conclusions (retrieved: http://www.ieep.eu/assets/947/IEEP_Biofuels_and_food_prices_June_2012.pdf).
14. D. Laborde (2011), "Assessing the land use change consequences of European biofuel policies", final report October 2011, International Food Policy Institute (IFPRI), Washington DC. (retrieved: http://www.ifpri.org/sites/default/files/publications/biofuelsreportec2011.pdf).
15. See also M. B. Fonseca et al. (2010), "Impacts of the EU biofuel target on agricultural markets and land use: A comparative modeling assessment", ed. A. Burrell, JRC-IPTS (retrieved: http://ec.europa.eu/energy/renewables/studies/doc/land_use_change/study_jrc_biofuel_target_iluc.pdf).
16. Commission staff working document – Impact assessment accompanying the document Proposal for a Directive of the European Parliament and of the Council amending Directive 98/70/EC relating to the quality of petrol and diesel fuels and amending Directive 2009/28/EC on the promotion of the use of energy from renewable Sources, SWD (2012) 343 final (17.10.2012) (retrieved: http://eur-lex.europa.eu/LexUriServ/LexUriServ.do?uri=SWD:2012:0343:FIN:EN:PDF).
17. Most of today's biofuels are produced from crops grown on agricultural land such as wheat and rapeseed. When agricultural or pasture land previously destined for the food, feed, and fiber markets is diverted to the production of biofuels, the non-fuel demand will still need to be satisfied. Although this additional demand can be met through intensification of the original production, bringing nonagricultural land into production (often through deforestation) elsewhere is also possible. It is in the latter case that land-use change occurs indirectly (hence the term "indirect land-use change"), which enhances the production of GHGs. While most biofuel feedstocks are being produced in the EU, the estimated indirect land-use change emissions are mostly expected to take place outside the EU, where the additional production is likely to be realized at the lowest cost. In the case that this production is realized through the use of additional land, its conversion could lead to substantial greenhouse gas emissions being released if high carbon-stock areas such as forests are affected as a result. Ibid. impact assessment (2012), p. 11. See further, European

Commission: COM (2010) 811 final, report from the Commission on indirect land-use change related to biofuels and bioliquids (retrieved: http://eur-lex.europa.eu/LexUriServ/LexUriServ.do?uri=COM:2010:0811:FIN:EN:PDF). Calculating this "indirect land-use change" (ILUC) effect is complicated, because it is based on economic models projecting behavior 10 or 20 years into the future. The numbers are different for different crops. But overall, when land-use effects are taken into account, most varieties of biodiesel turn out to produce more emissions than bioethanol – and often more than fossil fuels. The effect wipes out more than two-thirds of the carbon emissions that Europe's renewable-energy policy was supposed to save by 2020. In the United States, the Environmental Protection Agency did take the land-use effect into account in 2010 when it set standards for which fuels count as renewable. But the European Commission has ducked the issue in the face of strong resistance from the biofuels industry and Europe's energy and agricultural sectors.

18. See for a an excellent analysis of the EU's impact assessment 2012 report and legislative proposal, C. Charles, I. Gerasimchuk, R. Bridle, T. Moerenhout, E. Asmelash, and T. Laan (2013), "Biofuels at what cost? A review of costs and benefits of EU biofuel policies", research report, International Institute for Sustainable Development, April 2013 (retrieved: http://www.iisd.org/gsi/sites/default/files/biofuels_subsidies_Eu_review.pdf).
19. See B. Kampman, A. van Grinsven, and H. Croezen (2012), "Sustainable alternatives for land-based biofuels in the European Union: Assessment of options and development of a policy strategy", Delft, CE Delft (retrieved: http://www.transportenvironment.org/sites/te/files/publications/CE_Delft_Sustainable_alternatives_land_based_biofuels.pdf). For a very critical assessment of the EU impact assessment and its alleged flaws, see M. A. Kropiwnicka (2013), "ActionAid International, the European Commission's renewable energy progress and biofuels sustainability reports 2013: A critical analysis of the coverage of land rights and socio economic impacts" (retrieved: http://www.actionaid.org/sites/files/actionaid/analysis_of_Ec_report_impacts_of_biofuels_for_actionaid_may_2013_final.pdf). This report was followed in May 2013 by an expose claiming that the EU failed to respect the principles of the Lisbon Treaty. This claim was all related to the specific and limited use of data and narrow time frames for analysis, ignoring the large recent body of research when it doesn't fit the EU overall intended direction with respect to biofuels and renewable energy and lack of consideration of social, economic, and developmental aspects in developing economies. See also, B. Kretschmer, B. Allen, D. Kieve, and C. Smith (2013), "The sustainability of advanced biofuels in the EU: Assessing the sustainability of wastes, residues and other feedstocks set out in the European Commission's proposal on indirect land use change (ILUC)", Biofuel ExChange Briefing 3, Institute for European Environmental Policy (IEEP), London. The ILUC directive referred to points at Proposal COM (2012) 595 final of 17.10.2012 for a Directive of the European Parliament and of the Council amending Directive 98/70/EC relating to the quality of petrol and diesel fuels and amending Directive 2009/28/EC on the promotion of the use of energy from renewable sources.
20. This also needs to be read in conjunction with the reporting in May 2013 on energy technologies and innovation. Communication from the Commission to the European Parliament, the Council, the European economic and social committee, and the committee of the regions, Energy Technologies and Innovation. COM (2013) 253 final (retrieved: http://ec.europa.eu/energy/technology/strategy/doc/comm_2013_0253_En.pdf).
21. See further on this, G. Pehnelt and C. Vietze (2012), "Uncertainties about the GHG emissions saving of rapeseed biodiesel", Jena Economic Research Papers, 2012–039 (retrieved: http://www.econstor.eu/bitstream/10419/70134/1/720430488.pdf).

22. D. Diop, M. Blanco, A. Flammini, M. Schlaifer, M. A. Kropiwnicka, and M. Mautner Markhof (2013) "Assessing the impact of biofuels production on developing countries from the point view of Policy Coherence for Development" (retrieved: http://ec.europa.eu/europeaid/what/development-policies/documents/biofuels_final_report_assessing_impact_of_Eu_biofuel_policy_pcd_22022013_En.pdf).
23. Report (COM [2013] 175 final) from the Commission to the European Parliament, the Council, the European Economic and Social Committee, and the Committee of the regions, "Renewable energy progress report" (retrieved: http://ec.europa.eu/energy/renewables/reports/doc/com_2013_0175_res_En.pdf), p. 10. In the run-up to this report, the report ActionAid (2012), "Adding fuel to the flame: The real impact of EU biofuels policy on developing countries" (retrieved: http://www.actionaid.org/sites/files/actionaid/adding_fuel_to_the_flame_actionaid_2013_final_0.pdf), indicated that the EU will most likely not include the fact that EU investors in recent years have acquired already 6 million hectares of land in Sub-Saharan Africa. Their estimate is further that the EU biofuel policy consumes the equivalent of feeding 185 million people for a year, and filling up a tank with gas equals 200 days of food for an average child. See further, T. Kaphengst, S. Wunder, and K. Timeus (2012), "The social dimension of EU biofuel policy", Ecologic Briefs on International Relations and Sustainable Development (retrieved: http://www.ecologic.eu/files/publications/1358406689/kaphengst_12_Ecologic_Brief_Biofuel.pdf); European Academies Science Advisory Council (2012), "The current status of biofuels in the European Union, their environmental impacts and future prospects" (retrieved: http://www.easac.eu/fileadmin/PDF_s/reports_statements/Easac_12_Biofuels_Complete.pdf); ActionAid (2012), "Fuel for thought: Addressing the social impacts of EU biofuels policies" (retrieved: http://www.actionaid.org/sites/files/actionaid/fuel_for_thought.pdf).
24. The study (to be published by the end of 2013) explores whether consumers are able to make informed choices, by looking into consumer understanding and the transparency of information. It is expected to generate recommendations on improving and harmonizing fuel labeling at the pump across EU member states. The study also tackles the issue of the availability of different fuels and retailers, and retail prices. This study is expected at the end of 2013, after the close of the manuscript of this book. See also, Commission staff working document, Annexes 1 and 2 of the Impact Assessment, Accompanying the document Proposal for a Council Regulation on the Bio-Based Industries Joint Undertaking COM (2013) 496 final, SWD (2013) 248 final, and SWD (2013) 247 final.
25. It should be noted that the central and Eastern European countries have a strong and powerful agricultural, biofuel, and renewable energy lobby in the EU sphere.
26. One of the most media-hyped reports was that of Prof. A. Gohin (June 2013), "Le changement d'affectation des sols induit par la consomation européene de biodiesel: une analyse de sensibilité aux évolutions des rendements agricoles", Working Paper smart – lereco nr. 13–07 (retrieved: http://prodinra.inra.fr/ft?id=%7B7858309A-82-DB-4D52–9B5C-966887EBDEC3%7D), who demonstrated that that only by varying one of the dozens of assumptions of ILUC models, the final result can vary by 80%.
27. Based on for example seaweed or certain types of waste, advanced biofuels must account for no less than 2% of consumption by 2020 but should not be achieved at the cost of depriving other industries of raw materials, destabilizing EU waste policy, overexploiting forests, or reducing biodiversity.
28. That is, the report on the proposal for a directive of the European Parliament and of the Council amending Directive 98/70/EC relating to the quality of petrol and diesel fuels and amending Directive 2009/28/EC on the promotion of the use of energy from renewable sources (COM (2012) 0595 – C7–0337/2012 – 2012/0288 (COD)) (retrieved: http://www.europarl.europa.eu/sides/getDoc.do?pubRef=-//EP//

NONSGML+REPORT+A7–2013 0279+0+DOC+PDF+V0//EN). Ms Lepage's report argues that public subsidies for food-based biofuels in the EU encourage their cultivation on land that could otherwise be used for food production. Some studies suggest that continuing with the current level of EU incentives for food-based biofuels would actually cancel out the cuts in greenhouse gas emissions achieved by switching to biofuels. Clearing land to plant food for biofuel releases the greenhouse gas carbon dioxide (CO_2) through plowing and can involve deforestation, which reduces the "carbon sinks" – the trees that absorb CO_2. Ms Lepage's report also warns of the social impact of such biofuel cultivation – known as "indirect land use change (ILUC)." "The indirect land use change effects are not only environmental but also social, and are placing additional pressure on land use, particularly in developing countries, which is having a negative impact on the food security of local people, in particular women," her report says. The EU's Joint Research Centre has calculated that scrapping the current biofuel incentives could lower the price of vegetable oil in the EU by 48% by 2020. S. Hélaine, R. M'barek, and H. Gay (2012), "Impacts of the EU biofuel policy on agricultural markets and land use" (retrieved: http://static.euractiv.com/sites/all/euractiv/files/a%20JRC%20report.pdf); as well as (JRC) R. Edwards, S. Szekeres, F. Neuwahl, and V. Mahieu (ed. G. de Santi) (2012), "Biofuels in the European context: Facts and uncertainties" (retrieved: http://www.fp7.org.tr/tubitak_content_files/306/JRC/dokumanlar/jrc_biofuels_report.pdf). That on top of the fact that a 2013 report (Global Food Security Report) from the House of Commons International Development Committee (IDC) has found that the increasing use of biofuels is undermining food security by forcing up the price of food crops (retrieved: http://www.publications.parliament.uk/pa/cm201314/cmselect/cmintdev/176/17602.htm).

7 Financing the Agriculture Firm

1. See also, "Financial guidelines for agricultural producers: Recommendations of the farm financial standards council" (revised 1997) (retrieved: http://agmarketing.extension.psu.edu/Business/PDFs/FinGuidAgProd.pdf).
2. For a first good read on lending, see: M. S. Fridson and F. Alvarez (2011), *Financial Statement Analysis: A Practitioners Guide*, Wiley & Sons, 4th ed. Chichester. See further on agricultural specifics and the role of credit, C. dell'Aquila and M. Eboli (2013), "Financing production with liquidity constraints: The role of trade credit in agro-food supply chains" (retrieved: http://ageconsearch.umn.edu/bitstream/154656/2/dell%E2%80%99Aquila.pdf); and on the capital structure for agri-firms, R. Aulová and T. Hlavsa (2013), "Capital structure of agricultural businesses and its determinants" (retrieved: http://ageconsearch.umn.edu/bitstream/152688/2/agris_on-line_2013_2_aulova_hlavsa.pdf).
3. Crop budgets that provide the actual cost figures to finance different crop packages can vary from one area to another and from one year to the next.
4. USAID/FS Share (2012), "Lending to the agricultural sector: A toolkit" (retrieved-manual: https://www.agrifinfacility.org/sites/agrifinfacility.org/files/FS%20Share_Toolkit%20Training%20Guide.pdf), p. 2. The toolkit promotes a standardization of lending processes and a move toward international best practices for agricultural lending by FIs to stimulate expansion of agribusiness finance. See also in depth, FAO (2010), "Agricultural value chain finance: Tools and lessons", Rome (retrieved: http://www.fao.org/docrep/017/i0846e/i0846e.pdf).
5. See, for example, G. Berg and M. Fuchs (2013), "Bank financing of SMEs in five Sub-Saharan African Countries: The role of competition, innovation, and the government", Policy Research Working Paper 6563 (retrieved: ,http://wwwwds.worldbank.org/servlet/WDSContentServer/WDSP/IB/2013/08/08/000158349_20130808113228/Rendered/PDF/WPS6563.pdf).

6. Source: adapted from ibid., note 4, pp. 31–33.
7. See further, M. Rosenzweig and C. Udry (2013), "Forecasting profitability", Economic Growth Center Discussion Paper 1029, Economics Department Working Paper 120, Department of Economics, Yale University (retrieved: http://www.econ.yale.edu/growth_pdf/cdp1029.pdf).
8. A full write-up of the product groups can be found at ibid., note 4, p. 34.
9. Source: ibid., note 4, p. 35.
10. See for a full range of product worksheets, including repayment terms, interest rates and fees, and normal loan ranges, ibid., note 4, pp. 36–41.
11. F. Hollinger, L. Rutten, and K. Kiriakov (2009), "The use of warehouse receipt finance in agriculture in transition countries", Working Paper presented at the World Grain Forum 2009, St. Petersburg/Russian Federation, June 6–7.
12. Source: ibid., note 4, p. 46.
13. See for a fully worked out example, ibid., note 4, pp. 46–47.
14. Ibid., note 4, p. 48.
15. See for an overview of the most common financial matrixes, ibid., note 4, annex GG.
16. Source: ibid., note 4, p. 60.
17. For a full review of the risk categories and evaluation, see ibid., note 4, annex X.
18. Example derived from ibid., note 4, p. 61.
19 For a full-fledged SWOT analysis of the farm, see ibid., note 4, p. 63.
20. For a somewhat structured approach, see ibid., note 4, annex S.

Type of collateral	Considerations
Residential/commercial real estate Reasonable liquidation value = 60% to 70% of value	• Who owns the property? • Would losing this property have a significant effect on the borrower (psychological value of property)? • Are property values in the district increasing or declining? • What is the condition of the property? Has the property recently been renovated?
Equipment Reasonable liquidation value = 40% to 50% of value	• Who legally owns the equipment? • How old is it? Is it in good condition? • Would there be a market for the equipment if the financial institution had to sell it? (Typically, the more specialized the equipment, the more difficult it is to find a buyer.) • How important is the equipment to the borrower's operations (psychological value of property)? Is it integral to the borrower's operations or a piece of equipment that is no longer used? • How difficult is it for the equipment to be moved?
Vehicles Reasonable liquidation value = 30% to 40% of value	• Who legally owns the vehicle? • How old is it? Is it in good condition? • How important is the equipment to the borrower's operations (psychological value of property)? • Can/should the vehicle be locked during the term of the loan? • Will there be insurance for the vehicle if the borrower will continue to use it?
Inventory Reasonable liquidation value = 0% to 10% of value – raw materials and work in progress 50% to 60% of value – finished products	• What is the average level of inventory that the business typically maintains? • How important is the inventory to the borrower's operations (psychological value)? Is this inventory that the borrower doesn't expect to sell? • Should the inventory be locked? • What is the shelf life of the inventory? Is there a risk of obsolescence during loan term? • How difficult would it be for the financial institution to sell this inventory?
Accounts receivable Reasonable liquidation value = 50% to 60% of value	• Used only for borrowers with stable sales, credit functions, and borrowers. • How difficult would it be for the financial institution to collect these receivables?

21. See for boilerplate crop templates, ibid., note 4, pp. 129–130 as well as crop budget examples, ibid., note 3, annex P.
22. See for example, ibid., note 4, annex Q.
23. See ibid., note 4, p. 99.
24. DFID (2004), "The importance of financial sector development for growth and poverty reduction", Policy Division Working Paper, Financial Sector Team, Department for International Development, London (retrieved: http://ageconsearch.umn.edu/bitstream/12886/1/fi04im01.pdf). J. D. Von Pischke (1991), "Finance at the frontier: Debt capacity and the role of credit in the private economy", Development Studies, World Bank, Washington, DC.
25. See for a full review the chapter on risk management in agriculture.
26. See extensively, R. P. Christen and D. Pearce (2006), "Managing risks and designing products for agricultural microfinance: Features of an emerging model" (retrieved: http://www.ifad.org/ruralfinance/pub/risks.pdf). Also, L. Rutten (2012), "Innovative agricultural finance and risk management: Strengthening food production and trade in the transition region", Working Paper, FAO Investment Center, in particular chapter 4; IFPRI (2010), "Innovations in rural and agricultural finance" (eds R. Kloeppinger-Todd and M. Sharma) (retrieved: http://www.ifpri.org/sites/default/files/publications/focus18.pdf). ESFIM (G. Onumah and G. Meijerink) (2011), "Innovative agricultural financing models", Policy Brief 6 (retrieved: http://www.esfim.org/wp-content/uploads/policy-brief6-english.pdf); GPFI (2012) (retrieved: http://www.gpfi.org/sites/default/files/documents/G20%20Innovative%20Agricultural%20SME%20Finance%20Models.pdf); GPFI (2011), "Scaling-up access to finance for agricultural SME" (retrieved: http://www.gpfi.org/sites/default/files/documents/G20_Agrifinance_Report%20%28FINAL%20ONLINE%29.pdf).

8 Farmland I: An Investible Asset Class?

1. See further, W. Anseeuw, M. Boche, T. Breu, M. Giger, J. Lay, P. Messerli, and K. Nolte (2012), "Transnational land deals for agriculture in the global south: Analytical report based on the land matrix database", CDE/CIRAD/GIGA, 2013, Bern/Montpellier/Hamburg (retrieved: http://www.oxfam.de/sites/www.oxfam.de/files/20120427_report_land_matrix.pdf). The Land Matrix figures confirm that there is a worldwide rush for land, although it has lost some of its initial pace in recent years. Many deals have been more than just speculation and strategic positioning. A large number of contracts have been signed and followed through by the implementation and start of operation of projects. It involves a large number of target countries with very different investment conditions, as well as a great variety of actors with differing investment motives. Each land deal has its own specific characteristics. Little is known about the short and expected long-term effects of these investments. Even with the data collated by the Land Matrix project, the picture of large-scale land acquisitions remains hard to decipher. Africa is the most targeted region, and the majority of reported acquisitions are concentrated in a few countries. Among those 11 "high-concentration" countries, 7 are African, namely Sudan, Ethiopia, Mozambique, Tanzania, Madagascar, Zambia, and DR Congo. In Southeast Asia, the Philippines, Indonesia, and Laos are particularly affected.

   Other observations are as follows:

   - Investors are targeting countries that are among the poorest, are poorly integrated into the world economy, have a high incidence of hunger, and have weak land institutions.

- Investors are competing for land with local farming communities.
- Forested areas are highly affected by land acquisitions.
- Investors have a tendency to target land with high yield gaps, good accessibility, and considerable population densities.
- Investment is coming from wealthier, food-importing countries and is intraregional in nature.
- Investors are both public and private actors.
- Partnerships are important for investors.
- The rush for land is being driven by long-term trends.
- Investors are acquiring water.
- Both food and nonfood crops are important, but investors are seeking flexibility to switch between them.
- Most projects are export oriented.
- Land-governance systems are often not adequate to deal with large-scale land investments.
- Governments are selling land that is used by smallholders.
- Acquisitions are rarely based on free, prior, and informed consent (FPIC), and there is limited but worrying evidence on evictions.
- Rates of compensation are often very low. For a limited number of cases, the Land Matrix has information on compensation (typically stated intentions).
- Some infrastructure (i.e., health and education), but little evidence on benefits from employment generation.

See also, "Land tenure and international investments: A report by the high level panel of experts" (2011), Committee on World Food Security, High Level Panel of Experts on Food Security and Nutrition, Rome (retrieved: http://www.fao.org/file-admin/user_upload/hlpe/hlpe_documents/HLPE-Land-tenure-and-international-investments-in-agriculture-2011.pdf). See for a good regional introduction, Land Commodities AG (2012), "Investing in Australian agriculture: An introduction" (retrieved: http://www.landcommodities.com/wp-content/Assets/Downloads/Investing%20in%20Australian%20agriculture%20-%20An%20introduction.pdf); as well as Land Commodities AG (2012), "The Australian agricultural market: An introduction for investors" (retrieved: http://www.landcommodities.com/wp-content/Assets/Downloads/The%20Australian%20agricultural%20market%20%E2%80%93%20An%20introduction%20for%20investors.pdf); and Land Commodities AG (2012), "The Australian Wheatbelt: An introduction for investors" (retrieved: http://www.landcommodities.com/wp-content/Assets/Downloads/The%20Australian%20Wheatbelt%20%20An%20introduction%20for%20investors.pdf). For a forest-specific analysis, see New Forests (2011), "New Forests' timberland investment outlook, 2011–2015" (retrieved: http://www.newforests.com.au/news/pdf/articles/MarketOutlook_NewForestsTimberlandInvestmentOutlook.pdf); and their follow-through in New Forests (2013), "New Forests' timberland investment outlook, 2013–2017" (retrieved: http://newforests.com.au/news/pdf/articles/NewForestsTimberlandInvestmentOutlook2013.pdf).

2. See also H. Stookey and P. de Lapérouse (2009), "Agricultural land investment: A bright spot in the 2009 investment landscape" (retrieved: http://www.highquest-partners.com/userfiles/files/AgLand_Investing.pdf), combined with the traditional forex and environmental- and social-standard concerns.
3. See extensively, FAO (2006), "Livestock's long shadow: Environmental issues and options", Rome (retrieved: ftp://ftp.fao.org/docrep/fao/010/a0701e/a0701e00.pdf).
4. R. Devine (2003), "La consommation des produits carnés", INRA (retrieved: http://www6.inra.fr/productions-animales/2003-Volume-16/Numero-5–2003/La-consommation-des-produits-carnes).

5. FAO (2008), "The state of food insecurity in the world 2008", Rome (retrieved: ftp://ftp.fao.org/docrep/fao/011/i0291e/i0291e00a.pdf).
6. FAO (2009), "The state of food insecurity in the world 2009", Rome (retrieved: ftp://ftp.fao.org/docrep/fao/012/i0876e/i0876e_flyer.pdf).
7. See further, Stern Review (2006), "The economics of climate change, part II: The impacts of climate change on growth and development" (retrieved: http://www.wwf.se/source.php/1169157/Stern%20Report_Exec%20Summary.pdf), pp. 67–73
8. P. Pinstrup-Andersen and R. Pandya-Lorch (1998), "Food security and sustainable use of natural resources: A 2020 vision", *Ecological Economics* 26(1): 1–10.
9. J. R. Webb and J. H. Ruben (1988), "The effect of alternative return measures on restricted mixed-asset portfolios", *Journal of the American Real Estate and Urban Economics Association* (16): 123–137.
10. See, for example, http://farmdocdaily.illinois.edu/2013/09/managers-more-stable-farmland-prices.html.
11. See for some compelling analysis and data sets, "Is there a bubble forming in US farmland values" (2012), Global AgInvesting, Research and Insights (retrieved: http://www.globalaginvesting.com/downloads/files/Is-There-a-Bubble-Forming-in-U-S-Farmland-Values-.pdf), and C. Gustke (2013), "Are farmland investments heading for a barn bubble" (retrieved: http://www.bbc.com/capital/story/20130604-headed-for-a-barn-bubble).
12. Farmers make their decision as to which crops they plant in the spring based on the only information they have available, last year's prices. If corn prices were abnormally high one year, farmers will devote more acreage to corn, thus restricting the supply of other crops, pushing those prices higher. This corn price–land price relationship means that what you think will happen to the price of corn over the next 25 years will determine where you think land values will be in the next 25 years. (Source: ibid., note 8, p. 4.)
13. See also T. W. Hertel (2010), "The global supply and demand for agricultural land in 2050:A perfect storm in the making?", GTAP Working Paper 63 (retrieved: https://www.gtap.agecon.purdue.edu/resources/download/5115.pdf).
14. See also, H. Stookey and P. de Lapérouse (2009) "Agricultural land investment" (retrieved: http://www.highquestpartners.com/userfiles/files/AgLand_Investing.pdf.
15. Hancock Agricultural Investment Group (2009), "From the website of Hancock Agricultural Investment Group, an MFC Global Investment Management Company" (retrieved: http://www.haig.jhancock.com).
16. W. Hardin and P. Cheng (2002), "Farmland investment under conditions of certainty and uncertainty", *Journal of Real Estate Finance and Economics*.
17. Includes: alfalfa, corn, soybeans, peanuts, potatoes, rice, vegetables, and wheat.
18. Includes: almonds, apples, cherries, cranberries, pistachios, walnuts, and wine grapes.
19. See D. Garner and W. Brittain (2012), "Farmland as an alternative investment asset class: Fundamentals – characteristics – performance – opportunities – risks", Asset Class Guide (DGC/ACG/FARM/002) (retrieved: http://www.dgcassetmanagement.com/sites/www.dgcassetmanagement.com/files/Farmland%20Investment%20Report%202012%20-%20DGC%20Asset%20Management.pdf).
20. S. Chen, W. Wilson, R. Larsen, and B. Dahl (2013), "Investing in agriculture as an asset class", Agribusiness and Applied Economics Report 711 (retrieved: http://ageconsearch.umn.edu/bitstream/147053/2/AAE711.pdf).

21. See for competing research:
    - B. M. Clark, J. D. Detre, J. D'Antoni, and H. Zapata (2012), "The role of an agribusiness index in a modern portfolio", *Agricultural Finance Review* 72(3): 362–380.
    - G. Colvin and T. M. Schroeder (2012), "Investors' guide to farmland" (retrievable: https://www.createspace.com/3861185).
    - Global AgInvesting (2013), "Global AgInvesting 2013", New York (retrieved: http://www.globalaginvesting.com).
    - R. E. Just and Q. Weninger (1999), "Are crop yields normally distributed?", *American Journal of Agricultural Economics* 81(2): 287–304.
    - H. M. Kaplan (1985), "Farmland as a portfolio investment", *Journal of Portfolio Management* 11(2): 73–78.
    - Kleinwort Benson Investors (2010), "AgriEquities: The hard edge over soft commodities", presentation to the Global AgInvesting Conference, Geneva, November 2010.
    - C. B. Moss, A. M. Featherstone, and T. G. Baker (1987), "Agricultural assets in an efficient multiperiod investment portfolio", *Agricultural Finance Review* 47: 82–94.
    - D. L. Pendell and A. M. Featherstone (2006), "Agricultural assets in an optimal investment portfolio", selected paper presented at the Western Agricultural Economics Association Annual Meeting, Anchorage, AK, June 28–30, 2006.
    - D. V. Vedenov (2008), "Application of copulas to estimation of joint crop yield distributions", Paper presented at American Agricultural Economics Association Annual Meeting, Orlando, FL, July 27–29.
    - W. W. Wilson (2012), "Global fundamentals to 2020: Dynamic changes in agriculture and implications for investments", Global AgInvesting, New York.
    - Y. Zhu, S. K. Ghosh, and B. K. Goodwin (2008), "Modeling dependence in the design of whole farm: A copula-based model approach", Paper presented at the American Agricultural Economics Association Annual Meeting, Orlando, FL, July 27–29.
22. A good starting point to judge those price evolutions globally and per region is "The land commodities global agriculture & farmland investment report 2009" (retrieved: http://www.farmlandinvestmentreport.com/Farmland_Investment_Report.pdf), as well as Savills (2011), "International farmland market bulletin 2011" (retrieved: http://www.aginvestconference.com/usa/savills.pdf), pp. 5–8.
23. See for some further insights, J. Henderson (2012), "Will farmland values keep booming?" (retrieved: http://www.kc.frb.org/PUBLICAT/ECONREV/PDF/2q08henderson.pdf).
24. For the latest annual overview of land values in the United States, see "USDA land values summary 2012" (retrieved: http://www.nass.usda.gov/Publications/Todays_Reports/reports/land0812.pdf).
25. For more state details and/or updated numbers, see http://www.fcc-fac.ca/en/Products/Property/FLV/Spring2013/index.asp.
26. Quebec experienced the highest average increase at 19.4%, followed by Manitoba at 13.9% and Ontario at 11.9%. Saskatchewan and Alberta experienced 9.7% and 7.2% average increases respectively, followed by Nova Scotia at 6.8%, Prince Edward Island at 5.7%, and British Columbia at 0.4%. Average farmland values were unchanged in New Brunswick and Newfoundland and Labrador.
27. See further, ibid., Savills (2011), "International farmland market bulletin 2011".
28. See D. Pletichová and Z. Gebeltová (2013), "Development of market prices of agricultural land within the conditions of the EU", *Papers in Economics and*

Informatics 5(3) (retrieved: http://ageconsearch.umn.edu/bitstream/157586/2/agris_on-line_2013_3_pletichova_gebeltova.pdf).

29. N. Ishchukova and L. Smutka (2013), "Comparative advantage: Products mapping of the Russian agricultural exports" (retrieved: http://ageconsearch.umn.edu/bitstream/157521/2/agris_on-line_2013_3_ischukova_smutka.pdf).
30. See further, R. Slaston and A. Lissitsa (2013), "Productivity in agriculture of Ukraine and other CIS", presentation at the International Agricultural Trade Research Consortium's (IATRC's) 2013 Symposium: Productivity and Its Impacts on Global Trade, Seville, Spain, June 2–4. Labor productivity is lagging, however.
31. On September 23, 2013, it was reported that China's Xinjiang Production and Construction Corps (XPCC) had reached an agreement with Ukrainian KSO Agro about a 50-year lease of 3 million hectares of Ukrainian farmland (about 5% of total Ukrainian land and 9% of total Ukrainian farmland), although the deal was denied later in the day by KSO Agro (the Warsaw-listed agricultural giant managing the agricultural land on behalf of the Ukrainian ministry of agriculture). Its reporting indicates that it is only working with its Chinese partners on a project to install drip-irrigation systems over an area of 3,000 hectares in Ukraine next year. "KSG Agro does not intend or have any right to sell land to foreigners, including the Chinese," the statement further indicated. China's XPCC could not be immediately reached for comment. Uncertainty remains up to the date of closure of the manuscript of this book about the real depth of the deal. Under the (alleged) deal between China's Xinjiang Production and Construction Corps, or XPCC, and KSG Agro, a Ukrainian agricultural company, crops and pigs raised in the eastern region of Dnipropetrovsk will be sold at preferential rates to two Chinese state-owned grain firms. The deal comes after Ukraine lifted a law barring foreigners from buying (i.e., leasing) Ukrainian land last year. As part of the deal, China's Export-Import Bank has given Ukraine a $3 billion loan for agricultural development. In exchange for its produce, Ukraine will receive seeds, equipment, a fertilizer plant (Ukraine imports about $1 billion worth of fertilizer every year), and a plant to produce a crop-protection agent. XPCC will help build a highway in Ukraine's Autonomous Republic of Crimea as well as a bridge across the Strait of Kerch, a transport and industrial center for the country. The (alleged) deal has reignited the debate about international land grabs, especially as this deal emerges months after a respected study indicates that that somewhere between 0.7% and 1.75% of the world's agricultural land is being transferred to foreign investors from local landholders. That's an area bigger than France and Germany combined. See for that, M. C. Rulli, A. Saviori, and P. D'Odorico (2013), B. L. Turner, ed. "Global land and water grabbing", *Proceedings of the National Academy of Sciences* 110(3) (retrieved: http://www.pnas.org/content/110/3/892.full.pdf+html). The same study further indicates that the per capita volume of grabbed water often exceeds the water requirements for a balanced diet and would be sufficient to improve food security and abate malnourishment in the grabbed countries. It is found that about $0.31 \times 1012\ m^3 \times y - 1$ of green water (i.e., rainwater) and up to $0.14 \times 1012\ m^3 \times y - 1$ of blue water (i.e., irrigation water) are appropriated globally for crop and livestock production in 47×106 ha of grabbed land worldwide (i.e., in 90% of the reported global grabbed land). The study further found that foreign investors frequently buy tracts of land that have plenty of freshwater, either from local rainfall or underground aquifers. This is often good agricultural land that isn't yet fully utilized. It was being used by local farmers without modern technology, without irrigation, and without fertilizer. After the

land is bought up, large commercial farms will move in and boost production to grow their own crops. A 2010 World Bank study (see note 59) found that about 37% of this "grabbed" land is used to grow food crops, 21% to grow cash crops, and 21% to grow biofuels.

32. Savills (2013), "Market survey UK agricultural land", p. 3 (retrieved: http://www.savills.co.uk/research_articles/141557/144127–0).
33. See also Josh Miller and Diane Martin (2013), "RICS/ RAU rural land market survey" (retrieved: http://www.rics.org/Global/RICS%20RAU%20Rural%20Market%20Survey%20-%202013%20H1_Lite.pdf).
34. Hovering around 100,000 acres in 2013. In contrast, in 1945 that was 8700,000 acres, in 1980 400,000 acres, and in 2000 around 200,000 acres (see ibid., Savills (2013), "Market survey agricultural land", p. 4.
35. Ibid., Savills (2013), p. 6.
36. See for further details, Savills (2012), "Market survey agricultural land" (retrieved: http://pdf.savills.com/documents/ALMS_2012_low%20res.pdf), p. 3.
37. See for further details, Savills (2012), "International farmland focus 2012" (retrieved: http://www.gatewaytosouthamerica.com/archivo/savills_international-farmland-focus.pdf).
38. See for a very good comprehensive report on agriculture in Africa and its role in the world agricultural marketplace, (W. Oluoch-Kosura and G. Sikei) Gates and Rockefeller Foundation funded Alliance for a Green Revolution in Africa (AGRA) (2013), "Africa agriculture status report" (retrieved: http://reliefweb.int/sites/reliefweb.int/files/resources/agrafinalaugust20akim.pdf).
39. See further, J. Johnson (2013), "More cases of land grabs in South Sudan" (retrieved: http://www.southsudannewsagency.com/opinion/articles/more-cases-of-land-grabs-in-south-sudan), and M. Rhode (2011), "Is South Sudan's largest land deal a land grab?" (retrieved: http://thinkafricapress.com/south-sudan/largest-land-deal-land-grab).
40. See further, Savills World Research (2013), "Spotlight on African farmland" (retrieved: http://www.gatewaytosouthamerica-newsblog.com/wp-content/uploads/2013/05/Spotlight_African+Farmland.pdf).
41. See for a good and recent overview, Savills (2012), "International farmland focus: Going with the grain" (retrieved: http://farmlandgrab.org/uploads/attachment/international-farmland-focus.pdf).
42. See, for example, G. Schnitkey (2013), "2013 county cash rents: Levels variability and 2014 cash rent decisions", University of Illinois (retrieved: http://farmdoc.illinois.edu/manage/newsletters/fefo13_17/fefo_13_17.pdf). The University of Illinois' Department of Agriculture uses newly released USDA data on average farmland cash rents to analyze the relationship between rental rates and agricultural production potential. Their analysis provides insights on such questions as:
    1. How do rental returns compare for different classes of land?
    2. Are different classes of land rationally priced bearing in mind rental returns, and is the variability in cash rents fully explained by farmland productivity?
    3. In which part of the quality spectrum are cash rents least explained by farmland productivity?
    4. What is the outlook for rental rates in 2014?
43. The Senate bill is S. 954. The House bill is H.R. 1947.
44. See further, R. de Leon, T. Garcia, G. Kummel, L. Munden, S. Murday, and L. Pradela (Munden project) (2013), "A data-driven examination of land tenure risk and

industrial concessions in emerging market economies" (retrieved: http://www.rightsandresources.org/documents/files/doc_6301.pdf). In particular, it addresses questions such as:

1. What are the financial risks associated with overlooking or underestimating land-tenure contestation, ranging from increased operating costs to project delays and outright abandonment of operations?
2. What are the main sources of land-tenure risk and how should investors deal with these risks in their acquisition due-diligence process?
3. To what extent would tenure conflicts and financial damage be avoidable if investors had a more thorough understanding of preexisting or customary local land rights?
4. To what extent could developing appropriate risk-management processes provide investors with a competitive edge in selecting EME investment targets?

45. Source: FAO (2013), www.fao.org.
46. *Arable land* is the land under temporary agricultural crops (multiple-cropped areas are counted only once), temporary meadows for mowing or pasture, land under market and kitchen gardens, and land temporarily fallow (less than five years). The abandoned land resulting from shifting cultivation is not included in this category. *Permanent crops* is the land cultivated with long-term crops which do not have to be replanted for several years (such as cocoa and coffee); land under trees and shrubs producing flowers, such as roses and jasmine; and nurseries (except those for forest trees, which should be classified under "forest").
47. However, there is a significant difference between the structure of the US market compared to that of other countries. Non-owner-operators, or in other words investors who do not themselves farm (including institutions), owned 29% of farmland in 2007, and the Corn Belt had a higher proportion at 38%. While investor ownership is significant in the United States, foreign ownership of cropping land is still low at 0.4% as of 2010, with 0.6% of pastureland owned by foreigners and 1% of all forestland. This contrasts with estimates that institutional ownership in countries such as Australia, Argentina, Brazil, and Canada is below 10% and in many cases below 2%. Former Soviet Union countries have differing land-ownership structures. In some cases it is only the government that is able to own land, and access to farmland is only available through leasehold tenure. Data source: USDA (2012), "Trends in US farm values and ownership", Global AgInvesting Research & Insight.
48. Source: FAO (2011); Global AgInvesting Research & Insight Estimates (2012); Macquarie Agricultural Funds Management (2012).
49. Source: FAO (2011); GMO (2011); HNRG (2012); Global AgInvesting Research & Insight Estimates (2012); Timberland Investment Resources (2012); Macquarie Agricultural Funds Management (2012).
50. Source: B. Wheaton and W. J. Kiernan (2012), "Farmland: An untapped asset class? Quantifying the opportunity to invest in agriculture (retrieved: http://www.globalaginvesting.com/ResearchInsights/detail?contentid=2665).
51. See further, P. Campbell (2011), "The future prospects of global arable land", Global Food and Water Crises Research Program (retrieved: http://www.futuredirections.org.au/publications/food-and-water-crises/53-the-future-prospects-for-global-arable-land.html). Also regarding the impact of foreign buyers and the state of the farmland market, C. Schaffnit-Chatterjee (2012), "Foreign investment in farmland: No low-hanging fruit", Deutsche Bank Research (retrieved: http://www.dbresearch.com/PROD/DBR_INTERNET_EN-PROD/PROD0000000000296807/Foreign+investment+in+farmland%3A+No+low-hanging+fruit.PDF).

- Further reading can include the following:W. Anseeuw et al. (2012), "Land rights and the rush for land", International Land Coalition, Rome.
- R. Arezki et al. (2011), "What drives the global land rush?" IMF Working Paper 251, November 2011.
- R. Arezki et al. (2012), "The global land rush", *Finance & Development*, March 2012.
- J. Auer et al. (2012), "Real assets: A sought-after investment class in times of crisis", Deutsche Bank Research, Frankfurt am Main, June 2012.
- L. Cotula and E. Polack (April 2012), "The global land rush", IIED briefing, International Institute for Environment and Development.
- O. De Schutter (2011), "How not to think of land-grabbing: Three critiques of large-scale investment in farmland", *Journal of Peasant Studies* 38: 2.
- O. De Schutter (2011), "The green rush: The global race for farmland and the rights of land users", *Harvard International Law Journal* 52(2).
- FAO (2012), "Voluntary guidelines on the responsible governance of tenure of land, fisheries and forests in the context of national food security", Food and Agriculture Organization of the United Nations, Rome.
- D. Giovannucci et al. (2012), "Food and agriculture: The future of sustainability", United Nations, New York.
- GRAIN (2011), "Extent of farmland grabbing for food production by foreign interests", December 2011.GRAIN (2012), "Land grab deals", January 2012.
- HighQuest Partners, US (2010), "Private financial sector investment in farmland and agricultural infrastructure", Working Paper 33, OECD.
- A. Litovsky and P. Villalpando (2012), "The land security agenda", Earth Security Initiative, London, March 2012.

52. Source and see for, in my liking, an excellent review of the Australian market: Land Commodities (2013), "Achieving alpha in farmland investing: A realistic look at investment returns in the Australian Wheatbelt" (retrieved: http://www.landcommodities.com/a-realistic-look-at-investment-returns).
53. Farmland markets are characterized by: A general lack of liquidity – average transaction sizes are large with a relatively small number of potential buyers.Variation in the level of liquidity – farmer sentiment, farm profitability, and in turn the price of farmland fluctuates depending on seasonal conditions.A general lack of detailed or reliable information.
54. See also C. Häberli (2012), "Foreign direct investment in agriculture: Land grab or food security improvement" (retrieved: http://www.ssrn.com/link/SIEL-2012-Singapore-Conference.html).
55. Gulf Research Center, "Food inflation in the GCC countries", GRC Report, May 2008; and Emirates Business (2009), "Arab nations urged to set up strategic cereal stock", February 4. GCC countries include Bahrain, Kuwait, Oman, Qatar, Saudi Arabia, and the United Arab Emirates (UAE).
56. The soft sides of these deals are ignored here as not fitting in the context of this book, but not the less important. For example, Human Rights Watch released a report in 2012 alleging that the Ethiopian government was forcibly relocating tens of thousands of people in order to lease land to foreign investors from China and the Gulf states. "The first round of forced relocations occurred at the worst possible time of year – the beginning of the harvest," the report said. "Government failure to provide food assistance for relocated people has caused endemic hunger and cases of starvation." See further, Human Rights Watch (2012), "Waiting here for death: Displacement and villagiation in Ethiopia's Gambella region" (retrieved: http://www.hrw.org/sites/default/files/reports/ethiopia0112webwcover_0.pdf). See also ECVC

and HOTL Alliance (2012), "Land concentration, land grabbing and people struggles in Europe" (retrieved: http://www.tni.org/briefing/update-land-concentration-land-grabbing-and-peoples-struggles-europe).

57. See for China-specific deals, International Institute for Sustainable Development (IISD), C. Smaller, Q. Wei, and L. Yalan (2012), "Farmland and water: China invests abroad" (retrieved: http://www.iisd.org/pdf/2012/farmland_water_china_invests.pdf).
58. Contract farming can be defined as an agreement between farmers and processing and/or marketing firms for the production and supply of agricultural products under forward agreements, frequently at predetermined prices. FAO (2001), "Contract farming: Partnerships for growth, Rome (retrieved: http://www.fao.org/docrep/014/y0937e/y0937e00.pdf).
59. R. Rama and J. Wilkinson (2008), "Foreign direct investment and agri-food value chains in developing countries: A review of the main issues", in FAO, *Commodity Market Review 2007–2008*, Rome (retrieved: ftp://ftp.fao.org/docrep/fao/010/a1487e/a1487e00.pdf).
60. C. Smaller and H. Mann (2009), "A thirst for distant lands: Foreign investment in agricultural land and water", International Institute for Sustainable Development (retrieved: http://www.fao.org/fileadmin/templates/est/INTERNATIONAL-TRADE/FDIs/A_Thirst_for_distant_lands.pdf).
61. Saudi Arabia, for example, established the "King Abdullah Initiative for Saudi Agricultural Investment Abroad," which includes credit facilities to Saudi investors in agriculture abroad. Hail Agricultural Development Corporation (HADCO), a Saudi company, invested in Sudan, with the government providing 60% of the funding. The private equity company Abraaj Capital and other United Arab Emirates (UAE) companies and institutions have already acquired 800,000 acres of farmland in Pakistan with the support of the UAE.
62. Ibid., note 60, pp. 7–8.
63. An extensive overview can be found in Smaller and Mann, pp. 9–13.
64. Ibid., p. 9.
65. Ibid., note 27.
66. Reference can be made to Smaller and Mann, pp. 13–16, for a full review of the issues at hand.
67. See further on that: K. Deininger and D. Byerlee, with J. Lindsay, A. Norton, H. Selod, and M. Stickler (2011), "Rising global interest in farmland: Can it yield sustainable and equitable benefits?" (retrieved: http://siteresources.worldbank.org/INTARD/Resources/ESW_Sept7_final_final.pdf), and African Development Bank, V. Castel, and A. Kamara (2009), "Foreign investments in Africa's agricultural land: Implications for rural sector development and poverty reduction", Development Research Brief 2/2009 (retrieved: http://www.afdb.org/fileadmin/uploads/afdb/Documents/Financial-Information/Foreign%20Investments%20in%20Africa%E2%80%99s%20Agricultural%20Land%20Implications%20for%20Rural%20Sector%20Development%20and%20Poverty%20Reduction.pdf). They indicate, "Even though foreign investment in agriculture in Africa offers opportunities for business partnerships and technology transfer, it can jeopardize the livelihoods of the poorest, notably by increasing food insecurity. These investments may also create tensions in farming communities, as experienced in Madagascar. In this context, it is important to recall the case of Central America. During the 19th Century coffee boom, the privatization of previously customary lands led to a rapid land concentration,

and was immediately followed by decades of conflict and civil war which greatly undermined the development of the region (*). African governments need to design appropriate legislations and mechanisms to benefit from foreign investment in agricultural land, while preserving the livelihoods and interests of the local population (*)". See further for that: V. Songwe and K. Deininger (2009), "Foreign investment in agricultural production: Opportunities and challenges", *Notes Agriculture and Rural Development – Land Policy and Administration* no 45, World Bank, January 2009 (retrieved: http://www.landcoalition.org/pdf/08_note45.pdf).

68. See for the most recent compilation of studies and results, FAO (2013), "Trends and impacts of foreign investment in developing country agriculture: Evidence from case studies", Rome (retrieved: http://www.fao.org/fileadmin/user_upload/newsroom/docs/Trends%20publication%2012%20November%202012.pdf).
69. See for the report, note 31.
70. Author of the book *The Land Grabbers: The New Fight Over Who Owns the Earth* (Boston: Beacon Press, 2012), pp. 336.
71. See for the full interview with *The Guardian* in 2012, http://www.theguardian.com/world/2012/may/20/fred-pearce-land-grab-interview. See also D. Bornstein (2013), "How climate change politics are ruining African agriculture" (retrieved: http://www.policymic.com/articles/61801/how-climate-change-politics-are-ruining-african-agriculture).

9 Farmland II: Capita Selecta

1. See also, Land Commodities Asset Management AG (2012), "Investment summary".
2. In order to properly assess how quickly farmland values may rise in the future and to what levels, answers would be required to a number of questions, such as (1) What factors might act to change the supply of agricultural commodities, and to what extent will this affect prices? (2) What levels of global economic growth will occur in the future, and to what extent will this affect demand? (3) To what extent will climate change, in particular, severe-weather events, affect yields in the future? (4) To what extent might oil-supply shortages increase demand for biofuels or raise input costs? (5) How much land is currently being used for agriculture, and how much more productive land is available for future cropland expansion? (6) At what rate is productive farmland being lost to urbanization, land degradation, water shortages, and other effects? (7) What constraints might there be to the utilization of any remaining land appropriate for agriculture? (8) What effect would constraints on further cropland expansion or losses of existing cropland have on farmland values? See further on all these questions: "The Land commodities global farmland and investment report 2009: A mid-term outlook" (retrieved: http://www.farmlandinvestmentreport.com/Farmland_Investment_Report.pdf).
3. Ibid., Land Commodities (2009), chapters 2 and 3.
4. And further, ibid., Land Commodities (2009), chapters 4–6.
5. See further, D. Tilman et al. (2002), "Agricultural sustainability and intensive production practices", *Nature* 418(6898): 671–677.
6. Tilman goes on to make explicit: "Most of the best quality farmland is already used for agriculture, which means that further area expansion would occur on marginal land that is unlikely to sustain high yields and is vulnerable to degradation." Ibid., Tilman et al. (2002). See further, K. G. Cassman (1999), "Ecological intensification

of cereal production systems: Yield potential, soil quality, and precision agriculture", *Proceedings of the National Academy of Sciences* 96: 5952–5959. A. Young (1999), "Is there really spare land? A critique of estimates of available cultivable land in developing countries", *Environment, Development and Sustainability* 1: 3–18.

7. See J. Alcamo et al. (2003), "Global estimates of water withdrawals and availability under current and future 'business-as-usual' conditions", *Hydrological Sciences Journal* 48(3): 339–348.
8. See further, United Nations Environment Programme Rapid Response Assessment (2009), "The environmental food crisis: The environment's role in averting future food crises" (retrieved: http://www.grida.no/_res/site/file/publications/FoodCrisis_lores.pdf). J. M. Faurès et al. (2000), "The FAO irrigated area forecast for 2030", FAO, Rome (retrieved: ftp://ftp.fao.org/agl/aglw/docs/fauresetalagadir.pdf).
9. See further on that, ibid., Land Commodities (2009), chapter 6.
10. See FAO (2008), "The state of food and agriculture 2008", FAO, Rome (retrieved: http://www.fao.org/docrep/011/i0100e/i0100e00.htm). The typical remedies are being questioned in terms of their longevity: N. Alexandratos (1999), "World food and agriculture: Outlook for the medium and longer term", *Proceedings of the National Academy of Sciences* 96: 5908–5914. V. W. Ruttan (1999), "The transition to agricultural sustainability", *Proceedings of the National Academy of Sciences* 96: 5960–5967. V. W. Ruttan (2002), "Productivity growth in world agriculture: Sources and constraints", *Journal of Economic Perspectives* 16(4): 161–184 (retrieved: http://pubs.aeaweb.org/doi/pdfplus/10.1257/089533002320951028). S. Postel (1999), *Pillar of Sand: Can the Irrigation Miracle Last?* (New York: Norton). P. M. Vitousek, H. A. Mooney, J. Lubchenco, and J. M. Melillo (1997), "Human domination of earth's ecosystems", *Science* 277: 494–499. S. R. Carpenter et al. (1998), "Nonpoint pollution of surface waters with phosphorus and nitrogen", *Ecological Applications* 8: 559–568.
11. See further, ibid., Land Commodities (2009), chapter 7. P. A. Sanchez (2002), "Soil fertility and hunger in Africa", *Science* 205(5562): 2019–2020. J. Schmidhuber and F. N. Tubiello (2007), "Global food security under climate change", *Proceedings of the Natural Academy of Sciences* 104(50): 19703–19708. Stern Review (2006), "The economics of climate change, part II: The impacts of climate change on growth and development", *Stern Review*, UK, pp. 67–73. D. B. Lobell et al. (2008), "Prioritizing climate change adaptation needs for food security in 2030", *Science* 319(5863): 607–610. W. R. Cline (2007), "Global warming and agriculture: Impact estimates by country", Center for Global Development and Peterson Institute for International Economics, Washington, DC. Long et al. (2006), "Food for thought: Lower-than-expected crop yield stimulation with rising CO_2 concentrations", *Science* 312(5782): 1918–1921. F. N. Tubiello and G. Fischer (2006), "Reducing climate change impacts on agriculture: Global and regional effects of mitigation, 2000–2080", *Technological Forecasting and Social Change* 74(7): 1030–1056. D. R. Cline (2007), "Global warming and agriculture: Impact estimates by country", Center for Global Development and Peterson Institute for International Economics, Washington, DC. Hanasaki et al. (2008), "An integrated model for the assessment of global water resources, part 1: Model description and input meteorological forcing", *Hydrology and Earth System Sciences* 12(4): 1007–1025. N. Hanasaki et al. (2008), "An integrated model for the assessment of global water resources, part 2: Applications and assessments", *Hydrology and Earth System Sciences* 12(4): 1027–1037 (retrieved: http://www.hydrol-earth-syst-sci.net/12/1027/2008/hess-12-1027-2008.pdf). C. De Fraiture et al. (2003), "Addressing the unanswered questions in global water policy: A

methodology framework", *Irrigation and Drainage* 52(1): 21–30. Y. Shen et al. (2008), "Projection of future world water withdrawals under SRES scenarios: Water withdrawal", *Hydrological Sciences Journal* 53(1): 11–33. M. W. Rosegrant and X. M. Cai (2002), "Global water demand and supply projections, part 2: Results and prospects to 2025", *Water International* 27(2): 170–182. H. Yang et al. (2003), "A water resources threshold and its implications for food security", *Environmental Science and Technology* 37(14): 3048–3054. S. Rost et al. (2008), "Agricultural green and blue water consumption and its influence on the global water system", *Water Resources Research* 44(9). J. Böhner and F. Lehmkuhl (2005), "Environmental change modelling for Central and High Asia: Pleistocene, present and future scenarios", *Boreas* 32(2): 220–231. X. Jianchu et al. (2007), "The melting Himalayas: Regional challenges and local impacts of climate change on mountain ecosystems and livelihoods", ICIMOD Technical Paper, Katmandu, Nepal. X. D. Liu and B. D. Chen (2000), "Climatic warming in the Tibetan Plateau during recent decades", *International Journal of Climatology* 20(14): 1729–1742. D. Pimentel et al. (2001), "Economic and environmental threats of alien plant, animal and microbe invasions", *Agriculture, Ecosystems and Environment* 84(1): 1–20. A. Rossman (2009), "The impact of invasive fungi on agricultural ecosystems in the United States", *Biological Invasions* 11(1): 97–107. D. Pimentel et al. (2005), "Update on the environmental and economic costs associated with alien-invasive species in the United States", *Ecological Economics* 52(3): 273–288. A. Rossman (2009), "The impact of invasive fungi on agricultural ecosystems in the United States", *Biological Invasions* 11(1): 97–107. Alig et al. (2004), "Projecting large-scale area changes in land use and land cover for terrestrial carbon analyses", *Environmental Management* 33(4): 443–456. Anderson et al. (2004), "Emerging infectious diseases of plants: Pathogen pollution, climate change and agrotechnological drivers", *Trends in Ecology and Evolution* 19(10): 535–544. J. Gan (2004) "Risk and damage of southern pine beetle outbreaks under global climate change", *Forest Ecology and Management* 191(1–3): 61–71. Pyke et al. (2008), "Current practices and future opportunities for policy on climate change and invasive species", *Conservation Biology* 22(3): 585–592. H. A. Mooney and R. J. Hobbs, eds (2000), *Invasive Species in a Changing World* (Washington, DC: Island Press). M. Kenis et al. (2009), "Ecological effects of invasive alien insects", *Biological Invasions* 11(1): 21–45. Sala et al. (2000), "Global biodiversity scenarios for the year 2100", *Science* 287(5459): 1770–1774. Gaston et al. (2003), "Rates of species introduction to a remote oceanic island", *Proceedings of the Royal Society of London Series B: Biological Sciences* 270(1519): 1091–1098. P. Pinstrup-Andersen and R. Pandya-Lorch (1998), "Food security and sustainable use of natural resources: A 2020 vision", *Ecological Economics* 26(1): 1–10. Bai et al. (2007), "Land cover change and soil fertility decline in tropical regions", *Turkish Journal of Agriculture and Forestry* 32(3): 195–213. Den Biggelaar et al. (2004), "The global impact of soil erosion on productivity", *Advances in Agronomy* 81: 1–95. J. Henao and C. Baanante (2006), "Agricultural production and soil nutrient mining in Africa", Summary of IFDC Technical Bulletin, Alabama, USA. Stoorvogel et al. (1993), "Calculating soil nutrient balances in Africa at different scales 1: Supra-national scale", *Fertilizer Research* 35(3): 227–235. M. Stocking (1986), "The cost of soil erosion in Zimbabwe in terms of the loss of three major nutrients", Consultant's Working Paper 3, Soil Conservation Programme, FAO Land and Water Division, Rome. R. Lal (1998), "Soil erosion impact on agronomic productivity and environment quality", *Critical Reviews in Plant Sciences* 17: 319–464. Maizel et al. (1998), "Historical interrelationships between population settlement and farmland in the conterminous United States, 1790 to 1990", in T. D. Sisk, ed., *Perspectives On the Land Use History of North America: A Context for Understanding Our Changing Environment*, US Geological Survey, Biological Resources Division, Biological Science Report, USGS/BRD/BSR 1998–2003. G. K. Klein (2001), "Estimating global land use change over the past 300 years: The HYDE database", *Global Biogeochemical Cycles* 15(2): 417–433. G. K. Klein (2005), "Three centuries of global population growth: A spatial referenced population density database for 1700–2000", *Population and Environment* 26(4): 343–367. D. Potere and A. Schneider (2007), "A critical look at representations of urban areas in global maps", *Geojournal* 69(1–2): 55–80. E. Stehfest et al. (2008), "Climate benefits of changing diet", *Climatic Change* 95: 83–102 (retrieved: DOI 10.1007/s10584-008-9534-6). P. E. Waggoner (1995), "How much land can ten billion people spare for nature? Does technology make a difference?", *Technology in Society* 17: 17–34. R. Manning (2000), *Food's Frontier: The Next Green Revolution* (New York: North Point Press). D. L. Plucknett (1993), "International agricultural-research for the next century", *Bioscience* 43: 432–440. L. A. Bruijnzeel (2004), "Hydrological functions of tropical forests: Not seeing the soil for the trees?", *Agriculture Ecosystems & Environment* 104(1): 185–228. Fischlin et al. (2007), "Ecosystems, their properties, goods, and services", in M. L. Parry, O. F. Canziani, J. P. Palutikof, P. J. van der Linden, and C. E. Hanson, eds., *Climate Change 2007: Impacts, Adaptation and Vulnerability*, Contribution of Working Group II to the Fourth Assessment Report of the Intergovernmental Panel on Climate Change (Cambridge: Cambridge University Press). Nepstad et al. (2007), "Mortality of large trees and lianas following experimental drought in an Amazon forest", *Ecology* 88(9): 2259–2269. W. R. Cline (2007), "Global warming and agriculture: Impact estimates by country", Center for Global Development and Peterson Institute for International Economics, Washington, DC. MNP (2006) (edited by A. F. Bouwman, T. Kram, and K. K. Goldewijk), "Integrated modelling of global environmental change: An overview of IMAGE 2.4", Netherlands Environmental Assessment Agency (MNP), Bilthoven, the Netherlands, and K. K. Goldewijk, A. Beusen, M. de Vos, and G. van Drecht (2011), "The HYDE 3.1 spatially explicit database of human induced land use change over the past 12,000 years", *Global Ecology and Biogeography* 20(1): 73–86 (retrieved: DOI 10.1111/j.1466-8238.2010.00587.x). K. K. Goldewijk, A. Beusen, and P. Janssen (2010), "Long term dynamic modeling of global population and built-up area in a spatially explicit way, HYDE 3.1", *The Holocene* 20(4): 565–573 (retrieved: DOI 10.1177/0959683609356587).

12. See further, A. K. Chapagain and A. Y. Hoekstra (2008), "The global component of freshwater demand and supply: An assessment of virtual water flows between nations as a result of trade in agricultural and industrial products", *Water International* 33(1): 19–32. J. Schmidhuber and P. Shetty (2005), "The nutrition transition to 2030: Why developing countries are likely to bear the major burden", *Acta Agriculturae Scandinavica, Section C: Economy* 2(3–4): 150–166. A. Rae (1998), "The effects of expenditure growth and urbanisation on food consumption in East Asia: A note on animal products", *Agricultural Economics* 18(3): 291–299. B. S. King, J. L. Tietyen, and S. S. Vickner (2000), "Consumer trends and opportunities", University of Kentucky, Lexington. B. Popkins, S. Horton, and S. Kim (2001), "The nutrition transition and prevention of diet-related chronic diseases in Asia and the Pacific", *Food and Nutrition Bulletin* 22(4: suppl.), United Nations University Press, Tokyo. P. Pingali (2004), "Westernization of Asian diets and the transformation of food systems: Implications for research and policy", ESA Working Paper 04-17, FAO, Rome.

Keyzer et al. (2005), "Diet shifts towards meat and the effects on cereal use: Can we feed the animals in 2030?" *Ecological Economics* 55(2): 187–202. Z. Y. Zhou, Y. R. Wu, and W. M. Tian (2003), "Food consumption in rural China: Preliminary results from household survey data", *Proceedings of the 15th Annual Conference of the Association from Chinese Economics Studies*, Australia. Henningsson et al. (2004), "The value of resource efficiency in the food industry: A waste minimization project in East Anglia, UK", *Journal of Cleaner Production* 12(5): 505–512. V. Smil (2000), *Feeding the World: A Challenge for the Twenty-First Century* (Cambridge, MA: MIT Press). Lundqvist et al. (2008), "Saving water: From field to fork – curbing losses and wastage in the food chain", SIWI Policy Brief (retrieved: http://www.siwi.org/documents/Resources/Policy_Briefs/PB_From_Filed_to_fork_2008.pdf). WRAP (2008), "Food waste report 2: The food we waste", WRAP, UK (retrieved: http://www.wrap.org.uk/downloads/The_Food_We_Waste_v2__2_.d3471041.5635.pdf).

13. See OECD-FAO (2008), "Agricultural outlook 2008–2017", OECD, Paris (retrieved: http://www.agri-outlook.org/dataoecd/54/15/40715381.pdf).
14. Ibid., OECD-FAO (2008).
15. Ibid., OECD-FAO (2008).
16. Ibid., OECD-FAO (2008).
17. Peak oil is the point in time when the maximum rate of global petroleum extraction is reached, after which the rate of production enters terminal decline, otherwise known as the "geological limit" of production. The concept of peak oil is based on the historically observed production rates of individual oil wells and the combined production rate of a field or region of related oil wells. The aggregate production rate from an oil field over time grows until the rate peaks, after which it declines, sometimes rapidly, until the field is depleted. This concept has been shown to be applicable to the sum of a nation's domestic production rate. Peak oil theory maintains that the same sort of behavior can similarly be applied to the global rate of petroleum production. Peak oil has the potential to outweigh all other factors, including climate change and food demand growth, exerting upward pressure on agricultural commodity prices. See further, Hirsch et al. (2005), "Peaking of world oil production: Impacts, mitigation, & risk management", Science Applications International Corporation, commissioned by the US Department of Energy. Hirsch et al. (2007), "Peaking of world oil production: Recent forecasts", Science Applications International Corporation, commissioned by the US Department of Energy. Peak oil theory was first proposed by M. King Hubbert. This distinct curve, known as "Hubbert's curve," is characteristic of oil production profiles throughout history and can be applied to individual oil fields, specific domains, or an oil-producing region (e.g., individual states of the United States) and to the sum of such regions (e.g., the United States as a whole). Peak theory mathematically models the pattern by which peak oil production (or exploitation of reserves) for a given region follows the peak of discovery (of those reserves). Underlying this pattern is the fact that larger oil fields are more likely to be discovered and exploited first. These are also the cheapest fields to exploit, as exploration and production costs per unit of oil are higher the smaller the field. The more mature an oil-producing region, the smaller the discovered fields and the higher the cost of production becomes.
18. See further, Banse et al. (2008), "Will EU biofuel policies affect global agricultural markets?", *European Review of Agricultural Economics* 35(2): 117–141. Fitzherbert et al. (2008), "How will oil palm expansion affect biodiversity?", *Trends in Ecology and*

Evolution 23(10): 528–545. Hirsch et al. (2005), "Peaking of world oil production: Impacts, mitigation, & risk management", Science Applications International Corporation, commissioned by the US Department of Energy. Hirsch et al. (2007), "Peaking of world oil production: Recent forecasts", Science Applications International Corporation, commissioned by the US Department of Energy.

19. See further, M. C. Heller and G. A. Keoleian (2000), "Life cycle-based sustainability indicators for assessment of the U.S. food system", Center for Sustainable Systems, University of Michigan, Ann Arbor, pp. 40–42. L. Horrigan, R. S. Lawrence, and P. Walker (2002), "How sustainable agriculture can address the environmental and human health harms of industrial agriculture", *Environmental Health Perspectives* 110(5). Green et al. (2005), "Sparing land for nature: Exploring the potential impact of changes in agricultural yield on the area needed for crop production", *Global Change Biology* 10(11): 1594–1605.
20. See in detail, S. Hornblower and K. Kelly (2013), "Demand from China: fact or fiction?" (retrieved: http://www.growmark.com/sites/Files/Documents/ChinaDemand-FactOrFiction.pdf). Presented in this paper are analyses demonstrating China's lack of influence on commodity prices. They show that (1) other countries have a stronger relationship between their commodity imports and world commodity prices than China has; (2) although prices of commodities traded in dollars have risen, US exports of commodities have not generally increased; (3) there is, for most commodities, no relationship between exports of US commodities to China and their respective prices; where loose relationships do exist, price movements are still more closely tied to other market participants than to China; and (4) many commodities that China does not buy have experienced the same volatility as commodities it does buy, indicating that other factors besides Chinese purchases are driving commodity prices.
21. H. M. Kaplan (1985), "Farmland as a portfolio investment", *Journal of Portfolio Management* 12: 73–78.
22. According to the Hancock Agricultural Investment Group (HAIG), part of the US investment management division of Manulife Financial, "during the period from 1941 to 2002, average farmland values increased by almost two percent more than the average rate of inflation over that time period." Hancock Agricultural Investment Group (2009), from the website of Hancock Agricultural Investment Group, an MFC Global Investment Management Company (retrieved: http://www.haig.jhancock.com).
23. J. R. Webb and J. H. Rubens (1988), "The effect of alternative return measures on restricted mixed-asset portfolios", *Journal of the American Real Estate and Urban Economics Association* (16): 123–137.
24. Source: National Council of Real Estate Investment Fiduciaries (2008). Note that the study period ended in 2007, prior to the recent crash in equity and commercial real-estate values. The most commonly used measure of risk-adjusted return is known as the Sharpe ratio. It is a measure of the excess return, or "risk premium," per unit of risk in an investment asset or trading strategy and is used to assess how well the return on an investment compensates the investor for the risk taken. The higher the Sharpe ratio for a particular asset, the higher the return earned per unit of risk an investor is exposed to when owning that asset.
25. Some comparative studies on farmland have been criticized for taking numbers derived from agricultural equity-type investments (i.e., data that include income from the commercial operation of a farm), thus making a direct comparison

between straight farmland ownership and other asset classes difficult. D. A. Lins, B. J. Sherrick, and A. Venigalla (1992), "Institutional portfolios: Diversification through farmland investment", *Journal of the American Real Estate and Urban Economics Association* 20(4): 549–571 (retrieved: http://www.areuea.org/publications/ree/articles/V20/REE.V20.4.4.pdf). The study concludes as follows: "The study used cash rents after property taxes to derive the income part of the returns on farmland for the period 1967–88 and showed that diversification enhances portfolio performance for institutional investors. The results were robust across wide variations in variance and annual returns to farmland. For the period 1967–88, farmland exhibited a higher return than that of stocks and bonds. Further, returns on farmland were negatively correlated with stocks and bonds and positively correlated with inflation. Thus investment in farmland not only was a good hedge against inflation but also provided diversification for those who included it in their portfolio. The implication is that, by including farmland in their portfolio, they may be able to reduce the possibility of shortfalls of their funds in times of higher inflation." Another more recent study, based on figures from the NCREIF Farmland Index, reported similar findings – farmland performed well during the study period (1991–2004) compared to other assets such as US investment-grade bonds (as represented by the Lehman Aggregate Bond Index), a diversified portfolio of the top 1,000 US large capitalization stocks and top 2,000 small-capitalization stocks (as represented by the Russell 3000 Index), as well as mainstream real estate (as represented by the NCREIF Property Index). See for further details, B. Howard (2005), "Farmland investing: An overview", Callan Investments Institute. A further study conducted in Saskatchewan, a prairie province that produces just under half of Canada's grain, compared farmland with a number of other asset classes found in a typical globally diversified investment portfolio held by a Canadian mutual fund investor. The study concluded that "the addition of farmland ownership would have enhanced financial performance across the portfolio for average or medium levels of risk. The financial gains from farmland are a result of its negatively correlated returns with equity markets. When added to an equity portfolio, the risk level is reduced while maintaining the same rate of return on investment." See further, M. J. Painter (2000), "Efficient investment in Saskatchewan farmland", University of Saskatchewan.

26. A study on US farmland conducted in 2002 compared the effects on portfolio efficiency of including farmland in a mixed-asset portfolio under market conditions of certainty and uncertainty. It concluded that in both certain- and uncertain-world models, farmland can be shown to improve portfolio efficiency. See further for details, W. Hardin and P. Cheng (2002), "Farmland investment under conditions of certainty and uncertainty", *Journal of Real Estate Finance and Economics* 25(1): 81–98.
27. See also, Towers Watson (2012), "Investing in private markets: The right strategy, the right geography and the right manager", pp. 35–43, for the agricultural and timber markets.
28. See further, ETF Database (2012), "The importance of agricultural commodities: Access through ETFs". Asset allocation in some of the ETF or indexes are Rogers International Commodity Index (RICI), Rogers International Commodity Index Agriculture (RICIA), Reuters Jeffries CRB Index, or the Dow Jones USB Commodity Index. The weightings are adjusted on a frequent basis and are therefore not highlighted here.

29. ETNs are unsecured debt obligations and have (in contrast with ETPs) no backing by underlying assets, contain default risk, and no tracking error (both also in contrast with ETPs)
30. Savills (2009), "Savills agricultural land market survey". What is most interesting about this particular study is the comparison of leading agricultural equities with direct ownership of rural land. Direct farmland ownership has outperformed the top 25% of agricultural equities over all time periods.
31. Farmland is the foundation of all agriculture and the most basic repository of wealth and value within the sector. The fortunes of companies can wax and wane as management and market conditions change or one competitor outperforms another. Farmland on the other hand is fixed and will always gain in value over the mid- to long term as demand continues to rise. Ibid., Land Commodities (2009), p. 126. L. Barras, O. Scaillet, and R. R. Wermers (2009), "False discoveries in mutual fund performance: Measuring luck in estimated alphas", *Journal of Finance* 65(1): 179–216, Swiss Finance Institute Research Paper 08–18 (retrieved: http://papers.ssrn.com/sol3/papers.cfm?abstract_id=869748).
32. For investors with pockets that are not so deep, a number of research studies suggest that if assets are selected according to certain criteria, diversification is less of a concern than in the case of other asset classes. This is due to the fact that many farms operate a mixed cropping system with rotation between crop types. This creates diversification at the farm-enterprise level and hedges against fluctuations in individual commodity prices. If such farms are specifically selected during prepurchase due diligence, diversification at the portfolio level becomes less of an issue. One of the dominant studies that looked at the importance of diversification in farmland portfolios concluded, "With respect to farmland investment and geographic diversification, the results [of the study] question the ability of an optimised mean-variance portfolio to provide substantial improvement in comparison to a naive portfolio. The marginal improvement in portfolio efficiency of an optimised farmland portfolio is not statistically significant." Source: W. Hardin and P. Cheng (2002), "Farmland investment under conditions of certainty and uncertainty", *Journal of Real Estate Finance and Economics* 25(1): 81–98. This is especially pertinent when considered in the context of historical data on asset managers' failure to outperform investment benchmarks (achieve "alpha"). Also capital preservation works in favor of farmland versus farm equities or commodity futures. Even if a farmland tenant is unable to make a rent payment, the investor still owns the underlying asset. This cannot be said of agricultural equities where the worst case scenario is the loss of 100% of capital, or agricultural commodities where the loss of capital has the potential, under certain circumstances, to exceed the amount invested. Preservation of capital invested in individual assets within a portfolio is one of the most important contributing factors to long-term performance across the portfolio as a whole.
33. A study conducted in the United States that assessed farmland values over a prolonged period from 1900 to 1994 demonstrated that values only show inconsistencies in the short term. It may not be possible for speculators with a short holding horizon to obtain systematic excess returns by trading in land markets (due to transaction costs); however, farmland markets are efficient, and prices are consistent with standard farmland valuation models (even allowing for transaction costs), if a longer-term investment horizon is accepted. See S. H. Lence and Douglas J. Miller (1999), "Transaction costs and the present value model of farmland: Iowa, 1900–1994", *American Journal of Agricultural Economics* 81: 257–272.

34. Other questions might involve, How can I ensure I am making a sustainable and responsible investment? Do I need to redesign my approach to risk when allocating to agriculture? How can I access opportunities in new geographies? How do I manage the inherent volatility in agriculture? How can I maximize commodities futures as an investment route? What do I need to know about performance in private equity? How can I optimize my wider portfolio with real asset allocations? Where should I allocate my assets along the value chain? Should I rethink my approach to investments in real assets?
35. See further, D. Lins, A. Kowalski, and C. Hoffman (1992), "Institutional investment diversification: foreign stocks vs U.S. farmland", in *Proceedings of Regional Research Committee NC-161*, Department of Agricultural Economics, Kansas State University, Manhattan, Kansas, February. C. B. Moss, A. M. Featherstone, and T. G. Baker (1987), "Agricultural assets in an efficient multi-period investment portfolio", *Agricultural Finance Review* 47: 82–94.
36. See further, Land Commodities (2009), p. 131.
37. See for a good overview of the components of an agricultural due diligence, ibid., Land Commodities (2009), p. 130.
38. See, T. H. Kuethe, N. Walsh, and J. Ifft (2013), "Farmland versus alternative investments before and after the 2008 financial crisis", *Journal of the ASFMRA* 2013, pp. 120–131 (retrieved: http://www.asfmra.org/wp-content/uploads/2013/06/384-Kuethe.pdf). As this study on US farmland shows, compared to competing investments, farmland has historically delivered relatively high mean returns with low variability and low correlation with financial markets, both before and after the 2008 financial crisis. But much of this outperformance has been due to record increases in land values in many parts of the United States. Land accounts for more than 80% of the total value of farm assets and is the principal source of collateral for farm-operating finance and the loans that underpin rising farmland prices. As the study points out, despite the historical appeal of the asset class, there are a number of external risks that could impact returns going forward, perhaps to an even greater extent than food demand and commodity prices. Examples given are rising interest rates, cutbacks in government subsidy and support programs, rolling back of ethanol mandates, and changes in tax policy.
39. See in detail, J. M. MacDonald, P. Korb, and R. A. Hoppe (2013), "Farm size and the organization of U.S. crop farming", USDA Economic Research. See also J. M. MacDonald (2011), "Why are farms getting larger? The case of the US" (retrieved: http://ageconsearch.umn.edu/bitstream/115361/2/MacDonald.pdf).
40. See G. Ibendahl and T. Griffin (2013), "The connection between cash rents and land values" (retrieved: http://www.asfmra.org/wp-content/uploads/2013/06/392-Ibendahl.pdf). This ratio is important because it helps indicate whether cash rents are a cost-effective way of controlling farmland relative to purchasing the land. Results indicate that cash rents lag behind changes in land prices when land prices are increasing but not when land prices are decreasing. However, this relationship does not always hold. The study addresses such questions as the following:
    - How have income returns from rent (as a percentage of land value) changed for investors entering the asset class for the first time?
    - Does farmland look fairly priced on a price/earnings basis relative to historical levels, including by comparison to the period immediately preceding the farmland crash of the 1980s?
    - Are cash rents a leader or follower of land-price changes?

- See further about the disconnect between farmland value and cash rents, C. Dobbins and K. Cook (2013), "Up again: Indiana's farmland market in 2013" (retrieved: https://www.agecon.purdue.edu/extension/pubs/paer/pdf/PAER8_2013.pdf). Despite the worst drought in over a decade, crop prices in Indiana have continued to rise, supported by favorable interest rates, strong farmland demand, and a limited supply of land. But despite increasing over the period, the rise in cash rents has lagged behind land prices, meaning lower rental returns for investors.

41. A research study that looked at just this issue concluded, "There are barriers to the flow of non-farm equity into farm real estate markets due to high transaction costs and illiquidity. These barriers create a segmented farm real estate market where compensation for risk on farmland investment is high relative to well-established secondary markets." A. N. Shiha and J.-P. Chavas (1995), "Capital market segmentation and U.S. farm real estate pricing", *American Journal of Agricultural Economics* 77(May): 397–407.
42. Or as Lord Jacob Rothschild who has invested heavily in farmland in recent years said, "Land might be cheap and plentiful in Russia, but if the price of wheat goes up, is your deed going to be honored?", *Fortune* magazine, June 16, 2009, "Betting the farm: As world population expands, the demand for arable land should soar. At least that's what George Soros, Lord Rothschild, and other investors believe" (retrieved: http://money.cnn.com/2009/06/08/retirement/betting_the_farm.fortune/index.htm?section=money_latest).
43. See in detail on the topic:
    - C. Binkley, C. Raper, and C. Washburn (1996), "Institutional ownership of U.S. timberland: History, rationale, and implications for forest management", *Journal of Forestry* 94(9): 21–28.
    - D. Burch and G. Lawrence (2009), "Towards a third food regime: Behind the transformation", Agriculture and Human Values 26(4): 267–279.
    - D. Burch and G. Lawrence (2013), "Financialization in agri-food supply chains: Private equity and the transformation of the retail sector", Agriculture and Human Values 30(2): 247–258.
    - B. Christophers (2010), "On voodoo economics: Theorising relations of property, value and contemporary capitalism", Transactions of the Institute of British Geographers 35: 94–108.
    - J. Coakley (1994), "The integration of property and financial markets", Environment and Planning A 26: 697–713.
    - H. Cochet and M. Merlet (2011), "Land grabbing and share of the value added in agricultural processes: A new look at the distribution of land revenues", Paper presented at the International Academic Conference on Global Land Grabbing, University of Sussex, April 6–8.
    - N. Colbran (2010), "The financialisation of agricultural commodity futures trading and its impact on the 2006–2008 global food crisis", Paper presented at the 3rd biennial Ingram Colloquium on International Law and Development, University of New South Wales, December 2.
    - S. Daniel (2012), "Situating private equity capital in the land grab debate", Journal of Peasant Studies 39(3–4): 703–729.
    - G. Epstein (2005), "Introduction: Financialization and the world economy", in G. Epstein, ed., *Financialization and the World Economy* (Cheltenham, UK: Edward Elgar Publishing).

- Haila (1988), "Land as a financial asset: The theory of urban rent as a mirror of economic transformation", *Antipode* 20(2): 79–101.
- G. Krippner (2005), "The financialization of the American economy", Socio-Economic Review 3(2): 173–208.
- G. Krippner (2011), Capitalizing on Crisis: The Political Origins of the Rise of Finance (Cambridge: Harvard University Press).
- Leyshon and N. Thrift (2007), "The capitalization of almost everything: The future of finance and capitalism", Theory, Culture & Society 24(7–8): 97–115.
- H. Magdoff and P. Sweezy (1987), Stagnation and the Financial Explosion (New York: Monthly Review Press).
- R. Martin (2002), Financialization of Daily Life (Philadelphia: Temple University Press).
- Merian Research and CRBM (2010), "The vultures of land grabbing: The involvement of European financial companies in large-scale land acquisition abroad", Regulate Finance for Development, London (retrieved: http://www.rinoceros.org/IMG/pdf/VULTURES-completo-2.pdf).
- R. Parenteau (2005), "The late 1990s' bubble: Financialization in the extreme", in G. Epstein, ed., Financialization and the World Economy (Cheltenham, UK: Edward Elgar Publishing).
- Wheaton and W. Kiernan (2012), "Farmland: An untapped asset class? Quantifying the opportunity to invest in agriculture", Food for Thought, Macquarie Agricultural Funds Management, December (retrieved: http://www.macquarie.com/dafiles/Internet/mgl/com/agriculture/docs/food-forthought/food-for-thought-dec2012-anz.pdf).

44. See extensively, M. Fairbairn (2013), "'Like gold with yield': Evolving intersections between farmland and finance", international conference, Yale University, September 14–15. "Food sovereignty: A critical dialogue conference paper", no. 6, forthcoming, *Journal of Peasant Studies*. Also forthcoming, M. Fairbarn (2014), "'Just another asset'?: Farmland in an age of global finance", PhD dissertation, University of Wisconsin–Madison, and Fairbairn, "Preaching to acquire: Constructing the global farmland boom" (publication date unknown).
45. M. Fairbarn (2013) comments (p. 18), "The financialization of farmland could also alter land market dynamics. If attempts at farmland securitization progress, it would become possible to buy or sell farmland almost instantaneously and for retail investors to acquire land as a financial asset. The increasing liquidity and volume of investment associated with securitization could greatly increase the volatility of farmland markets. Though increased volatility translates into the possibility of higher profits for speculators, it would not necessarily be welcome to those more staid farmland investors that were drawn to the sector for the steady, predictable returns. However, these investors – many of the pension funds and others employing a 'own-lease out' strategy – could also contribute to changing land market dynamics."
46. Ibid., Fairbarn, p. 17.
47. "Their focus is a carefully timed purchase and subsequent disposal" (2011), InvestAg Savills. "International Farmland Market Bulletin", Investagsavills.com (retrieved: http://www.investag.co.uk/Bulletin2011.pdf). Ibid., Fairbarn, p. 18: "The idea of entering into land ownership with an 'exit strategy' in place would thoroughly confound most of the world's farmers, for whom hanging on to their land is a primary objective."

10 Farmland III: Land-Expectation Value and Timberland Valuation

1. See further, D. Aakr (2010), "Agricultural land valuation", presentation, NDSU Extension Service, and Dr. Berdár Béla (2011), "Agricultural land valuation", presentation. He distinguishes between the valuing of land, plantation, and forests and the special situations of tree stands, mature stands, and young tree stands.
2. See further, K. Viitanen, S. Hannelius, and M. Airaksinen (2006), "Valuation guidance of forest properties within valuation standards needed?" (retrieved: http://www.fig.net/pub/fig2006/papers/ts76/ts76_01_viitanen_Etal%20_0746.pdf).
3. See in detail also for other (minor) methods not covered in this chapter, J. Leech and I. Ferguson, technical eds. (2012), "A standard for valuing commercial forests in Australia", Association of Consulting Foresters of Australia, Division of the Institute of Foresters of Australia, Canberra (retrieved: http://www.forestry.org.au/kcfinder/upload/files/A%20standard%20for%20valuing%20forests%20in%20Australia%20ACFA%20July%202012.pdf). Also, Code of Practice Valuation of Commercial Forest Plantations (Ireland) (draft 2012), working group: D. Little, J. Phelan, T. McDonald, and H. Phillips (retrieved: http://www.agriculture.gov.ie/media/coford/content/eventspresentations/events2012/Code%20of%20Practice%20Valuation%20of%20Commercial%20Forest%20Plantations%20Draft%20%2002-10-2012.pdf).
4. Agricultural land is defined as land that is one or more of the following:
   1. Cropland of at least 20 gross acres
   2. Ten or more gross acres of permanent crop
   3. Grazing land with a minimum carrying capacity of 40 animal units, and containing an economically feasible number of units
   4. Land devoted to high-density use in the production of commodities
   5. Land devoted to use in the processing of cotton necessary for marketing
   6. Land devoted to use in the processing of grapes for marketing

   Other definitions used are:
   - *Animal init*: A unit of measure to classify grazing land according to its carrying capacity, i.e., ability of a parcel of land to graze a certain number of animal units over a period of time (usually a year, but sometimes expressed on a monthly basis) without injurious effect upon the natural vegetative cover of the land. In the state of Arizona, the following equivalent measures have been classed as one animal unit:
   1. One mature beef animal of 1,000 pounds
   2. One and one-quarter horses
   3. Five head of sheep
   4. Five head of goats
   5. Five head of ratites
   - *Crop value*: The value of the vine, tree, or plant over and above the land value. The add-on value as permanent crops reach maturity or the value of the harvest from the field crop.
   - *Fallow land*: Land capable of being farmed but due to the lack or cost of water or participation in a crop-rotation program is left idle. Its value is the same as irrigated lands.
   - *Agricultural property*: Property used for agronomy, horticulture, or animal husbandry, producing an agricultural crop or commodity.
   - *Farm property qualification*: The type of crop being raised on a farmstead relates directly to the qualification for farm property. Field crops (grain, alfalfa, sugar beets,

lettuce, safflower, etc.) require 20 acres or more. For permanent crops (citrus, nuts, grapes, and fruit), 10 acres or more are normally required.

- *Headquarters land*: Land used for storage and farm/ranch improvements (barns, sheds, corrals, seasonal employees housing) and valued the same as crop or grazing land.
- *Irrigated land*: Land cleared, leveled, and ditched for the application of water and growing of plants (includes the roads, ditches, well sites, ponds, and turn and skip rows as well as the cropped acres).
- *Nonqualifying rural property*: Land used for residential pleasure, development, speculative, or recreational purposes, classed and valued according to its primary use.
- *Qualifying agricultural property*: Land must be in active production for seven out of the last ten years prior to application as agricultural land for qualification as agricultural property. Failure to farm the property due to severe drought conditions may extend the idle period on a year-to-year basis as initiated by the cognizant assessor, and as approved by the Department of Revenue.
- *Qualifying ranch property*: Natural grazing land must have a minimum annual carrying capacity of 40 animal units per year to qualify as ranch property.
- *Undeveloped land*: Raw acreage that is not used as part of farming acreage and does not meet the criteria for classification as wasteland. Value is based upon comparable properties in the locale of similar size, quality, and use potential.
- *Unit (annual) carrying capacity*: That which the land will naturally support. A measurement of the animal units allowed by the Arizona Land Department for negotiation of grazing leases of state land.
- *Wasteland*: Land that cannot be converted to an economically beneficial use, including such land areas as river bottoms, sand hills, rock outcroppings, sand washes, and soil salinity areas. Value is based on its contribution to the farming operation. In valuing ranch or grazing property, wasteland is not segregated and valued separately.

5. See M. R. Langemeijer (2013), "Impact of crop insurance indemnity payments on cash rent and land values", Department of Agricultural Economics, Purdue University, selected paper prepared for presentation at the AAEA Crop Insurance and Farm Bill Symposium, Louisville, KY, October 8–9, 2013 (retrieved: http://ageconsearch.umn.edu/bitstream/156836/2/Impact%20of%20Crop%20Insurance%20Indemnity%20Payments%20-%20Paper%20with%20Tables%20and%20Figures.pdf). Very little impact was identified.
6. See further, P. Riera and G. Signorello, eds (2013), "Good practice guidelines for the non-market valuation of forest goods and services", EFS (retrieved: http://news.efi.int/files/attachments/Network%20news/1cost_E45_guidelines.pdf), with comprehensive literature list, pp. 78–80.
7. Indicative only as it will evolve over time due to changes in the marketplace and credit supply
8. See T. J. Straka and S. H. Bullard, "Land expectation value calculation in timberland valuation" (retrieved: http://www.timbertax.org/getstarted/appraisal/papers/pdf/ajoct96.PDF), from which text and examples are derived. See further, S. Chang (1990), "1 comment ll", *Forest Science* 36(1): 177–179. L. S. Davis and K. N. Johnson (1987), *Forest Management* (New York: McGraw-Hill), pp. 408–443. B. B. Foster (1986), "Evaluating pre-commercial timber", *Forest Farmer* 46(2): 20–21. B. B. Foster (1986), "An alternative method for evaluating pre-commercial timber", *The Consultant* 31(2): 29–34. W. D. Klemperer (1987), "Valuing young timber scheduled for future harvest", *The Appraisal Journal* 55(4): 535–547. I. Straka and J. Waluing (1991), "Stands

of pre-commercial timber", *Real Estate Review* 21(2): 92–96. B. P. Vicary (1988), "Appraising pre-merchantable timber", *The Consultant* 33(3): 56–59. P. Feichtinger and K. Salhofer (2011), "CEPS: The valuation of agricultural land and the influence of government payments", Factor Markets Working Paper 10. The latter study gives an overview of the theoretical foundations, empirical procedures, and derived results of the literature identifying determinants of land prices. Special attention is given to the effects of different government support policies on land prices. Empirical studies tend to use a broad range of variables to explain land values and are systematized into six categories in order to investigate the influence of different measures of government support on land prices. Results reveal a significantly higher rate of capitalization for decoupled direct payments and a significantly lower rate of capitalization for agri-environmental payments, as compared to the rest of government support. Furthermore, the results show that taking theoretically consistent land rents (returns to land) and including nonagricultural variables like urban pressure in the regression implies lower elasticities of capitalization. In addition, we find a significant influence of the land type, the data type, and estimation techniques on the capitalization rate. See further, B. Johnson (2012), "Agricultural land valuation and economic updates" (retrieved: http://agecon.unl.edu/c/document_library/get_file?uuid=03c5f957-1e16-4888-b36e-d0e94e6fa17f&groupId=2369805&.pdf).

9. See IVSC (2012), "Exposure draft: The valuation of forests" (retrieved: http://www.ivsc.org/sites/default/files/Forestry%20TIP%20Exposure%20Draft_0.pdf). It is expected, according to the IVSC, that in Q4 of 2013 (after the close of the manuscript of this book), this will, after public consultation, result in a technical information paper (TIP) on the topic. The comments can be consulted here: http://www.ivsc.org/comments/valuation-commercial-forests-exposure-draft#overlay-context=.
10. The information contained in this section does not constitute accounting or tax-related advice, and counsel needs to be sought on individual matters.
11. The "Guide to Costings", as issued by the Central Association of Agricultural Valuers, together with figures produced by other independent institutions, provides useful models to establish costs.
12. See for an extensive report on the issue, R. Meade, G. Fiuza, and A. Lu (2009), "Forest and forest land valuation: How to value forests and forest land to include the carbon costs and benefits" (retrieved: http://ageconsearch.umn.edu/bitstream/97127/2/2009_1_ETS%20impact%20forest%20value_Meade%20etal%5B1%5D.pdf), particularly chapters 3 and 4. New Zealand was the first country in the world to introduce an "all sectors, all gases" carbon scheme in 2008. Both illustrations have been derived from the report.
13. See further, FAO (2013), "Climate change guidelines for forest managers", FAO Forestry Paper 172 (retrieved: http://www.fao.org/docrep/018/i3383e/i3383e.pdf).

11 Advanced Technologies and Agriculture: Genomics, Nanotechnology, and Beyond

1. See further,
   - D. R. Bentley (2006), "Whole-genome re-sequencing", *Current Opinion in Genetics & Development* 16: 545–552.
   - G. Bernardi, E. O. Wiley, H. Mansour, M. R. Miller, G. Orti, D. Haussler, S. J. O'Brien, O. A. Ryder, and B. Venkatesh (2012), "The fishes of Genome 10K", *Marine Genomics* 7: 3–6.

- Y. Dong, M. Xie, Y. Jiang, et al. (2013), "Sequencing and automated whole-genome optical mapping of the genome of a domestic goat (Capra hircus)", *Nature Biotechnology* 31: 135–141 (retrieved: http://www.nature.com/nbt/journal/v31/n2/full/nbt.2478.html).
- B. Fan, Z.-H. Du, D. M. Gorbach, and M. F. Rothschild (2010), "Development and application of high-density SNP arrays in genomic studies of domestic animals", *Asian-Australasian Journal of Animal Science* 23: 833–847 (retrieved: http://www.ajas.info/Editor/manuscript/upload/23-109.pdf).
- FAO (2001), "Glossary of biotechnology for food and agriculture" (retrieved: http://www.fao.org/biotech/biotech-glossary/en) (available in English, French, Spanish, Chinese, Arabic, Russian, Polish, Serbian, Vietnamese, and Kazakh).
- FAO (2007), "Marker-assisted selection: Current status, and future perspectives in crops, livestock, forestry and fish", edited by E. P. Guimarães, J. Ruane, B. D. Scherf, A. Sonnino, and J. D. Dargie (retrieved: http://www.fao.org/docrep/010/a1120e/a1120e00.htm).
- FAO (2008), "Biotechnology as a toolbox to study and monitor agricultural genetic resources", side event to the thirteenth meeting of the Subsidiary Body on Scientific, Technical and Technological Advice (SBSSTA) of the Convention on Biological Diversity, FAO Headquarters, Rome, February 22 (retrieved: http://www.fao.org/biotech/docs/sbstta.htm).
- FAO (2011a), "Introduction to molecular biology and genetic engineering", Module A, by O. Brandenberg, Z. Dhlamini, A. Sensi, K. Ghosh, and A. Sonnino, in the five-module *Biosafety Resource Book* (retrieved: http://www.fao.org/docrep/014/i1905e/i1905e00.pdf).
- FAO (2011b), "Biotechnologies for agricultural development", *Proceedings of the FAO International Technical Conference on Agricultural Biotechnologies in Developing Countries: Options and Opportunities in Crops, Forestry, Livestock, Fisheries and Agro-industry to Face the Challenges of Food Insecurity and Climate Change* (ABDC-10) (retrieved: http://www.fao.org/docrep/014/i2300e/i2300e00.htm).
- C. Feuillet, J. E. Leach, J. Rogers, P. S. Schnable, and K. Eversole (2011), "Crop genome sequencing: Lessons and rationales", *Trends in Plant Science* 16: 77–88.
- R. D. Fleischmann et al. (1995), "Whole-genome random sequencing and assembly of Haemophilus influenzae Rd", *Science* 269: 496–512.
- M. A. M. Groenen et al. (2012), "Analyses of pig genomes provide insight into porcine demography and evolution", *Nature* 491: 393–398 (http://www.nature.com/nature/journal/v491/n7424/full/nature11622.html).
- E. C. Hayden (2013), "Gene sequencing leaves the laboratory: Maturing technology speeds medical diagnoses", *Nature* 494: 290–291.
- B. Hayes and M. Goddard (2010), "Genome-wide association and genomic selection in animal breeding", *Genome* 53: 876–883.
- B. Hayes, H. A. Lewin, and M. Goddard (2013), "The future of livestock breeding: Genomic selection for efficiency, reduced emissions intensity, and adaptation", *Trends in Genetics* 29(4): 206–214.
- E. S. Lander (2011), "Initial impact of the sequencing of the human genome", *Nature* 470: 187–197.
- P. Lidder and A. Sonnino (2011), "Biotechnologies for the management of genetic resources for food and agriculture", Background Study Paper 53, FAO Commission on Genetic Resources for Food and Agriculture (retrieved: http://www.fao.org/docrep/meeting/022/mb387e.pdf).

- C. McSweeney and R. Mackie (2012), "Micro-organisms and ruminant digestion: State of knowledge, trends and future prospects", Background Study Paper 61, FAO Commission on Genetic Resources for Food and Agriculture (retrieved: http://www.fao.org/docrep/016/me992e/me992e.pdf).
- P. L. Morrell, E. S. Buckler, and J. Ross-Ibarra (2012), "Crop genomics: Advances and applications", *Nature Reviews Genetics* 13: 85–96.
- Murphy, D. J. (2011), *Plants, Biotechnology, and Agriculture* (Oxfordshire, UK: CABI Press).
- D. B. Neale and A. Kremer (2011), "Forest tree genomics: Growing resources and applications", *Nature Review Genetics* 12: 111–122.
- O. W. Nelson and G. M. Garrity (2012), "Genome sequences published outside of Standards in Genomic Sciences, July–October 2012", *Standards in Genomic Sciences* 7: 131–149 (retrieved: http://standardsingenomics.org/index.php/sigen/article/view/sigs.3416907/814).
- OIE (2012), "The application of biotechnology to the development of veterinary vaccines", Guideline 3.3 (adopted by OIE World Assembly in May 2010) of the *Manual of Diagnostic Tests and Vaccines for Terrestrial Animals*, World Organization for Animal Health (OIE) (retrieved: http://www.oie.int/international-standard-setting/terrestrial-manual/access-online).
- N. L. Quinn, A. P. Gutierrez, B. F. Koop, and W. S. Davidson (2012), "Genomics and genome sequencing: Benefits for finfish aquaculture", in Z. Muchlisin, ed., *Aquaculture* (retrieved: http://www.intechopen.com/books/aquaculture/genomics-and-genome-sequencing-benefits-for-finfish-aquaculture).
- D. A. Relman (2011), "Microbial genomics and infectious diseases", *New England Journal of Medicine* 365: 347–357 (retrieved: http://www.nejm.org/doi/pdf/10.1056/NEJMra1003071).
- W. W. Soon, M. Hariharan, and M. P. Snyder (2013), "High-throughput sequencing for biology and medicine", *Molecular Systems Biology* 9: 640–654 (retrieved: http://www.nature.com/msb/journal/v9/n1/full/msb201261.html).

2. See further, http://www.fao.org/biotech/biotech-forum.
3. See G. Heard (2013), "Nitrogen fix innovation" (retrieved: http://www.theland.com.au/news/agriculture/cropping/general-news/nitrogen-fix-innovation/2670128.aspx). Professor Edward Cocking, director of the University of Nottingham's Centre for Crop Nitrogen Fixation, has developed a new method of inserting nitrogen-fixing bacteria into plant roots (by placing nitrogen-fixing bacteria into the roots of a cereal plant, through a technology dubbed N-Fix). The work occurred after he found nitrogen-fixing rhizobium in sugarcane, which he found had the ability to colonize other major crop plants. This now could mean that nonlegume plants will be able to fix atmospheric nitrogen. The find would help ease pressure on world food security, particularly given that nitrogen fertilizer is produced from nonrenewable fossil fuels. The process relies simply on a seed coating and does not need genetic modification of the plant.
4. See http://www.ncbi.nlm.nih.gov/genome/browse and the Genomes OnLine Database (GOLD, http://www.genomesonline.org).
5. See detailed per organism and category, FAO (2013), "Impacts of genomics and other 'omics' for the crop, forestry, livestock, fishery and agro-industry sectors in developing countries" (retrieved: http://www.fao.org/docrep/017/aq145e/aq145e.pdf), chapter 2.

6. See in extension, ibid., note 4, chapter 3, including examples of applications.
7. "Nanoscale science and engineering for agriculture and food systems", Department of Agriculture, United States, 2003.
8. See further for a good introduction, A. Jones, J. Nye, and A. Greenberg (2012), "Nanotechnology in agriculture and food technology" (retrieved: http://ice.chem.wisc.edu/NanoDecisions/PDF/Agriculture.pdf). C. I. Moraru et al. (2003), "Nanotechnology: A new frontier in food science", *Food Technology* 57: 24–29.
9. T. Joseph and M. Morrison (2006), "Nanotechnology in agriculture and food" (retrieved: ftp://ftp.cordis.europa.eu/pub/nanotechnology/docs/nanotechnology_in_agriculture_and_food.pdf), p. 2.
10. See also, M. C. DeRosa et al. (2010), "Nanotechnology in fertilizers", *Nature Nanotechnology* 5(2): 91.
11. See for example on electrospinning and more (replacement of plastic), A. Johnson (2006), "Agriculture and nanotechnology" (retrieved: http://www.tahan.com/charlie/nanosociety/course201/nanos/AJ.pdf), pp. 1–2. See further, WHO (2012) "Nanotechnology in food and agriculture", meeting report (retrieved: http://www.fao.org/fileadmin/templates/agns/pdf/topics/Nano_Report_Final_20120625.pdf).
12. See further, C. Silvestre, D. Duraccio, and S. Cimmino (2011), "Food packaging based on polymer nanomaterials", *Progress in Polymer Science* 36: 1766–1782.
13. The concept of precision farming is extensively described in the book. The discussion here is limited to the implications of nanotechnology.
14. See further for applications, ibid., T. Joseph and M. Morrison (2006), pp. 8–10.
15. See in detail, N. M. Manalili, M. A. Dorado, and R. van Otterdijk (2011), "Appropriate food packaging solutions in developing countries", FAO, Rome (retrieved http://www.fao.org/docrep/015/mb061e/mb061e00.pdf). Also K. Marsh and B. Bugusu (2007), "Food packaging: Roles, materials, and environmental issues", *Journal of Food Science* 72(3): 39–55.
16. See further for applications, ibid., T. Joseph and M. Morrison (2006), pp. 10–11.
17. See ibid., T. Joseph and M. Morrison (2006), p. 11.
18. See for a full review of systems in place, FAO (2013), "State of the art on the initiatives and activities relevant to risk assessment and risk management of nanotechnologies in the food and agriculture sectors", FAO/WHO technical paper (retrieved: http://www.fao.org/docrep/018/i3281e/i3281e.pdf).
19. See in detail, J. Lua and M. Bowles (2013), "How will nanotechnology affect agricultural supply chains?", *International Food and Agribusiness Management Review* 16(2) (retrieved: http://ageconsearch.umn.edu/bitstream/148580/2/20120066.pdf), focusing on these areas where most advancement still is to be made. See further,
    - Z. Akbari, T. Ghomashchi, and S. Moghadam (2007), "Improvement in food packaging industry with biobased nanocomposites", *International Journal of Food Engineering* 3(4): 24.
    - A. Arora and G. W. Padua (2010), "Review: Nanocomposites in food packaging", *Journal of Food Science* 75(1): R43–R49 (retrieved: DOI 10.1111/j.1750-3841.2009.01456.x).
    - European Food Safety Authority (EFSA) (2012), "Annual report of the EFSA Scientific Network of Risk Assessment of Nanotechnologies in Food and Feed for 2012", European Food Safety Authority, Parma, Italy.
    - B. A. Magnuson, T. S. Jonaitis, and J. W. Card. (2011), "A brief review of the occurrence, use, and safety of food-related nanomaterials", *Journal of Food Science*, 76(6): R126–R133 (retrieved: DOI 10.1111/j.1750-3841.2011.02170).

- J. M. Wilkinson (2002), "The important role of micro & nanotechnology in the creation of new products and the difficulties faced by SME's in acquiring and exploiting these disruptive technologies", *IEMC-2002: IEEE International Engineering Management Conference* 1, 2: 64–66.
- Yu Liu, Xinghui Li, M. R. Dokmeci, and M. L. Wang. (2011), "Carbon nanotube sensors integrated inside a microfluidic channel for water quality monitoring: Sensors and smart structures technologies for civil, mechanical, and aerospace systems 2011", edited by M. Yun, C. B. Giurgiutiu, V. Lynch, and J. P. Tomizuka.

20. T. Sporleder and M. A. Boland (2011), "Exclusivity of agrifood supply chains: Seven fundamental economic characteristics", *International Food and Agribusiness Management Review* 14(5): 27–51.
21. See in detail, FAO (2007), "Market-assisted selection: Current status and future perspectives in crops, livestock, forestry and fish", edited by Elcio P. Guimarães, John Ruane, Beate D. Scherf, Andrea Sonnino, and James D. Dargie (retrieved: ftp://ftp.fao.org/docrep/fao/010/a1120e/a1120e.pdf).
22. See further, J. Ruane and A. Sonnino (2011), "Agricultural biotechnologies in developing countries and their possible contribution to food security", *Journal of Biotechnology* 156: 356–363 (retrieved: http://www.fao.org/docrep/015/an111e/an111e00.pdf).
23. FAO (2001). "Glossary of biotechnology for food and agriculture" (retrieved: http://www.fao.org/biotech/biotech glossary/en).
24. See further, FAO (2011), "Genetically modified crops (part 4)", Rome (retrieved: http://www.fao.org/docrep/015/i2490e/i2490e04d.pdf).
25. C. James (2011), "Global status of commercialized biotech/GM crops: 2011", International Service for the Acquisition of Agri-biotech Applications (retrieved: http://www.isaaa.org/resources/publications/briefs/43/executivesummary/default.asp).
26. P. Nowicki, L. Aramyan, W. Baltussen, L. Dvortsin, R. Jongeneel, I. P. Domínguez, C. v. Wagenberg, N. Kalaitzandonakes, J. Kaufman, D. Miller, L. Franke, and B. Meerbeek (2010), "Study on the implications of asynchronous GMO approvals for EU imports of animal feed products", report prepared for the Directorate-General for Agriculture and Rural Development, European Commission (retrieved: http://ec.europa.eu/agriculture/analysis/external/asynchronous-gmo-approvals/full-text_En.pdf).
27. EMA (2012), "Searchable database of medicines", European Medicines Agency(retrieved: http://www.ema.europa.eu/ema/index.jsp?curl=pages/includes/medicines/medicines_landing_page.jsp&mid), and N. Vàzquez-Salat and L. M. Houdebine (2012), "Will GM animals follow the GM plant fate? Transgenic research" (retrieved: http://www.springerlink.com/content/qq3gj7717p83250j/fulltext.pdf).
28. E. Hallerman (2004), "Glofish, the first GM animal commercialized: Profits amid controversy", *Information Systems for Biotechnology*, June 2004, pp. 8–10 (retrieved: http://www.isb.vt.edu/articles/jun0405.htm).
29. See further, FAO (2011), "Current status and options for biotechnologies in food processing and in food safety in developing countries", ABDC-10 proceedings, chapter 5 (retrieved: http://www.fao.org/docrep/014/i2300e/i2300e00.htm).

    FAO (2011), "Current status and options for livestock biotechnologies in developing countries", ABDC-10 proceedings, chapter 5 (retrieved: http://www.fao.org/docrep/014/i2300e/i2300e00.htm).
30. OIE (2012), "The application of biotechnology to the development of veterinary vaccines", Guideline 3.3 (adopted by OIE World Assembly in May 2010) of the

Manual of Diagnostic Tests and Vaccines for Terrestrial Animals, World Organisation for Animal Health (OIE) (retrieved: http://www.oie.int/international-standard-setting/terrestrial-manual/access-online).

31. Further details about these kinds of vaccines, including examples of their commercial application, are provided in ibid., OIE (2012).
32. Phillips McDougall (2011), "The cost and time involved in the discovery, development and authorization of a new plant biotechnology derived trait", a consultancy study for Crop Life International (retrieved: http://www.croplife.org/PhillipsMcDougallStudy).
33. M. Lusser, T. Raney, P. Tillie, K. Dillen, and E. Rodríguez-Cerezo (2012), "International workshop on socio-economic impacts of genetically modified crops co-organised by JRC-IPTS and FAO: Workshop proceedings", JRC Scientific and Technical Report EUR 25265 (retrieved: http://www.fao.org/docrep/015/ap016e/ap016e.pdf).
34. See further, A. J. Stein and E. Rodríguez-Cerezo (2009), "The global pipeline of new GM crops: Implications of asynchronous approval for international trade", JRC Technical Report EUR 23486 (retrieved: http://ipts.jrc.ec.europa.eu/publications/pub.cfm?id=2420).
35. N. Vàzquez-Salat and L. M. Houdebine (2012), "Will GM animals follow the GM plant fate?", *Transgenic Research*, in press (retrieved: http://www.springerlink.com/content/qq3gj7717p83250j/fulltext.pdf).
36. P. Kanowski (2012), "Genetically-modified trees: Opportunities for dialogue", a scoping paper for The Forests Dialogue (TFD), Gland, Switzerland, October 17–18, 2012 (retrieved: http://environment.yale.edu/tfd/dialogue/genetically-modified-trees/genetically-modified-trees-scoping-dialogue1).
37. Ibid., Vàzquez-Salat (2012).
38. A. Maxmen (2012), "Politics holds back animal engineers", *Nature* 490: 318–319.
39. FDA (2010), "Materials from the AquAdvantage salmon meeting of the Veterinary Medicine Advisory Committee", US Food and Drug Administration, Rockville, United States, September 19–20, 2010 (retrieved: http://www.fda.gov/AdvisoryCommittees/CommitteesMeetingMaterials/VeterinaryMedicineAdvisoryCommittee/ucm201810.htm), and ibid., Maxmen (2012).
40. C. McSweeney and R. Mackie (2012), "Micro-organisms and ruminant digestion: State of knowledge, trends and future prospects", CGRFA Background Study Paper 61, Commission on Genetic Resources for Food and Agriculture (retrieved: http://www.fao.org/docrep/016/me992e/me992e.pdf).
41. J. Ruane, A. Sonnino, and A. Agostini (2010), "Bioenergy and the potential contribution of agricultural biotechnologies in developing countries", *Biomass and Bioenergy*, 34: 1427–1439 (retrieved: http://www.fao.org/docrep/article/al311e.pdf).
42. The most recent one took place in March 2013, and the feedback report can be found at FAO (2013), "All messages from the FAO 2013 e-mail conference on 'Impacts of genomics and other "omics" for the crop, forestry, livestock, fishery and agro-industry sectors in developing countries'" (retrieved: http://www.fao.org/fileadmin/user_upload/biotech/docs/conf19msgs.pdf). See also, FAO (2012/2013), "GMOs in the pipeline: Looking to the next five years in the crop, forestry, livestock, aquaculture and agro-industry sectors in developing countries" (retrieved: http://www.fao.org/docrep/016/ap109e/ap109e00.pdf), and its summary report (retrieved: http://www.fao.org/docrep/017/ap998e/ap998e.pdf).
43. See, for example, Brian G. Sims and Josef Kienzle (2006), "Farm power and mechanization for small farms in Sub-Saharan Africa", FAO, Rome (retrieved: ftp://ftp.

fao.org/docrep/fao/009/a0651e/a0651e00.pdf). The term "mechanization" is used to describe tools, implements, and machinery applied to improving the productivity of farm labor and of land; it may use either human, animal, or motorized power, or a combination of these. In practice, therefore, it involves the provision and use of all forms of power sources and mechanical assistance to agriculture, from simple hand tools, to draft-animal power, to mechanical power technologies. Mechanization is a key input in any farming system. It aims to achieve the following:

- increased productivity per unit area due to improved timeliness of farm operations;
- an expansion of the area under cultivation where land is available, as it often is in Sub-Saharan Africa;
- accomplishment of tasks that are difficult to perform without mechanical aids;
- improvement of the quality of work and products; and
- a reduction of drudgery in farming activities, thereby making farm work more attractive.

Mechanization systems are categorized into human, animal, and mechanical technologies. Based on the source of power, the technological levels of mechanization have been broadly classified as hand-tool technology, draft-animal technology, and mechanical-power technology. The constraints to increased farm production are due, to a large extent, to three factors:

- an excessive reliance on human power,
- the low productivity of human labor, and
- a decrease in the labor available.

(Ibid., p. xii) and the summary of issues and challenges to the adoption of different forms of mechanization, such as draft-animal power and tractors (ibid., p. xv).

44. See for a general overview, D. Sunding and D. Zilberman (2001), "The agricultural innovation process: Research and technology adoption in a changing agricultural sector", in B. L. Gardner and G. C. Rausser, eds, *Handbook of Agricultural Economics*, vol. 1, chapter 4, pp. 207–261 (New York: Elsevier). Also, W. E. Huffman (2009), "Technology and innovation in world agriculture: Prospects for 2010–2019" (retrieved: http://www2.econ.iastate.edu/research/webpapers/paper_13060_09007.pdf).
45. See further, C. Pray, D. Gisselquist, and L. Nagarajan (2011), "ASTI, Private investment in agricultural research and technology transfer in Africa", Conference Working Paper 13 (retrieved: http://www.ifpri.org/sites/default/files/publications/pray.pdf). Imported innovations in machinery, pesticides, fertilizers, poultry, and plant varieties have been very important to the development of modern agriculture in Africa. These technologies are now primarily brought in by private agricultural input industries and by some processing industries. Private-sector R&D is still quite limited in Sub-Saharan Africa, with the exception of South Africa. It is concentrated in the maize seed industry and in the processing and plantation subsectors. The evidence presented suggests that governments can encourage the introduction of more private technology by continuing to liberalize: allowing local and foreign firms to enter, providing firms with a stable policy and regulatory environment, strengthening Intellectual Property Rights (IPR), and not taxing agriculture. When barriers to investment, importation, and the introduction of technology fall, private firms will introduce appropriate technology even to the smallest markets. Other factors that could increase markets and stimulate research are the reduction of barriers to regional trade in fertilizer, seed, and other agricultural inputs, and removal of badly designed input subsidies that channel input trade through government tenders rather than markets. Further relaxation – or, as a second best, regional harmonization – of

technical regulations on agriculture could have a big impact on the pace of cultivar introduction. Many of the surveyed companies commented on this.

46. See further: N. Beintema and H. Elliott (2009), "Setting meaningful investment targets in agricultural research and development: Challenges, opportunities and fiscal realities", Paper prepared for the Expert Meeting on "How to Feed the World in 2050", FAO, Rome, June 24–26, 2009 (retrieved: ftp://ftp.fao.org/docrep/fao/012/ak978e/ak978e00.pdf). The Agricultural Science and Technology Indicators (ASTI) initiative provides frequent updates and analysis on agricultural R&D around the world.
47. See also, K. Vanangamudi, S. Kalaivani, M. Vanangamudi, G. Sasthri, A. Selvakumari, and P. Srimathi (2010), "Seed quality enhancement: Principles and practices", Scientific Publishers Journals Department.
48. See, L. Nalley, M. Popp, and Z. Niederman (2013), "Embedded seed technology and greenhouse gas emissions reductions: A meta-analysis", *Journal of Agricultural and Applied Economics* 45(3): 523–535 (retrieved: http://ageconsearch.umn.edu/bitstream/155428/2/jaae453ip14.pdf).

12 Challenges in Agricultural Production and Natural-Resources Management

1. See extensively and in particular chapter 4 (pp. 94–132) from which this chapter has been extensively drawn: N. Alexandratos and J. Bruinsma (2012), "World agriculture towards 2030/2050: The 2012 revision", ESA Working Paper 12–03, FAO Rome (retrieved: http://www.fao.org/fileadmin/templates/esa/Global_persepctives/world_ag_2030_50_2012_rev.pdf).
2. Since total agricultural production is measured by weighing individual products with average international prices, the price-based index of the volume of production grows faster than aggregates expressed in physical units or using a calorie-based index as diets change away from staples to higher-value commodities.
3. Ibid., note 1, p. 95, which includes a breakdown in regional differences in (expected and historical) growth.
4. Based on (gross imports + gross exports/2)/production.
5. For the full data set, see ibid., note 1, p. 98.
6. G. Fischer, H. van Velthuizen, and F. Nachtergaele (2011), "GAEZ v3.0: Global Agro-ecological Zones Model documentation", mimeo, IIASA, Luxemburg (retrieved: http://www.iiasa.ac.at/Research/LUC/GAEZv3.0/gaez2010-Flyer_1final1.pdf).
7. G. Fischer, E. Hisznyik, S. Prieler, and D. Wiberg (2010), "Scarcity and abundance of land resources: Competing land uses and the shrinking resource base", Thematic Report 2, prepared for SOLAW, FAO, 2011.
8. See for the full reporting, ibid., note 1.
9. Derived from ibid., note 1, p. 105.
10. G. Fischer, H. van Velthuizen, M. Shah, and F. Nachtergaele (2002), "Global agro-ecological assessment for agriculture in the 21st century: Methodology and results", RR-02-002, IIASA, Luxemburg, and FAO, Rome (retrieved: http://www.iiasa.ac.at/Research/LUC/SAEZ/pdf/gaez2002.pdf).
11. J. Cohen (1995), *How Many People Can the Earth Support?* New York: Norton.
12. A. Young (1999), "Is there really spare land? A critique of estimates of available cultivable land in developing countries", *Environment, Development and Sustainability* 1: 3–18.

13. N. Alexandratos, ed. (1995), *World agriculture: Towards 2010*, an FAO study (Chichester, UK: Wiley) (retrieved: http://www.fao.org/docrep/V4200E/V4200E00.htm#Contets).
14. See ibid., note 1, pp. 106–107.
15. See ibid., note 1, pp. 109–114.
16. See in more detail, ibid., note 1, p. 116. They further elaborate on the estimation of the pressure of irrigation on water resources.
17. See ibid., note 1, p. 117, and table 4.11 on page 118.
18. See note 6.
19. See ibid., note 1, pp. 120–121.
20. See World Bank (2008), "World development report 2008: Agriculture for development", World Bank, Washington, DC, pp. 66–69; J. Bruinsma, ed. (2003), *World Agriculture: Towards 2015/2030 – An FAO Perspective*, Earthscan, London, and FAO, Rome (retrieved: http://www.fao.org/fileadmin/user_upload/esag/docs/y4252e.pdf), pp. 146–148. J. Bruinsma (2011), "The resource outlook to 2050: By how much do land, water use and crop yields need to increase by 2050?", chapter 6 in P. Conforti, ed., *Looking ahead in World Food and Agriculture: Perspectives to 2050* (Rome: FAO). Also literature mentioned in notes 6 and 7 (retrieved: http://www.fao.org/docrep/014/i2280e/i2280e06.pdf).
21. See ibid., note 1, pp. 124–126.
22. See for an application in the cereal production area, ibid., note 1, p. 126, pointing at a 40% to 50% increase in production due to fertilizer use.
23. S. Daberkow, J. Poulisse, and H. Vroomen (2000), "Fertilizer requirements in 2015 and 2030", FAO, Rome (retrieved: ftp://ftp.fao.org/agl/agll/docs/fertreqrev.pdf).
24. See for an application for maize, C. Frink, P. Waggoner, and J. Ausubel (1998), "Nitrogen fertilizer: Retrospect and prospect", paper presented at the NAS colloquium, Plants and Population: Is there Time?, Irvine, CA, December 5–6.
25. R. Socolow (1998), "Nitrogen management and the future of food: Lessons from the management of energy and carbon", paper presented at the NAS colloquium, Plants and Population: Is there Time?, Irvine, CA, December 5–6.
26. See ibid., note 1, p. 128.
27. See ibid., note 1, p. 129.
28. See J. Henao and C. Baanante (1999), "Nutrient depletion in the agricultural soils of Africa", Vision 2020 Brief 62, IFPRI, Washington, DC. See further, P. Gruhn, F. Goletti, and M. Yudelman (2000), "Integrated nutrient management, soil fertility, and sustainable agriculture: Current issues and future challenges", Food, Agriculture, and the Environment Discussion Paper 32 (retrieved: http://www.ifpri.org/sites/default/files/pubs/2020/dp/2020dp32.pdf), as well as O. E. Omotayo and K. S. Chukwuka (2009), "Soil fertility restoration techniques in Sub-Saharan Africa using organic resources", *African Journal of Agricultural Research* 4(3): 144–150 (retrieved: http://www.academicjournals.org/ajar/pdf/pdf%202009/Mar/Omotayo%20and%20Chukwuka.pdf), and "Guidelines on nitrogen management in agricultural systems" (2008), Training Course Series 29, IAE Agency, Vienna (retrieved: http://www-pub.iaea.org/MTCD/publications/PDF/TCS-29_PDF/TCS-29.pdf).
29. See, J. Syers, A. Johnston, and D. Curtin (2008), "Efficiency of soil and fertilizer phosphorus use", FAO Fertilizer and Plant Nutrition Bulletin 18, FAO, Rome.
30. See ibid., note 1, p. 130.
31. See ibid., note 1, p. 130.
32. See ibid., note 1, p. 131.

33. See ibid., note 1, p. 132.
34. See for a schematic overview per region and commodity, ibid., note 1, p. 133.
35. See, for example, R. Townsend and T. W. Hertel (2011), "The global supply and demand for agricultural land in 2050: A perfect storm in the making?", *American Journal of Agricultural Economics* 93(2): 259–275, and T. W. Hertel (2013), "Land, environment and climate: Contributing to the global public good", background paper prepared for the Results Meeting on Aid, Environment and Climate Change, organized by the United Nations University-WIDER, Stockholm, June 4, 2013 (retrieved: https://geoshareproject.org/resources/78/download/Land_as_a_Global_Public_Good_-__05-24-2013.pdf), as well as T. W. Hertel and D. B. Lobell (2012), "Agricultural adaptation to climate change in rich and poor countries: Current modeling practice and potential for empirical contributions", GTAP Working Paper 72 (retrieved: https://www.gtap.agecon.purdue.edu/resources/download/6112.pdf).

13 Sustainability of Agricultural Productivity Growth

1. See extensively, "Sustainable agricultural productivity growth and bridging the gap for small-family farms" (2012), Interagency Report to the Mexican G20 Presidency, p. 21 and further (retrieved: http://www.oecd.org/tad/agricultural-policies/50544691.pdf).
2. Because of the public interest, it has also gained traction on the political agenda. See "Sustainable agricultural productivity growth and bridging the gap for small-family farms" (2012), Interagency Report to the Mexican G20 Presidency (retrieved: http://unctad.org/en/PublicationsLibrary/sucmisc2012d1_En.pdf).
3. K. O. Fuglie (2012), "Productivity growth and technology capital in the global agricultural economy", in K. O. Fuglie, S. L. Wang, and V. E. Ball, eds, *Productivity Growth in Agriculture: An International Perspective* (Oxfordshire, UK: CAB International).
4. See for further details, FAOSTAT. Similar trends as the one indicated for wheat have been reported for maize and rice with similar magnitude in delta since the early 1960s.
5. FAO (2011), "Save and grow: A policymaker's guide to the sustainable intensification of smallholder crop production", Rome (retrieved: http://www.fao.org/docrep/014/i2215e/i2215e.pdf).
6. C. Nellemann, M. MacDevette, T. Manders, B. Eickhout, B. Svihus, A. G. Prins, and B. P. Kaltenborn, eds (2009), "The environmental food crisis: The environment's role in averting future food crises", A UNEP rapid response assessment, Norway, United Nations Environment Programme and GRID-Arendal (retrieved: http://www.grida.no/files/publications/FoodCrisis_lores.pdf).
7. OECD (2012), "OECD environmental outlook to 2050", OECD Publishing (retrieved: http://www.oecd.org/env/indicators-modelling-outlooks/49846090.pdf).
8. OECD (2012), "Environmental performance of agriculture at a glance" (retrieved: http://www.oecd.org/greengrowth/sustainable-agriculture/40953155.pdf).
9. UN (2001), "World economic and social survey 2011: The great green technological transformation", Department of Economic and Social Affairs (retrieved: http://www.un.org/en/development/desa/policy/wess/wess_current/2011wess.pdf).
10. IPCC, M. L. Parry, O. F. Canziani, J. P. Palutikof, P. J. van der Linden, and C. E. Hanson (2007), Climate Change 2007: Impacts, Adaptation and Vulnerability, Contribution of Working Group II to the Fourth Assessment Report of the Intergovernmental Panel on Climate Change (Cambridge, UK: Cambridge University Press). D. B. Lobell, W. Schlenker, and J. Costa-Roberts (2011), "Climate trends and global

crop production since 1980", Science 333(6042): 616–620 (retrieved: DOI 10.1126/ science.1204531, or www.ncbi.nlm.nih.gov/pubmed/21551030). G. C. Nelson, M. W. Rosegrant, J. Koo, R. Robertson, T. Sulser, T. Zhu, and C. Ringler (2009), Climate change: Impact on agriculture and costs of adaptation", IFPRI, Washington, DC (retrieved: dx.doi.org/10.2499/0896295354 or www.ifpri.org/publication/climate-change-1). C. Müller, W. Cramer, W. L. Hare, and H. Lotze-Campen (2011), "Climate change risks for African agriculture", Proceedings of the National Academy of Sciences 108(11): 4313–4315. R. Wassmann, G. C. Nelson, S. B. Peng, K. Sumfleth, S. V. K. Jagadish, Y. Hosen, and M. W. Rosegrant (2010), "Rice and global climate change", in Rice in the Global Economy: Strategic Research and Policy Issues for Food Security, p. 22 (Los Banos, Philippines).

11. J. A. Burney, S. J. Davis, and D. B. Lobell (2010), "Greenhouse gas mitigation by agricultural intensification", *Proceedings of the National Academy of Sciences* 107(26): 12052–12057. SREX (2012), "Managing the risks of extreme events and disasters to advance climate change adaptation (SREX)", Special Report of the Intergovernmental Panel on Climate Change.
12. Royal Society (2009), "Reaping the benefits: Science and the sustainable intensification of global agriculture", RS Policy Document 11/09, London.
13. OECD (2011), "OECD green growth studies: Food and agriculture", OECD Publishing (retrieved: dx.doi.org/10.1787/9789264107250-en).
14. J. Alston (2010), "The benefits from agricultural research and development, innovation, and productivity growth", OECD Food, Agriculture and Fisheries Working Papers 31, OECD Publishing (retrieved: dx.doi.org/10.1787/5km91nfsnkwg-en).
15. S. Fan, ed. (2008), *Public Expenditures, Growth, and Poverty: Lessons from Developing Countries* (Baltimore, MD: Johns Hopkins University Press). FAO (2012), "State of food and agriculture: Investment in agricultural for food security".
16. K. O. Fuglie (2012), "Productivity growth and technology capital in the global agricultural economy", in K. O. Fuglie, S. L. Wang, and V. E. Ball, eds (2012), "Productivity growth in agriculture: An international perspective (Oxfordshire, UK: CAB International).
17. Ibid., note 7.
18. World Bank (2012), *Agricultural Innovation Systems: An Investment Sourcebook* (Washington, DC: World Bank) (retrieved: http://siteresources.worldbank.org/INTARD/Resources/335807–1330620492317/9780821386842.pdf).
19. OECD (2012), "Improving agricultural knowledge and innovation systems: OECD conference proceedings", OECD Publishing (retrieved: http://www.iadb.org/intal/intalcdi/PE/2012/09772.pdf).
20. OECD (2013), "Agricultural innovation systems: A framework for analysing the role of the government", OECD Publishing (retrieved:http://search.oecd.org/officialdocuments/publicdisplaydocumentpdf/?cote=TAD/CA/APM/WP(2012)19/FINAL&docLanguage=En). Innovation policy is central to improving AIS. The public sector plays a major role in the provision of knowledge infrastructure (e.g., communication technology, databanks, centers of technology convergence) and the financing of basic research, or research with long-term and public-good aspects, in particular for agriculture and natural-resource management. The public sector remains a major funder and performer of R&D for agriculture. Governments also encourage innovation activities in the private sector, including by fostering knowledge markets through IPR protection, engaging in public–private partnerships (PPP), providing information and sharing the outcomes of public research (spillovers), and providing direct or indirect financial incentives.

21. See, S. Kimura and C. Le Thi (2013), "Cross country analysis of farm economic performance", OECD Food, Agriculture and Fisheries Papers 60, OECD Publishing (retrieved: http://www.oecd-ilibrary.org/docserver/download/5k46ds9ljxkj.pdf?expires=1372759420&id=id&accname=guest&checksum=A3C78924115F508AAF764E18A0058BBC).
22. K. Andersen, G. Rausser, and J. Swinnen (2013), "Political economy of public policies: Insights from distortions to agricultural and food markets", Policy Research Working Paper WPS 6433 (retrieved: http://www-wds.worldbank.org/external/default/WDSContentServer/IW3P/IB/2013/05/02/000158349_20130502090516/Rendered/PDF/wps6433.pdf).
23. OECD (2011), "Fostering productivity and competitiveness in agriculture", OECD Publishing(retrieved:http://browse.oecdbookshop.org/oecd/pdfs/product/5111131e.pdf).
24. K. Z. Chen, K. Flaherty, and Y. Zhang (2012), "China: Recent developments in agricultural research", ASTI Country Note, International Food Policy Research Institute, Washington, DC (retrieved: http://www.asti.cgiar.org/pdf/China-Note.pdf).
25. N. M. Beintema and G.-H. Stads (2011), "African agricultural R&D in the new millennium", IFPRI (retrieved: www.ifpri.org/publication/african-agricultural-rd-new-millennium).
26. N. M. Beintema and G.-H. Stads (2008), "Measuring agricultural research investments: A revised global picture", ASTI Background Note, International Food Policy Research Institute, Washington, DC (retrieved: www.asti.cgiar.org/pdf/Global_revision.pdf).
27. R. G. Echeverría and N. M. Beintema (2009), "Mobilizing financial resources for agricultural research in developing countries: Trends and mechanisms", Global Forum for Agricultural Research, Rome (retrieved: http://www.fao.org/docs/eims/upload//261898/mobilizing%20financial%20resources%20for%20AR4D.pdf).
28. N. Beintema and H. Elliott (2009), "Setting meaningful investment targets in agricultural research and development: Challenges, opportunities, and fiscal realities", Paper prepared for the expert meeting, How to Feed the World in 2050, Food and Agricultural Organisation of the United Nations, Rome, June 24–26 (retrieved: ftp://ftp.fao.org/docrep/fao/012/ak978e/ak978e00.pdf).
29. T. Mogues and S. Benin, eds (2012), "Public expenditures for agricultural and rural development in Africa", Routledge Studies in Development Economics.
30. Foresight (2011), "The future of food and farming: Final project report", Government Office for Science, London (retrieved: http://www.bis.gov.uk/assets/foresight/docs/food-and-farming/11–546-future-of-food-and-farming-report.pdf).
31. FAO (2012), "State of food and agriculture: Investment in agricultural for food security" (retrieved: http://www.fao.org/docrep/017/i3028e/i3028e.pdf).
32. K. Anderson, ed. (2009), "Distortions to agricultural incentives from a global perspective 1955–2007", World Bank (retrieved: http://siteresources.worldbank.org/INTTRADERESEARCH/Resources/544824–1163022714097/Five_decades_of_distortions_0309rev.pdf).
33. OECD (2012), "Development aid reaches an historic high in 2010" (retrieved: http://www.oecd.org/dac/stats/developmentaidreachesanhistorichighin2010.htm).
34. Ibid., note 1, p. 21 and further.
35. See further, J. van der Burg (2012), "Raising seed quality: What is in the pipeline?" (retrieved: https://www.seedtest.org/upload/cms/user/5vanderBurg.pdf) distinguishing between physical, physiological, and genetic quality. Categories of technology

include testing and research (which includes X-ray, image analysis (crack detection), CF (chlorophyll fluorescence)-analyzer, Q2 technology, ethanol assay (ethanol as a measure for seed quality or ethanol as marker for seed deterioration, germination energy, and ethanol production move inversely and molecular technologies (flow cytrometry and immunobeds, proteomics and metabolomics) as well as enhancement (priming [electron treatment], sorting, and sanitation – via aerated steam).

36. See also, S. Ali, A. R. Khan, G. Mairaj, M. Arif, M. Fida, and S. Bibi (2008), "Assessment of different crop nutrient management practices for yield improvement" (retrieved: http://www.cropj.com/Microsoft%20Word%20-%20Sajid%20ali%20main%20150–157.pdf). L. Xiaoyan, P. He, and J. Jin (2011), "A long-term analysis of factors to improve nutrient management for winter wheat production in China", *Better Crops* 95(3): 16–18 (retrieved: http://www.ipni.net/publication/bettercrops.nsf/0/FE8645500795AF158525797C0078FF20/$FILE/Better%20Crops%202011–3%20p16–17.pdf). N. D. Mueller, J. S. Gerber, M. Johnston, D. K. Ray, N. Ramankutty, and J. A. Foley (2012), "Closing yield gaps through nutrient and water management", *Nature*, MaxMillan Publishers. (retrieved: doi: 10.1038/nature11420).
37. See in detail, J. B. Passioura and J. F. Angus (2010), "Improving productivity of crops in water-limited environments", in D. L. Sparks, ed., *Advances in Agronomy*, vol. 106 (Burlington: Academic Press), pp. 37–75.

14 Commodities I: Derivatives Markets

1. The Commitment of Traders report is published each week on Friday and reports the outstanding positions of different categories of traders. The reports can be found here: http://www.cftc.gov/marketreports/commitmentsoftraders/index.htm.
2. The Bank for International Settlements reports the open positions on a six-month basis in its semiannual OTC derivatives statistics. The most recent data can be found here and include data to mid-2012 (Table 22A under other commodities): http://www.bis.org/statistics/otcder/dt21c22a.pdf. The volumes to the end of 2012 were released in May 2013 (http://www.bis.org/publ/otc_hy1305.pdf). The next update is due November 15, 2013, after the close of the manuscript of this book (http://www.bis.org/publ/rpfx13.htm).
3. In the extended Chapter 15 we will discuss potential other classifications of agricultural commodity-market participants.
4. M. W. Masters (2009), "Testimony before the commodity futures trading commission", technical report, Commodity Futures Trading Commission (retrieved: http://www.cftc.gov/ucm/groups/public/@newsroom/documents/file/hearing080509_masters.pdf), and Scott H. Irwina and Dwight R. Sandersb (2011 revised), "Testing the masters hypothesis in commodity futures markets, Energy Economics 34(1): 256–269 (retrieved: http://www.econ.iastate.edu/sites/default/files/mastershypothesis_0.pdf). S. H. Irwin (2013), "Commodity index investment and food prices: Does the 'masters hypothesis' explain recent price spikes?" (retrieved: http://www.farmdoc.illinois.edu/irwin/research/Irwin_IAAE_Paper.pdf).
5. G. Soros (2008), "Testimony before the U.S. Senate Commerce Committee Oversight Hearing on FTC advanced rulemaking on oil market manipulation", Washington, DC, June 4, 2008.
6. M. Radetzki (2008), *A Handbook of Primary Commodities in the Global Economy* (Cambridge: Cambridge University Press).

7. D. Greely and J. Currie (2008), "Speculator, index investors and commodity prices", Goldman Sachs – Commodities.
8. These so-called index traders will be drawn from both previous noncommercial and commercial categories. Coming from the former, there will be managed funds, pension funds, and other institutional investors that generally seek exposure to commodity prices as an asset class in an unleveraged and passively managed manner using a standardized commodity index. Coming from the second category, there will be entities whose positions predominantly reflect hedging of OTC transactions (swap dealers, holding long futures positions to hedge short OTC commodity-index exposure, opposite institutional traders such as pension funds).
9. Nonetheless, a warning issued by the CFTC clarifies that there are still significant limitations to the new data, for instance to the extent that traders may engage in different types of activity: producers may decide to engage in swap activities, and investors classified among swap dealers may also be involved in commercial activities.
10. United Nations Conference on Trade and Development (2011), "Price formation in financialized commodity markets: The role of Information".
11. Barley, cocoa, crude oil, maize, sugar, and wheat.
12. Including the United Nations Conference on Trade and Development (2011), "Price formation in financialized commodity markets: The role of Information".
13. IOSCO (2010), "Task force on commodity futures markets", Report to the G-20, Technical Committee of the International Organization of Securities Commissions, No. OR08/10, Madrid, November, p. 5.
14. Ibid., IOSCO, p. 6.
15. A recent example, in the wider commodity space, of a persisting differential is the wide gap between Brent crude oil prices and West Texas Intermediate (WTI) oil prices, which exceeded $15 per barrel in early February 2011. The price of New York Mercantile Exchange (NYMEX) WTI, the leading oil contract in the world, has been significantly below Brent crude futures at ICE, although the two are similar in quality. This can be explained by various factors. The NYMEX WTI contract envisages physical delivery in Cushing, Oklahoma. Inventories in Cushing are soaring, recently reaching a peak of 38.3 million barrels (G. Meyer [2011], "Oil price divergence takes crude dealers down different track", *Financial Times*, February 24) as a result of increased oil production both in North Dakota and Canada. As pipelines deliver oil to Cushing from the north and the south, but cannot transport oil from Cushing (see, for example, IntercontinentalExchange, undated), inventories there keep on rising, whereas demand from nearby refineries does not keep up, which depresses the price. Any market participant wishing to engage in arbitrage would have to move the oil from Cushing to the Gulf of Mexico for shipment to Europe. This is costly and takes time. There has been some arbitrage between Cushing and the Gulf of Mexico to exploit higher prices on the coast, but the price differential between WTI and Brent has persisted.
16. FAO (2009), "The state of food and agriculture 2009: Livestock in the balance", FAO, Rome (retrieved: http://www.fao.org/docrep/012/i0680e/i0680e.pdf).
17. R. K. Kaufmann (2011), "The role of market fundamentals and speculation in recent price changes for crude oil", *Energy Policy* 39(1): 105–115.
18. See also, B. W. Brorsen (2012), "Discussion: Agricultural commodities and agribusiness stocks as financial assets", *Journal of Agricultural and Applied Economics* 44(3): 397–399.

19. H. Zapata, J. D. Detre, and T. Hanabuchi (2012), "Historical performance of commodity and stock markets", *Journal of Agricultural and Applied Economics* 44: 339–358.
20. G. Schnitkey and C. Kramer (2012), "Performance of publically-traded agricultural companies since 2000: Construction of the AgIndex", *Journal of Agricultural and Applied Economics* 44: 359–370.
21. S. H. Irwin and D. R. Sanders (2012), "Financialization and structural change in commodity futures markets", *Journal of Agricultural and Applied Economics* 44: 371–396.
22. S. H. Irwin and D. R. Sanders (2011), "Index funds, financialization, and commodity futures markets", *Applied Economic Perspectives and Policy* 33: 1–31.
23. Selected paper prepared for presentation at the Southern Agricultural Economics Association Annual Meeting, Orlando, FL, February 2–5, 2013 (retrieved: http://ageconsearch.umn.edu/bitstream/143080/2/Determining%20the%20Nature%20of%20Dependency%20between%20Agribusiness%20and%20Non-Agribusiness%20Stocks.pdf).
24. C. Revoredo-Giha and M. Zuppiroli (2013), "Effectiveness of hedging within the high price volatility context", Land Economy Working Paper Series 69 (retrieved: http://ageconsearch.umn.edu/bitstream/142546/2/leergworkingpaper69.pdf).
25. S. H. Lence (2009), "Do futures benefit farmers?", *American Journal of Agricultural Economics* 91(1): 154–167.
26. M. Bohl and P. Stephan (2012), "Does futures speculation destabilize spot prices? New evidence for commodity markets", available at SSRN 1979602, papers.ssrn.com.
27. Ibid., note 24.
28. C. Carter (1984), "An evaluation of pricing performance and hedging effectiveness of the barley futures market", *Western Journal of Agricultural Economics* 9(1): 1–13.
29. M. G. Castelino (1989), "Basis volatility: Implications for hedging", *Journal of Financial Research* 12(2): 157–172.
30. E. Fama (1970), "Efficient capital markets: A review of theory and empirical work", *Journal of Finance* 25(2): 383–417.
31. Ibid., note 28.
32. R. W. Gray (1961), "The search for a risk premium", *Journal of Political Economy,* 69(3): 250–260.
33. P. H. Cootner (1960), "Returns to speculators: Telser versus Keynes", *Journal of Political Economy* 68(4): 396–404.
34. Cootner's empirical research on US wheat futures statistics found that, on the average, prices fell from May to October–November and rose steadily thereafter. "In short, as the crop came and the movement into commercial channels reached a peak, prices fell. As the crop was consumed, hedges were lifted and prices rose" (Cootner, 1960, p. 401). Carter applied a Cootner-type trading routine, assuming speculators were short, in the Winnipeg barley market for October and November and in the CBOT corn market for September, October, and November. The period hypothesized by Carter is shorter (two or three months) than the one used by Cootner.
35. D. R. Sanders and M. R. Manfredo (2004), "Comparing hedging effectiveness: An application of the encompassing principle", *Journal ofAgricultural and Resource Economics* 29(1): 31–44.
36. D. Lien and Y. K. Tse (2002), "Some recent developments in futures hedging", *Journal of Economic Surveys* 16(3): 357–395.

37. M. Hernandez, R. Ibarra, and D. Trupkin (2012), "How far do shocks move across borders? Examining volatility transmission in major agricultural futures markets", selected paper prepared for presentation at the International Association of Agricultural Economists (IAAE) Triennial Conference, Foz do Iguaçu, Brazil, August 18–24, 2012, based on the extensive study, M. A. Hernandez, R. Ibarra, and D. R. Trupkin (2011), "How far do shocks move across borders? Examining volatility transmission in major agricultural futures markets", Discussion Paper 1109, International Food Policy Research Institute (IFPRI), Washington DC (retrieved: http://ageconsearch.umn.edu/bitstream/122511/2/Hernandez.pdf).
38. In the case of soybeans, both China and Japan also show important cross-volatility spillovers.
39. This study evaluates the level of interdependence and volatility transmission in major agricultural exchanges between the United States (Chicago, Kansas), Europe (France, United Kingdom), and Asia (China, Japan). In particular, they examine the dynamics and cross-dynamics of volatility across futures markets for three key agricultural commodities: corn, wheat, and soybeans. The period of analysis is 2004–2009 for corn and soybeans and 2005–2009 for wheat. They follow a multivariate GARCH approach that allows us to evaluate whether there is volatility transmission across exchanges, the magnitude and source of interdependence (direct or indirect) between markets, and ultimately how a shock or innovation in a market affects volatility in other markets. This study is relevant for a number of reasons: First, it provides (as the first study) an in-depth analysis of volatility transmission across several important exchanges of different agricultural commodities. Second, and contrary to previous related studies, they account for the potential bias that may arise when considering agricultural exchanges with different closing times. They synchronize their data by exploiting information from markets that are open to derive estimates for prices when markets are closed. Third, the sample period allowed them to examine whether there have been changes in the dynamics of volatility due to the recent food-price crisis of 2007–2008, a period of special interest with unprecedented price variations. Finally, they applied different MGARCH specifications to analyze in detail the cross-market dynamics in the conditional volatilities of the exchanges.
40. C. Gardebroek and M. Hernandez (2012), "Do energy prices stimulate food price volatility? Examining volatility transmission between US oil, ethanol and corn markets", selected paper prepared for presentation at the Agricultural and Applied Economics Association's 2012 AAEA Annual Meeting, Seattle, Washington, August 12–14, 2012 (retrieved: http://ageconsearch.umn.edu/bitstream/122476/2/Gardebroek.pdf).

15 Commodities II: Trading Agricultural Commodities

1. Adapted from and based on CBOT Agricultural Markets (2006), "CME commodity trading manual: An introduction to trading CBOT agricultural futures and options" (retrieved: http://gpvec.unl.edu/files/Futures/CME%201%20Commodity%20Trading%20Manual.pdf). All examples in this chapter are drawn from or adapted from CBOT Agricultural Markets (2006), "CME Commodity trading manual: An introduction to trading CBOT agricultural futures and options". Also the structure of the first part of this chapter is built based on the structure of the same trading manual. See further, CBOT (2006), "Agricultural markets: An introduction to trading

CBOT futures and options" (retrieved: http://www.kisfutures.com/CBOTIntroductiontoTradingCBOTAgFutures&Options.pdf). W. D. Purcell (1991), Agricultural Futures and Options: Principles and Strategies (New York: Macmillan), p. 375.

2. See for monthly updated review of prices, http://usda.mannlib.cornell.edu/MannUsda/viewDocumentInfo.do?documentID=1002. Often includes historical reviews of pricing patterns and correlations between livestock and grains, grain and fertilizers, etc.
3. CME (2013), "A trader's guide to futures" (retrieved: http://www.cmegroup.com/education/files/a-traders-guide-to-futures.pdf).
4. See further, "CME Globex Reference Guide" (2012) (retrieved: http://www.cmegroup.com/globex/files/GlobexRefGd.pdf).
5. A good introduction could be C. A. Carter (2012), "Futures and options: An introduction", Rebel text, p. 358.
6. CME (2006), "An introduction to futures and options" (retrieved: http://www.cmegroup.com/files/intro_fut_opt.pdf).
7. See further, J. B. Bittman (2008), *Trading and Hedging with Agricultural Futures and Options*, Wiley Trading (Hoboken, NJ: Wiley), p. 353. Specifically for the grain market, E. Kub (2012), "Mastering the grain markets: How profits are really made" (CreateSpace Independent Publishing Platform), p. 288. R. A. Brock (T. Lambert ed.) (2005), "Grain trading: Basics of fundamental and technical analysis", Brock Associates, p. 120. From an industry perspective, D. Morgan (2005), *Merchants of Grain: The Power and Profits of the Five Giant Companies at the Center of the World's Food Supply* (IUniverse) p. 424. B. Yoder (2004), *Mastering Futures Trading: An Advanced Course for Sophisticated Strategies*, McGraw-Hill Trader's Edge Series (New York: McGraw-Hill), p. 288. M. C. Thomsett (2013), Getting started in advanced options (Hoboken, NJ: Wiley), p. 300. CBOE, the Options Institute (W. Gaynor) (1998), "Mastering options strategies: A step-by-step guide to understanding profit & loss diagrams" (retrieved: http://www.cboe.com/learncenter/pdf/MasteringOptionsStrategies.pdf). Options University (2004), "Module 4: Advanced options strategies" (retrieved: https://optionsmastery.s3.amazonaws.com/PDFs/ModuleIV.pdf).
8. See further, CME (2007), "An overview of commodity futures for hedgers" (retrieved: http://www.kisfutures.com/CMEOverviewofCMECommodityFuturesforHedgers.pdf).
9. See further, G. Angell (1990), *Agricultural Options: Trading Puts and Calls in the New Grain and Livestock Futures Markets* (Brightwaters, NY: Windsor Books), p. 230.
10. F. Mattos, P. Garcia, and J. M. E. Pennings (2008), "Dynamic decision making in agricultural futures and options markets", Proceedings of the NCCC-134 Conference on Applied Commodity Price Analysis, Forecasting, and Market Risk Management, St. Louis, MO (retrieved: http://ageconsearch.umn.edu/bitstream/37605/2/confp09–08.pdf).
11. See also, CME (2010), "An introduction to trading dairy futures and options" (retrieved: http://www.cmegroup.com/trading/agricultural/files/AC-226_DairyProductBroch_final.pdf).
12. NYBOT/ICE (2004), "Agricultural futures and options" (retrieved: http://www.danielstrading.com/resources/education/exchange-publications/nybot/agricultural-futures-and-options.pdf).
13. CME (2012), "Options on corn futures rulebook" (retrieved: http://www.cmegroup.com/rulebook/CBOT/II/10A/10A.pdf). CME (2013), "Short-dated new crop options" (retrieved: http://www.cmegroup.com/trading/agricultural/files/short-dated-new-crop-options-contract-specs.pdf).

14. CME (2013), "Wheat futures and options" (retrieved: http://www.cmegroup.com/trading/agricultural/files/fact-card-wheat.pdf). ICE (2013), "ICE durum wheat futures and options" (retrieved: https://www.theice.com/publicdocs/futures_canada/ICE_Durum_Wheat_white_paper.pdf). ICE (2013), "ICE milling wheat futures and options" (retrieved: https://www.theice.com/publicdocs/futures_canada/ICE_Milling_Wheat_white_paper.pdf).
15. See for a detailed analysis of regional variations in wheat production practices, costs, and yields, G. Vocke and M. Ali (2013), "U.S. wheat production practices costs and yields: Variations across regions". From an investment perspective, it highlights the importance of geographical diversification and the extent to which regional and farm-to-farm differences in cropping practices, yields, and the costs of land, labor, and capital assets can affect farm profits. It considers important questions such as:
    1. How have production costs, wheat yields, and wheat enterprise profits changed in recent decades in light of new technology and production practices?
    2. Which cost items account for most of the variation in total production costs?
    3. Given current production costs, what is the required breakeven wheat price given regional variations in expected yields?
    4. What proportion of farms' wheat operations are profitable under certain production price, grain price, and yield scenarios?
16. CBOT (2004), "Grain and soybean futures and options" (retrieved: http://agmarketing.extension.psu.edu/Commodity/PDFs/grainsoybean_future.pdf). CME (2013), "Soybeans options rulebook" (retrieved: http://www.cmegroup.com/rulebook/CBOT/II/11A/11A.pdf).
17. CME (2010), "CME group livestock futures and options: Introduction to underlying market fundamentals" (retrieved: http://www.cmegroup.com/trading/agricultural/files/LivestockFundamentals.pdf).
18. King Commodities (2010), "Hedging with livestock futures" (retrieved: http://www.kingcommodities.com/res/default/hedginglivestock.pdf). CME (2009), "Self-study guide to hedging with livestock options and futures" (retrieved: http://www.cmegroup.com/trading/agricultural/files/AC-215_SelfStuy_GuideNYMEX.pdf). CME (2007), "Strategies for CME livestock futures and options" (retrieved: http://gpvec.unl.edu/files/Futures/CME%203%20Hedging%20Strategies%20for%20Livestock%20Futures%20&%20Options.pdf).
19. J. D. Sartwelle and J. Mintert (2002), "Hedging using livestock futures" (retrieved: http://ag.arizona.edu/arec/wemc/cattlemarket/Hedgingusinglivestockfuture.pdf).
20. CME (2011), "Lean hogs futures and options" (retrieved: http://www.cmegroup.com/trading/agricultural/files/Lean-Hog-Futures-Options.pdf).
21. See CME (2010), "Understanding dairy price volatility", webinar (retrieved: http://accordent.powerstream.net/008/00102/071129fdsa/msh.html), as well as E. V. Jesse and R. A. Cropp (2010), "Futures and options trading in milk and dairy trading: A guidebook for dairy producers", Working Paper. CME (n.d.), "An introduction to dairy futures and options" (retrieved: http://www.cmegroup.com/trading/agricultural/files/AC-226_DairyProductBroch_final.pdf), as well as the speculative position limits that can be taken (by CFTC regulation 150.2) (retrieved: http://www.cmegroup.com/trading/agricultural/files/Spec_Limits_and_Hedge_Fund_Exemptions_070708.pdf). See also, E. Jesse and B, Cropp (2009), "Futures and options: Trading milk and dairy products; a guidebook for dairy producers" (retrieved: http://future.aae.wisc.edu/publications/A3732.pdf).

22. NYBOT/ICE (2004), "Cotton futures and options" (retrieved: http://insigniafutures.com/Files/NYBOT_Cotton.pdf). The NYBOT was acquired in 2007 by ICE (IntercontinentalExchange) and known as ICE futures US, who also have operations in Europe as ICE Europe. The most recent contract specifics per agricultural commodity can be found at www.theice.com. ICE trades the cotton nr. 2. See for an introduction to cotton futures, B. K. Bennett (2009), "Introduction to cotton futures" (retrieved: http://www.cottoninc.com/fiber/AgriculturalDisciplines/AgriculturalEconomics/Cotton-Futures-Options/IntroductionToCottonFutures.pdf). G. J. Power and J. R. C. Robinson (2009), "Cotton futures dynamics: Structural change, index traders and the returns to storage", Paper presented at the NCCC-134 Conference on Applied Commodity Price Analysis, Forecasting and Market Risk Management, St. Louis, MO, April 20–21, 2009 (retrieved: http://ageconsearch.umn.edu/bitstream/53044/2/confp10–09.pdf).
23. NYSE Euronext Liffe (2008), "Coffee futures and options summary" (retrieved: http://www.plantureux.com/PDF/liffe_robusta.pdf). NYBOT (2004), "Coffee futures and options" (retrieved: http://www.plantureux.com/PDF/liffe_robusta.pdf), and ICE (2012), "Coffee 'C' futures" (retrieved: https://www.theice.com/publicdocs/ICE_Coffee_Brochure.pdf).
24. CME (2012), "Cleared OTC swaps on agricultural commodities" (retrieved: http://www.cmegroup.com/education/files/cleared-otc-swaps-on-commodity-indexes.pdf). These indexes have been designed to track the futures prices of products representing energy, agricultural, and metals products with a transparent, rule-based system for choosing and weighting the products. The engineering of investment-funding vehicles that track commodity futures indexes has provided investors with the means for gaining exposure to commodities that offers better potential to capture the full benefits of the asset class. Cleared OTC commodity index swaps offer OTC market participants the full benefits of central clearing, including (1) a decrease in counterparty credit risk; (2) a decrease in operational and legal risks that customers face in OTC trading; (3) customer funds and positions are held in sequestered accounts, separate from the assets of each customer's clearing firm; (4) mark-to-market margining; and (5) capital efficiencies for offsetting positions and multilateral netting.
25. CME (2009), "An introductory guide to random length lumber futures and options" (retrieved: http://www.cmegroup.com/trading/agricultural/files/AC-243_RanLenLumberBrochure.pdf).
26. NYBOT/ICE (2004), "Sugar future and options" (retrieved: http://www.gofutures.com/pdfs/sugar-futures-options.pdf). ICE handles sugar contracts nr. 11.
27. ICE (2012), "Cacao futures" (retrieved: https://www.theice.com/publicdocs/ICE_Cocoa_Brochure.pdf).
28. ICE (2012), "Frozen concentrated orange juice" (retrieved: https://www.theice.com/publicdocs/ICE_FCOJ_Brochure.pdf).
29. ICE (2012), "Canola futures" (retrieved: https://www.theice.com/publicdocs/ICE_Canola_Brochure.pdf).
30. www.cme.com or www.cbot.com and www.usda.com. See further, CME commodity products (2007), "An overview of commodity futures and options". C. Garner (2010), *A Trader's First Book on Commodities* (Upper Saddle, NJ: FT Press), p. 244.
31. CME (1998), *Commodity Trading Manual*, 9th ed. (New York: Routledge), p. 325.
32. See, for example, CME (2011), "Feeder cattle futures and options chartbook" (retrieved: http://www.cmegroup.com/trading/agricultural/files/Feeder_Cattle_Chartbook.

pdf). CME (n.d.), "25 proven strategies for trading options on CME group futures" (retrieved: http://www.cmegroup.com/education/25_proven_strategies/CME-113_21brochure_SIDE_SR.pdf).

33. See, for example, CME white paper (2012), "Structuring minimum price contracts with short-dated new crop options" (retrieved: http://www.cmegroup.com/trading/agricultural/files/structuring-minimum-price-contracts-with-short-dated-new-crop-options.pdf).
34. CME (2011), "An introduction to cattle feeding spreads ('cattle crush')" (retrieved: http://www.cmegroup.com/trading/agricultural/files/AC-378_CattleFeeding WhitePaper_r2.pdf).
35. C. Sorensen (1999), "Seasonality in agricultural commodity futures" (retrieved: http://openarchive.cbs.dk/bitstream/handle/10398/7146/soerensen_seasonality_wp9914.pdf?sequence=1). M. Richter and C. Sorensen (2003), "Stochastic volatility and seasonality in commodity. Futures and options: The case of soybeans" (retrieved: http://www.wiwi.uni-frankfurt.de/finance/common/fiwikolloq/archiv/03_WS/sorensen.pdf). K. Ovavarin and N. Meade (2010), "Mean reversion and seasonality in GARCH of agricultural commodities", International Conference on Applied Economics – ICOAE 2010 (retrieved: http://kastoria.teikoz.gr/icoae2/wordpress/wp-content/uploads/articles/2011/10/064.pdf), pp. 573–581.
36. CME (2008), "Managed futures: Portfolio diversification strategies" (retrieved: http://insigniafutures.com/Files/Portfolio_Diiversification_Opportunities.pdf).
37. ICE (2013), "ICE futures block trades FAQ" (retrieved: https://www.theice.com/publicdocs/futures_us/exchange_notices/Block_Trade_FAQ.pdf).
38. See for a recent overview of the full product category in the agricultural space, "CME Leading Products Guide" (2013), for agricultural futures (pp. 5–6) and agricultural options (pp. 7–8) (retrieved: https://www.cmegroup.com/trading/files/leading-products-guide.pdf).
39. All examples are derived from CBOT Agricultural Markets (2006), "An introduction to trading CBOT agricultural futures and options", pp. 8–10.
40. CME commodity products (2005), "Options on CME dairy products: A self-study guide"; all examples in this section are derived from this particular document. Other applications for different agricultural commodities can be found at CME (2012), "Agricultural products: Self-study guide to hedging with grain and oilseed futures and options" (retrieved: http://www.cmegroup.com/trading/agricultural/files/AC-216_HedgersGuideNewBoilerplate.pdf). CME (2009), "Commodity products: Self-study guide to hedging with livestock futures and options" (retrieved: http://www.cmegroup.com/trading/agricultural/files/AC-215_SelfStuy_Guide-NYMEX.pdf); CME (2008), "Managing price risk with grain and oilseeds futures and options" (retrieved: http://www.cmegroup.com/trading/agricultural/files/AC-211_UpdateManaginPriceRiskGrainsSP_r1.pdf).
41. See in detail, M. J. Lombardi and F. Ravazzolo (2013), "On the correlation between commodity and equity returns: Implications for portfolio allocation", BIS Working Papers 420 (retrieved: www.bis.org/publ/work420.pdf). They compared the weekly performances of the MSCI World and the S&P Goldman Sachs Commodity Index. The tipping point was September 5, 2008, the day that Lehman Brothers failed. A possible explanation with the period before is that since 2008, more financial investors have started investing in commodities, while before the market was predominantly taken by users/producers and hedgers (i.e., specialists). These specialists trade on fundamental analysis, which is commodity and position specific, like

drought, demand/supply relations, etc. These are different market movers that macroeconomic variables that move the stock market. Or put differently, the arrival of non-specialist financial investors (both in terms of volume and number), who behave differently than specialists, that is, they behave on indicators on which they also trade in the stock market, makes both asset classes now behave in sync. A potential observation with respect to this piece of research is that correlation, as always, doesn't mean causality. That however, does not change the implications for a portfolio but ignores other potential drivers as the role of the Chinese economy in commodities. The Goldman Sachs index consists of 75% energy commodities, which might compromise the view. However, the index is most often used for commodity-ETFs, etc., and thus can move markets significantly now that 50% or more of volume in the market comes from ETFs. Potentially, the correlation may decline as the aftermath of the financial crisis in the market eases and financial investors withdraw from the commodity markets. Up until Q3 of 2013 (close of the manuscript), there were no indications in the agricultural commodities markets of such a thing happening, and the author's expectation is that it is unlikely to happen in the years to come – to a large degree since the fundamental for agri-commodities are positive across the board (but with the usual variations across agri-commodities and time periods).

16 Commodities III: Speculation on (Agricultural) Commodity Derivatives and Financialization of Commodity-Price Formation

1. B. Lallemand (2012), "Investing not betting: Making financial markets serve society", Finance Watch, in particular chapter 4.
2. For a recent bibliography, see M. Henn from WEED (2013 updated), "Evidence on the negative impact of commodity speculation by academics, analysts and public institutions" (retrieved: http://www2.weed-online.org/uploads/evidence_on_impact_of_commodity_speculation.pdf). For a discussion on the conclusions from some of the main papers, see SOMO, R. van Tilburg, and M. Vander Stichele (2011), "Feeding the financial hype: how excessive financial investments impact agricultural derivatives markets" (Retrieved: http://somo.nl/publications-en/Publication_3726), pp. 28–34. M. W. Masters (2010), "Ending excessive speculation in commodity markets: legislative options", Testimony before the Committee on Homeland Security and Governmental Affairs, United States. See further, L. Gutierrez (2012), "Speculative bubbles in agricultural commodity markets", *European Review of Agricultural Economics*, pp. 1–22. K. Tang and W. Xiong (2012), "Index investment and the financialization of commodities", Financial Analyst Journal 68(6): 54–74. Y. Tse and M. Williams (2013), "Does index speculation impact commodity prices?", *Financial Review* 48(3): 365–383. M. Nissanke (2012), "Commodity market linkages in the global financial crisis: Excess volatility and development impacts", *Journal of Development Studies* 48(6): 732–750. P. Adämmer, M. T. Bohl, and P. M. Stephan (2011), "Speculative bubbles in agricultural prices" mimeo, Westphalian Wilhelminian University of Münster, Münster. A. Silvennoinen and S. Thorp (2013), "Financialization, crisis and commodity correlation dynamics", *Journal of International Financial Markets, Institutions and Money* 24: 42–65. L. Baldi, M. Peri, and D. Vandone (2011), "Price discovery in agricultural commodities: The shifting relationship between spot and futures prices". S. Basak and A. Pavlova (2013), "A model of financialization of commodities" London Business School; Centre for Economic Policy Research (CEPR) Research Paper. J. A.

Frankel and A. K. Rose (2010), "Determinants of agricultural and mineral commodity prices" Harvard University & University of California, Berkeley Working paper. B. J. Henderson, N. D. Pearson, and L. Wang (2012), "New evidence on the financialization of commodity markets" University of Illinois at Urbana-Champaign, University of Massachusetts, and the FDIC's 22nd Annual Derivatives and Risk Management Conference Working Paper. Y. Tse (2012), "The relationship among agricultural futures, ETFs, and the US stock market" 20(2), Article 2.

3. There is one exception in the financial asset space where the impact is similar to the commodity markets' impact. That is the impact that CDSs (credit default swaps) have on the cost of funding of a sovereign. Credit default swaps are insurance-like derivatives that compensate the holder of, for example, treasury bonds in case the sovereign defaults on the bond. Each sovereign has a credit rating, reflecting the intrinsic creditworthiness of the paper it issues. So, in basic terms there is no real problem. Everybody who owns an asset can get insurance in case the asset gets destroyed or in case of financial assets becomes nonperforming. The cost of that insurance will be based on the intrinsic risk there is in the asset (creditworthiness, are you a solid driver, did you build your home in an earthquake- or flooding-sensitive area, etc.?). Now, in the insurance world, in order to get insurance, you need to have what they call "an insurable interest," or put differently, you need to have an interest for the object not to be destroyed, burned down, or become nonperforming. That makes sense: taking fire insurance on your neighbor's home with no equity in that house yourself, the unthinkable might happen sooner or later. In the CDS market it is possible to buy CDS products even if you don't own the asset (T-bond) against which default it insures. The more buyers of that CDS, the higher the price of the insurance and the more the market will perceive the asset to be risky (although that intrinsically is not the case). We call that transaction a naked CDS as there is no underlying insurable interest in the hands of the CDS buyer. But once perception is created about a certain government being risky, reality will follow, as over time that government will have to pay more interest on its newly issued bonds, pushing up its interest burden, deteriorating public finance, and ultimately impacting (lowering) the credit rating of the sovereign. It will become a vicious circle. The only objective was to speculate against a certain government, and having the tools to create perception will help materialize that reality.
4. Testimony of Michael W. Masters, managing member, Portfolio Manager Masters Capital Management LLC before the Commodities Futures Trading Commission (2010) (retrieved: http://www.nefiactioncenter.com/PDF/masters_testimony_25mar2010.pdf).
5. The most recent data can be found here: http://www.usda.gov/oce/commodity/wasde/latest.pdf.
6. A. Dwyer, J. Holloway, and M. Wright (2012), "Commodity market financialisation: A closer look at the evidence", Reserve Bank of Australia, March quarter, 2012, 65–78.
7. For further details and analysis, see B. Scott (2011) "Barclays PLC and agricultural commodity derivatives" (retrieved: http://www.wdm.org.uk/sites/default/files/Brett%20Scott%20%20BarCap%20Report%20March%202011.pdf); T. Jones (2010), "The great hunger lottery: How banking speculation causes food crisis", WDM (retrieved: http://www.wdm.org.uk/sites/default/files/hunger%20lottery%20report_6.10.pdf); M. Worthy (2011), "Broken markets: How financial market regulation can help prevent another global food crisis", WDM (http://www.wdm.org.uk/sites/default/files/Broken-markets.pdf).

8. For example, IFF Commodities Task Force submission to the G20 (2011), "Financial investment in commodities markets: potential impact on commodity prices & volatility", September.
9. See, J. M. Riley (2013), "Extension's role in commodity marketing education: Past, present, and future", *Journal of Agricultural and Applied Economics* 45(3): 537–555 (retrieved: http://ageconsearch.umn.edu/bitstream/155429/2/jaae453ip15.pdf).
10. J. S. Netz (1995), "The effect of futures markets and corners on storage and spot price variability", American Journal of Agricultural Economics 77(1): 182–193; M. Powers (1970), "Does futures trading reduce price fluctuations in the cash markets?", American Economic Review 60(3): 460–464; G. Taylor and R. Leuthold (1974), "The influence of futures trading on cash cattle price variations", Food Research Institute Studies 13(1): 29–35; S. J. Turnovsky (1979), "Futures markets, private storage, and price stabilization", Journal of Public Economics 12: 301–327; B. W. Brorsen, C. M. Oellermann, and P. L. Farris (1989), "The live cattle futures market and daily cash price movement", Journal of Futures Market 9(4): 273–282; C. L. Gilbert (1989), "Futures trading, storage, and price stabilization", Review of Futures Markets 8(2): 152–176. Discussion, by Jean-Marie Viaene, pp. 177–179.
11. R. J. Shiller (1990), "Speculative prices and popular models", *Journal of Economic Perspectives* 4(2): 55–66; T. Gehrig and L. Menkhoff (2004), "The use of flow analysis in foreign exchange: Exploratory evidence", *Journal of International Money and Finance* 23(4): 573–594.
12. B. R. Holt and S. H. Irwin (2000), "The effect of futures trading by large hedge funds and CTAS on market volatility", No. 18935, from NCR-134 Conference on Applied Commodity Price Analysis, Forecasting, and Market Risk Management, Chicago, IL, April 17–18, 2000.
13. P. K. Clark (1973), "A subordinated stochastic process model with finite variance for speculative prices", *Econometrica* 41(1): 135–355.
14. B. J. De Long, A. Shleifer, L. H. Summers, and R. J. Waldmann (1990), "Noise trade risk in financial markets", *Journal of Political Economy* 98(4): 703–738.
15. IMF (2009), "What explains the rise in food price volatility?", *World Economic Outlook*, October 2009, chapter 1, pp. 61–63, Box 1.7.
16 Ibid., note 42.
17. More recently, part of the empirical literature tackled the issue from a different perspective, focusing on a precise aspect of financialization. For instance, Phillips and Yu (P. C. B. Philips and J. Yu [2010], "Dating the timeline of financial bubbles during the subprime crisis", Cowles Foundation DP No. 1770, Yale University) examined the migration of price bubbles across equity, bond, currency, and commodity markets and tend to confirm the presence of a price bubble in crude oil in mid-2008 (while no bubbles are detected for agricultural commodities). Buyuksahin and Robe (B. Buyuksahin and M. Robe [2010], "Speculators, commodities and cross-market linkages", technical report, Commodity Futures Trading Commission), using CFTC proprietary data, investigated the determinants of the conditional correlation between commodities and equity prices and conclude that commodity-index traders (CIT hereafter) did not influence this measure of co-movement across markets at all, while hedge funds, operating simultaneously on equity and commodity-derivative markets, have a sizeable and persistent impact. Nonetheless they fail to explain why the conditional correlation goes up in 2009, when hedge-fund activity slowly declined, leaving the task of finding an explanation for future research. Mou (Y. Mou [2010], "Limits to arbitrage and commodity index investments: Front-running the Goldman roll", technical report, Columbia Business School) instead finds that

CIT rolling activity – commodity-index traders often roll contracts forward a few days before the expiration, on to the next nearby contract – has a significant and sizeable effect on rolling yields. Finally on the same point, Singleton (K. Singleton [2011], "Investor flows and the 2008 boom/bust in oil prices", mimeo.) shows that quarterly changes in the spread position held by money managers positively affects the average price return.

18. N. Boyd, B. Buyuksahin, J. Harris, and M. Haigh (2010), "The prevalence, sources and effects of herding", technical report, US Commodity Futures Trading Commission; N. Boyd, B. Buyuksahin, J. Harris, and M. Haigh (2009), "The impact of hedging on futures prices", technical report, US Commodity Futures Trading Commission; C. Brunetti and B. Buyuksahin (2009), "Is speculation destabilizing?" technical report, US Commodity Futures Trading Commission; B. Buyuksahin and J. Harris (2009) "The role of speculators in the crude oil futures market", technical report, US Commodity Futures Trading Commission; B. Buyuksahin and M. Robe (2009), "Commodity traders' positions and energy prices: Evidence from the recent boom-bust cycle", technical report, US Commodity Futures Trading Commission; B. Buyuksahin and M. Robe (2010), "Speculators, commodities and cross-market linkages", technical report, US Commodity Futures Trading Commission.
19. See for references, FEFAC (2011), "G-20 ministers of agriculture meeting on price volatility of agricultural commodities/viewpoint of the European feed industry", June 21; A. Heap and A. Tonks (2008), "Citi commodities strategy: Great bulks of fire IV", April 7, 2008, p. 1; N. M. Arjun, B. Singer, et al. (2008), "$100 oil reality, part 2: Has the super-spike end game begun?", Goldman Sachs Global Investment Research, May 5, p. 12; A. Awad, W. Canaday, et al. (2008), "Financial investors fueling commodities boom", Greenwich Associates, May 2008, p. 1.
20. The beta of an asset is a measure of the relative volatility of that asset and the volatility of the market. A beta of zero means that the asset's return is not sensitive to the market's returns.
21. K. Tang and W. Xiong (2010), "Index investing and the financialization of commodities", technical report, Princeton University.
22. For a complete description of this phenomenon, the reader can refer to M. W. Masters (2008),"Testimony before the Committee on Homeland Security and Governmental Affairs, United States Senate, June 24, 2008".
23. They provide liquidity by buying and selling futures contracts. That means nonfinancial operators in this market can find counterparties relatively cheaply in order to enter into contracts. If there were too few financial participants in the market, liquidity would be too low, with the possibility of large seasonal price movements. An excessively large number of financial participants could also result in wide price fluctuations. This is the case when these parties assume that past developments carry information on future price movements, giving rise to trend chasing. This will result in buying after prices rise and selling after prices fall.
24. See in detail, M. Lagi, K. Z. Bertrand, and Y. Bar-Yam (2011), "The food crises and political instability in North Africa and the Middle East" (retrieved: http://arxiv.org/pdf/1108.2455.pdf); M. Lagi, Y. Bar-Yam, K. Z. Bertrand (2011), "The food crises: A quantitative model of food prices including speculators and ethanol conversion" (retrieved: http://arxiv.org/pdf/1108.2455.pdf); M. Lagi, Y. Bar-Yam, and K. Z. Bertrand (2012), "The food crises: Predictive validation of a quantitative model of food prices including speculators and ethanol conversion" (retrieved: http://necsi.edu/research/social/foodprices/update/food_prices_update.pdf).

25. D. Bicchetti and N. Maystre (2012), "The synchronized and long-lasting structural change on commodity markets: Evidence from high frequency data", UNCTAD Discussion Paper 208, UNCTAD, Geneva (retrieved: http://unctad.org/en/PublicationsLibrary/osgdp2012d2_En.pdf) or D. Bicchetti and N. Maystre (2013), "The synchronized and long-lasting structural change on commodity markets: Evidence from high frequency data", *Algorithmic Finance* 2: 3–4, 233–239 (retrieved: DOI 10.3233/AF-13028).
26. See also, International Food Policy Research Institute (2008), "High food prices: The what, who, and how of proposed policy actions" (retrieved: http://www.ifpri.org/sites/default/files/publications/foodpricespolicyaction.pdf).
27. Or to be precise, causality refers to the fact that something causes an effect, or the action of causing an effect; correlation is the relationship in which two or more things are mutual or complementary, or one thing is caused by another.
28. IFF (2011), "Financial investment in commodity markets: Potential impact on commodity prices and volatility" (retrieved: http://www.iif.com/download).
29. See further, "Price volatility in food and agricultural markets: Policy responses" (2011, June 2), policy report including contributions by FAO (Food and Agricultural Organisation of the UN), IFAD (International Fund for Agricultural Development), IMF, OECD, UNCTAD (United Nations Conference on Trade and Development), WFP (World Food Program), the World Bank, the WTO (World Trade Organization), FPRI (International Food Policy Research Institute), and the UN HLTF (UN High Level Task Force on Food Security).
30. UNCTAD (2012), "Don't blame the physical markets: Financialization is the root cause of oil and commodity price volatility", Policy Brief 25 (retrieved: http://unctad.org/en/PublicationsLibrary/presspb2012d1_En.pdf).
31. Institute of International Finance (2011), "IIF Commodities Task Force submission to the G20: Financial investment in commodity markets: Potential impact on commodity prices and volatility", September 2011.
32. A. Silvennoinen and S. Thorp (2010), "Financialization, crisis and commodity correlation dynamics", Research Paper 267, Quantitative Finance Research Centre, University of Technology Sydney.
33. UNCTAD (D. Bicchetti and N. Maystre) (2012), "The synchronized and long-lasting structural change on commodity markets: Evidence from high frequency data", Discussion Paper 208 (retrieved: http://unctad.org/en/PublicationsLibrary/osgdp2012d2_En.pdf).
34. Ibid., UNCTAD (2012), pp. 16–17. See further, UNCTAD (J. Mayer) (2009), "The growing interdependence between financial and commodity markets", Discussion Paper 195 (retrieved: http://unctad.org/en/Docs/osgdp20093_En.pdf). UNCTAD (P. Fajarnes) (2011), "An overview of major sources of data and analysis relating to physical fundamentals in international commodity markets", Discussion Paper 202 (retrieved: http://unctad.org/en/Docs/osgdp20112_En.pdf). UNCTAD (N. E. Maurice and J. Davis) (2011), "Unraveling the underlying causes of price volatility in world coffee and cocoa commodity markets", UNCTAD Special Unit on Commodities working paper series on commodities and development, Discussion Paper 1 (retrieved: http://unctad.org/en/PublicationsLibrary/suc-miscDP01_En.pdf).
35. UNCTAD (2009), "The global economic crisis: Systemic failures and multilateral remedies", United Nations publication, Sales No. E.09.II.D.4, New York and Geneva. UNCTAD (2011a), "Price formation in financialized commodity markets: The role of information", United Nations publication, UNCTAD/GDS/2011/1, New York and

Geneva. UNCTAD (2011b), "Trade and development report, 2011: Post-crisis policy challenges in the world economy", United Nations publication, Sales No. E.11.II.D.3, New York and Geneva. UNCTAD (2012), "Trade and development report 2012: Policies for inclusive and balanced growth", United Nations publications, Sales No. E.12.II.D.6, New York and Geneva.

36. B. Algieri (2012), "Price volatility, speculation and excessive speculation in commodity markets: Sheep or shepherd behavior?" ZEF-Discussion Papers on Development Policy 166, Center for Development Research, Bonn, May 2012, p. 37.
37. See further, C. Arnade and G. Vocke (2013), "Investigating the divergence in wheat prices", ERS USDA, selected poster prepared for presentation at the Agricultural & Applied Economics Association's 2013 AAEA & CAES Joint Annual Meeting, Washington, DC, August 4–6, 2013 (retrieved: http://ageconsearch.umn.edu/bitstream/149694/2/ArnadeVockeAAEAR.pdf). This study investigates whether importers of US wheat form an integrated market or a series of segmented markets. Two market-integration tests are applied: one based on equilibrium price relationships and one based on disequilibrium price relationships. With the exception of a few importers, both tests tend to agree. This study also seeks to explain the reasons for the wide divergence in prices paid by importers of US wheat. They find that factors such as the class of wheat that is purchased and the size of annual purchases influence pricing. They also find that there continues to be large persistent difference in the prices paid by importers.
38. See particularly on price discovery in commodity markets, UNCTAD (2011), "Price formation in financialized commodity markets: The role of information", United Nations, which includes an extensive further literature list.
39. I. Piot-Lepetit and R. M'Barek, eds. (2011), "Methods to analyse agricultural commodity price volatility", Springer Science+Business Media LLC (retrieved: DOI 10.1007/978-1-4419-7634-5_2).
40. To the extent not mentioned yet, B. Buyuksahin and M. A. Robe (2011), "Speculators, commodities and cross-market linkages", mimeo. B. Fattouh, L. Kilian, and L. Mahadeva (2012), "The role of speculation in oil markets: What have we learned so far?", CEPR Discussion Paper 8916, available at VoxEU.org. G. Gorton and K. G. Rouwenhorst (2006), "Facts and fantasies about commodity futures", *Financial Analysts Journal* 62(2): 47–68. D. Sheppard (2011), "NYMEX oil trade 45 percent computer-driven", Reuters, March 3. H. R. Stoll and R. E. Whaley (2010), "Commodity index investing and commodity futures prices", *Journal of Applied Finance* 20(1): 7–46. H. R. Stoll and R. E. Whaley (2011), "Commodity index investing: Speculation or diversification?", *Journal of Alternative Investments* 14(1): 50–60. K. Tang and W. Xiong (2011), "Index investment and financialization of commodities", Princeton University; A. Borin and V. Di Nino (2012), "The role of financial investments in agricultural commodity derivatives markets", Working Paper 849, Banca D'Italia, January 2012.
41. See for that in my earlier writings, L. Nijs (2011), *Shaping Tomorrow's Marketplace: Investment Philosophies for Emerging Markets and a Semi-Globalized World* (Euromoney Books, 2011), pp. 256.
42. For a quick refresher on Granger causality tests, J. Lung-Lin (2008), "Notes on testing causality", Institute of Economics, Academia Sinica Department of Economics, National Chengchi University, 2008 (retrieved: http://faculty.ndhu.edu.tw/~jlin/files/causality.pdf). The Granger causality test is essentially a statistical hypothesis test for determining whether one time series is useful in forecasting another. That

in contrast to regressions that merely reflect correlation. Or a little bit more complicated: the Granger causality test, which is based on Wald statistics, investigates whether the estimated coefficients on the lagged values of a particular regressor are jointly statistically different from zero. The null hypothesis is that the considered coefficients are all equal to zero.

43. C. L. Gilbert (2010a), "Commodity speculation and commodity investments", in *Commodity Market Review*, pp. 25–46, FAO, Rome; C. L. Gilbert (2010b), "Speculative influences on commodity futures prices 2006–2008", UNCTAD DP No. 197; S. H. Irwin and D. Sanders (2010), "The impact of index swap funds on commodity futures markets", technical report, OECD Food, Agriculture and Fisheries Working Paper 27; A. Borin and V. Di Nino (2012), "The role of financial investments in agricultural commodity derivatives markets", Working Paper 849, Banca D'Italia, January 2012.
44. IMF (2006), "Has speculation contributed to higher commodity prices?", *World Economic Outlook*, September 2006, chapter 5, pp. 153–156, Box 5.1.
45. S. Basak and A. Pavlova (2011), "Asset prices and institutional investors", AFA, Denver Meetings Paper.
46. See, M. Tothova (2011), "Main challenges of price volatility in agricultural commodity markets", in I. Piot-Lepetit and R. M'Barek, eds, *Methods to Analyse Agricultural Commodity Price Volatility*, Springer Science+Business Media (retrieved: DOI 10.1007/978-1-4419-7634-5_2, C _). See also S. Grosche (2011), "Methods to analyze agricultural commodity market volatility" University Bonn Working paper. S.-H. Poon and C. W. J. Granger (2003), "Forecasting volatility in financial markets: A review", *Journal of Economic Literature* 41: 478–539. Although volatility has always been a feature of agricultural commodity markets, the evidence suggests that volatility has increased at least in some commodity markets. There seems to be an overlap between periods of high prices and increased volatility. Volatility peaks also seem to coexist with decreased stocks. Persistence of volatility points to uncertainty in developments of market fundamentals coupled with structural and monetary policy. Higher price volatility means higher costs of managing risks (such as higher margins on futures contracts and higher premiums for crop revenue insurance). However, with increasing biofuels production, a tightened interdependence between crude oil and commodity markets can be expected which could result in increased transmission of crude-oil price volatility into agricultural commodity markets. It is likely that higher costs of risk mitigation would eventually translate into higher consumer prices. Increased volatility highlights the presence of greater uncertainty on the market. Two broad sets of policies could be employed: (1) those that target volatility itself, such as price and supply controls, and (2) those that deal with the effects of price volatility while letting markets work, such as risk-management instruments, safety nets, and so forth. Policies based on price and supply controls do not appear to have an impressive precedent and reduce market signals. Policies mitigating volatility or the effects of volatility should aim to address uncertainties and focus on risk management while keeping markets working. It remains impossible to capture future price variability. However, elements from the past that resulted in past and present variability remain active.
47. See also Borin and Nino ibid., and the long literature list included on pp. 27–30 and 43–48.
48. See M. Almánzar, M. Torero, and K. von Grebmer (2013), "Futures commodities prices and media coverage", ZEF. Discussion Papers on Development Policy 178,

Center for Development Research, Bonn, May 2013, pp. 34 (retrieved: http://www.zef.de/fileadmin/webfiles/downloads/zef_dp/zef_dp_178.pdf).

49. M. T. Bohl and P. M. Stephan (2013), "Does futures speculation destabilize spot prices? New evidence for commodity markets", *Journal of Agricultural and Applied Economics* 45(4): 595–616 (retrieved: http://ageconsearch.umn.edu/bitstream/157413/2/jaae608.pdf).
50. See X. L. Etienne, S. H. Irwin, and P. Garcia (2013), "Bubbles in food commodity markets: Four decades of evidence", Paper prepared for presentation at the Conference on Understanding International Commodity Price Fluctuations, International Monetary Fund, Washington, DC, March 20–21, 2013
51. DG Markt (2010), "Towards more responsibility and competitiveness in the European financial sector" (retrieved: http://ec.europa.eu/internal_market/finances/docs/leaflet/financial_services_En.pdf).
52. Retrieved: http://ec.europa.eu/agriculture/analysis/tradepol/worldmarkets/high_prices_En.pdf and http://ec.europa.eu/agriculture/analysis/tradepol/worldmarkets/outlook/2011_2020_En.pdf, for the 2011–2020 agricultural outlook of the EU.
53. Document COM (2010), "484 final" (retrieved: http://ec.europa.eu/internal_market/financial-markets/docs/derivatives/20100915_proposal_En.pdf).
54. Retrieved: http://eur-lex.europa.eu/LexUriServ/LexUriServ.do?uri=OJ:L:2012:201:0001:0059:EN:PDF.
55. They were very much guided by the IOSCO report, "Principles for the regulation and supervision of commodity derivatives markets" (retrieved: http://www.iosco.org/library/pubdocs/pdf/IOSCOPD358.pdf).
56. See for more details, "Draft technical standards under the Regulation (EU) No 648/2012 of the European Parliament and of the Council of 4 July 2012 on OTC derivatives, CCPs and trade repositories" (retrieved: http://www.esma.europa.eu/system/files/2012-600_0.pdf). In 2013 that was followed by two Q7A reports: "ESMA 2013-685, questions and answers: Implementation of the Regulation (EU) No 648/2012 on OTC derivatives, central counterparties and trade repositories (EMIR) (retrieved: http://www.esma.europa.eu/system/files/2013-685_qa_iI_on_Emir_implementation_final_for_publication_20130604.pdf) and "ESMA 2013-1080, implementation of the Regulation (EU) No 648/2012 on OTC derivatives, central counterparties and trade repositories (EMIR) (retrieved: http://www.esma.europa.eu/system/files/2013-1080_qa_iiI_on_Emir_implementation.pdf).
57. A thematic overview of the swap/derivatives chapter of the Dodd-Frank Act can be found here: http://us.practicallaw.com/3-502-8950.
58. Final rules can be found here: http://www.cftc.gov/ucm/groups/public/@lrfederalregister/documents/file/2011-20337a.pdf.
59. Consider a bread baker in Kansas City. He buys thousands of pounds of flour each year. Because of this year's droughts, though, he's worried that America's wheat supply will shrink, driving up flour prices in the next 12 months. So, to hedge against a price spike, he buys $10,000 in over-the-counter flour swaps today to guarantee that next year he'll be able to purchase up to 10,000 pounds of flour for $1 per pound. In 12 months, if the price of flour is dramatically higher, he'll have saved himself a huge amount of money. The price also could be lower, of course, but that's the risk he runs to guarantee he won't go bankrupt. Nevertheless, this new round of Dodd-Frank rules would impose a slew of new requirements on these instruments. Buyers and sellers would have to trade them on approved exchanges, and they would need to have a certain level of capital on hand to trade. Reporting rules regarding profits

and losses would be ratcheted up. There would also be mandatory "clearing," meaning swap trades would be required to run through certified middlemen. (Retrieved: http://www.agweek.com/event/article/id/20275.)

60. See further, UNCTAD (2011), *Price Formation in Financialized Commodity Markets: The Role of Information*, in particular chapter 4 (retrieved: http://unctad.org/en/Docs/gds20111_En.pdf).
61. The growth of domestic demand in emerging economies has been an important driver of growth for commodities markets. Cross-border trade liberalization has increased the effect of competition on commodities production costs and so made "traditional" subsidy programs ineffective and/or too costly. New developments on the supply side, such as new unconventional sources of natural gas or the new co-products of corn processing (e.g., biofuels), have also been stimulating cross-border trade in new markets. Cross-border competition has come with the price of higher short-term volatility, though, which is coupled with the effects of government subsidy programs that have supported artificial prices in several commodities and have increased incentives to invest in new more efficient technologies to reduce energy consumption in metal production or harvested areas for crops, for example. Growing links between commodities markets and international trade have intensified the effects of government actions such as export bans. Most notably, direct market interventions in an open-market model with international trade are unable to create incentives to tackle underlying problems of market structure. When the fiscal capacity of a country is reduced, the market has to face sudden adjustments with highly volatile patterns. For instance, in agricultural and soft-commodities markets, where the opportunity costs of land use are high (e.g., US wheat farms) or too low (e.g., sugar plantations in Brazil), public investments in new technologies for innovative applications and infrastructures, respectively, might be a preferable alternative to subsidies. They might favor more efficient allocation of the land if the market itself is unable to rebalance due to such transaction costs.
62. J. Frankel (2006), "The effect of monetary policy on real commodity prices", in John Y. Campbell, ed., *Asset Prices and Monetary Policy* (Chicago: University of Chicago Press).
63. G. Gorton and K. G. Rouwenhorst (2006), "Facts and fantasies about commodity futures", Working Paper 10595, National Bureau of Economic Research (NBER), March.
64. See the subchapter in the chapter on trading agricultural commodities for recent research evidencing that commodity futures are no longer a hedge for a portfolio (since the 2008 crisis).
65. M. Radetzki (2006), "The anatomy of three commodity booms", *Resources Policy* 31(1): 56–64.
66. A. Heap (2005), "China: The engine of a commodities super cycle", Citygroup Global Markets Inc., Smith Barney.
67. UNCTAD (2005), "Trade and development report, 2005, chapter II: Income growth and shifting trade patterns in Asia", United Nations publication, New York and Geneva.
68. Commodity investment can also take the form of holding physical stocks, but this is generally considered profitable only for precious metals, or buying shares in enterprises that produce commodities. However, the correlation between a firm's share price and the price of the underlying commodity may be low, inter alia be-

cause of the additional layer of management and company risk that may swamp the underlying commodity risk.

69. In the S&P GSCI, weights are based on five-year averages of relative world production quantities; energy products usually account for about two-thirds of the total index. In the DJ-UBSCI, weights are also based on five-year averages but rely primarily on the relative amount of trading activity of a particular commodity; weights are limited to 15% for individual commodities and to one-third for entire sectors in order to allow for a greater degree of diversification across commodities.
70. B. Scherer and L. He (2008), "The diversification benefits of commodity futures indexes: A mean variance spanning test", in F. J. Fabozzi, R. Füss, and D. G. Kaiser, eds, *The Handbook of Commodity Investing* (Hoboken, NJ: Wiley), 241–265.
71. A long position is a market position that obligates the holder to take delivery (i.e., to buy a commodity), in contrast with a short position, which obligates the holder to make delivery (i.e., to sell a commodity). The aggregate of all long open positions is equal to the aggregate of all short open positions. For individual traders, net long positions are total long positions minus total short positions.
72. *Backwardation*: Market situation where futures prices are progressively lower with rising maturities. (In some cases the term is used to describe a situation where the futures price is below the expected future spot price.) *Contango*: Market situation where futures prices are progressively higher with rising maturities. (In some cases the term is used to describe a situation where the futures price exceeds the expected future spot price.)
73. Notional amount refers to the value of the underlying commodity. However, traders in derivatives markets do not own or purchase the underlying commodity; hence notional value is merely a reference point based on underlying prices. The limited transparency of data on OTC markets is underlined by the fact that the category "other commodities" accounted for roughly 40% of the total OTC exposure in the late 1990s but now constitutes 80% to 90%.
74. More precisely, among the types of firms engaged in business activities that can be hedged and therefore classified as "commercial" by the CFTC are merchants, manufacturers, producers, and commodity swaps and derivative dealers. The CFTC classifies as "noncommercial" all other traders, such as hedge funds, floor brokers and traders, and nonreporting traders (i.e., those traders whose positions are below the reporting thresholds set by an exchange).
75. These 12 commodities are feeder cattle, live cattle, cocoa, coffee, cotton, lean hogs, maize, soybeans, soybean oil, sugar, Chicago wheat, and Kansas wheat. The CFTC explains that these index-trader data should ultimately be considered as estimates because, for example, "some traders assigned to the Index Trader category are engaged in other futures activity that could not be disaggregated... Likewise, the Index Trader category will not include some traders who are engaged in index trading, but for whom it does not represent a substantial part of their overall trading activity" (CFTC, Commitments of Traders, Explanatory Notes, retrieved: http://www.cftc.gov/MarketReports/CommitmentsofTraders/ExplanatoryNotes/index.htm).
76. S. H. Irwin and D. R. Sanders (2010), "The impact of index and swap funds on commodity futures markets: Preliminary results", Working Paper 27, OECD Food, Agriculture and Fisheries, Paris.
77. Source: Ibid., note 76.
78. Asymmetric exposure to leverage is one factor that may create greater risk. For example, when highly leveraged financial investors with long positions face margin

calls because they are subject to adverse price movements, they may not be able to pay this additional margin unless they liquidate their position. At the same time, commercial participants with short positions may prefer holding their positions until expiry and accept physical delivery. In this situation, a sudden and rapid selling pressure ignited by financial investors will not be contained by commercial participants. As a result, prices will move rapidly and excessively. Leverage may be a particularly important issue on OTC markets because these markets are characterized by a high degree of concentration and much higher leverage ratios than usually observed on futures exchanges. According to the International Swaps and Derivatives Association (2010), only about 60% of commodity derivatives trades are collateralized, compared with over 90% of credit derivatives trades. The low level of collateralization of the former implies higher leverage ratios. Although the relationship between OTC markets, futures exchanges, and spot markets is not entirely clear, the high concentration and high leverage ratios in OTC markets pose systemic risks to commodity markets, and to financial stability more generally.

79. See D. Valiante (CEPS Task Force reports) (2013), "Price formation in commodity market: Financialization and beyond" (retrieved: http://www.ceps.eu/ceps/dld/8241/pdf).
80. Ibid.
81. Source: Ibid.
82. For a heat map per (agricultural) commodity, see ibid., Valiante, pp. 8–9. See also for a detailed analysis per agricultural commodity:
    - Wheat market: pp. 169–190.
    - Corn market: pp. 191–213.
    - Soybean market: pp. 214–229.
    - Sugar market: pp. 230–246.
    - Cacao markets: pp. 246–261.
    - Coffee market: pp. 262–280.
83. Ibid., Valiante.
84. See extensively on the topic ibid., note 1, in particular chapter 4, subchapter on herding behavior.
85. Uncertainty in decision making may be a defining characteristic of commodity markets. This is because (1) medium- and longer-term commodity supply-and-demand conditions are subject to considerable uncertainty, for example because of unknown depletion rates of nonrenewable resources and unknown effects of climate change on agricultural production; and (2) inventory data, which provide valuable signals for short-term price expectations, suffer from significant measurement errors (G. Gorton, F. Hayashi, and K. G. Rouwenhorst [2007], "The fundamentals of commodity futures returns", Working Paper 13249, National Bureau of Economic Research [NBER], July; M. S. Khan [2009], "The 2008 oil price 'bubble'", Policy Brief 09/19, Peterson Institute for International Economics, Washington DC, August.); and (3) data on current global commodity supply-and-demand conditions are published with long time lags and are frequently revised. Therefore, even well-informed traders must formulate price expectations on the basis of partial and uncertain data.
86. Source: S. Bikhchandani and S. Sharma (2001), "Herd behavior in financial markets", *IMF Staff Papers* 47(3): 279–310, and A. Shleifer and L. H. Summers (1990), "The noise trader approach to finance", *Journal of Economic Perspectives* 4(2): 19–33.
87. C. L. Gilbert (2010), "Commodity speculation and commodity investment", *FAO Commodity Market Review*, 2009–2010, 26–46.

88. K. Tang and W. Xiong (2010), "Index investment and financialization of commodities", Princeton University, Working Paper 16385, National Bureau of Economic Research, Cambridge, MA, September.
89. C. L. Gilbert (2010), "Speculative influences on commodity futures prices 2006–2008", UNCTAD Discussion Paper 197, March.
90. Ibid.
91. See further, chapter 4 ibid., note 1, for a more in-depth review of the research conducted and their findings.

17 The Global Water Challenge

1. See B. Pearce (2007), *When the River Runs Dry. Water: The Defining Crisis of the 21st Century* (Boston: Beacon Press), p. 336. S. Solomon (2011), *Water: The Epic Struggle for Wealth, Power, and Civilization* (New York: Harper Perennial), p. 624. C. Fishman (2012), *The Big Thirst: The Secret Life and Turbulent Future of Water* (New York: Free Press), p. 402. A. Prudhomme (2011) *The Fate of Fresh Water in the 21st Century* (New York: Scribner), p. 448. M. Black, J. King, and C. Lacey (2009), *The Atlas of Water: Mapping the World's Most Critical Resource*, 2nd ed. (Berkeley: University of California Press), p. 129. M. De Villier (2001), *Water: The Fate of Our Most Precious Resource* (Boston: Mariner Books), p. 368.
2. OECD (2012), *OECD: Environmental Outlook to 2050: The Consequences of Inaction*, rev. ed., OECD Publishing, chapter 5, "Water". For a summary see http://www.oecd.org/env/indicators-modelling-outlooks/49910023.pdf.
3. See further, OECD (2012), "OECD environmental outlook to 2050: The consequences of inaction" (retrieved: http://www.keepeek.com/Digital-Asset-Management/oecd/environment/oecd-environmental-outlook-to-2050_9789264122246-en#page4); OECD (2013), *Water Scarcity for Better Lives*, OECD Publishing, p. 171; OECD (2013), *Water and Climate Change Adaptation: Policies to Navigate Uncharted Water*, OECD Studies on Water, OECD Publishing (retrieved: http://www.keepeek.com/Digital-Asset-Management/oecd/environment/water-and-climate-change-adaptation_9789264200449-en#page1; OECD (2012), *Meeting the Water Challenge*, OECD Publishing, p. 172; OECD (2012), *Water Quality and Agriculture: Meeting the Policy Challenge*, OECD Publishing, p. 156; OECD (2011), *Water Governance in OECD Countries: A Multi-level Approach*, OECD Publishing, p. 244.
4. To illustrate, according to the World Resources Institute (WRI), Germany has renewable freshwater resources totaling close to 1,900 m^3 per person per year. In Egypt, the reading is slightly less than 800 m^3, and in Canada over 91,000 m^3 per person per year; the world average totals roughly 8,550 m^3, which is why, on paper, there is sufficient water on Earth even if the world population does continue to increase. In most countries, the available water resources are not in full use; Germany's water-industry association, the Bundesverband der Energie- und Wasserwirtschaft (BDEW), says the figure for Germany is little more than 20%. The renewable freshwater resources correlate closely with the levels of precipitation in a given country. But there are exceptions.
5. E. Heymann, D. Lizio, and M. Siehlow (2010), "World water markets: High investment requirements mixed with institutional risk", Deutsche Bank Research (retrieved: http://www.dbresearch.com/PROD/DBR_INTERNET_EN-PROD/PROD000000000 0258353.PDF).

6. Think about India as an example, where 90% of the annual precipitation falls during the summer monsoon and where the same country is experiencing serious droughts during other parts of the year. E. Heymann et al. (2007), "450 bn reasons to invest in India's infrastructure", Deutsche Bank Research, *Current Issues*, November 28. Other examples are Spain and the observed differential between northern and southern Spain.
7. See further, S. "Ban" Banerjee and B. A. Obembe (2013), "Econometric forecasting of irrigation water demand conserves a valuable natural resource", *Journal of Agricultural and Applied Economics* 45(3): 557–568 (retrieved: http://ageconsearch.umn.edu/bitstream/155430/2/jaae453ip16.pdf). Natural causes (such as droughts), nonnatural causes (such as competing uses), and government policies limit the supply of water for agriculture in general and irrigating crops in particular. Under such reduced water-supply scenarios, existing physical models reduce irrigation proportionally among crops in the farmer's portfolio, disregarding temporal changes in economic and/or institutional conditions. Hence, changes in crop mix resulting from expectations about risks and returns are ignored. A method is developed that considers those changes and accounts for economic substitution and expansion effects.
8. See C. Schaffnit-Chatterjee (2009), "The global food equation: Food security in an environment of increasing scarcity", Deutsche Bank Research, *Current Issues* (retrieved: http://www.dbresearch.com/PROD/DBR_INTERNET_EN-PROD/PROD0000000000247631.PDF).
9. Ibid., note 2 and 5, pp. 4–6.
10. Source: GRID – Arendal (retrieved: http://www.grida.no/graphicslib/detail/trends-in-global-water-use-by-sector_f83a).
11. Ibid., note 2 and 5, p. 7.
12. World Bank (2006), "The challenge of reducing non-revenue water (NRW) in developing countries", Washington, DC.
13. Inadequate funding structures, mismanagement, ring-fencing of agricultural markets, inefficiencies in the distribution of foodstuffs or their waste are further reasons for food shortages in some countries. Further, unsustainability of production techniques and soil degeneration are on that list.
14. Leaky pipes lead to water losses of 50% or more in irrigation farming worldwide. With many conventional procedures (e.g., surface or sprinkler irrigation, flooding), a large part of the water is lost to evaporation before it reaches the plants in the first place. Besides, the risk of soil salination is very considerable with these procedures because surface water that comes from, say, rivers has a higher salt content than rainwater. If the fields are irrigated with surface water for a fairly long period, the salt content of the soil increases. Part of the water evaporates, and the salt is deposited on the surface. If this process occurs in countries with little natural precipitation, the salt ultimately fails to be washed out of the soil (Heymann et al., p. 9)
15. Genetic engineering and biotechnology offer opportunities to reduce the water requirements of cultivated plants so that they can also be grown in dry regions (Heymann et al., p. 10)
16. J. Liu, T. W. Hertel, F. Taheripour, T. Zhu, and C. Ringler (2013), "Water scarcity and international agricultural trade", selected paper prepared for presentation at the Agricultural & Applied Economics Association's 2013 AAEA & CAES Joint Annual Meeting, Washington, DC, August 4–6, 2013 (retrieved: https://www.gtap.agecon.purdue.edu/resources/download/6527.pdf).

17. Trust in the existence of comparative advantages. Countries in which the production of certain agricultural products is only possible at very high costs and/or with very high water use ought to import these products instead (Heymann et al., p. 10). Countries do however do not want to rely on other countries for their food supply. Notable is the example of Saudi Arabia where large desalination plants guarantee agricultural activities fueled by the availability of oil. Better and/or cheaper (at least more efficient) would be to export the oil and import the required food levels.
18. Ibid., note 2 and 5, p. 8. Discussed in more detail, pp. 9–11.
19. Ibid., note 7. The World Bank estimates the loss at 35% in many emerging economies, often going to 50% of total consumption. The reasons for these inefficiencies are the age and poor maintenance of the systems, which in turn are linked with funding bottlenecks and political mismanagement. A further reason for the infrastructure shortcomings is that the authorities responsible for the water supply in cities and municipalities frequently lack the required expertise, are insufficiently motivated, or are understaffed (Heymann et al., p. 12).
20. Rising prices would have to be charged for water consumption for additional or commercial purposes. Consumption-related prices – instead of flat rates – are a key signal to discourage the wasting of water (Heymann et al., p. 13).
21. Ibid., Heymann et al., p. 13.
22. Apart from pathogenic organisms, the high concentration of nutrients in wastewater can accelerate the growth of algae. Some types of algae have a toxic effect on the human organism. Wastewater can contaminate drinking-water resources and soil either at particular spots or by diffusion through the system (Heymann et al., p. 14).
23. A reduction of the nutrient emissions into bodies of water is required and urgent. This could be done by ensuring that the nutrients from the wastewater are separated and subsequently recycled. Moreover, this would be a way of reducing the costs of procuring nutrients. The most important lever, though, probably lies in raising the efficiency of fertilization (Heymann et al., p. 14).
24. Two main advantages emerge: first, the reduction of emissions into bodies of soil and water, and second, the recycling of recoverable resources.
25. Household wastewater consists of three main components (gray water, yellow water, and brown water) with diverse properties. While brown water (feces contaminated) contains most of the organic substances (substrate), yellow water (urine contaminated) contains nearly all the soluble nutrients, such as nitrogen, phosphorous, potassium, etc. The type of wastewater found most frequently is gray water (household wastewater excluding yellow and brown water), which shows very low nutrient loads. Following treatment, gray water may be channeled back into the household water cycle, into a receiving watercourse (e.g., river), or be allowed to drain back into the soil for groundwater recharge. The material substances can then be recovered from the yellow and brown water; the remaining water can be released into a receiving watercourse (Heymann et al., p. 15).
26. Ibid., note 2 and 5, pp. 15–16.
27. See FAO (2008), "Water and the rural poor: Interventions for improving livelihoods in Sub-Saharan Africa", Rome.
28. Ibid., Heymann et al., p. 18.
29. Ibid., Heymann et al., p. 18.
30. See further, UN (2009), "Water in a changing world", London, Paris.
31. See Heymann et al., pp. 19–20.

32. Ibid., Heymann et al., p. 20.
33. *Types of Seawater Desalination*
    - The electrodialysis process enables pure water to be extracted from seawater by electrochemical means. To do so, seawater is introduced into electrolysis cells. These cells contain numerous chambers formed by alternating cation- and anion-impermeable membranes. Applying an electric current to them produces concentrated salt water on the one hand and desalinated water on the other. This procedure is not very widespread yet since the membranes and the water to be desalinated have to meet high quality standards. Moreover, blockage of the membranes is also proving to be a problem in practice.
    - In the reverse-osmosis process, pure water is won by creating filtration pressure. In this case, the water molecules of the seawater are pressed through a membrane, whereas the salt ions are unable to pass through. Concentrated saltwater is formed on one side of the membrane, while desalinated water collects on the other. However, before beginning the reverse-osmosis process it is necessary to pretreat the seawater to remove foreign substances that may block or damage the membrane. The process can be applied particularly efficiently when upgrading surface water with a low salt concentration. In practice, seawater can be desalinated at a cost of less than one euro per cubic meter of water.
    - In multistage flash distillation, seawater is purified by thermal means. In this case the seawater is heated and steam is produced. The steam is subsequently liquefied in a condenser which thus produces freshwater. The process is very energy intensive and is therefore applied mainly in countries with substantial resources of primary energy (see ibid., Heymann et al., p. 21).
34. Ibid., Heymann et al., pp. 19–21.
35. Estimated at 400 to 500 billion USD globally and annually. The investment requirements in the water sector are thus of a magnitude similar to that of the market volume (by revenue), which is uncommon in most markets. However, this can be explained by the fact that revenues are kept down artificially because of low or absent pricing (Heymann et al., p. 23).
36. For example Israel with its permanent water shortage has a large set of (private) tech companies looking at a range of water solutions.
37. Ibid., Heymann et al., p. 25.
38. Resemblance with the energy market cannot be denied.
39. Deutsche Bank for example created a ranking. See Heymann et al., pp. 25–27.
40. A write-up of the two sectors and their respective characteristics and financial performance (public equities) can be found in Summit Global Management (2012), "The case for water equity investing 2012: Value and opportunity in an economic environment" (retrieved: http://www.summitglobal.com/documents/SummitCaseforWaterInvesting2012-a030512.pdf), pp. 10–13 and Summit Global Management (2012), "2012 water market review" (retrieved: http://www.summitglobal.com/documents/Maxwell2012WaterMarketReview-a030912.pdf).
41. See further, J. R. Diego, C. van den Berg, and A. McMahon (2012), "Investing in water infrastructure: Capital, operations and maintenance", water papers, World Bank, November (retrieved: http://water.worldbank.org/sites/water.worldbank.org/files/publication/water-investing-water-infrastructure-capital-operations-maintenance.pdf), and further:
    - V. Alavian et al. (2009), Water and Climate Change: Understanding the Risks and Making Climate-Smart Investment Decisions (Washington, DC: World Bank).

- B. C. Bates, Z. W. Kundzewicz, S. Wu, and J. Palutikof, eds (2008), "Climate change and water", IPCC Technical Paper 6, IPCC, Geneva.
- Danilenko, E. Dickson, and M. Jacobsen (2010), "Climate change and urban water utilities: Challenges and opportunities", water note, World Bank, Washington, DC.
- M. Fay, et al. (2010), "Infrastructure and sustainable development in post crisis growth and development", World Bank, Washington, DC.
- D. François, A. Correljé, and J. Groenewegen (2010), "Cost recovery in the water supply and sanitation sector: A case of competing policy objectives?", Utilities Policy 18(3): 135–141.
- See also the extensive literature list in Rodriguez et al. (2012), pp. 41–44.

42. Netbase Software is owned by Crowder Consulting.
43. Owned by General Electric.
44. "EPA water recycling and reuse: The environmental benefits".
45. For some examples, see Goldman Sachs (2013), "Water: Emerging risks & opportunities" (retrieved: http://www.goldmansachs.com/our-thinking/our-conferences/water-conference/water-summit-white-paper.pdf), pp. 6–7.
46. For example, "USGS water use in the United States" (http://ga.water.usgs.gov/edu/wateruse-total.html). The breakdown is 49% for thermoelectric power; 5% domestic, aquaculture, mining, and livestock; 4% industrial; 31% irrigation; and 11% public supply. These numbers seem to contrast with the FAO (supra in this chapter) numbers where 70% of consumption is allocated to agriculture. It is very likely that the denominator (total volume of water supply) is calculated differently, although the difference remains significant.
47. See further, "Desalination and Water Purification Technology Roadmap" (2003), Desalination & Water Purification Research & Development Program Report 95 (retrieved: http://www.usbr.gov/research/AWT/reportpdfs/report095.pdf).
48. See FAO (2006), "Water desalination for agricultural applications", Proceedings of the FAO Expert Consultation on Water Desalination for Agricultural Applications, Rome, April 26–27, 2004, land and water discussion paper 5, edited by J. Martínez Beltrán and S. Koo-Oshima (retrieved: ftp://ftp.fao.org/agl/aglw/docs/lwdp5_E.pdf).
49. See ibid., FAO (2004); J. Martínez Beltrán, S. Koo-Oshima, and P. Steduto, "Desalination of saline waters"; and Koussai Quteishat "Water desalination technologies and costs – state of the art", pp. 5–19. And ibid., note 48.
50. See further ibid., FAO (2004), D. B. Burnett and C. A. Jones, "Freshwater resources from desalination of wastewater and oilfield-produced brine", pp. 29–37.
51. See further ibid., FAO (2004), J. A. Medina, "Feasibility of water desalination for agriculture", pp. 37–45.
52. See, for example, Y. Dreizin, A. Tenne, and D. Hoffman (2007), "Integrating large scale seawater desalination plants within Israel's water supply system", in Y. Dreizin et al. *Desalination*, pp. 1–18. The paper quantifies the cost-and-benefit consequences of these factors, thereby establishing their relative weight, importance, and significance. Cost consequences relate not only to the effect each factor had on desalinated water costs at their inlets to the national or regional water supply grids, but also to its effect on overall investments and operating costs related to expanding the entire water supply system to meet projected increases in demand, including seasonal, multiseasonal, and local storage capacities, distribution line sizes, pumping energy requirements, etc., and to dealing with deteriorating groundwater quality, including

rehabilitation of salinized and/or contaminated wells, etc. Benefits included factors such as potable-water-supply reliability and quality enhancement, expanded and environmentally safer water-reuse potential, etc. As will be shown, the challenge was to create a master plan that accounts for all of these factors and optimizes their overall cost-benefit ratio both short and long term.

53. See for the current state of the desalination industry, existing and planned projects, etc., Abdulrahman Al-Ibrahim (2012), "Investment opportunities in the desalination sector in the Kingdom of Saudi Arabia" (retrieved: http://www.saudiarabia-jccme.jp/forumpdf/6-4.pdf).
54. Membrane processes include reverse osmosis (RO) and electrodialysis (ED). Whereas ED is suitable for brackish water, RO can be used for both brackish water and seawater.
55. See further, J. Labre (2002), "Desalination market: Demand approach. Matching desalinated water and conventional resources", Ondeo Workshop, Paris; World Bank (n.d.), "Renewable energy desalination" (retrieved: http://water.worldbank.org/sites/water.worldbank.org/files/publication/water-wpp-Sun-Powered-Desal-Gateway-Meeting-MENAs-Water-Needs_2.pdf).
56. See ibid., FAO (2004), pp. 40–41.
57. See extensively on this topic, J. Mateo-Sagasta and J. Burke (2008), "Agriculture and water quality interactions: A global overview", SOLAW Background Thematic Report TR08 (retrieved: http://www.fao.org/fileadmin/templates/solaw/files/thematic_reports/TR_08.pdf).
58. See further N. T. Carter (2013), "Desalination and membrane technologies: Federal research and adoption issues", Congressional Research Service 7–5700, R40477, pp. 12–15 (retrieved: http://www.fas.org/sgp/crs/misc/R40477.pdf).
59. A. Subramani, "Energy minimization strategies and renewable energy utilization for desalination: A review", *Water Research* 45(5): 1907–1920.
60. H. B. Park et al., "Highly chlorine-tolerant polymers for desalination", *Angewandte Chemie* 120: 6108–6113.
61. H. Tanaka and Y. Ho, "Global trends and performance of desalination research", *Desalination and Water Treatment* 25.
62. R. L. McGinnis and M. Elimelech (2007), "Energy requirements of ammonia carbon dioxide forward osmosis desalination", *Desalination* 207(1–3): 370–382.
63. A. Bennett (2011), "Innovation continues to lower desalination costs", *Filtration+Separation*, July–August.
64. Ibid., Carter (2013), p. 6.
65. H. Cooley et al. (2006), "Desalination, with a grain of salt: A California perspective", Pacific Institute. Only 7% of the existing desalination capacity uses seawater as its source. More than half of US desalinated water is from brackish sources. Another 25% is river water treated for use in industrial facilities, power plants, and some commercial applications. Globally, seawater desalination represents 60% of the installed desalination capacity. The Middle East, Algeria, Spain, and Australia are leading in the installation of new desalination capacity, with Saudi Arabia and the United Arab Emirates leading in annual production of desalinated water (J. Hughes [2009], "Seawater desalination leads response to global water crisis", AWWA Streamlines).
66. S. Chaudry (2003), "Unit cost of desalination", California Desalination Task Force, California Energy Commission.
67. A survey of municipal desalination facilities in Texas found that the cost for brackish desalination ranged from $410 to $847 per acre-foot, and for seawater desalination

ranged from $1,168 to $1,881 per acre-foot (J. Arroyo and S. Shirazi [2009], "Cost of water desalination in Texas", Texas Water Development Board, Austin, TX, p. 6 [retrieved: http://www.twdb.texas.gov/innovativewater/desal/doc]). Water produced from proposed seawater desalination facilities in California is estimated to range from $1,900 to $3,000 per acre-foot (H. Cooley and N. Ajami [2012], "Key issues for desalination in California: Cost and financing", Pacific Institute, November, p. 5 [retrieved: http://www.pacinst.org/reports/desalination_2013/financing_final_report.pdf]).

68. J. McEvoy and M. Wilder (2012), "Discourse and desalination: Potential impacts of proposed climate change adaptation interventions in the Arizona-Sonora border region", *Global Environmental Change* 22.
69. Several reports in the last decade have aimed to inform the path forward for US desalination research. The first was the 2003 "Desalination and Water Purification Technology Roadmap". The National Research Council reviewed the road map in a 2004 report, "Review of the Desalination and Water Purification Technology Roadmap", which called for a strategic national research agenda. That NRC committee published a report in 2008, "Desalination: A National Perspective". It concluded that research should focus on reducing the cost of desalination and that substantial further cost savings are unlikely to be achieved through incremental advances in the commonly used desalination technologies, such as reverse osmosis. In 2010, the Water Research Foundation, WateReuse Foundation, and Sandia National Laboratories published a report on how to implement the 2003 roadmap (Water Research Foundation, WateReuse Foundation, and Sandia National Laboratories [2010], "Implementation of the National Desalination and Water Purification Technology Roadmap" [retrieved: http://www.sandia.gov/water/docs/DesalImplementRoadmap1-26-2010_c_web.pdf]).
70. See ibid., Carter (2013), p. 8.
71. M. Elimelech and W. A. Phillip (2011), "The future of seawater desalination: Energy, technology, and the environment", *Science* 333: 712–717.
72. Pretreatment is necessary in order to avoid fouling and harm to the reverse-osmosis membranes.
73. Ibid., Subramani (2011).
74. A. Bennett, "Innovation continues to lower desalination costs", *Filtration+Separation*, July/August. Packaging of pre-engineered membrane-based desalination plants also has reduced the upfront capital costs for some desalination applications.
75. J. McEvoy and M. Wilder (2012), "Discourse and desalination: Potential impacts of proposed climate change adaptation interventions in the Arizona-Sonora border region", *Global Environmental Change* 22.
76. A major benefit of co-location is using the cooling water from the power plant for desalination; this water has been warmed by the power plant, which reduces the energy requirements for desalinating it. Also, the desalination facility may avoid construction costs by sharing intake and discharge facilities.
77. Ibid., McEvoy et al. (2012).
78. For example, M. S. Miranda and D. Infield (2002), "A wind-powered seawater reverse-osmosis system without batteries", *Desalination* 153; D. Weiner et al. (2001), "Operation experience of a solar- and wind-powered desalination demonstration plant", *Desalination* 137.
79. European Union (2008), "ADIRA handbook: A guide to desalination system concepts", Euro-Mediterranean Regional Program for Water Management (MEDA), SBN

978-975-561-311-6 (retrieved: http://wrri.nmsu.edu/conf/conf11/2008_adira_handbook. pdf). Desalination requires a considerable amount of energy. Seawater desalination via MSF consumes typically 80.6 kWh of heat energy (290 MJ thermal energy per kg) plus 2.5 to 3.5 kWh of electricity per cubic meter of water, while large-scale RO requires only about 3.5 to 5.0 kWh of electricity per cubic meter. Currently, the global production of about 65.2 million cubic meters per day of desalinated water involves the use of at least 75.2 TWh per year, which equals about 0.4% of the global electricity consumption. The cost of desalination has been decreasing over the last years down to USD 0.5 cubic meters, while market prices for desalinated water are typically between 1 and 2 USD per cubic meter.

80. See further, "Water desalination using renewable energy, including up-to-date data and technical considerations regarding the different technologies and applications" (retrieved: http://www.dlr.de/tt/Portaldata/41/Resources/dokumente/institut/system/projects/MENA_REGIONAL_WATER_OUTLOOK.pdf). Fichtner (2011), "MENA regional water outlook part II: Desalination using renewable energy, final report", Germany (retrieved: http://www.medrc.org/download/twb/FICHT-6911691-v3-Task_1-Desalination_Potential.pdf). V. Gude et al. (2010), "Renewable approaches for desalination", *Renewable & Sustainable Energy Reviews* 14: 2641–2654. J. Koschikowski (2011), "Water desalination: When and where will it make sense?" presentation at the 2011 annual meeting of the American Association for the Advancement of Science, Fraunhofer Institute for Solar Energy Systems (ISE) (retrieved: http:// ec.europa.eu/dgs/jrc/downloads/jrc_aaas2011_Energy_water_koschikowski.pdf). Pike Research (2010), "Research report, executive summary: Desalination technology markets – global demand drivers, technology issues, competitive landscape, and market forecasts", Pike Research. WIPO (2011), "Patent landscape report on desalination technologies and the use of alternative energies for desalination" (retrieved: http://www.wipo.int/export/sites/www/patentscope/en/programs/patent_landscapes/documents/patent_landscapes/948–2E-WEB.pdf). Irena (2012), "Water desalination using renewable energy technology brief", IEA-ETSAP and IRENA Technology Brief I12 (retrieved: http://www.irena.org/DocumentDownloads/Publications/IRENA-ETSAP%20Tech%20Brief%20I12%20Water-Desalination.pdf). See for further reporting World Bank Desalination Training and Research Center, http://www.medrc.org/index.cfm?area=research&page=Final_Reports.
81. For a comparative cost analysis for common renewable desalination, see IEA-ETSAP and IRENA Technology Brief I12, p. 13–14 and summary tables pp. 21–24. Further, M. Papapetrou et al. (2010), "Roadmap for the development of desalination powered by renewable energy: Promotion of renewable energy for water desalination" (retrieved: http:// wrri.nmsu.edu/conf/conf11/prodes_roadmap_online.pdf). European Union (2008), "ADIRA handbook".
82. National Research Council, *Desalination: A National Perspective*, 2008.
83. Fluoride is low in seawater and is further depleted by desalination; communities can choose to add fluoride to treated water consistent with their health goals.
84. J. Cotruvo (2006), "Health aspects of calcium and magnesium in drinking water", *Water Conditioning and Purification*, June. Remineralization would also help reduce the corrosivity of desalinated water on piping.
85. D. A. Caron et al. (2009), "Harmful algae and their potential impacts on desalination operations off southern California", *Water Research*. Coastal algal blooms known as red tides were the subject of a 2012 expert workshop ("Red tide and HABs: Impact on desalination plants", Expert Workshop, Muscat, Sultanate of Oman, February

2012 (retrieved: http://www.medrc.org/index.cfm?area=about&page= expert_workshop_download). According to this article, "There are no published reports on the effectiveness of reverse osmosis for removing dissolved algal toxins from seawater. Some of these toxin molecules (e.g., domoic acid) are near the theoretical molecular size of molecules rejected by reverse osmosis membranes, but experimental studies are required to validate the effective [*sic*] of this process on toxin removal."

86. N. Voutchkov (2011), "Management of desalination plant concentrate", SunCam (retrieved: http://s3.amazonaws.com/suncam/ npdocs/113.pdf).
87. Ibid., Carter (2013), p. 12.
88. A. Fried and B. Serio (2012), "Water industry segment report", *Desalination*, p. 6. Including comprehensive data sets for the USA and the MENA region (country-by-country analysis) (retrieved: http://www.wtcsd.org/Resources/WEMI/Segment%2520Report/120730%2520Desalination%2520segment%2520report.pdf). See also, World Bank (2012), "Mena development report Renewable Energy Desalination: An emerging solution to close the water gap in the Middle East and North Africa" (retrieved: http://water.worldbank.org/sites/water.worldbank.org/files/publication/water-wpp-Sun-Powered-Desal-Gateway-Meeting-MENAs-Water-Needs_2.pdf).

18 Future Agricultural Dynamics

1. OECD-FAO (2013) "Agricultural outlook 2013–2022" (retrieved: http://www.oecd.org/site/oecd-faoagriculturaloutlook/highlights-2013-EN.pdf). The fully documented outlook database, including historical data and projections, is available through the OECD-FAO joint internet site, www.agri-outlook.org. Data indicated in the text have been selected from the document and/or the database referred to.
2. Actual prices can be consulted at any time at www.agri-outlook.org.
3. "Global aquaculture production fishery statistical collections", FAO, Rome.
4. Ibid., note 1, p. 15.
5. Ibid., note 1, p. 19.
6. Ibid., note 1, p. 25.
7. Ibid., note 1, p. 31.
8. This analysis involves performing multiple simulations that draw alternative values from distributions for key driving variables and analyzes their impacts on selected important outcomes for variables such as prices, production, trade, etc. Partial stochastic analysis does not attempt to forecast the implications of all possible uncertainties for future market outcomes. Instead, it allows the policy maker to select specific sources of uncertainty and to quantify the likely range of variation around the deterministic baseline values that derive from these identifiable sources of uncertainty (see ibid., note 1, p. 43). The methodology is detailed in A. Burrell and Z. Nii-Naate (2013), "Partial stochastic analysis with the European Commission's version of the AGLINK-COSIMO model", JRC Scientific and Technical Reports, European Commission, JRC76019 (received: http://ftp.jrc.es/EURdoc/JRC76019.pdf).
9. Findings of the research, ibid., note 1, pp. 45–48.
10. This can be done by analyzing a subset of the simulated solutions in which the values of one or more drivers lie in a particular range. A limitation of this approach is that, if a specific risk is defined very narrowly, it may result in a subset of runs that is too small to be reliable (see further ibid., note 1, p. 48).
11. Ibid., note 1, p. 43.

Index

Printed and bound by CPI Group (UK) Ltd, Croydon, CR0 4YY